X-Ray CT

Hiroyuki Toda

X-Ray CT

Hardware and Software Techniques

Springer

Hiroyuki Toda
Mechanical Engineering
Kyushu University
Fukuoka, Japan

ISBN 978-981-16-0589-5 ISBN 978-981-16-0590-1 (eBook)
https://doi.org/10.1007/978-981-16-0590-1

This Springer imprint is published by the registered company Springer Nature Singapore Pte Ltd.
The registered company address is: 152 Beach Road, #21-01/04 Gateway East, Singapore 189721, Singapore

Preface

Opportunities for the utilization of X-ray tomography in industry and scientific research have increased in recent years. Its various applications have been used for the assessment of the performance, quality, reliability, dimensions, and morphology of industrial products, components, and materials made of metal, ceramic, and polymers in a variety of manufacturing industries. Furthermore, cutting-edge scientific research that uses X-ray tomography has become increasingly common in synchrotron radiation facilities. However, the knowledge, information, techniques, and learning structures necessary for this type of work have not been covered in detail in present-day university education in engineering departments. With this in mind, there have been many requests for a technical book that comprehensively illuminates these topics with the inclusion of applied examples.

There have been many technical books and texts published relating to X-ray tomography; however, the majority have focused on the fields of medicine and dentistry.

To the best of the authors' knowledge, publication focusing on industry and scientific research has been limited to the English-language material published from Oxford University Press. Indeed, the author contributed a chapter to that body of work; collaborations among multiple authors, however, often encourage individuals to devote themselves to expanding on their own specializations, a process that does not lend itself to the creation of a technical book that exhaustively covers all the necessary techniques in the field.

This book was written for researchers and graduate students working in industry, research institutes, and academia on topics relating to various structural and functional material research and development, as well as engineers involved in the development, setup, and production of industrial products and components. In this sense, it was intended to act as a "bible" of sorts, covering both the soft and hard aspects of X-ray tomography, from foundations to applications, and physics to engineering.

Finally, I would like to extend my heartfelt gratitude to all those who have understood and aided me in the publication of this textbook, particularly Dr. Yoshio Suzuki, formerly at SPring-8 and Dr. Masami Tomizawa at Toshiba IT and Control Systems Co. for reviewing this entire book; Mr. Masahito Natsuhara of the Japanese Inspection Instruments Manufacturers Association (Shimadzu Co.), who has provided various advice and assistance. I would also like to thank specially appointed Assistant Professor Kazuyuki Shimizu of Kyushu University, who provided support with data analysis; graduate student Hiro Fujiwara; and Secretary Yumi Yuasa, who assisted with some of the drawings. Finally, I would like to thank my family, who have tolerated over a year and a half of my sleepless and restless nights while working on this book.

Fukuoka, Japan
February 2021

Hiroyuki Toda

Acknowledgements The present publication received a grant from JSPS Grants-in-aid for Scientific Research JP20HP5222.

The original version of the book was revised: The copyright holder name has been changed. The correction to the book is available at https://doi.org/10.1007/978-981-16-0590-1_11

Contents

Chapter 1
History and Present

1.1 History of Medical X-ray Tomography

1.1.1 Early Years

When the German scientist Wilhelm Roentgen discovered X-rays in 1895, the first demonstration he conducted in front of an audience was to show the bones in his hand with a transmission image. People immediately recognized that materials like soft tissue, bones, and metals were distinguishable with high contrast if X-rays were used, noting their effectivity for clinical purposes. For this reason, vehicles with X-ray facilities were applied in the treatment of wounded soldiers within 20 years of these discoveries during World War I. X-ray imaging spread across the medical disciplines within the blink of an eye.

At the same time, some individuals recognized that material interiors could be observed in 3D using X-rays. The Bohemian mathematician Johann Radon published his theory on image reconstruction in 1917 [1]. However, as he was purely a mathematician, it perhaps never occurred to him that it could be applied to X-ray computed tomography (CT). Approximately half a century afterward, Allan Macleod Cormack formulated several mathematical principles for image reconstruction in 1963 [2]. Afterward, the English engineer Godfrey Hounsfield developed a commercial-grade X-ray CT scanner, shown in Fig. 1.1 [3]. Cormack and Hounsfield together were awarded the Nobel Prize in Physiology or Medicine in 1979 for their achievements. During this time, Bocage proposed a cross-sectional imaging technique in 1921 by translating X-rays and film simultaneously, in parallel, and in opposite directions [4]. Prior to this, the idea of only moving the X-ray source was already observed in 1914. This type of technique has been referred to as "planigraphy." Subsequently, Grossman developed a device that moves the beam source and film in a circular arc, as shown in Fig. 1.2 [5]. The first use of the term "tomograph" has been attributed to him. The etymology of this term comes from the ancient Greek "tomos" (meaning slice) and the suffix "-graph," which means photography. His device was commercially sold by the German company Zanitas in 1934. This method sets up a *region*

H. Toda, *X-Ray CT*,
https://doi.org/10.1007/978-981-16-0590-1_1

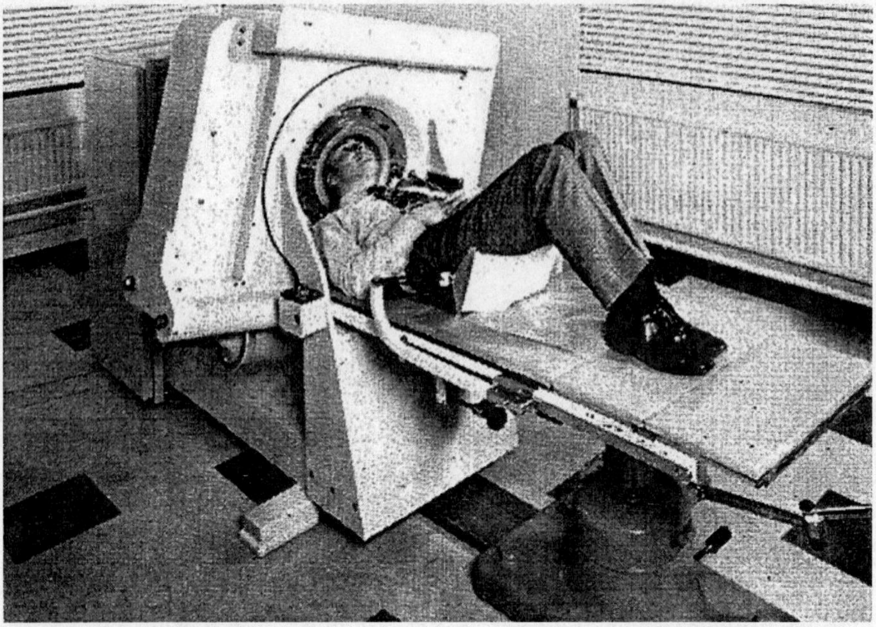

Fig. 1.1 X-ray CT scanner developed by Hounsfield and set up at Atkinson Morley Hospital in Wimbledon [3]

of interest in a given area of the subject and blurs all planes other than that set as the observation subject so the obtained image is entirely different from what is obtained through modern X-ray CT. Meanwhile, in Japan, Shinji Takahashi of what was then known as the Tohoku Imperial University began research on "X-ray rotational imaging methods" around 1945 [6]. His research also included X-rays passing through a *slit* and irradiating in the vertical axial direction, while simultaneously rotating the film, thereby obtaining a sinogram [6]. After imaging, the drawing paper on top of a rotating table and the obtained X-ray image were aligned as shown in Fig. 1.3; cross-sections were then drawn with a pen and paper. This can be considered pioneering research into the prototype of present-day X-ray CT scanners during an analog era where film needed to be used for imaging.

1.1.2 Conditions After the 1970s

Hounsfield first used an X-ray CT scanner on a human body on October 1st, 1971, to inspect a cerebral tumor in Atkinson Morley Hospital in London. Operational X-ray CT scanners for medical use surpassed 10,000 units by 1980; there are now too many to count. This dissemination of this device was also due to its highly effective performance. According to Hsieh, the increasingly high speed of X-ray tomography

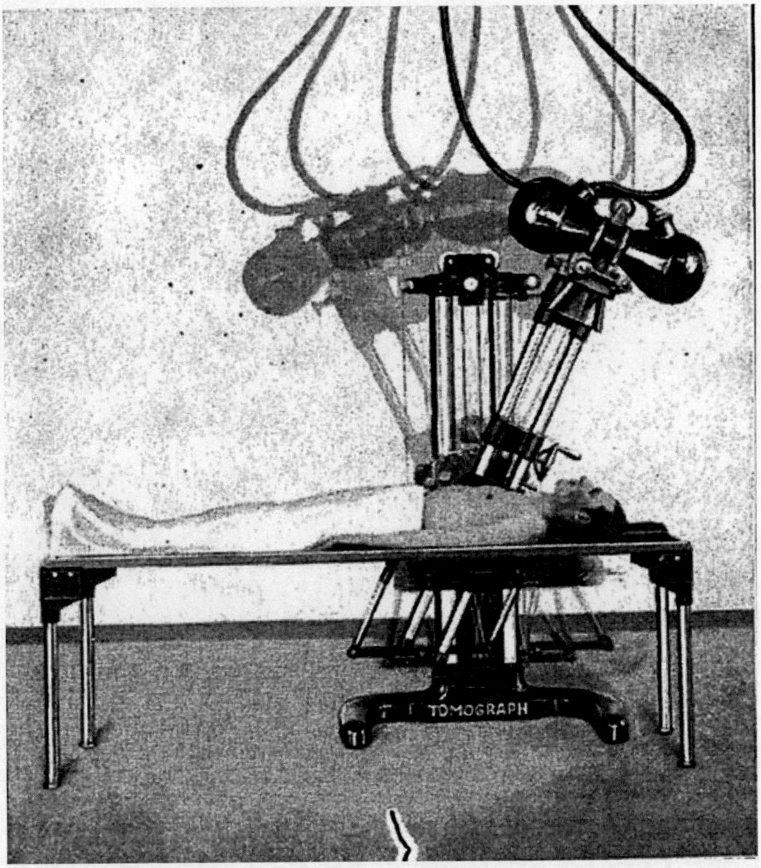

Fig. 1.2 X-ray tomography device proposed by Grossman [5]

has enabled it to reduce its scanning time by half every 2.3 years, similar to Moore's law regarding the integration density of semiconductors, as stated by the eponymous co-founder of the U.S. company Intel (Fig. 1.4) [7]. This shows that X-ray tomography has developed alongside other improvements seen in digital techniques. Furthermore, *spatial resolution* has improved from over 1.5 mm in its initial development to below 1 mm in the 1980s, with lesions measuring several hundred μm now becoming detectable in the present day [8]. Because of these technical developments, X-ray tomography techniques have rapidly advanced radiation diagnosis and its significant contributions to the advancement of the medical sciences are well known.

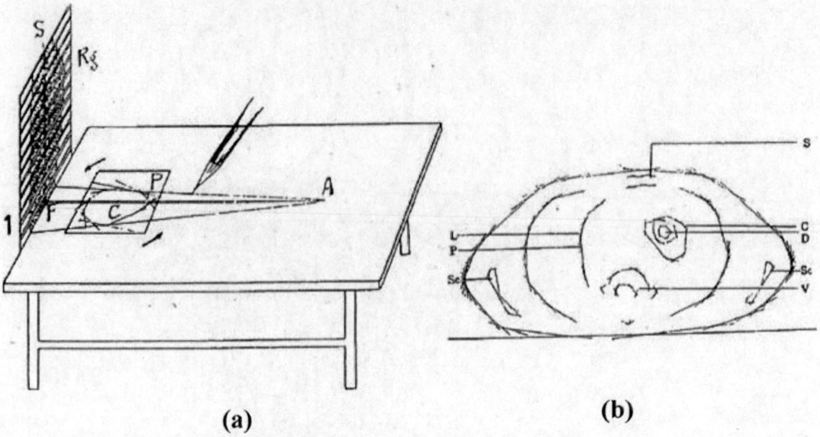

Fig. 1.3 **a** Schematic shown in the manuscript published by Shinji Takahashi. After X-ray tomography imaging, the drawing paper on top of a rotating table and the obtained X-ray image are aligned, and the cross-sections are drawn using a pen and paper [6]; **b** shows the hand-drawn cross-sectional image [6]. The letters in the figure are defined as follows: S: clavicle; C: cavity; D: necrotic tissue; Sc: shoulder bone; V: vertebrae; L: lung; and P: pleura.

Fig. 1.4 Acceleration observed for the amount of time required for a single X-ray tomogram [7]

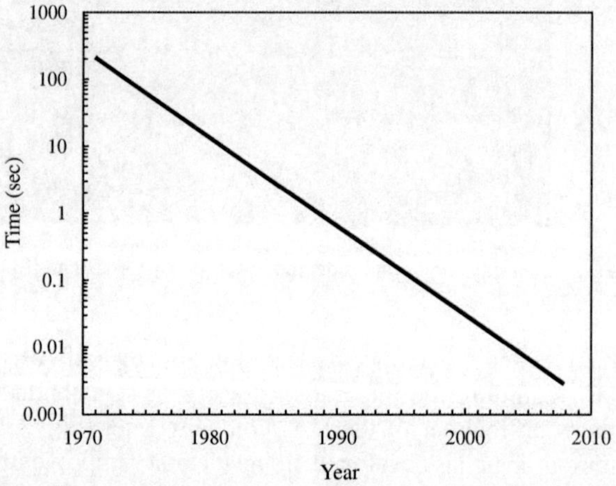

1.2 History and Present of X-ray Tomography for Non-medical Use

In his Nobel Lecture, Cormack recounted an episode of a request from the Swiss Federal Institute for Snow and Avalanche Research, during his early research years when his developments had received no attention at all [9]. Many people were cognizant of the potential applications of being able to see objects in 3D since the early years of X-ray tomography. Seven years after the brain inspections at Atkinson

Table. 1.1 Classifications of X-ray tomography

	Classification	Object size	Irradiation	X-ray energy	Spatial resolution	Examples of needs
	Medical use	Large (Human body)	Low dose needed	40–60 keV	0.1–1 mm	High speed, high definition, low dose, VR, etc.
Industrial and scientific uses	Microtomography	Small (small components, microstructure, defects, etc.)	No limit	A few ~100 keV	0.1 – 10 μm	High resolution, quantitativeness, phase contrast, element selectivity, etc.
	Others	Large (large components, whole products)	No limit	~1 MeV	Under 0.1–1 mm	Defect detection, dimensional measurement, etc.

Morley Hospital in 1978 for example, Reimers et al. from the German organization BAM conducted 3D observations of wooden materials, ceramics, plastics, and archaeological artifacts using medical X-ray CT scanners [10].

A history of the advances in X-ray tomography techniques for non-medical use is explored below, primarily from the perspective of spatial resolution and its improvements. Table 1.1 shows a broad classification of X-ray tomography based on its applications. There is no unequivocal definition of the term "microtomography" shown in the table, however, this generally refers to applications where spatial resolution or pixel size is below 50 μm, reaching 1 μm in some cases. The classifications shown in Table 1.1 are primarily for expediency and can vary widely according to the chemical composition of the subject. Therefore, care must be taken as the overlap of the three items is considerably large. Figure 1.5 shows an overview of the size and weight of X-ray CT-imaged subjects for each applied field. The upper limit of spatial resolution is approximately 1/500–1000th of the size of the imaging domain (generally, spatial resolution decreases one to several times that amount).

Figure 1.6 shows the number of original manuscripts with the term "*microtomography*" in their keyword list or abstract when searching in a database, (SCOPUS) plotted by year of publication. Manuscripts in Fig. 1.6 include those from both X-ray tomography at synchrotron radiation facilities and industrial X-ray CT scanners. There have been 2,461 manuscripts published from the 1980s to 2016. The number of manuscripts has rapidly increased since 2000 and it can be observed that its rate of increase has accelerated from 2004 onwards. Figure 1.7 shows the spatial resolution values in the abstracts of these papers. These include both applied research and instrument development research. Perhaps because the spatial resolution of X-ray tomography is usually kept low, there does not seem to be a clear relationship between spatial resolution and year at first glance. However, as seen with the dotted

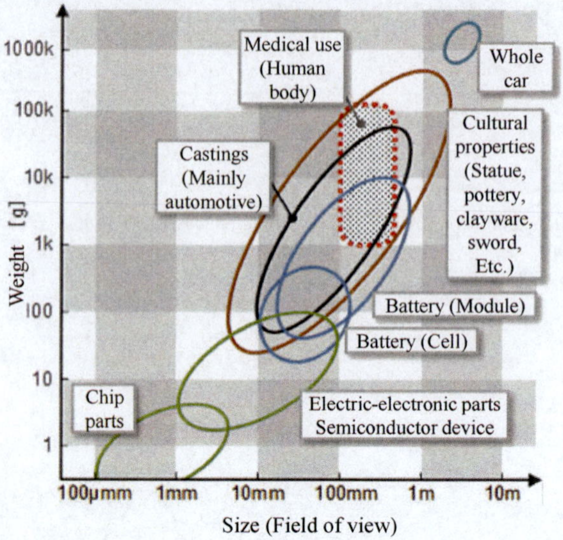

Fig. 1.5 CT imaging subject size and weight, and their relationship with applied fields [11]

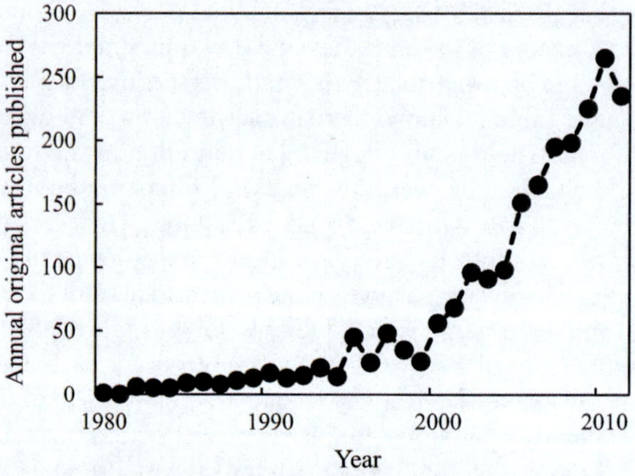

Fig. 1.6 Number of original manuscripts with the term "microtomography" when searching in the reference database SCOPUS; medical and pharmaceutical papers have been excluded

line in Fig. 1.7, connecting the upper limit of the spatial resolution shows the champion data for each year. Virtually all of these papers report synchrotron radiation research. An overview of its advances is presented below.

1.2.1 *Synchrotron Radiation X-ray Microtomography*

A year after the first report on microtomography by Elliot et al. [12], theoretical investigation results using synchrotron radiation for X-ray tomography were published by

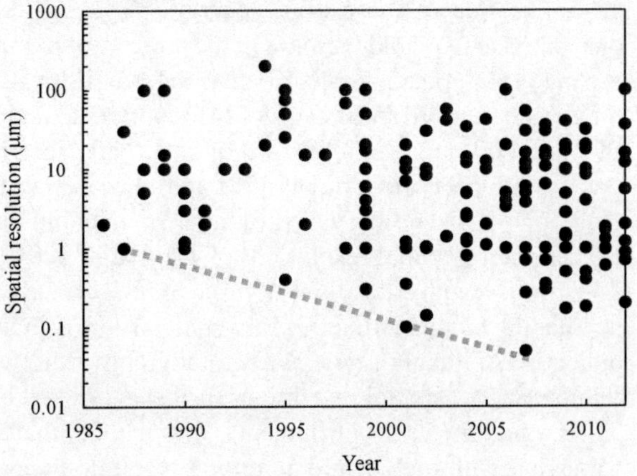

Fig. 1.7 Spatial resolution values present in the abstracts of search results in Fig. 1.6 and plotted by publication year

Grodzins from MIT in the U.S. [13] and by Cho et al. in Korea the following year in 1984 [14]. Grodzins, in particular, showed the conditions under which a 1-μmspatial resolution could be obtained. Similarly, Thompson et al. from the U.S. reported actual synchrotron radiation experiments in 1984 [15]. These experiments were conducted at the Stanford *Synchrotron Radiation* Lightsource (SSRL), wherein the heart of a pig was visualized at a low spatial resolution above 1 mm using a monochromatic X-ray *fan-beam* and detector that arranged Si (Li) elements in a single dimension. Research pursuing higher resolutions was subsequently conducted by various international research groups in the 1980s, including Hirano, Usami, Suzuki et al. from Hitachi, Ltd. [16–21]. Flannery et al. from Exxon in the U.S. achieved a 10-μm spatial resolution by combining a high-resolution *scintillator* with a cooled 330 × 512-pixel charge-coupled device *(CCD) camera* and optical lens at the National Synchrotron Light Source (NSLS) at Brookhaven National Laboratory in the U.S. [16, 17]. Aside from the specifications, the composition of the experimental setup is virtually identical to that used currently. Shortly afterward, Kinney et al. from Lawrence Livermore National Laboratory in the U.S. used a similar setup at SSRL and achieved a spatial resolution of 20 μm first with a *bending magnet* beamline [18] and then of 10 μm with a *wiggler* beamline the following year [19]. Virtually concurrent with this, the Hitachi group tested the direct detection of X-ray images using high-resolution X-ray imaging tubes at the High Energy Accelerator Research Organization Photon Factory (KEK-PF) [20, 21]. This group was able to obtain a 3D image of SiC fiber-reinforced aluminum and meteorites using 21 keV X-ray energies and achieved a spatial resolution equivalent to that of Flannery and Kinney [20, 21].

Incidentally, increasing the spatial resolution without reducing the *signal-to-noise ratio* (S/N ratio) requires the dramatic increase of *light quantity* (photon count per unit time/surface area) of X-rays emitted onto the specimen. For this, the scientific community had to wait for the introduction of so-called third-generation synchrotron radiation facilities such as APS (U.S., beginning operations in 1996), ESRF (France, beginning operations in 1994), and SPring-8 (Japan, beginning operations in 1997).

The application of these third-generation synchrotron radiation facilities largely contributed to the rapid increase in the number of publications from 2000 onwards as seen in Fig. 1.6. For example, Koch et al. from ESRF [22], Wang et al. from APS [23], and Uesugi et al. from SPring-8 [24] reported spatial resolutions of 1μm between 1998 and 2001. Furthermore, although slightly inferior in comparison, Beckmann et al. from DESY in Germany [25] and Dowd et al. from NSLS reported a spatial resolution of 2 μm in 1999 [26], and Stampanoni et al. from SLS in Switzerland [27] reported a spatial resolution of 1.4 μm in 2004. This spatial resolution value of 1μm was the physical limit of *projection-type X-ray tomography*, which does not use imaging elements like *Fresnel zone plates* or mirrors. Afterward, the practical application of imaging-type X-ray tomography (discussed later) was seen in various facilities and its spatial resolution further improved by over an order of magnitude [28]. Furthermore, an environment is now in place that allows for the daily application of imaging with high spatial/temporal resolution and contrast imaging such as *fast tomography* and *phase-contrast imaging*, which apply the radiation characteristics of high brightness and coherence.

Phase-contrast imaging will be touched upon in Chaps. 2 and 5. The publication of the principles of *phase-contrast imaging* by the Dutch scientist Frits Zernike, who used visible light and was later awarded the Nobel Prize in Physics, was extremely early in 1942 [29]. The subsequent application of X-rays to phase-contrast imaging by Bonse and Hart occurred over 20 years later [30]. It was over 30 years after this that Momose et al. at Hitachi Ltd. applied this to X-ray tomography observations in living tissue [31].

1.2.2 X-ray Tomography Using an Industrial X-ray CT Scanner

The first report on microtomography was thought to be produced by Elliot et al. of the U.K. in 1982 [12]. They passed the X-rays generated from an X-ray tube through a *pinhole* and achieved a high spatial resolution of 15 μm [12]. Figure 1.8 shows an image obtained at that time, in which a snail shell was somehow visualized. In the same year, Burnstein et al. from American Science and Engineering Inc. introduced an X-ray CT scanner that used a linear accelerator (linac) and applied it to a rocket engine (Fig. 1.9) [32, 33]. This is a visualization example of a large component at a low spatial resolution (2 mm) using high-energy X-rays (~15 MeV). Seguin et al. from the same company were able to achieve a spatial resolution of 50 μm in 1985 using X-ray tubes [34]. Feldkamp et al. of Ford in the U.S. also proposed a cone-beam reconstruction in 1984 [35]. Afterward, they used a microfocus X-ray tube with a focal spot size of 50 μm to achieve a spatial resolution of 100–125 μm [36]. Meanwhile, albeit through a two-dimensional projection, Aoki et al. from Tsukuba University achieved a spatial resolution of 10 μm using enlarged projections from X-rays, radiatively expanding from the microfocus X-ray source, published in

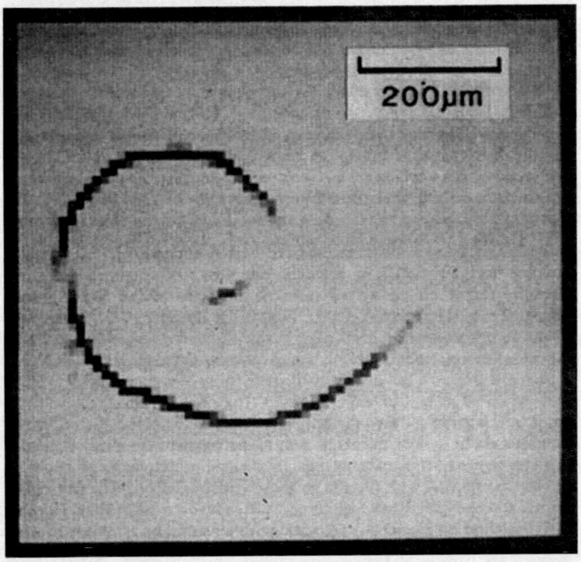

Fig. 1.8 First application example of microtomography by Elliot et al. [12]. Cross-section of a tropical freshwater snail shell. The swirl of the shell is somehow visible

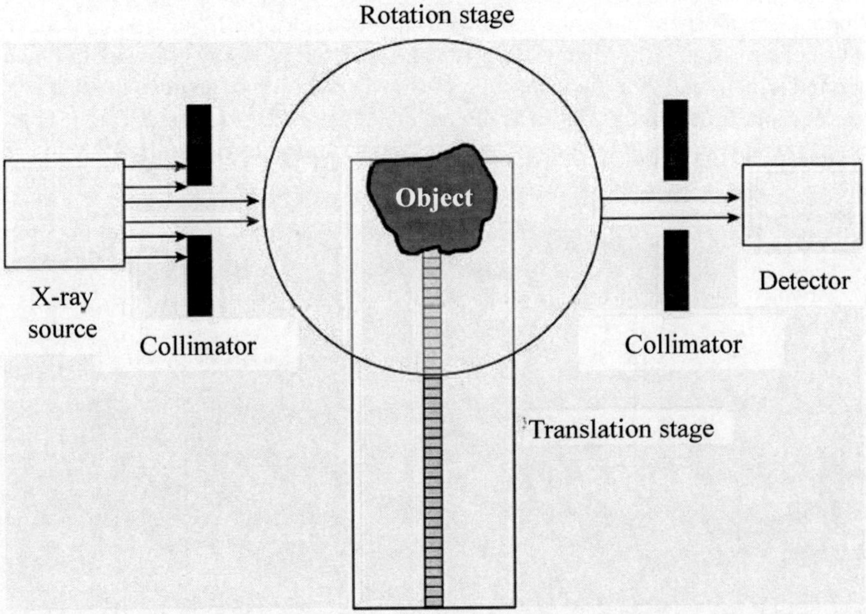

Fig. 1.9 Schematic of X-ray CT scanner used for a rocket engine of American Science and Engineering, Inc. [32, 33]

1988 [37]. This was the same spatial resolution level achieved using the synchrotron radiation of the time. Under such a context, the American Society for Nondestructive Testing held their first conference on industrial X-ray CT scanners in Seattle in 1989. Reports were given on industrial X-ray CT scanners, which used cone-beam methods developed by Vickers and others. The Lawrence Livermore National Laboratory and Sandia National Laboratories of the U.S. actively reported on the development of industrial X-ray CT scanners during this time.

In parallel, more companies began to test and manufacture industrial X-ray CT scanners. Hopkins et al. of Scientific Management Systems, Inc. of the U.S. used beam sources that applied radioisotopes such as ^{137}Cs and ^{192}Ir and reported on X-ray CT scanners that could use higher X-ray energies (300–600 keV) than those of medical tomography (40–60 keV), as well as its application to steel [38]. The spatial resolution in these cases was similar to that in medical tomography devices, at approximately 1–2 mm. Furthermore, reports were seen during this period that purposefully combined industrial X-ray CT scanners with ^{241}Am, which can only achieve low gamma-ray energies, and compared them with medical CT scanners [39]. Afterward, the aforementioned American Science and Engineering, Inc. produced what was at the time the world's largest X-ray CT scanner in 1984 to inspect rocket engines with a diameter of several meters [32, 33]. At around the same time, Toshiba Co. first produced a high-energy X-ray CT scanner specializing in steel specimen usage in Japan in 1982 (Fig. 1.10: tube voltage, 420 kV) [40]. This was used by the Nippon Steel Corporation at the time, with case examples of stainless-steel porosity and steel corrosion [40]. Meanwhile, Hitachi Ltd. used an X-ray CT scanner manufactured with a linac for the inspection of Mazda automobile engines in 1991. The external appearance of this device is illustrated in Fig. 1.11. Additional unique application examples include the manufacturing of an X-ray CT scanner by Hitachi Ltd.

Fig. 1.10 First high-energy X-ray CT scanner developed in Japan in 1982 by Toshiba [40] (courtesy of Junichi Iwasawa, Toshiba IT Control System)

Fig. 1.11 X-ray CT scanner for the inspection of automobile engines manufactured by Hitachi Ltd. in 1991 (courtesy of Katsutoshi Sato, Hitachi Ltd.)

for the inspection of burning behavior in a fuel assembly, which was set up in the Joyo fast reactor in Oarai city in Ibaraki Prefecture and is currently still in operation.

With regards to microtomography , a microfocus X-ray CT scanner manufactured by the research, "elemental technique development of a ceramics turbine for coal gasification", which was sponsored in FY 1988 by what was at the time the Japanese Ministry of International Trade and Industry, was set up and used for defect detection. A commercially available X-ray CT scanner was also first available in 1994 [41]. An X-ray CT scanner for microtomography was produced in Japan in 1999 by Shimadzu Co. and Toshiba Co. Figure 1.12 shows the external appearance of an X-ray CT scanner from that time.

1.2.3 Advances in Elemental Techniques

W. Edward Chamberlain of Temple University published information on image intensifiers in 1941. Westinghouse first achieved its development into a product in 1952 and this was achieved in Japan by Toshiba Co. and Shimadzu Co. in 1956. However, the complete digitalization of the detector did not occur until much later. In the present day, image intensifiers are used in combination with a CCD camera.

It was in 1969 that Boyle and Smith of Bell Labs in the U.S. invented the CCD camera, for which they were awarded the Nobel Prize. The Nobel Prize in Physics was awarded to them in 2009 for their publication in the Bell System Technical Journal in 1970. Afterward, the U.S. company Fairchild Imaging first sought to

Fig. 1.12 X-ray CT scanner for microtomography manufactured by Shimadzu Co. in 1999 (courtesy of Masahito Natsuhara of Shimadzu Co.)

commercialize CCD cameras in 1973. There was a demand at that time for detectors in space telescopes and unmanned probes. The scientific applications of CCD camera development saw significant advances at this time. Combining CCD cameras, which appeared almost simultaneously alongside X-ray tomography techniques, with a scintillator and lens enabled the digital recording of X-ray images.

In 1995, a *flat panel detector*, which was a high-performance large-surface-area X-ray detector , was developed. Meanwhile, Wanlass invented the complementary metal–oxide–semiconductor (CMOS) camera in 1963. However, the actual practical application of *CMOS cameras* occurred primarily in the 1990s alongside advances in semiconductor micro-fabrication techniques. The commercial availability of the Fastcam MOS camera by Photron Limited as a high-speed camera also occurred in the1990s. CMOS cameras , which had a global shutter functionality that could activate an electronic shutter on all pixels at once, subsequently became commercially available in the 2000s. Furthermore, Fairchild Imaging in the U.S., PCO in Germany, and Andor Technology PLC in the U.K. developed a series of high-performance CMOS cameras for scientific measurements, called sCMOS cameras , in the 2010s. Currently, CMOS cameras have seen widespread use as detectors with a high *frame rate* , high resolution, and wide *dynamic range* due to their advancement in overcoming high readout noise. Following the period of X-ray films, which obtained images through photographic emulsion exposure, various devices that can store images as digital information were developed; these advancements are considered key to the development of X-ray tomography techniques.

Regarding X-ray sources, Feinfocus, founded in 1982, manufactured a *micro-focus X-ray source* on a commercial basis. Objects could be placed closer to an

X-ray source by developing microfocus capabilities, which contributes to a higher spatial resolution. Shortly after this, Hamamatsu Photonics succeeded in developing a microfocus X-ray source in Japan in 1992 [42]. Microfocus X-ray sources are an essential technique in realizing microtomography, which is essential in industrial X-ray CT scanners. In the present day, X-ray sources can be selected among synchrotron radiation facilities, X-ray tubes, and linacs, according to the application of interest. Alternatives continue to increase due to the development of X-ray CT scanners using desktop-type radiation devices by the National Institute of Advanced Industrial Science and Technology and Hitachi Ltd. in projects started by the Japanese Ministry of Economy, Trade, and Industry in 2013 [43].

Furthermore, the development of computers touched upon in Sect. 1.1.2 is another important element that supports the progress of X-ray tomography. Early-stage experiments by Hounsfield took 2.5 h to conduct image reconstruction of a single cross-section with a large computer. The introduction of microcomputers in the 1970s, which coincided with the development of X-ray tomography, as well as its dissemination in the 1980s, greatly improved the computational environment for X-ray tomography. By the 2000s, personal computers were used to conduct X-ray CT scanner reconstructions and 3D visualizations.

Advances in X-ray tomography were achieved with the support of various elemental technologies and continues to mature technically. Currently, industrial X-ray CT scanners can also conduct what is referred to as nanotomography, where even higher spatial resolution and higher contrast 3D imaging is possible. Furthermore, high-functionality X-ray CT scanners (e.g. industrial X-ray CT scanners, which are used for *in-vivo tomography* and morphological measurements) can also be used according to the application of interest.

There have been trends toward standardization, as can be seen by the German guidelines for dimensional measurements based on X-ray tomography in 2011 [44, 45]. Furthermore, X-ray CT scanners have been set up on manufacturing lines for products such as circuit boards since approximately 2010, which have led to large improvements in quality control. This application seeks to reduce the time needed for scanning and conducting full inline inspections.

Meanwhile, development of application technologies has also progressed, going beyond simply visualizing the physical interior structure in 3D using X-ray tomography and calculating its size, morphology, and spatial distribution, but quantitatively calculating the physical quantities that are key to elucidating the phenomena observed in mechanical and materials engineering [46, 47]. For example, this may include mechanical engineering values such as strain, stress intensity factor, and the J-integral, as well as materials engineering values such as the crystallographic orientation of a polycrystalline metal, defects, and chemical compositions. Superior verification of these physical objects, which goes beyond simple observations of their interior, can thus be assured.

References

1. J. Radon, Berichte Sächsischen Akademie Der Wissenschaft **69**, 262–279 (1917)
2. A.M. Cormack, J. Appl. Phys. **34**, 2722–2727 (1963)
3. G.N. Hounsfield, Br. J. Radiol. **46**, 1016–1022 (1973)
4. A.E.M. Bocage, French Patent No. 536464 (1921)
5. G. Grossman, Br. J. Radiol. **8**, 733–751 (1935)
6. S. Takahashi, Jpn. J. Clin. Radiol. **21**, 1037–1045 (1976)
7. J. Hsieh, *Computed Tomography Principles, Design, Artifacts, and Recent Advances* (Wiley Interscience, Bellingham, Washington, 2009).
8. Y. Hirao, Technique Systematization Investigatory Report of the National Museum of Nature and Science, No. 12, Medical-Use X-Ray CT Techniques Systematization Investigatory Report, Center of the History of Japanese Industrial Technology, (2008), 93.
9. A. Cormack: Nobel Lecture, (1979). (https://www.nobelprize.org/nobel_prizes/medicine/laureates/1979/cormack- lecture.pdf)
10. P. Reimers, H. Heidt, J. Stade, H.-P. Weise, Materialprüfung **22**, 214–217 (1980)
11. M. Tomizawa, T. Yamamoto, J. Inst. Electr. Eng. Jpn. **136**, 755–758 (2016)
12. J.C. Elliot, S.D. Dover, J. Microsc. **126**, 211–213 (1982)
13. L. Grodzins, Nucl. Instrum. Methods **206**, 541–545 (1983)
14. Z.H. Cho, K.S. Hong, O. Nalcioglu, Nucl. Instrum. Methods Phys. Res. **227**, 385–392 (1984)
15. A.C. Thompson, J. Llacer, L.C. Finman, E.B. Hughes, J.N. Otis, S. Wilson, H.D. Zeman, Nucl. Instrum. Methods Phys. Res. **222**, 319–323 (1984)
16. B.P. Flannery, H.W. Deckman, W.G. Roberge, K.L. D'Amico, Science **237**, 1439–1444 (1987)
17. K.L. D'Amico, H.W. Deckman, J.H. Dunsmuir, B.P. Flannery, W.G. Roberge, Rev. Sci. Instrum. **60**, 1524–1526 (2016)
18. J.H. Kinney, Q.C. Johnson, M.C. Nichols, U. Bonse, R.A. Saroyan, R. Nusshardt, R. Pahl, Rev. Sci. Instrum. **59**, 196–197 (1988)
19. J.H. Kinney, Q.C. Johnson, M.C. Nichols, U. Bonse, R.A. Saroyan, R. Nusshardt, R. Pahl, Rev. Sci. Instrum. **60**, 2471–2474 (1989)
20. Y. Suzuki, K. Usami, K. Sakamoto, H. Shiono, H. Kohno, Jpn. J. Appl. Phys. **27**, L461–L464 (1988)
21. T. Hirano, K. Usami, K. Sakamoto, Rev. Sci. Instrum. **60**, 2482–2485 (1989)
22. A. Koch, C. Raven, P. Spanne, A. Snigirev, J. Optical Soc. Am. A. **15**, 1940–1951 (1998)
23. Y. Wang, F.D. Carlo, I. Foster, J. Insley, C. Kesselman, P. Lane, G. Laszewski, D.C. Mancini, I. McNulty, M.-H. Su, B. Tieman: Proc. SPIE 3772, Developments in X-Ray Tomography II, (1999), pp 318–327
24. K. Uesugi, Y. Suzuki, N. Yagi, A. Tsuchiyama, T. Nakano, Nucl. Instrum. Methods Phys. Res. A **467–468**, 853–856 (2001)
25. F. Beckmann, U. Bonse, Proc. SPIE 3772, Developments in X-Ray Tomography II, Denver, Colorado, (1999), pp 179–187
26. B.A. Dowd, G.H. Campbell, R.B. Marr, V.V. Nagarkar, S.V. Tipnis, L. Axe, D.P. Siddons: Proc. SPIE 3772, Developments in X-Ray Tomography II, (1999), pp 224–236
27. S. Heinzer, T. Krucker, M. Stampanoni, R. Abela, E. Meyer, A. Schuler, P. Schneider, R. Muller: Proc. SPIE 5535, Developments in X-Ray Tomography IV, (2004), pp 65–76
28. Y. Suzuki, H. Toda, *Advanced tomographic methods in materials research and engineering.* (Oxford University Press, 2008), pp 181–201
29. F. Zernike, Physica **9**, 686–698 (1942)
30. U. Bonse, M. Hart, Appl. Phys. Lett. **6**, 155–156 (1965)
31. A. Momose, T. Takeda, Y. Itai, K. Hirano, Nat. Med. **2**, 473–475 (1996)
32. P. Burnstein, R.C. Chase, R. Mastronardi, T. Kirchner, Mater. Eval. **40**, 1280–1284 (1982)
33. P. Burnstein, P.J. Bjorkholm, R.C. Chase, F.H. Seguin, Nucl. Instrum. Methods Phys. Res. **221**, 207–212 (1984)
34. F.H. Seguin, P. Burnstein, P.J. Bjorkholm, F. Homburger, R.A. Adams, Appl. Opt. **24**, 4117–4123 (1985)

35. L.A. Feldkamp, L.C. Davis, J.W. Kress, J. Opt. Soc. Am. A **1**, 612–619 (1984)
36. L.A. Feldkamp, G. Jesion, Rev. Prog. Quant. Nondestruct. Eval. **5A**, 555–566 (1986)
37. S. Aoki, M. Oshiba, Y. Kagoshima, Jpn. J. Appl. Phys. **27**, 1784–1785 (1988)
38. F.F. Hopkins, I.L. Morgan, H.D. Ellinger, R.V. Klinksiek, G.A. Meyer, J.N. Thompson, IEEE Trans. Nucl. Sci. **28**, 1717–1720 (1981)
39. W.B. Gilboy, J. Foster, M. Folkard, Nucl. Instrum. Methods **193**, 209–214 (1982)
40. I. Taguchi, S. Nakamura, J. Iron Steel Inst. Japan **71**, 1685–1691 (1985)
41. M. Stauber, R. Müller: Osteoporosis, Volume 455 of the series Methods In Molecular Biology, (2008), pp 273–292
42. M. Hirano, J. Jpn. Soc. Non-Destruct. Insp. **58**, 219–222 (2009)
43. Industrial Structure Council/Industrial Technical Environment Sub-Committee, *Research Development/Innovation Sub-Committee, Assessment Working Group: Manufacturing Revolution Program with Three-Dimensional Contrast Imaging at Its Core* (Development of Assessment Standard Techniques for Next-Generation Three-Dimensional Interior/Exterior Measurements), Project Technical Assessment Result Report, 2017).
44. VDI/VDE 2617–13: VDI, Duesseldorf, 2011. (2011).
45. VDI/VDE 2630–1.3: VDI, Duesseldorf, 2011. (2011)
46. H. Toda, M. Kobayashi, Y. Suzuki, A. Takeuchi, K. Uesugi, J. Jpn. Soc. Non-Destruct. Insp. **58**, 433–438 (2009)
47. M. Kobayashi, Y. Ohkawa, H. Toda, K. Uesugi, Y. Suzuki, A. Takeuchi, J. Jpn. Inst. Met. Mater. **77**, 375–384 (2013)

Chapter 2
Fundamentals of X-Ray Imaging

As shown in Fig. 2.1, X-rays are electromagnetic waves with a short wavelength, or, in other words, high energy. Soft X-rays with a wavelength of approximately 0.1–10 nm (X-ray energies corresponding to this wavelength are 0.124–12.4 keV) cannot penetrate thick substances; hence, they are used for imaging organisms or polymers made of light elements such as silicon, carbon, oxygen, hydrogen, and nitrogen. Meanwhile, the wavelengths of hard X-rays are even shorter and are smaller than the size of an atom (~0.1 nm). For this reason, hard X-rays tend to transmit through thick objects as well. Furthermore, the 2–6 keV X-ray energy range between soft and hard X-rays have been recently separately classified as "tender X-rays." Their free path in the air is short but they indicate X-ray energy domains used by crystal spectrometers in synchrotron radiation facilities.

Incidentally, the relationship in Fig. 2.1 between the X-ray wavelength λ and energy E can be calculated using the well-known equation shown below:

$$E = h\nu = \frac{hc}{\lambda} \tag{2.1}$$

Here, h is the Planck constant (6.63×10^{-34} Js), ν is the frequency, and c is the speed of light (3×10^8 m/s). A convenient rule-of-thumb is that X-ray energy in units of keV can be roughly calculated by dividing 1.24 by the wavelength λ in units of nm. The value of 1 eV corresponds to the amount of energy obtained when an electron is accelerated with a voltage of 1 V, which refers to a value of 1.6×10^{-19} J in SI units. Furthermore, X-ray wavelengths are often expressed in nm but wavelengths shorter than soft X-rays may sometimes use Å as well (1 nm = 10 Å).

As shown in Fig. 2.2, a number of configurations can be considered when conducting imaging with X-rays. First, (c) is rarely used in the material and mechanical engineering disciplines. As discussed later, using the (a) type, it is difficult to separate the effects of density and atomic number. Meanwhile, usage of the (b) type allows for the calculation of sample-interior density distributions [1]. The two phenomena of X-ray absorption and phase shift are discussed below, keeping in

H. Toda, *X-Ray CT*,
https://doi.org/10.1007/978-981-16-0590-1_2

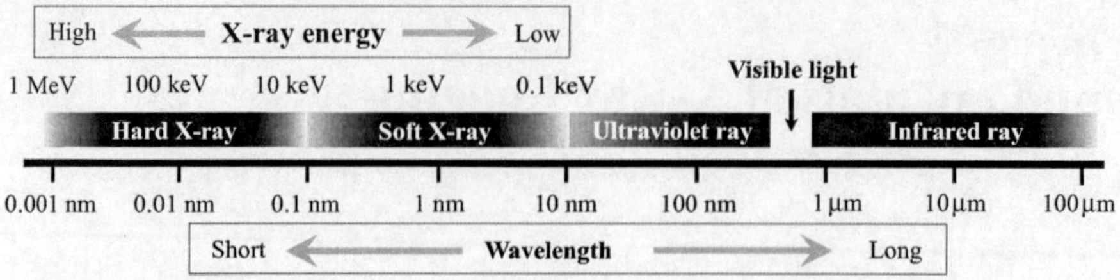

Fig. 2.1 Relationship between the wavelength and energy of electromagnetic waves (including X-rays)

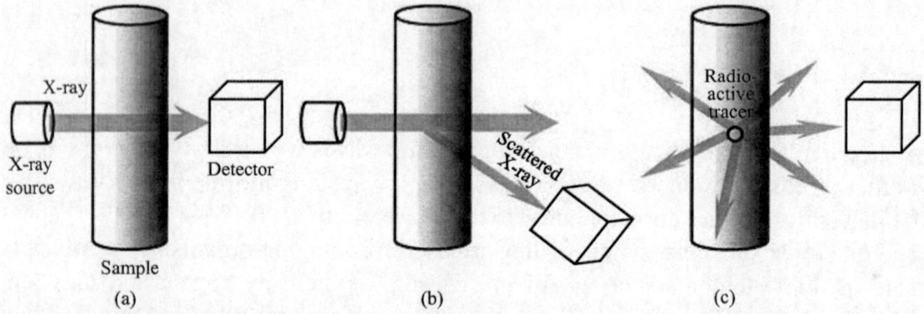

Fig. 2.2 Schematic of imaging with X-rays: **a** acquisition of the transmission image, **b** detection of the scattered Xrays, and **c** detection of gamma rays, etc. emitted from the sample interior

mind the transmission-type X-ray imaging in Fig. 2.2a, which is most commonly used. These correspond to the two techniques of *absorption contrast tomography* and phase-contrast tomography, respectively. These are explained in order in this chapter.

Note that a distinction may be made between X-ray absorption and attenuation depending on the text used. For example, as discussed later, the photoelectric effect may exhibit the physical absorption of X-rays but coherent scattering does not. However, if an imaging experiment is established where an X-ray source and detector are placed in front of and behind the sample, respectively, a decrease in the X-ray intensity after transmitting through a sample can be observed, despite coherent scattering occurring. Therefore, both X-ray scattering and absorption can be interpreted as X-ray absorption. With this in mind, this text will express all cases where either X-ray absorption or scattering is dominant as X-ray absorption.

2.1 Absorption Contrast

2.1.1 Absorption Coefficient and Contrast

(1) Absorption Coefficient

As shown in Fig. 2.3, we assume a simple case where *monochromatic X-rays* obtained from locations such as a synchrotron radiation facility are emitted onto a sample with thickness L with uniform density and chemical composition. The X-ray intensities before and after transmission are set as I_0 and I, respectively. I is defined as follows according to the *Lambert-Beer law*:

$$I = I_0 e^{-\mu L} \tag{2.2}$$

Here, μ is the *linear absorption coefficient*, which is dependent on X-ray wavelength as well as sample density and atomic number. It is often expressed not in SI units but as [cm^{-1}]. X-rays are absorbed by the sample with even slight differences in thickness, so I will always be smaller than I_0. The linear absorption coefficient value can be determined as shown in Eq. (2.2) if X-ray intensities in the absence and presence of the sample are experimentally calculated. Furthermore, the magnitude of X-ray absorption may vary by location if the linear absorption coefficient is not uniform throughout the sample interior and areas with variable linear absorption coefficients are present in its interior. Light–dark contrasts that correspond to this appear in the transmission image and therefore, the existence and shapes of these domains can be identified. This is the foundation of absorption contrast imaging. Equation (2.2) is not accurate in cases where the electron density in a specimen is not uniform, or in other words, where the X-ray is either *refracted* or reflected on the boundary surface when transmitting through a second phase, particles, or void. As an example, this corresponds to the use of X-ray beams with high *spatial coherence*

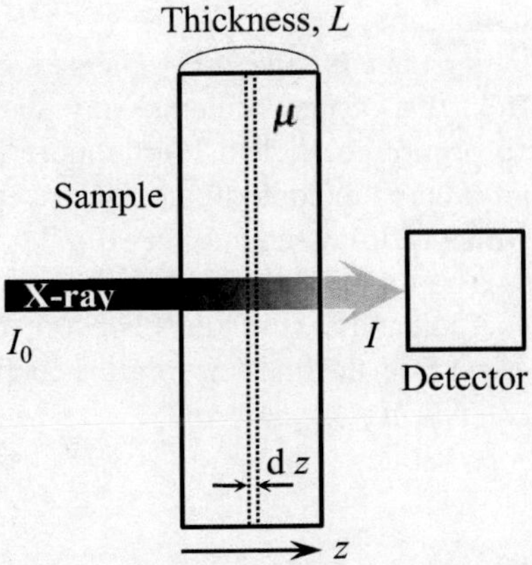

Fig. 2.3 Monochromatic X-ray with intensity I_0 emitted onto a sample of thickness L, with a uniform density and chemical composition. We consider the X-ray intensity I after transmission through the sample

(lateral coherence, transverse coherence) discussed in Sect. 2.2.2 (3), such as when using synchrotron radiation facilities or X-ray tubes with an extremely small focal spot size.

We consider a differential thickness dz at a distance z from the sample surface onto which the X-ray was emitted, as shown in Fig. 2.3. If the X-ray intensity at position z is set as I (z), the X-ray attenuation will be proportional to μ dz, thereby yielding Eq. (2.3). Solving for this differential equation by setting the boundary condition I (0) = I_0 yields Eq. (2.2).

$$\frac{dI}{I(z)} = -\mu dz \tag{2.3}$$

If the linear absorption coefficient is not uniform throughout the sample interior, the linear absorption coefficient at position z can be set as μ (z) and integration that follows the X-ray path can be used to write Eq. (2.2) as shown below:

$$I = I_0 e^{-\int_0^L \mu(z)dz} \tag{2.4}$$

Furthermore, X-rays from X-ray tubes are not monochromatic and have an energy distribution. Integrations of the X-ray energy spectra need to be conducted in these cases, as shown in the following equation.

$$I = \int_E I_0(E) e^{-\int_0^L \mu(z,E)dz} dE \tag{2.5}$$

Here, I_0 (E) and μ (z, E) are the incident X-ray intensity and linear absorption coefficient as functions of X-ray energy, respectively. Equations (2.4) and (2.5) are one-dimensional but the linear absorption coefficient has a distribution in the x–y plane vertical to the z-axis, so this is generally imaged using a two-dimensional detector. The sample is imaged from multiple directions while being rotated, and an X-ray tomography image reconstruction algorithm is used to reconstruct a 3D distribution of the linear absorption coefficient. This is the basic principle of X-ray tomography.

Incidentally, the X-ray energy spectra after transmission through the object vary from that of the incident X-ray due to the X-ray energy dependency of the linear absorption coefficient. Furthermore, lower-energy X-rays are more rapidly attenuated after being incident to an object, generating the artifact known as beam hardening, which is discussed later in Sect. 7.6.1.

The linear absorption coefficient is also dependent on the object density ρ, so a coefficient that is not dependent on density can be defined by removing the density effect from the linear absorption coefficient. This is referred to as the *mass absorption coefficient* μ_{m}.

$$\mu_m = \frac{\mu}{\rho} \tag{2.6}$$

For example, as the same element may have large differences in density depending on whether it is a gas, liquid, or solid, and structural materials may have voids of various sizes, the mass absorption coefficient can often be useful. The units of the mass absorption coefficient are generally [cm^2/g].

Other than these, the *atomic absorption coefficient* μ_a and the *molar absorption coefficient* μ_M are also used. Their units are [cm^2/atom] and [cm^2/mol], respectively. The following relationship exists between each of these absorption coefficients.

$$\mu = \mu_m \rho = \mu_a \rho_a = \mu_a \rho \left(\frac{N_A}{A} \right) = \mu_M \left(\frac{\rho}{A} \right) \tag{2.7}$$

Here, A is the atomic weight [g/mol], N_A is Avogadro's number (6.02×10^{23} atoms/mol), and ρ_a is the atomic density [atoms/cm^3]. The variable μ_a is also known as the *atomic cross-section* and is generally represented by σ. The atomic cross-section is the probability of a photon interacting with a single atom. The extent of X-ray absorption can be assessed by multiplying the atomic cross-section with an atomic number of unit surface area. Returning to the transmission image in Fig. 2.3, the following equation can be obtained when expressing the event number W, which recounts the number of times an X-ray absorption event occurs in a region of differential thickness:

$$W = I(z)\rho_a dz \mu_a = I(z)\mu dz \tag{2.8}$$

In cases of alloys or composites where multiple types of atoms are mixed together, the probability of absorption by the constituent jth element is usually calculated as a sum. In other words, the mass absorption coefficient of alloys and composites μ_m^{mix} is the total sum of cross-sections per unit mass. Equation (2.9) is also known as the *rule of mixture*.

$$\mu_m^{mix} = \sum_{i=1}^{j} \mu_m^i x_i \tag{2.9}$$

Here, $\mu_m{}^i$ is the mass absorption coefficient of the ith element and x_i is its mass fraction. The rule of mixture states that the mass absorption coefficient for an X-ray energy that is sufficiently distant from the *absorption edge* (discussed later) can be calculated with an error of less than a few percent or 2% for X-ray energies above 10 keV and 1 keV, respectively [2].

(2) How to Investigate the Absorption Coefficient

Values such as the linear absorption coefficient for various elements at various X-ray energies can be obtained from handbooks or databases [3–6]. A Ref. [3] that is used frequently is the Physical Measurement Laboratory at the National Institute of Standards and Technology in the U.S. Detailed data on the absorption edge, absorption coefficient, and cross-section of elements and compounds with an atomic number

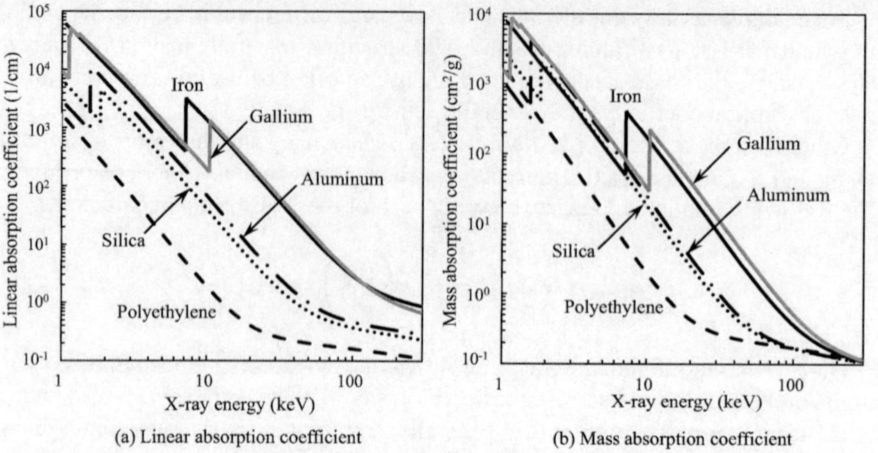

Fig. 2.4 X-ray energy dependencies of the linear absorption coefficient and mass absorption coefficient of the three representative compositional materials (polymers, ceramics, metals) [3]. High-density polyethylene was shown as a polymer, silica as a ceramic, and aluminum as a metal. Gallium is also shown as a reference due to its use in the 3D imaging of grain boundaries discussed in Sect. 5.5.1

over 90 can be obtained; for example, X-ray energy dependencies. Reference [5] is from the Center for X-ray Optics at the Lawrence Berkeley National Laboratory, where detailed data such as the complex index of refraction and transmissivity of elements or compounds can be obtained. Other than this, data on the absorption edge or transmissivity can be easily looked up using smartphone apps, which can be readily used during experiments [6].

Figure 2.4 shows the X-ray energy dependencies of the linear absorption coefficient and mass absorption coefficient using databases from reference [3]. Representative structural materials such as high-density polyethylene, silica, aluminum, and iron were included in Fig. 2.4. The absorption coefficient rapidly decreased as energy became higher. Figure 2.4 shows that the absorption coefficient jumped at around the 7-keV point for iron; this is the absorption edge, which corresponds to the *binding energy* of the electrons in the innermost shell (K-shell in this case). The K absorption edge, which is the absorption edge corresponding to the K-shell of each element, is summarized in Table 2.1 [4]. As an extremely rough approximation, the energy at the K absorption edge is proportional to Z^2. Furthermore, materials with a large Z exhibit large X-ray absorption but can also be inverted in certain sections due to the presence of the absorption edge. The X-ray energy dependencies of the absorption coefficient include the linear domain of the low-energy section (e.g. approximately below 30 keV for aluminum) and flat domain of the high-energy section (approximately over 100 keV for aluminum). This is due to the transition of the predominant absorption mechanism discussed in Sect. 2.1.2. The X-ray energy range often used in X-ray microtomography is thought to be between 5 to 50 keV but this range includes the transition domains of absorption edges, linear domains, and

Table 2.1 Values for coefficients *C* and *D*, as well as Z/A, used in the Victoreen formula (Eq. (2.10)) for representative elements. C_1 and D_1 are values below the K absorption edge; C_2 and D_2 are values above the K absorption edge [4]. Values for the K absorption edge are shown for each element in the table as well

Element	C_1	D_1	K absorption edge (keV)	C_2	D_2	Z/A
Be	0.365	0.00213				0.4438
C	1.22	0.0142				0.4995
N	2.05	0.0317				0.4997
O	3.IS	0.0654				0.5000
Mg	11.3	0.539				0.4934
Al	14.4	0.803				0.4818
Si	18.2	1.10				0.4984
Ca	55.8	7.56				0.4990
Ti	75.6	12.3	4.96	5.15	0.153	0.4593
V	86.9	15.1	5.46	6.14	0.203	0.4514
Cr	99.0	18.2	5.99	7.24	0.268	0.4614
Mn	112	22.3	6.54	8.51	0.344	0.4550
Fe	126	27.2	7.11	9.95	0.433	0.4655
Co	141	33.2	7.71	11.6	0.535	0.4581
Ni	158	40.1	8.33	13.4	0.651	0.4769
Cu	176	48.3	8.98	15.6	0.779	0.4564
Zn	195	57.7	9.66	17.8	0.937	0.4589
Ga	216	68.6	10.37	20.2	1.13	0.4446
Mo	555	336	20.00	56.2	7.73	0.4377

flat domains, depending on the element. Furthermore, the absorption coefficients of various materials converge above 20–100 keV, hence, contrast becomes difficult to obtain when imaging is conducted. This is also discussed in Sect. 2.1.2.

The mass absorption coefficient can also be estimated from the empirical equation known as the *Victoreen formula* [7]:

$$\mu_m = \frac{\mu}{\rho} = C\lambda^3 - D\lambda^4 + \frac{\sigma_{KN} N_A Z}{A} \tag{2.10}$$

Here, *C* and *D* are coefficients, which vary according to the atomic number *Z*, and σ_{KN} is the *Klein-Nishina cross-section*, which is used as a correction when the wavelength is short [8]. The *C*, *D*, and $\sigma_{\mathrm{KN}} N_{\mathrm{A}}$ values necessary for the calculation of Eq. (2.10) are shown for representative elements in Tables 2.1 and 2.2 [4]. Care must be taken as *C* and *D* values can vary around the absorption edge. Comparisons of the mass absorption coefficient for aluminum between 5–60 keV, between the calculated value from Eq. (2.10) and Ref. [3] database showed that its differences were just slightly over 5% at a maximum.

Table 2.2 $\sigma_{KN}N_A$ values for the Victoreen formula (Eq. (2.10)) [4]

Energy (keV)	$\sigma_{KN}N_A$ (cm^2)
62.0	0.329
41.3	0.350
31.0	0.362
24.8	0.368
20.7	0.374
17.7	0.378
15.5	0.381
13.8	0.383
12.4	0.385
11.3	0.386
10.3	0.388
9.5	0.389
8.9	0.390
8.3	0.391
7.8	0.392
7.3	0.393
6.9	0.394
6.5	0.394
6.2	0.394
5.9	0.395
5.6	0.395
5.4	0.395
5.2	0.395
5.0	0.396
4.8	0.396
4.6	0.396

From Eq. (2.10) and Table 2.1, the mass absorption coefficient value has a monotonically increasing tendency with an increasing atomic number. In other words, absorption contrast becomes more difficult to obtain when the atomic numbers are close to each other in value or when the absorption coefficients of alloys or composites calculated in Eq. (2.9) are close in value to each other even if the atomic numbers are not. Conversely, when neutrons are used, elements with extremely large mass absorption coefficients—such as hydrogen, boron, lithium, cadmium, and gadolinium—are observed across a wide range of atomic numbers. Accordingly, elements that have high mass absorption coefficients for X-rays (e.g. lead or gold) may have relatively low neutron absorption. This is because X-rays interact with the electrons on the outer edges of the atomic nucleus, whereas neutrons interact with the nucleus and are absorbed/scattered. With this in mind, the complementary use of X-rays and neutrons can allow for the imaging of various elements with sufficient contrast.

2.1.2 Various X-Ray Absorption Processes

There are various mechanisms by which an object absorbs X-rays, which vary according to the X-ray energy used. The contribution of various X-ray absorption mechanisms can be summed together as shown in Eq. (2.11), which allows us to discern the X-ray absorption of the object.

$$\mu = \mu_a \rho \left(\frac{N_A}{A}\right) = \left(\mu_a^{pe} + \mu_a^{incoh} + \mu_a^{coh} + \mu_a^{pair} + \mu_a^{trip} + \mu_a^{ph.n}\right) \rho \left(\frac{N_A}{A}\right) \tag{2.11}$$

Here, μ_a^{pe}, μ_a^{incoh}, μ_a^{coh}, μ_a^{pair}, μ_a^{trip}, and $\mu_a^{ph.n}$ are the atomic cross-sections that correspond to *photoelectric absorption*, *incoherent scattering*, *coherent scattering*, *pair production*, *triplet production*, and photonuclear absorption, respectively. Here, coherent scattering refers to cases where there is a relationship with the X-ray phase before and after X-ray scattering. In coherent scattering, the electrons in the object exposed to the same incident X-ray, all vibrate along the same phase, so the scattered X-rays produced by these electrons also reach the same phase and interfere with each other. Meanwhile, in incoherent scattering, there is no relationship with the X-ray phase before and after X-ray scattering. X-ray scattering can be classified further into *elastic scattering* and *inelastic scattering*; the former is when the X-ray energy does not change before and after scattering and the latter is when the X-ray energy does change.

The variables, μ_a^{trip} and $\mu_a^{ph.n}$ in Eq. (2.11) are insignificantly small for X-ray energies below 20 MeV [9]. Here, X-ray imaging, which uses absorption contrasts, can generally be approximated as shown below:

$$\mu = \left(\mu_a^{pe} + \mu_a^{incoh} + \mu_a^{coh} + \mu_a^{pair}\right) \rho \left(\frac{N_A}{A}\right) \tag{2.12}$$

The most important X-ray scattering process is inelastic incoherent scattering. This was named Compton scattering, after the eponymous American physicist who received the Nobel Prize for his work in Physics. Coherent scattering can be classified between scattering due to free and unrestrained *charged particles*, also known as *Thomson scattering*, and scattering due to orbital electrons tightly bound to the nucleus, also known as *Rayleigh scattering*.

Details on interactions between X-rays and atoms are left to others [2], and the following discussion will focus on these processes in order. This will support better understanding of the fundamentals of absorption contrast tomography.

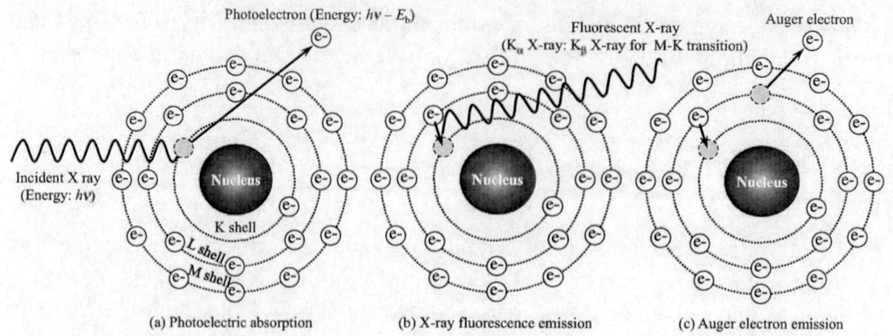

Fig. 2.5 Schematic of the photoelectric absorption process and the corresponding processes of X-ray fluorescence emission and Auger electron emission

(1) Photoelectric Absorption

As shown in Fig. 2.5a, the incident X-ray transfers all of its energy to the electron when the X-ray energy is greater than the binding energy E_b of electrons in places like the K-shell, and the electron is ejected from the atom. This phenomenon is referred to as the *photoelectric effect*. Albert Einstein received the Nobel Prize in 1921 for his explanation of this effect. The K-shell electron binding energy is also referred to as the K absorption edge; the value of each element is shown in Table 2.1. The ejected electron has high kinetic energy and is referred to as a *photoelectron*. The photoelectron kinetic energy E_{pe} is expressed with the *Einstein relation*:

$$E_{pe} = h\nu - E_b \tag{2.13}$$

X-rays with energies lower than the K absorption edge can only eject electrons from the L- or M-shells. Electrons can be ejected from the K-shell once the X-ray energy exceeds that of the K absorption edge and the absorption coefficient increases by an order of magnitude. The absorption edge energies of a number of elements subject to X-ray imaging are included as a reference in Table 2.3 [3]. Among these, for example, the L_I absorption edge corresponds to the 2 s-orbital of the L-shell, the L_{II} and L_{III} absorption edges correspond to the 2p-orbital of the L-shell, etc. The absorption edge energy of the L-shell ranges from one order of magnitude smaller than the K-shell energy to even lower. Furthermore, photoelectric absorption is more likely to occur if the incident X-ray and electron binding energies are closer to each other in value.

Incidentally, the absorption coefficient of X-rays due to photoelectric absorption can be expressed as follows:

$$\mu_{pe} = k\rho Z^{\alpha}\lambda^{\beta} \tag{2.14}$$

Table 2.3 Absorption edge energy values for representative elements [3] (keV)

Element	K	L_I	L_{II}	L_{III}	M_I	M_{II}	M_{III}	M_{IV}	M_V
C	0.28	0.020	0.0064						
O	0.53	0.024	0.0071	0.0071					
Mg	1.30	0.089	0.051	0.051					
All	1.56	0.12	0.073	0.073	0.0084				
Si	1.84	0.15	0.099	0.099	0.011	0.0051			
Ti	4.97	0.56	0.46	0.46	0.060	0.035	0.035		
Fe	7.11	0.85	0.72	0.71	0.093	0.054	0.054	0.0036	0.0036
Cu	8.98	1.10	0.95	0.93	0.12	0.074	0.074	0.0016	0.0016

Here, $\alpha \approx 4–5$, $\beta \approx 3–3.5$, and k is a constant. The cross-section due to photoelectric absorption is proportional to the 4th or 5th power of the atomic number Z, so the X-ray absorption extent can vary widely with the element type, which is extremely useful for X-ray imaging. Furthermore, Eq. (2.14) clarifies how materials with large atomic numbers can be used for shielding. A well-known application example of the photoelectric effect is *photoelectron spectroscopy*, which obtains electron energy distributions by studying photoelectron energy and spin conditions. Another well-known application example is the *photoelectron microscope*, which enlarges and images photoelectrons using an electron optics system. This enables imaging at high spatial resolutions (several dozen nm) in the soft X-ray domain with synchrotron radiation [10]. Furthermore, the usage of circularly polarized light in the soft X-ray domain allows for the observation of magnetic domain structures in ferromagnetic materials using X-ray magnetic circular dichroism [11].

An electron vacancy remains in the spot from which the electrons were ejected with photoelectric absorption. This is an unstable condition, so electrons from outer shells such as the L-shell transit to fill this vacancy, as shown in Fig. 2.5b. The binding energies for the electron in the orbitals before and after the transition are different, so an X-ray that has the amount of energy equivalent to this difference is emitted. This is called *fluorescence*. *X-ray fluorescence* energy is dependent on the atom that produced the fluorescence. The usage of these as high-sensitivity chemical analysis methods is well known.

Meanwhile, as shown in Fig. 2.5c, the difference in binding energy generated to fill the vacancy is reduced to the ground state energy through the emission of the electron. The emitted electron is referred to as the *Auger electron*. The Auger electron has element-specific energy and is used for chemical analysis referred to as *Auger electron spectroscopy*. The *mean free path* of Auger electrons is extremely short, so they are used for surface analysis.

The probability that a fluorescent X-ray is emitted through the relaxation process is known as the *fluorescence yield* ω [12]. In contrast, the probability that an Auger electron is emitted is $1 - \omega$.

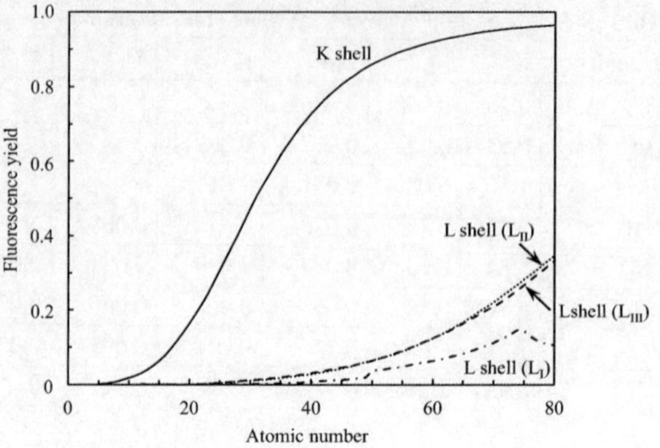

Fig. 2.6 Fluorescence yield for each element [12]

$$\omega = \frac{1}{1 + a_y Z^{-4}} \tag{2.15}$$

Here, $a_y = 1.12 \times 10^{6}$. From Eq. (2.15), atoms with a large atomic number emit a fluorescent X-ray, whereas the Auger effect is predominant in light elements. For example, the fluorescence yields of aluminum, iron, and tin are 0.025, 0.29, and 0.85, respectively [12]. The fluorescence yields due to the electrons, including the L-shell, of each element are shown in Fig. 2.6 [12]. Figure 2.6 shows that the fluorescence yield in the L-shell is extremely low, relative to others.

(2) *Compton Scattering*

We first consider an incident X-ray with an energy higher than the binding energy of an orbital electron. The photon collides with a weakly-bound outer-shell electron, the electron with kinetic energy is ejected as a *recoil electron*, and the scattering X-ray energy simultaneously becomes lower (i.e. its wavelength increases) in a phenomenon referred to as the *Compton effect*. Figure 2.7 shows a schematic of this effect. In this case, the energy and momentum are preserved. Using this, the change in wavelength $\Delta\lambda$ due to Compton scattering can be determined as shown below:

$$\Delta\lambda = \frac{h}{m_e c}(1 - cos\theta) \tag{2.16}$$

Here, m_e is the electron mass and θ is the scattering angle. From Eq. (2.16), changes in the wavelength increase as the scattering angle becomes a larger back-scattering value.

The Compton effect becomes predominant over the photoelectric effect at high X-ray energies. Furthermore, low-energy X-rays have a deflection angle θ, which exceeds 90°, and its *back-scattering* tendencies increase. Meanwhile, high-energy X-rays have a deflection angle θ, which drops below 90°, and its *forward scattering* tendencies increase. Due to their low binding energy, outer-shell electrons have a

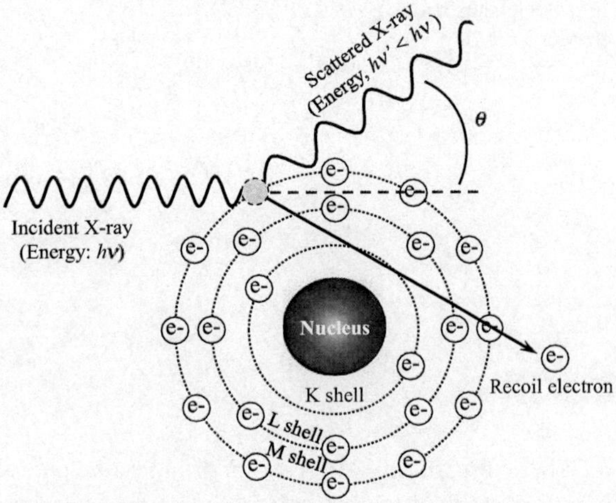

Fig. 2.7 Schematic of compton scattering

relatively low amount of scattering X-ray energy loss. For this reason, scattering X-rays can continue to interact with the atom, and can be the cause of negative influences on image quality during X-ray imaging. Meanwhile, recoil electrons act as free electrons, move to other atomic locations and serve a vacancy-filling role.

Even with the same non-elastic scattering, *Raman scattering* occurs within tightly-bound inner-shell electrons of light elements. Raman scattering is known to occur in the visible light spectrum but is also used in *X-ray Raman scattering*, which uses high-brilliance X-rays due to synchrotron radiation, to analyze the electron configuration of light elements using X-rays.

The probability of the Compton effect occurrence is dependent only on the electron density and not on the atomic number of the elements involved. For this reason, the atomic cross-section only shows a weak dependence on the atomic number Z of a physical material; in other words, only low contrast can be achieved with Compton scattering.

(3) Coherent Scattering

Coherent scattering is the interaction between low-energy (below 10 keV) X-rays and the material. The X-ray energies in these cases are lower than the binding energy of the electron. There are two types of coherent scattering: Thomson scattering, and Rayleigh scattering. The difference between them is in how tightly bound the electron is to the atom. During coherent scattering, the incident X-ray induces resonant vibration in the electron. As a result, the atom emits an X-ray with the exact wavelength as the incident X-ray in the opposite direction, as shown in Fig. 2.8. Therefore, there is no energy exchange between the X-ray and the material. In these cases, the scattering X-ray angle distribution is known to deviate towards the front. Furthermore, the phase relationship between the incident and scattering X-rays is retained.

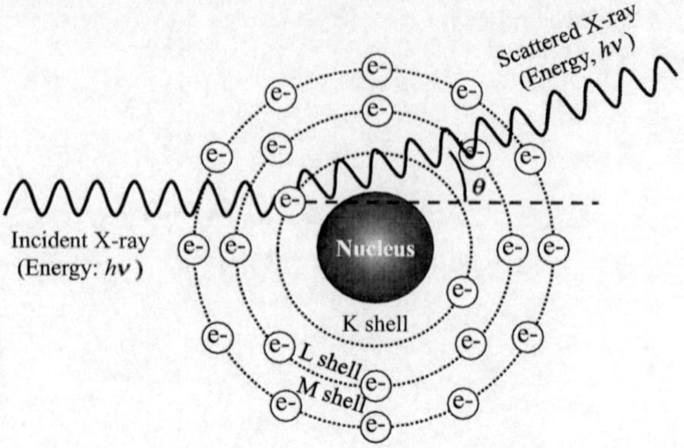

Fig. 2.8 Schematic of coherent scattering

Coherent scattering is not as important as the photoelectric effect, Compton scattering, and pair production in X-ray imaging, however, its presence should be recognized. The reason for this is because the scattered X-ray follows a path that is different from the incident X-ray, which generates a blur in the X-ray image. Coherent scattering induces X-ray *diffraction* in crystals. The diffraction angle in this case is dependent on the X-ray wavelength and crystallographic lattice spacing. Meanwhile, wavelength increases with Compton scattering; hence, there is no interference with the incident X-ray and no effect on the measurement of interplanar spacing using X-ray diffraction.

In Thomson scattering, the cross-section does not depend on X-ray energy. Meanwhile, for Rayleigh scattering, the cross-section is roughly proportional to the second power of the atomic number of the material. Rayleigh scattering becomes insignificantly small above 100 keV in materials with a small atomic number.

(4) Pair Production

Pair production becomes possible when the incident X-ray energy is over 1.022 MeV. This corresponds to the sum of the electron and positron static energies (m_e c^2 = 511 keV for each). As shown in Fig. 2.9, the incident X-ray interacts with the electric field of the nucleus and is annihilated; most of the energy of the X-ray photon becomes the kinetic energy of the positron and electron. The reaction cross-section of a single atom is roughly proportional to the second power of the atomic number of the material. Pair production cannot be ignored in X-rays whose high energies exceed 10 MeV. The generated positron is unstable and is known to become annihilated upon interaction with an electron in a material (i.e. *positron annihilation*).

Furthermore, although the probability is low at less than half that of pair production, recoil electrons are produced upon interaction with the electron field when the incident X-ray energy is above 2.044 MeV, resulting in triplet production.

(5) Photonuclear Reaction

The photonuclear reaction refers to interactions between high-energy X-rays and the nucleus and is the phenomenon in which the total energy of the X-ray photon is

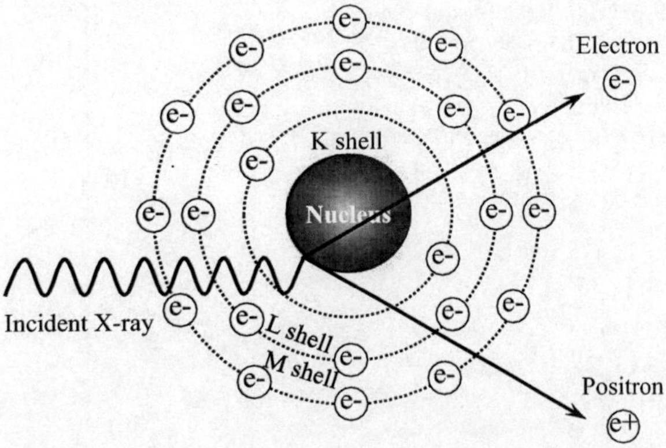

Fig. 2.9 Schematic of pair production

transferred to the nucleus and where the photon is annihilated. Alongside this process, neutrons, protons, π-mesons, etc. are all emitted. Generally, this reaction occurs with extremely high X-ray energies that exceed 7–20 MeV and is not important for X-ray imaging.

(6) Dominant Absorption/Scattering Processes

As shown in Eq. (2.12), the contributions to the mass absorption coefficient from photoelectric absorption, incoherent scattering, coherent scattering, and pair production (excluding triplet production and photonuclear reactions) are investigated. Figures 2.10, 2.11, 2.12 and 2.13 show the X-ray energy dependencies of the mass absorption coefficients of carbon, aluminum, iron, and zirconium using the Ref. [3]

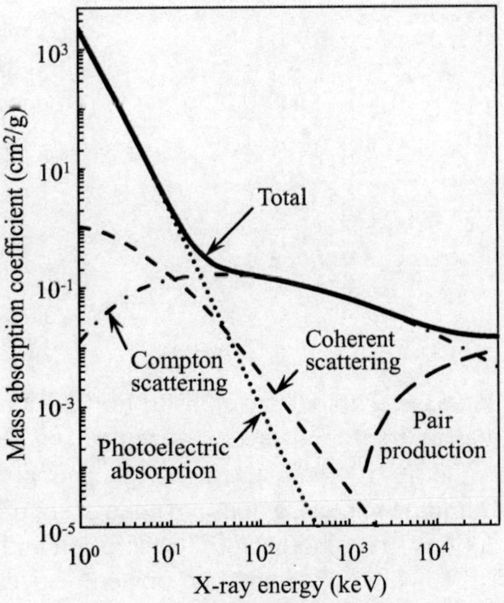

Fig. 2.10 Relationship between mass absorption coefficient and X-ray energy for carbon [3], drawn to show the contributions of the various interactive effects of the X-ray and the material

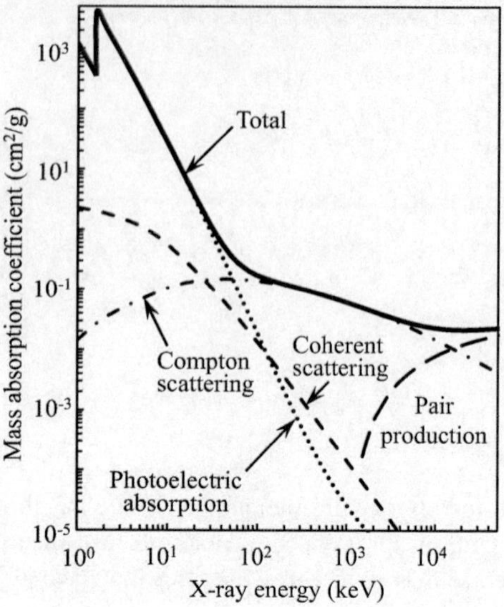

Fig. 2.11 Relationship between mass absorption coefficient and X-ray energy for aluminum [3], drawn to show the contributions of the various interactive effects of the X-ray and the material

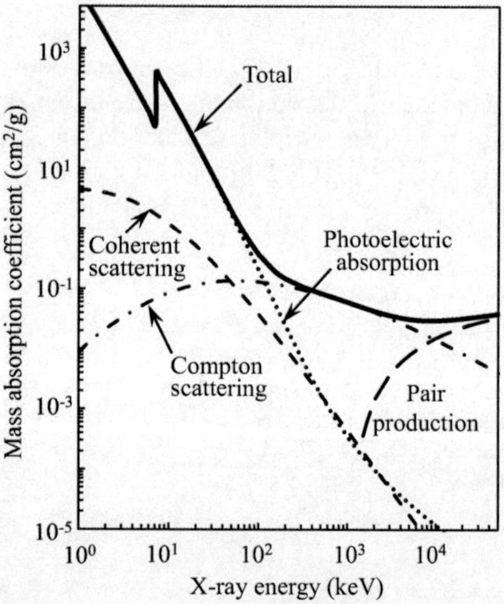

Fig. 2.12 Relationship between mass absorption coefficient and X-ray energy for iron [3], drawn to show the contributions of the various interactive effects of the X-ray and the material

database. Photoelectric absorption is dominant for each material for X-ray energies up to around several dozen to 100 keV. Below this range of values, Compton scattering is dominant; above it, pair production is dominant. Absorption due to Compton scattering does not change significantly with X-ray energy, whereas contributions from photoelectric absorption and pair production increase in proportion to the 4th–5th power and 2nd power, respectively, of Z, as seen in Sect. 2.1.2 (1) and

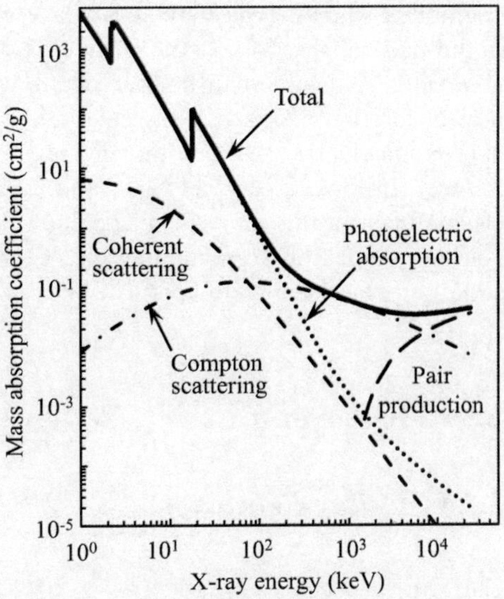

Fig. 2.13 Relationship between mass absorption coefficient and X-ray energy for zirconium [3], drawn to show the contributions of the various interactive effects of the X-ray and the material

(4). For this reason, the Compton scattering-dominant range becomes narrower with larger *Z*. A clear X-ray energy dependency, shown in Eq. (2.14), can be seen in the domain in which photoelectric absorption is dominant. This suggests that X-ray energy modulation is important to obtaining favorable imaging conditions.

Figure 2.14 shows changes in the mass absorption coefficient with a number of X-ray energies in elements ranging from light elements to heavy elements like mercury, using the same Ref. [3] database. X-rays with energies of 10 keV show large changes in the mass absorption coefficient particularly in the range of atomic numbers around 15. Meanwhile, there is very little change in the mass absorption

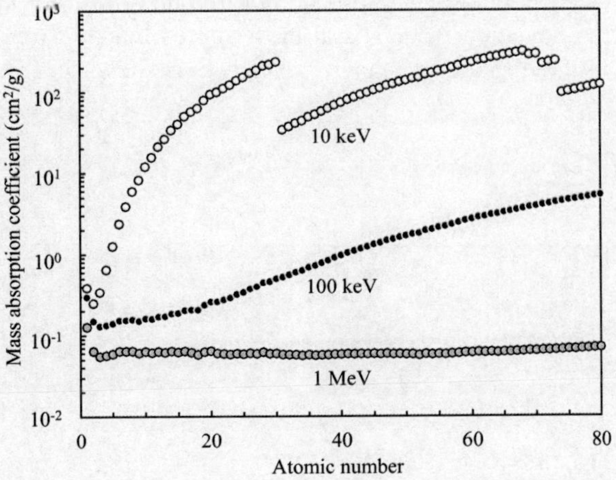

Fig. 2.14 Atomic number dependency of mass absorption coefficients at representative X-ray energies used in X-ray tomography [3]

coefficient when energies are at 100 keV. Similarly, there are virtually no changes in the mass absorption coefficient with 1-MeV energy X-rays. This indicates that contrast becomes more difficult to obtain with higher energy in X-ray imaging that uses absorption contrast. X-ray energies ranging from several dozen keV to below approximately 100 keV are known to be optimal for obtaining sufficient contrast. As seen in Figs. 2.10, 2.11, 2.12 and 2.13, photoelectric absorption is generally predominant in this range. The specific selection of the used X-ray energies must consider the factors for obtaining the necessary S/N ratio and spatial resolution, in addition to the strength of contrast. These will be summarized in Chap. 6.

2.2 Phase Contrast

2.2.1 *X-Ray Refraction*

(1) Snell's Law

High-school physics classes teach *Snell's law*, which expresses the relationship between the incidence and refraction angles. This is the relational expression between the incident angle θ_1 and angle of refraction θ_2:

$$\cos\theta_1 = n\cos\theta_2 \tag{2.17}$$

Figures shown in high-school textbooks are similar to Fig. 2.15a, wherein medium 1 is assumed to be air and medium 2 is assumed to be water or transparent plastic and where the incident light is depicted as visible light. The refraction indices n of water, polymers, and glass in the visible light spectrum often have values of 1.3–1.8. The frequency of X-rays is higher than the resonant frequency associated with electron binding conditions. For this reason, n becomes smaller than 1. However, the deviation from 1 is at most approximately 10^{-5}. For this reason, a schematic of the diffraction behavior of X-rays is shown in Fig. 2.15b. Incident X-rays at an

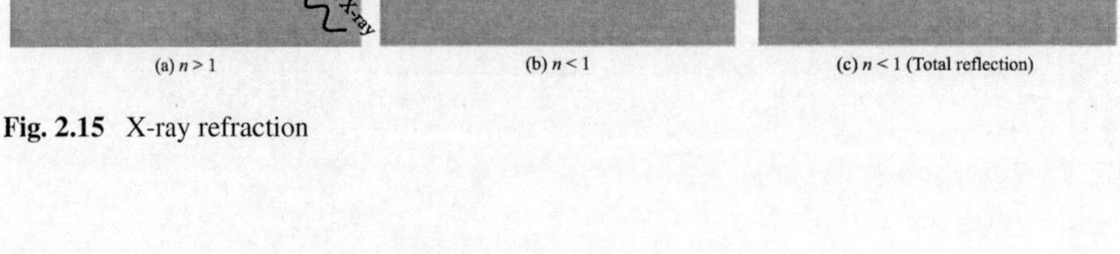

Fig. 2.15 X-ray refraction

angle orthogonal to the surface do not experience refraction and smaller θ_1 values result in larger refraction. When θ_1 is smaller than the *critical angle* θ_c, the X-ray undergoes *total reflection*, as shown in Fig. 2.15c. The total reflection of X-rays was reported by Compton as early as 1923 [13]. When $\theta_2 = 0$ in Eq. (2.17) and substituted into Eq. (2.18) (shown below) assuming that $\beta = 0$, $\theta_c \approx \sqrt{2\delta}$ (δ is discussed later). The θ_c value takes low values in the order of mrad, with values for silicon at approximately 3.1 mrad and 1.6 mrad for incident X-rays at energies of 10 keV and 20 keV, respectively.

In the case of Fig. 2.15c, X-rays can be focused if medium 2 is made into a gradual concave surface. This is the basic principle of an *X-ray condenser mirror* often used in X-ray imaging in synchrotron radiation facilities. X-ray beams with a given width are covered if the incident X-ray angle is too shallow, requiring a longer X-ray mirror. For this reason, it is advantageous to acquire as large a critical angle as possible. As seen in Eq. (2.19), discussed later, δ is proportional to the density of the surface material. With this in mind, heavy elements like platinum, rhodium, and nickel are used to obtain a larger critical angle for the X-ray mirror surface. For example, an adhesive layer is placed in a silicon substrate whose surface has been sufficiently polished and coated with a thin metallic film that is approximately 30–100 nm thick. Typically, the surface roughness of the mirror is controlled to approximately 0.5–1 nm and below.

Figure 2.16 shows the X-ray energy dependencies of the critical angle of the total reflection in three material types [5]. Increased X-ray energies result in a smaller critical angle of the total reflection. Furthermore, the size of the critical angle greatly varies according to the material. Points where the critical angle of total reflection greatly decrease correspond to the K absorption edge of each element. Here, we consider the X-ray mirror being fixed to a given incident angle and an incident X-ray with a given energy distribution present. As shown in Fig. 2.16, the high-energy X-ray above a given energy passes through the object without total reflection. This aspect

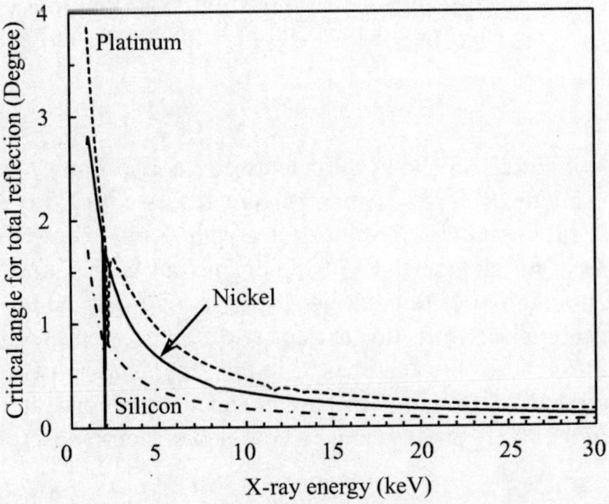

Fig. 2.16 Critical angle for the total reflection of X-rays in three material types of different densities [5]. The vertical axis corresponds to X-ray energy. Platinum (21.5 g/cm^3 density) and nickel 8.9 g/cm^3) are used as the surface layer of an X-ray mirror; silicon (2.3 g/cm^3 density) is used as its substrate

is applied to use X-ray mirrors as a *filter* that cuts high-energy X-rays. For example, *higher-order radiation* is present in X-rays that have undergone spectral diffraction with a *monochromator*. Higher-order radiation can be cut using X-ray mirrors for various experiments in which higher-order radiation may influence measurements.

Furthermore, during the total reflection of an X-ray, the penetration depth of an X-ray, which should ordinarily have high transmission capabilities into an object, can be controlled to the nm order. This can be used to selectively analyze only scattering X-rays from near the surface.

(2) Complex Refractive Index

Generally, the refractive index in cases where an X-ray advancing through the object interior is absorbed by the object is expressed with the *complex refractive index* $\hat{n}$. The variable $\hat{n}$ is shown below based on the free-electron approximation:

$$\hat{n} = 1 - \delta + i\beta \tag{2.18}$$

$$\delta = \frac{N_a Z \rho e^2 \lambda^2}{2\pi A m_e c^2} \tag{2.19}$$

$$\beta = \frac{\mu\lambda}{4\pi} \tag{2.20}$$

Here, N_a is Avogadro's number, A the mass number, ρ the density, and $e^2/(m_e c^2)$ the classic electron radius. The real component δ expresses the refraction and phase shift; the *phase shift* of X-rays is dependent on δ. However, $Z/A \approx 0.5$ for many materials, so δ becomes proportional to the density. From this, information relating to the density of the material interior can be obtained from phase-contrast imaging. Furthermore, the imaginary component β expresses X-ray absorption and is simply proportional to the mass absorption coefficient. However, care must be taken as the above equation does not hold near the absorption edge. From Eqs. (2.14), (2.19), and (2.20), δ and β are dependent on X-ray energy as shown below:

$$\beta \propto E^{-4}, \delta \propto E^{-2} \tag{2.21}$$

Figure 2.17 shows the calculated δ and β for a variety of light and heavy elements using the Ref. [5] database. δ/β increases with X-ray energy but decreases with soft X-rays. Actual δ/β values for a wide X-ray energy range are approximately 1000 for light elements, 100 for elements around aluminum, and above 10 for iron or zirconium. For this reason, stronger contrasts can be obtained in X-ray imaging by using phase shifts due to density differences instead of linear absorption coefficient differences between phases. In this way, phase-shift contrasts can be used in almost all types of structural and functional materials and, at the very least, they are effective in the X-ray energy range where photoelectric absorption is predominant.

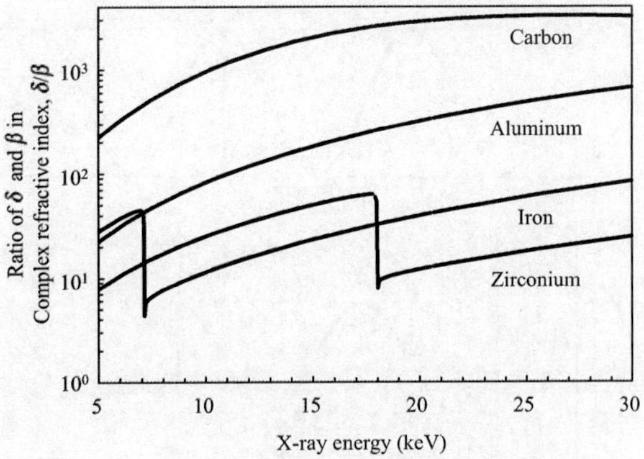

Fig. 2.17 Ratio of complex refractive index coefficients δ and β plotted against X-ray energy [5]; index showing how much more effective refraction- or phase-based contrast is relative to absorption contrast

There is a rich amount of information in phase-contrast imaging that cannot be obtained with absorption contrast. On the other hand, there are phenomena that can be seen by absorption contrast but not phase contrast, due to the usage of the absorption edge. These can be used in a complementary manner. However, simply measuring the X-ray intensity may not yield phase information. Some form of conversion between the phase, which interferes with a reference wave that acts as a standard, and intensity is needed to obtain phase contrast. For this, X-rays, which are spatially coherent to some extent, are needed. Furthermore, special setups discussed in Chap. 5 are needed. The reality is that there are difficulties in the implementation of phase-contrast imaging and analysis of the obtained images. On the other hand, even when intending to simply conduct absorption contrast imaging, internal structure size may be misestimated due to the convolution of phase effects, false recognition of unrealistic structures, and mischaracterization of material properties based on the absorption coefficient. These will be touched on in the implementation methods of Sect. 5.2 and the application examples in Sect. 6.6.3. Furthermore, Sect. 2.2.2 (3) will discuss what coherence is.

2.2.2 *X-Ray Phase Shift*

(1) Phase Shift

Phase-contrast imaging is a method that creates grayscale images by measuring phase shifts in X-rays passing through an object. Specific descriptions of this technique are provided in Chap. 5 and a further look into X-ray phase shifts will be conducted here.

When assuming a scenario where a monochromatic X-ray passes through an object with thickness L as shown in Fig. 2.18, both X-ray absorption-based attenuation and phase shift Φ occur concurrently. Here, "monochromatic" refers to a single wavelength with no wavelength distribution. The X-ray intensity before and after

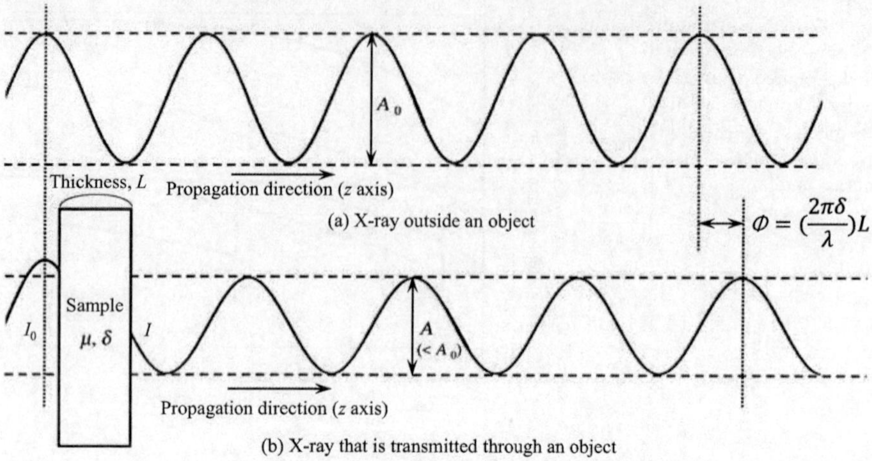

Fig. 2.18 Schematic showing the changes in amplitude A and phase shift Φ after a monochromatic X-ray with intensity I_0 is emitted toward an object with a thickness L and is transmitted through it

transmitting through the object is as shown in Eq. (2.1), but expressing X-ray attenuation with its ratio $\frac{I}{I_0}$ allows for a similar expression with X-ray amplitude before and after transmission as shown below:

$$\frac{I}{I_0} = \left(\frac{A}{A_0}\right)^2 \tag{2.22}$$

Meanwhile, a phase shift is generated, as the X-ray propagation speed in the object is different from its speed in a vacuum. Figure 2.19 shows an example in two dimensions instead of one: a two-dimensional schematic of the relationship between the X-ray *wavefront* and object. As previously mentioned, the refractive index of X-rays is less than 1, so the wavefront of X-rays after transmitting through an object is expanded slightly forward [14]. The phase shift expresses the amount of wavefront deformation similar to that shown in Fig. 2.19. The propagation direction of the X-ray at this point is in a direction orthogonal to the wavefront in Fig. 2.19. Therefore, adjacent waves are superimposed and interfere with one another in areas where the wavefront is curved. Interference fringes due to *Fresnel* or *Fraunhofer diffraction* can be observed depending on the distance from the object to the detector. This is explained in Sect. 2.2.2 (2).

Incidentally, the phase shift can be expressed by the complex refractive index δ from Eq. (2.18) and X-ray wavelength, as shown as follows:

$$\Phi = \left(\frac{2\pi\delta}{\lambda}\right)L \tag{2.23}$$

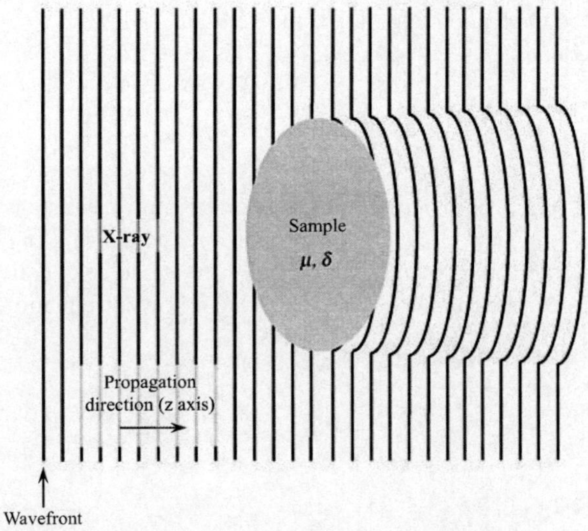

Fig. 2.19 Schematic showing changes in the wavefront of a monochromatic X-ray emitted toward an object and after transmitting through it

This equation is as shown below when δ is not homogeneous within the object interior:

$$\Phi = \left(\frac{2\pi}{\lambda}\right) \int_0^L \delta(z) dz \tag{2.24}$$

Therefore, the phase shift is the projection of δ that shows how much the real component of the complex refractive index decreases from 1, in the z-axis direction, corresponding to the propagation direction of the X-ray. This is identical to when the local linear absorption coefficient in Eq. (2.4) was projected in the pathway of the X-ray, when the linear absorption coefficient was not homogeneous within the object interior. The smallest detectable phase shift has been shown to be around 5% (0.05 rad) [15]. δ has a distribution in the x–y plane orthogonal to the z-axis, so a 3D δ distribution can be reconstructed if the specimen is quantitatively measured from multiple angles while rotated. This is the basic principle of phase-contrast X-ray tomography.

The equation below focuses on electrons and relates δ to compounds and alloys of a variety of elements [16]:

$$\delta = \frac{r_e \lambda^2}{2\pi} \sum_k \rho_a^k (Z_k + f_k) \tag{2.25}$$

Here, $r_{\mathrm{e}} = \mathrm{e}^{\,2}/(m_e c^2)$ is the classic electron radius, f_{k} is the real component of the correction factor for the atomic scattering factor, and the subscript k of each parameter indicates the kth element. The term f_k can be ignored if the vicinity of the absorption edge is excluded, which allows δ to be expressed by the following equation:

$$\delta = \frac{r_e \lambda^2 \rho_e}{2\pi} \tag{2.26}$$

Here, ρ_e is the electron density. δ simply corresponds to the electron density and, as shown in Eq. (2.19), is proportional to density. In contrast, elemental concentration mapping using the absorption edge can be conducted near the absorption edge, as touched upon in Chap. 5. The following equations can also be obtained from Eqs. (2.24) and (2.25) [16]:

$$\Phi = \int \sum\nolimits_k \rho_a^k p_k dx \tag{2.27}$$

$$p_k = r_e \lambda (Z_k + f_k) \tag{2.28}$$

Here, p_k corresponds to the atomic absorption coefficient μ_a in the case of absorption contrast and equal to the X-ray phase shift cross-section. As a result, the difference between absorption and phase can be traced back to the difference between p_k and μ_a. The ratio of p_k and μ_a here is dependent on the X-ray energy and atomic number. This ratio takes a large value of 100–1000 for light elements in particular, as has already been confirmed in Fig. 2.17.

(2) X-ray Diffraction and Imaging

When the wavefront is deformed as shown in Fig. 2.19, adjacent waves are partially superimposed and interfere with one another. This is referred to as the *diffraction fringe*. Generally, the diffraction fringe is clearly observed in waves that have transmitted through interfaces or surfaces. Furthermore, the diffraction fringe pattern varies depending on the distance between the detector and object. In other words, a large detector-object distance results in Fraunhofer diffraction, whereas short distances result in Fresnel diffraction. Each of these phenomena are applied in X-ray imaging, as discussed later; in other words, a small *coherence length* results in a binary fringe pair, whereas a large coherence length results in multiple diffraction fringes appearing on both sides of the interface. In contrast, an extremely small coherence length eliminates the diffraction fringe.

As shown in Fig. 2.20, we assume two-dimensional imaging using a coherent monochromatic X-ray. This is the assumption for synchrotron radiation. At this time, the X-ray amplitude $\psi(x, y)$ at the detector position when Fresnel diffraction occurs is as follows [17]:

$$\psi(x, y) = \frac{i \exp(ikz)}{\lambda z} \iint_q (X, Y) \exp\left[\frac{ik}{2z}\{(x - X)^2 + (y - Y)^2\}\right] dXdY \tag{2.29}$$

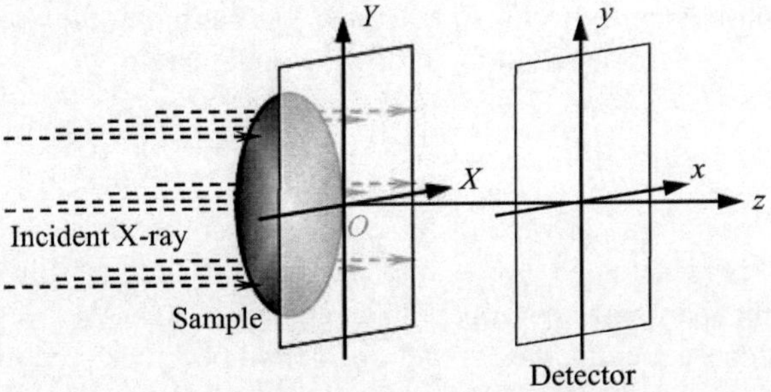

Fig. 2.20 Assumed case where a coherent monochromatic X-ray is emitted toward an object and the transmitted X-ray is observed on a two-dimensional detector. The X-ray intensity at the (*x*, *y*) plane that shows the detector plane located at a distance z from the (*X*, *Y*) plane immediately behind the sample is discussed

Here, k is the wavenumber and $k = \lambda/2\pi$. The term $q(X, Y)$ is the X-ray amplitude distribution immediately after transmitting through the sample ($z = 0$). The distance from the object to the detector is expressed with the coordinate system shown in Fig. 2.20 for Eq. (2.29), which was approximated with a fourth-order Taylor series expansion (Fresnel approximation). Furthermore, the wave function can be obtained by a simple correction with the magnification M in cases where X-rays are emitted from point light sources and not as *planar waves*, as would be the case when X-ray tubes are used [17]. The following equation shows Fraunhofer diffraction, which shows the same conditions up to the third order [17]:

$$\psi(x, y) = C \iint q(X, Y) exp\left\{\frac{ik}{z}(xX - yY)\right\} dXdY \tag{2.30}$$

Here, C is a constant. The inside of this integration is a Fourier transform of $q(X, Y)$.

Generally, when a structural size of interest inside the object is set as D, Fresnel diffraction appears when the sample and detector are relatively close to each other, as shown below:

$$z < \frac{D^2}{\lambda} \tag{2.31}$$

However, Fresnel diffraction does not appear when the detector is extremely close to the sample. This is referred to as the near field. In this case, only X-ray absorption contrast can be used. Furthermore, $F = \frac{D^2}{\lambda z}$ is referred to as the *Fresnel number*. This is a non-dimensional number that indicates the interface between refraction, which is a phenomenon of geometrical optics, and diffraction, which is a phenomenon of wave optics [18]. A geometrical optics approximation is made when $F \gg 1$, and refraction is considered predominant [18]. Meanwhile, multiple diffraction fringes are generated in the vicinity when $F \ll 1$ [18]. Furthermore, the *radius of the first Fresnel zone* r_{FZ} is expressed with the following equation:

$$r_{FZ} = \sqrt{\lambda z} \tag{2.32}$$

The $F \ll 1$ condition corresponds to when D is considerably smaller than the radius of the first Fresnel zone. For example, $F \approx 13$ when the sample and detector are placed 30 mm apart with an X-ray with an energy of 20 keV ($\lambda = 0.062$ nm) and $D = 5$ μm, and consequently, the Fresnel diffraction becomes predominant.

Figure 2.21 is an imaging simulation result where samples that include both an X-ray absorbing and non-absorbing structure are placed between 0.1 and 1000 mm from the detector [19]. Figure 2.21b–f are the Fresnel numbers, which correspond to 2500, 250, 25, 2.5, and 0.25, respectively. The X-ray non-absorbing structure is not visible when $F \gg 1$ and z is extremely small, but the interface becomes gradually highlighted as the distance between the sample and detector increases; its presence can be clearly visualized. Furthermore, multiple fringes can be seen when $F \approx 1$ in (e). The image is so disordered for $F \ll 1$ in (f) that the original structure is no longer even visible. Black and white diffraction fringes can be seen in the vicinity of the object interior structure in Fig. 2.21c and d ($F \gg 1$). The imaging process that applies this phenomenon is referred to as *refraction-contrast imaging*. In refraction-contrast

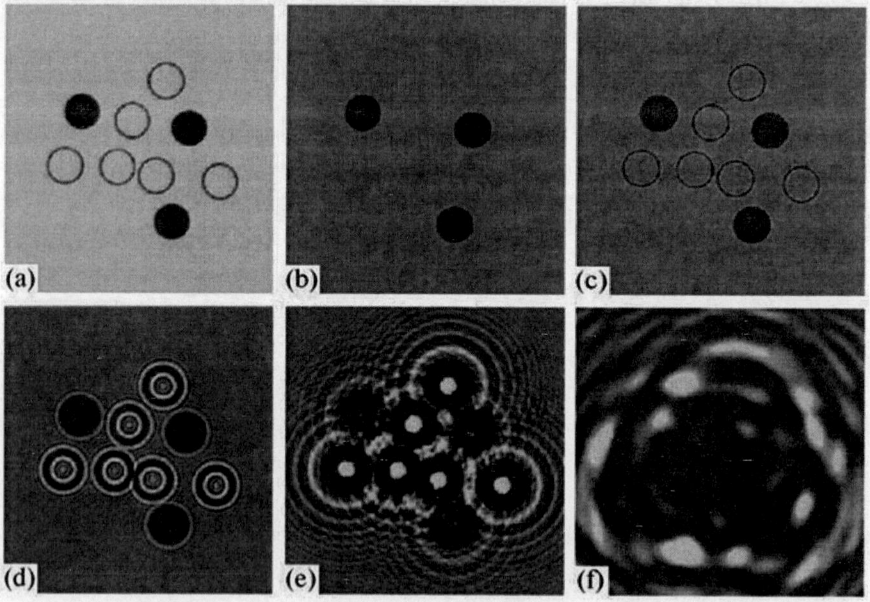

Fig. 2.21 **a** Image of object obtained with a 12.4 keV monochromatic X-ray. This includes three black circles that absorb X-rays and six white circles that do not (each has a diameter of 5 μm); **b–f** show imaging simulations with camera-detector distances of 0.1 mm, 1 mm, 10 mm, 100 mm, and 1000 mm, respectively [19]

imaging, the structure of the object interior (e.g. heterogeneous phases, particles, or voids) is mostly retained in the image. The interface is then highlighted as a result of intense black and white fringes appearing on these structures and the interface at the base. For this reason, these domains are also known as interface detection domains. The deflection angle increases when the incident angle into the interface is relatively shallow and the fringe appears even more clearly. Furthermore, image contrast is dependent on the phase shift. With refraction contrast imaging, we can anticipate the increasing effects of microstructure detection capabilities and visualization effects of objects that cannot be seen at all with absorption contrast by simply adjusting the distance between the sample and detector and not having to use any specialized devices.

Figure 2.22 shows visualizations of silicon particles dispersed in an aluminum alloy using monochromatic X-rays with an energy of 20 keV from synchrotron radiation [20]. The relatively small black silicon particle (diameter of approximately 5 μm) corresponds to $F \approx 13$ in the cross-sectional image on the top-right of Fig. 2.22. For example, the atomic number difference between aluminum and silicon is 1 and, as shown in the top-left cross-sectional image in Fig. 2.22, the silicon particles cannot be distinguished when the sample and detector are close in distance. The term $r_{FZ} \approx 1.4$ μm when the distance is 30 mm in Fig. 2.22; the silicon particle morphology

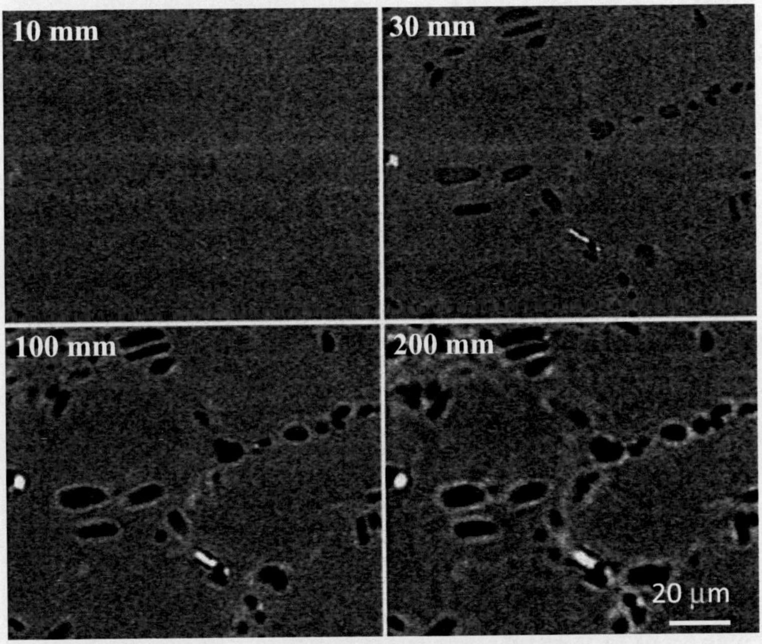

Fig. 2.22 Changes in a virtual 3D cross-section where the sample-detector distance was varied from 10 to 200 mm [20]. The sample was an Al-7% Si alloy

and size information is not lost while still being able to detect the edge when r_{FZ} approaches the spatial resolution at a distance of 20–30 mm.

Spatial coherence needs to be somewhat high to use refraction contrast imaging. Synchrotron radiation usage is typical for these purposes. However, refraction contrasts can be used even with X-ray tubes if those with a sufficiently small focal point size are used [21, 22]. Furthermore, the effect of the angle of view from the X-ray source is superimposed even when synchrotron radiation with an extremely small X-ray *divergence angle* is used in addition to X-ray refraction at the interface. For example, the angle of view is 10 μrad when the effective size of the X-ray source is 500 μm and the distance from the X-ray source and specimen is set to 50 m. This generates approximately 0.3 μm of divergence for a sample-detector distance of 30 mm. In practice, the angle of view can be considerably larger than this. In such cases, care must be taken as the quantitative aspect of the linear absorption coefficient and size is lost because of the fringe due to X-ray refraction, and image magnification due to the X-ray emission.

Meanwhile, Fraunhofer diffraction is generated when the sample and detector are placed considerably apart from one another.

$$z > \frac{D^2}{\lambda} \tag{2.33}$$

When calculating the integration of Eq. (2.30), it can be observed that $\psi(x, y)$ is dependent on $sinc\left(\frac{kD}{2z}\right)$ (here, $c(x) = \frac{sin(x)}{x}$). This amplitude distribution no longer corresponds to the object structure and is instead a Fourier transform of the X-ray amplitude distribution generated from the object. Furthermore, X-ray intensity is proportional to $sinc^2\left(\frac{kD}{2z}\right)$. X-ray intensity is observed in X-ray imaging, so the square of the absolute Fourier transform value is observed in the case of Fraunhofer diffraction. For this reason, information relating to the object cannot be recovered with simple inverse transformations.

Incidentally, when the observed object structure is close to the radius of the first Fresnel zone, the fringe generated in the interface includes phase modulation information due to the object. In these cases, the image changes significantly with multiple sample-detector distances. Cloetens et al. from ESRF proposed the *X-ray holography* method, wherein phase information is recovered by acquiring images with multiple distances and solving for the intensity propagation equation, which formulates the X-ray wavefront deformation. This has been used in synchrotron radiation research [23].

(3) Coherence

The ease of appearance and clarity of the interference fringes discussed in Sect. 2.2 can be understood by assessing the coherence of X-rays. Increasing coherence can allow for the usage of various X-ray imaging techniques, including phase-contrast imaging, X-ray holography, coherent X-ray diffraction microscopy (CXDM) [24], and wave-front shearing interferometry [25]. Coherence is also important when implementing *imaging-type X-ray tomography* [26]. Meanwhile, increased coherence results in the generation of *speckle noise* [27], which can conversely make X-ray imaging difficult. Coherence assessments are important to quantitatively determine and improve these aspects.

Discussions in this chapter include explanations where monochromatic X-rays and planar waves, or in other words 100% completely coherent conditions, are assumed. However, actual X-rays cannot be completely monochromatic. Furthermore, light sources cannot be placed in an infinite direction and are located within finite distances. Although the term *point light source is used*, the regions from which X-rays are emitted comprise some degree of size. In reality, X-rays, which are neither coherent nor incoherent instead sitting somewhere in the middle, are often used. Here, we apply the concept of coherence length and assess its extent.

Incidentally, double-slit interference experiments conducted by Young [28] are a familiar teaching method used to teach the wave nature of light in high-school physics classes. Calculating the optical path difference of the light coming out of the two slits considering the distance between the slit and screen enables the elucidation of conditions under which light waves mutually strengthen each other. However, high-school textbooks do not reference the conditions under which interference fringes appear during Young's experiments. In actuality, interference fringes either disappear or are unclear depending on the light source size, double-slit interval, distance from the light source to the slit, and the light wavelength.

These are considered using Fig. 2.23. A monochromatic X-ray with the same wavelength and whose wavefront is completely aligned is emitted from a position at a distance of D, propagating in the direction with just a microscale angle difference $\Delta\theta$; the slit is assumed here. Moving from position A, where the wavefront peaks of both X-rays are aligned and following the wavefront to position B, which is shifted in the horizontal direction by only a distance l_s, both waves are misaligned by half a wavelength and are antiphase. At position C, which is further shifted in the horizontal direction by only a distance l_s, the waves once again have the same phase; in other words, when X-ray interference occurs in this way, the distance range at which this interference occurs in the plane orthogonal to the propagating direction of light is defined as the coherence length relating to spatial coherence. From Fig. 2.23, l_s can be defined as follows:

$$l_s = \frac{\lambda}{2\Delta\theta} \tag{2.34}$$

When the distance from the observed plane and light source is set as R, $\Delta\theta = \frac{D}{R}$ and l_s is expressed as follows.

$$l_s = \frac{\lambda R}{2D} \tag{2.35}$$

Disruption of the light wavefront generally becomes smaller with increased distance from the small light source and the amplitude and phase in the direction

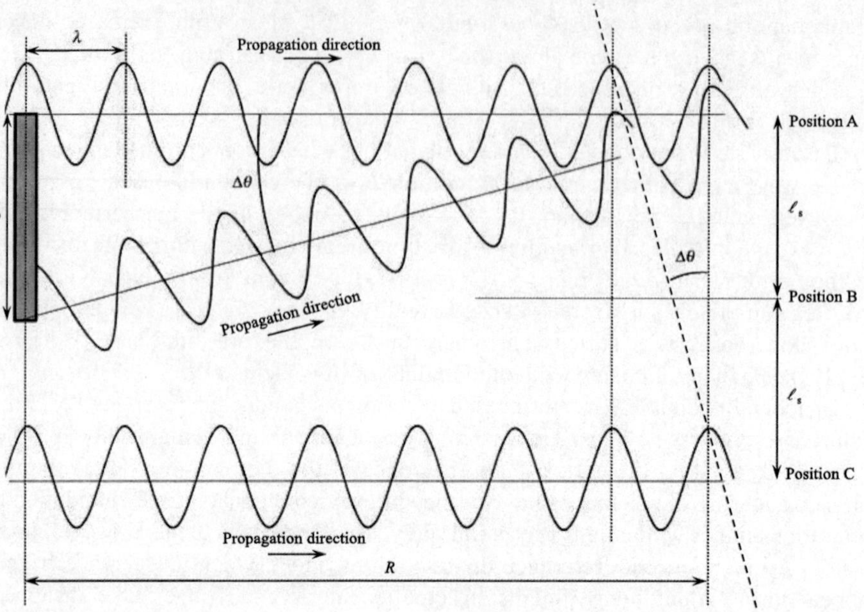

Fig. 2.23 Schematic illustrating a case where X-rays with the same wavelength and aligned wavefront are emitted from a position separated by distance D and propagate with the only difference being a microscale angle $\Delta\theta$. By position B, shifted transversely from position A along the wavefront by distance l_s only, waves become antiphase; they return to inphase at position C, shifted from position B transversely by distance l_s. From this, the coherence length l_s relating to spatial coherence is defined

orthogonal to the X-ray propagation direction are well-aligned. This can be expressed with spatial coherence fluctuations by using the measure known as coherence length from Eq. (2.35). Furthermore, X-rays with lower energy are known to obtain higher spatial coherence. Incidentally, the coherence lengths when under the coherent and incoherent conditions discussed in the beginning are infinite and 0, respectively.

Slits or pinholes can be inserted and its insertion positions can be seen as provisional light sources when the spatial coherence is insufficient. The spatial coherence and intensity in these cases can be controlled through the position and opening size of the installed slits or pinholes. Other than this, the light source can be made to ostensibly be smaller using Bragg diffraction with crystals such as silicon [29]. Furthermore, nonsymmetrical crystals can be used to virtually place light sources at a further distance [29]. In contrast, speckle noise due to defects or the heterogeneity of optical elements, beryllium windows, etc. can occur when spatial coherence is too high. Using graphite-coated boards, paper, or sandpaper in these cases can reduce spatial coherence and control speckle noise if these materials are rotated and inserted in the X-ray pathway [29].

Synchrotron radiation is beneficial in that it can increase spatial coherence. To quantitatively determine the coherence length, we can use SPring-8 as an example.

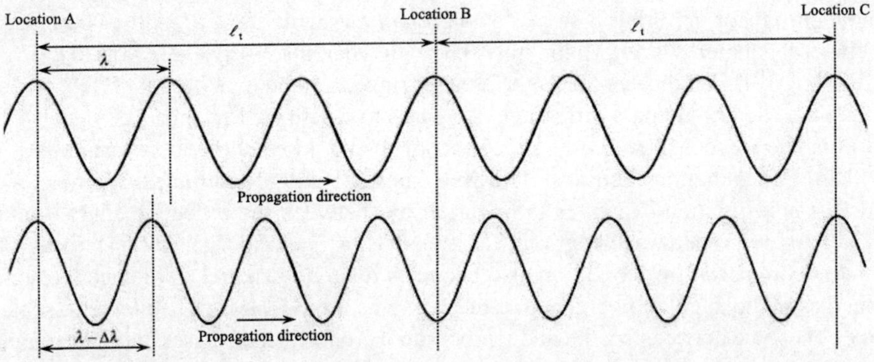

Fig. 2.24 Schematic illustrating a case where X-rays with a wavelength difference of only $\Delta\lambda$ are propagated in the entirely identical direction. X-rays that were in-phase at position A became antiphase to each other after propagating by a distance of l_t and in-phase for a second time after further propagating by a distance of l_t. From this, a coherence length l_t relating to temporal coherence is defined

The size of the light source in the vertical direction when the X-ray energy is 10 keV at the standard undulator light source of SPring-8 is around 16 μm [30]. The spatial coherence in these cases is approximately 390 μm when the distance from the light source to the experimental hatch is set to 100 m. Using long-/medium-length beamlines such as BL20XU is advantageous when taking the distance from the light source to the observed plane. Meanwhile, the beam size in the horizontal direction is relatively large at around 0.6 mm. Thus, it is sometimes necessary to insert a slit that controls the beam width only in the horizontal direction [31].

Next, we consider a case where X-rays propagating in the same direction have only slightly varying wavelengths, as shown in Fig. 2.24. This is identical to the case of spatial coherence; this time, however, we consider the phase shift in the X-ray propagating direction. From position A where both X-ray wavefront peaks are aligned, the X-rays at position B where the waves have only propagated by a distance l_t, are misaligned by half a wavelength and are antiphase. At position C, where the waves have further propagated by a distance l_t, the waves are once again in-phase; in other words, the extent to which X-rays interfere with each other within the wavelength difference range when X-rays emitted from the same light source interfere is expressed with temporal coherence. When $l_t = N\lambda$ for the X-ray in the top row of Fig. 2.24, $l_t = \left(N + \frac{1}{2}\right)(\lambda - \Delta\lambda)$ for the X-ray in the bottom row and the coherence length l_t relating to *temporal coherence/longitudinal coherence* is as expressed below:

$$l_t = \frac{\lambda^2}{2\Delta\lambda} \tag{2.36}$$

This is also estimated using synchrotron radiation as an example, similar to the case of spatial coherence. Monochromatic X-rays obtained from the double crystal monochromator in SPring-8 has a *monochromaticity* with $\lambda/\Delta\lambda$ around 10,000 [32]. Temporal coherence is approximately 0.31 μm when the X-ray energy of 20 keV is considered. This indicates that interference fringes can be experimentally observed when the X-ray light path difference decreases to a value below this.

Finally, we consider X-ray spectroscopy using monochromators available at synchrotron radiation facilities. The well-known Bragg condition $2dsin\theta = n\lambda$ for X-ray diffraction expresses the relationship between the reflective angle θ and wavelength λ. When wavelengths are screened using materials like silicon crystals at a monochromator, this condition also becomes related to spatial coherence through changes in the angular divergence of the X-ray beam [33]. It is not suitable to separate coherence between spatial and temporal coherence in such cases and it becomes necessary to use descriptions with *mutual coherence*, which integrate these two. Please refer to other texts [33] for further details.

References

1. T.T. Truong, M.K. Nguyen, INTECH Open Access Publisher (2012).
2. D.F. Jackson, D.J. Hawkes, Phys. Rep. **70**, 169–233 (1981)
3. Physical Measurement Laboratory (PML), National Institute of Standards and Technology (NIST): X-Ray Form Factor, Attenuation and Scattering Tables, NIST, U.S. Commerce Department, USA, (2011). https://www.nist.gov/pml/x-ray-form-factor-attenuation-and-scattering-tables (Accessed March 2017)
4. International Tables for X-Ray Crystallography, Vol. 3, Physical and chemical tables, ed. by C.H. Macgillavry, G.D. Rieck, D. Reidel Publishing Company, (1985). (published online (https://xdb.lbl.gov/xdb.pdf). Accessed August 2017).
5. The Center for X-Ray Optics (CXRO), Lawrence Berkeley National Laboratory (LBNL), X-ray interactions with matter calculator. https://henke.lbl.gov/optical_constants/. Accessed March 2017
6. X-ray Utilities, Sam's X-rays - Applications for Synchrotrons. https://www.sams-xrays.com/xru. Accessed March 2017
7. J.A. Victoreen, J. Appl. Phys. **20**, 1141–1147 (1949)
8. O. Klein, Y. Nishina, Zeitschrift Für Physik **52**, 853–868 (1928)
9. H. Kato, Jpn. J. Radiol. Technol. **70**, 684–691 (2014)
10. K. Horiba, Y. Nakamura, N. Nagamura, S. Toyoda, H. Kumigashira, Rev. Sci. Instrum. **82**, 113701 (2011)
11. B.T. Thole, P. Carra, F. Sette, G. van der Laan, Phys. Rev. Lett. **68**, 1943–1946 (1992)
12. M.O. Krause, J. Phys. Chem. Ref. Data **8**, 307–327 (1979)
13. A.H. Compton, Phil. Mag. **45**, 1121–1131 (1923)
14. A. Momose, J. Jpn. Soc. Synchrotron Radiat. Res. **10**, 23–35 (1997)
15. P. Cloetens, M. Pateyron-Salomé, J.Y. Buffiére, G. Peix, J. Baruchel, F. Peyrin, M. Schlenker, J. Appl. Phys. **81**, 5878–5886 (1997)
16. A. Momose, Jpn. J. Appl. Phys. **44**, 6355–6367 (2005)
17. J.M. Cowley, *Diffraction Physics*, 3rd edn. (North Holland, 1995)
18. N. Yagi, Y. Suzuki, Med. Imaging Technol. **24**, 380–384 (2006)
19. V.V. Lider, M.V. Kovalchuk, Crystallogr. Rep. **58**, 769–787 (2013)
20. T. Hidaka, H. Toda, M. Kobayashi, K. Uesugi, T. Kobayashi, J. Jpn. Inst. Light Metals **58**, 58–64 (2008)
21. S.W. Wilkins, T.E. Gureyev, D. Gao, A. Pogany, A.W. Stevenson, Nature **384**, 335–338 (1996)

22. D. Gao, A. Pogany, A.W. Stevenson, S.W. Wilkins, Imaging Ther. Technol. **18**, 1257–1267 (1998)
23. P. Cloetens, W. Ludwig, J. Baruchel, D. Van Dyck, J. Van Landuyt, J.P. Guigay, M. Schlenker, Appl. Phys. Lett. **75**, 2912–2914 (1999)
24. Y. Nishino, T. Ishikawa, J. Jpn. Soc. Synchrotron Radiat. Res. **19**, 3–14 (2006)
25. Y. Suzuki, J. Jpn. Soc. Synchrotron Radiat. Res. **18**, 75–83 (2005)
26. Y. Suzuki, H. Toda, *Advanced tomographic methods in materials research and engineering*, ed. by J. Banhart (Oxford University Press, 2008), Section 7.1.
27. M. Awaji, Y. Suzuki, A. Takeuchi, H. Takano, N. Kamijo, S. Tamura, M. Yasumoto, Nucl. Instrum. Methods Phys. Res. Sect. A **845**, 467–468 (2001)
28. E. Shigemasa, M. Yabashi, J. Jpn. Soc. Synchrotron Radiat. Res. **19**, 33–40 (2006)
29. A. Momose, J. Jpn. Soc. Synchrotron Radiat. Res. **20**, 43–49 (2007)
30. H. Takano, Y. Kohmura, J. Jpn. Soc. Synchrotron Radiat. Res. **19**, 314–322 (2006)
31. Y. Suzuki, Jpn. J. Optics **42**, 303–308 (2013)
32. Y. Suzuki, A. Takeuchi, H. Takenaka, I. Okada, X-Ray optics and instrumentation, (2010), article ID 824387, 6 pages, doi:https://doi.org/10.1155/2010/824387.
33. H. Yamasaki, T. Ishikawa, J. Jpn. Soc. Synchrotron Radiat. Res. **20**, 18–25 (2007)

Chapter 3
3D Image Reconstruction

Image reconstruction refers to the process by which images are restored to their original higher-order dimensions after X-rays are transmitted through a two- or three-dimensional object and the transmitted X-rays are measured as one- or two-dimensional observable values.

The introduction to Johann Radon's 1917 paper "*On the determination of functions from their integrals along certain manifolds*" (translated from German to English by Dr. R. Lohner at the Georgia Institute of Technology) poses the questions, "is every line function that satisfies suitable regularity conditions obtainable in a plane?" and "if this is the case, is the point function uniquely determined and how is it found?", then subsequently provides mathematical proofs [1]. This was at a time where the word "tomography" did not even exist and two-dimensional detectors or computers were not used. However, after this was shown, there have been discussions for three dimensions and higher, calling to mind Radon's foresight in these subjects.

Over half a century passed and medical X-ray CT scanners began to see practical use in the 1970s, after which the practical application of various X-ray CT scanner mechanisms and development of multiple corresponding image reconstruction methods began to take hold. These include a number of strengths and weaknesses, including the extent of X-ray exposure on the human body, the difficulty of mathematical solutions, spatial resolution, temporal resolution, noise, artifacts, and the complexity of mechanisms and devices. Details on the wide-ranging techniques cannot be entirely covered in this chapter alone. Furthermore, the strict mathematical derivations of image reconstruction algorithms have been omitted here due to limited space. For this reason, please see specialist texts for reference and more detailed elaboration on these aspects [2–7]. This chapter will outline the fundamentals of several image reconstruction approaches currently in use in X-ray tomography at synchrotron radiation facilities and industrial X-ray CT scanners, clarifying both their proper use and points of caution in image reconstruction.

H. Toda, *X-Ray CT*,
https://doi.org/10.1007/978-981-16-0590-1_3

3.1 Projection Data

3.1.1 Basic Calculation Method

Simple methods like those used to acquire a single two-dimensional cross-sectional image from the early days of medical X-ray tomography are described so that complex 3D image reconstruction can be easily understood. For this reason, an overview that effectively follows the chronological history of X-ray tomography development is included.

Figure 3.1 shows the medical X-ray CT scanner system, which at the time was already in use, introduced by Hounsfield in his lecture when awarded the Nobel Prize [8]. X-ray tomography had shown rapid progress within 8 years of his first using X-ray CT scanners in clinical inspections in 1971. Figure 3.1a shows how the detector and *X-ray tube* are situated on both sides of the human body and translate in the same direction, thereby covering the person, while simultaneously rotating around the body, in what is called a *translate/rotate system*. Using traditional classifications, this is also referred to as a *second-generation CT*. Hounsfield created a detector out of 30 detection element groups and had an imaging time of 18 s. The X-ray expands from several to approximately a dozen degrees to cover the thirty detection elements. This is referred to as a *narrow fan beam*. The traditionally classified first-generation CT systems had one or two detectors and used a collimated *pencil beam*. Scans required approximately 300 s in these cases. Figure 3.1b is referred to as a *rotate/rotate system*. The *fan beam* (X-ray beam that expands in a fan-like way) emitted from the X-ray tube transmits through the human body and enters the arc-arranged detectors facing the radiation source. According to Hounsfield's explanations, there were 300–500

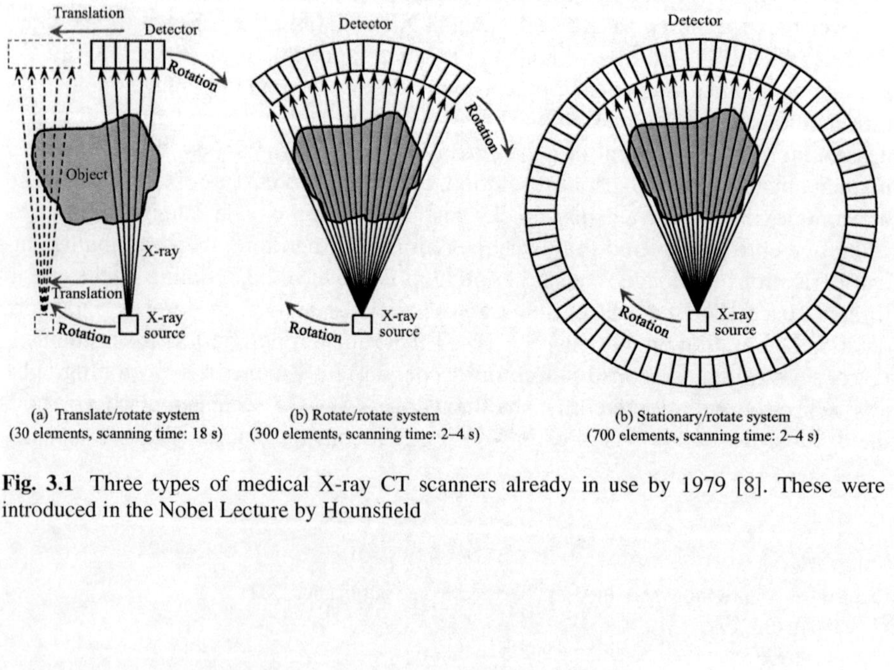

Fig. 3.1 Three types of medical X-ray CT scanners already in use by 1979 [8]. These were introduced in the Nobel Lecture by Hounsfield

detector elements and the scanning time was within 3 s. This was the so-called *third-generation CT*. Finally, Fig. 3.1c is referred to as a *stationary/rotate system*, and the fan beam is irradiated while the X-ray tube rotates on the inside of the many detection elements placed along the periphery. This is *the fourth-generation CT*. According to Hounsfield's explanations, there were 700–1000 detector elements and scans were completed within 3 s. Incidentally, *the fifth-generation CT* deflects the electron beam with deflecting coils and generates X-rays by hitting it on a ring target; high-speed scans are achieved by continuously rotating the X-ray generation positions.

These methods were used to acquire a single cross-section. Information on the human body in the longitudinal direction was acquired by scanning repeated images where the bed on which the patient was sitting was moved slightly and then stopped to acquire multiple two-dimensional images (the so-called step scan). Figure 3.2 shows the shape of the X-ray beam when it is used to seek 1D information—in other words, it shows the shape of the beam when seeking the two-dimensional distribution of the linear absorption coefficient of a given cross-section from a certain transmitting X-ray intensity line profile at a given projection angle.

Meanwhile, the *helical scan* system of X-ray CT scanners, referred to as the *sixth-generation CT*, was introduced in 1989 [9, 10] with advantages including being able to rapidly take 3D images by continuously moving the bed, as shown in Fig. 3.3. In this case, projection images are not obtained from every incident direction at every position in the subject interior, so interpolations from projection data above and below the slice are necessary for image reconstruction. For this reason, this is also known as the interpolated reconstruction method. This method is not frequently used in X-ray tomography at synchrotron radiation facilities or industrial X-ray CT scanners, so

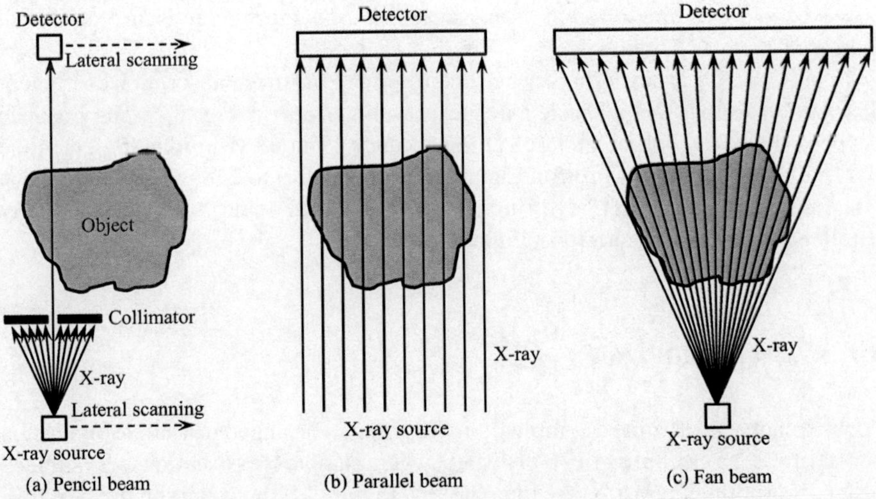

Fig. 3.2 X-ray beam shape when determining two-dimensional information (linear absorption coefficient distribution in a virtual cross-section) from one-dimensional information (line profile of transmitted X-ray intensity at a given projection angle)

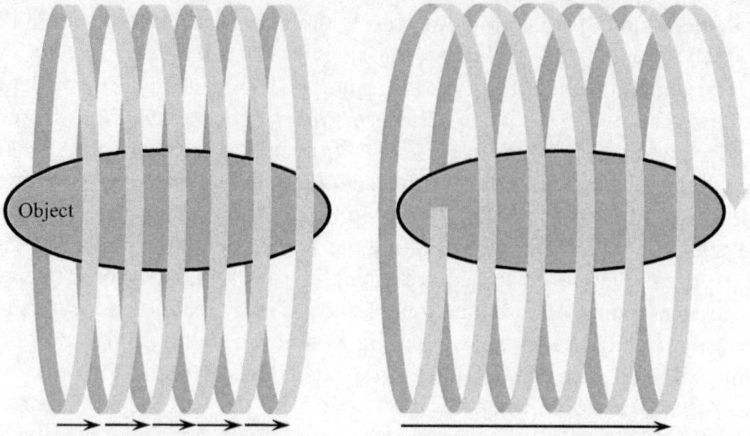

Fig. 3.3 3D image acquisition methods for helical scan system X-ray CT scanners for medical use (also referred to as sixth-generation CT) and prior systems. Path of the X-ray source is depicted

we omit this detail here. For further information please see referenced works [2–7, 11]. With the introduction of flat-panel detectors in the 1990s, the so-called cone-beam CT began to see practical use. This is referred to as the *seventh-generation CT*, which is a method that uses two-dimensional detectors to directly measure a two-dimensional mapping of a transmitted X-ray intensity at a given projection angle and creates a 3D image made of several hundred to several thousand two-dimensional cross-sections in a single round of measurements. This technique is also used with industrial X-ray CT scanners.

Finally, the X-ray beam shape when determining a 3D linear absorption coefficient distribution from a 2D projected image is summarized in Fig. 3.4. The distance between the X-ray source and detector can range from 22 (Bio-imaging beamline BL07 at the Kyushu Synchrotron Light Research Center) to 248 m (high-resolution imaging beamline BL20XU at SPring-8); hence, an X-ray beam that appears virtually parallel can be used, as shown in Fig. 3.4a.

3.1.2 Projection Data

For simplicity, we assume a combination of a detector arranged in a single dimension and a parallel monochromatic X-ray beam, where sample cross-sections are obtained using absorption contrast X-ray tomography. Figure 3.5 shows a schematic of this. In X-ray tomography, either the sample is fixed and the detector and X-ray source will rotate while this positional relationship is maintained or the X-ray source is fixed and the sample rotates. Generally, the former corresponds to medical X-ray

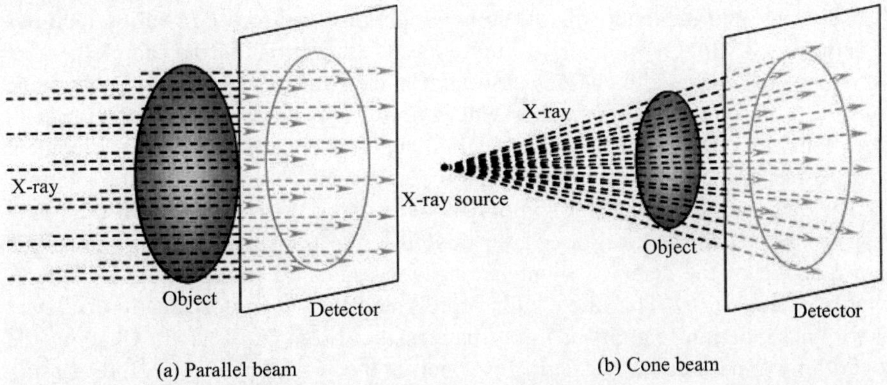

Fig. 3.4 Representative X-ray beam shape when determining 3D information (3D linear absorption coefficient distribution in a virtual cross-section) from 2D information (two-dimensional mapping of transmitted X-ray intensity at a given projection angle)

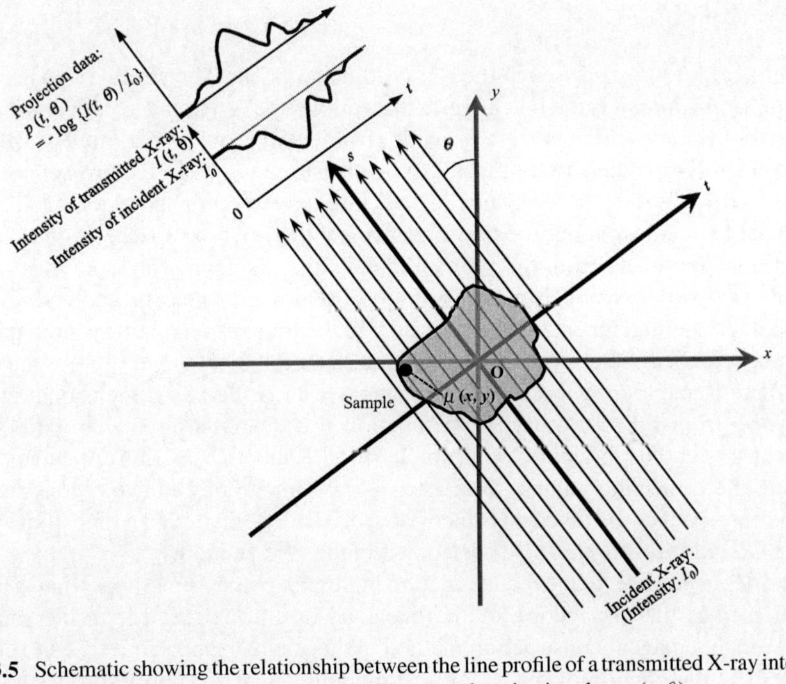

Fig. 3.5 Schematic showing the relationship between the line profile of a transmitted X-ray intensity $I(t)$ at a given projection angle θ and given position t, and projection data $p(t, \theta)$

CT scanners, and the latter to X-ray tomography at synchrotron radiation facilities or industrial X-ray CT scanners. In these cases, all sections of the sample need to be within the beam width and detector field-of-view range. Furthermore, the image must be acquired from X-rays that are incident to all positions of the sample from all directions; 180° is sufficient for the rotation angle when a parallel beam is incident to a sample. The coordinate origin O in Fig. 3.5 refers to the center of this rotation.

The data measured in X-ray tomography consist of the intensity, *I*, of the X-ray that has passed through an object. In this case, the transmitted X-ray intensity *I* is a function of the detector element's array coordinate *t* and projection angle θ, expressed as $I\ (t,\ \theta)$. The line profile represented by $I\ (t,\theta)$ is typically displayed as a high-gradation (e.g. 16-bit) grayscale image. From Eq. (2.4) in Chap. 2, the logarithm of the ratio of $I\ (t,\theta)$ to I_0, which is the X-ray intensity before passing through an object is expressed as follows:

$$p(t,\theta) = -log\frac{I(t,\theta)}{I_0} = \int_{-\infty}^{\infty} \mu(tcos\theta - ssin\theta, tsin\theta + scos\theta)ds \tag{3.1}$$

Here, $\mu\ (x,\ y)$ is expressed in the *t-s* coordinate system. The term $p\ (t,\ \theta)$ is called the *projection data*; it is the line integral that follows the X-ray beam path of the local linear absorption coefficient $\mu\ (x,\ y)$ inside an object. It is obvious from Eq. (3.1) that the incident X-ray intensity I_0 must also be measured to obtain the projection data. Tomography is a method for evaluating the internal structure and external shape of an object by reconstructing the two-dimensional distributions of the original linear absorption coefficient from the projection data and displaying it in the form of an image. As two-dimensional images become a 3D image when stacked and layered on top of each other, Fig. 3.5 can be considered the foundation of X-ray tomography. The mathematical solution to this problem is *Radon's theorem*, as presented by John Radon, mentioned in Chap. 1. Equation (3.1) is the two-dimensional *Radon transform* from function $\mu\ (x,\ y)$ to function $p\ (t,\ \theta)$. Measurements with X-ray CT scanners can be thought of as a single set of Radon transforms collected while varying the projection angles. Similarly, X-ray tomography image reconstruction can be perceived as the inverse problem of analytically obtaining an object image by *inverse Radon transform*. This is discussed in the next section.

The description of image reconstruction in this chapter assumes that what is being constructed is the distribution of the linear absorption coefficient in the sample. However, the image reconstruction method in this chapter can also be used to calculate the 3D distribution of other physical quantities that are not uniform inside an object, such as the phase shift described in Sect. 2.2.2 (1), so long as the local values can be measured and recorded from many directions along the X-ray path as projected data.

3.2 Fundamentals of Image Reconstruction

3.2.1 Radon Transform and Radon Space

The linear equation with an inclination of angle θ from the x-y coordinate system and with a distance t from the origin in the t-axis direction is expressed as follows:

$$t = xcos\theta + ysin\theta \tag{3.2}$$

With this, a delta function can be used to express the Radon transform of Eq. (3.1) as follows.

$$p(t, \theta) = \iint_{-\infty}^{\infty} \mu(x, y)\delta(xcos\theta + ysin\theta - t)dxdy \tag{3.3}$$

The p (t, θ) space is referred to as a *Radon space*.

As an example, Fig. 3.6 shows a schematic where the projection data of the black circular area in the object from Fig. 3.5 are depicted in Cartesian and polar coordinates. The parallel beam projection is placed radiatively with the origin at the center for polar coordinates. If there were 180° worth of projection data at this point, the Radon space would be covered without any gaps. Similar to the black circle in Fig. 3.5, the trajectory of the interior structure placed at a distance from the center of rotation is a trigonometric function in Cartesian coordinates and a circle

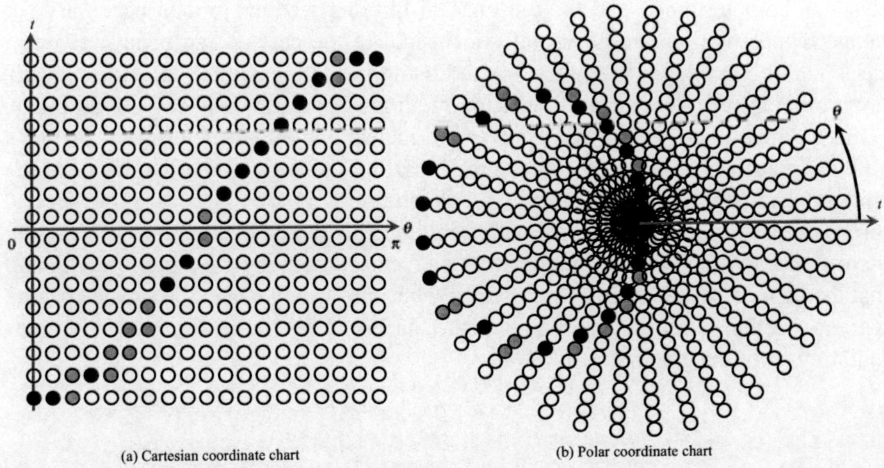

Fig. 3.6 Schematic of a Radon transform of the projection data $p(t, \theta)$ when the trajectory of the black circular area shown within the object in Fig. 3.5 was imaged from all directions with 16 detector elements; **a** shows a sinogram, shown in Cartesian coordinates and where θ is the horizontal axis and t the vertical axis; and **b** shows the transform in polar coordinates, where t is the radius

that passes the origin in polar coordinates. The Cartesian coordinate case in Fig. 3.6a in particular is referred to as a *sinogram* based on the sinusoid sine curve. Structures with larger distances from the center of rotation have a larger sine wave amplitude in the Radon space expressed in Cartesian coordinates, whereas the radius of the circle becomes larger in polar coordinates. In the case of polar coordinates, care must be taken so that the origin does not become an undefined singularity. The Radon transform has characteristics of symmetry ($p(t, \theta) = p(-t, \theta + \pi)$), where inverted imaging produces the same image; periodicity ($p(t, \theta) = p(t, \theta + 2n\pi)$), where a single interval returns the function to its original value; and linearity, where the Radon transform of the sum of two images multiplied by coefficients ($a f_1(x, y) + b f_2(x, y)$, where a and b are constants) is equal to the sum of both Radon transforms ($a p_1(t, \theta) + b p_2(t, \theta)$). Furthermore, when rotating an original image by an angle φ and projecting point (x, y) to point $(x', y') = (x \cos\varphi - y \sin\varphi, x \sin\varphi + y \cos\varphi)$, Radon transform would only move φ in the angular direction. This is easy to understand when using Fig. 3.6.

Figure 3.7 shows a virtual cross-section and Radon space obtained by imaging Al-Cu alloys with synchrotron radiation. The white point-like structures here are particles of intermetallic compounds that have a chemical composition of Al_2Cu [12]. As shown in Fig. 3.7, cases where there are multiple structures in the object interior appear as a series of superimposed sine waves with different amplitude and phase values in Cartesian coordinates and as multiple superimposed circles that pass through the origin with polar coordinates.

A sinogram expresses the Radon transform of $\mu(x, y)$ in the form of an image, as well as the data necessary for the reconstruction of $\mu(x, y)$. This is useful as the X-ray tomography imaging results can be easily confirmed. For example, parameters that are important for X-ray tomography imaging (e.g. center of rotation, magnification, etc.) can be determined and the presence of missing wedges in data necessary for image reconstruction can be visually inspected in the case of fan beams. Even in cases where 3D images are reconstructed using two-dimensional detectors like X-ray tomography at synchrotron radiation facilities or industrial X-ray CT scanners, a sinogram can be used to express a two-dimensional data structure that includes all data in each line of the detector in the horizontal direction, and the utility of the sinogram is therefore the same. For example, the sinogram in Fig. 3.7c shows several thread-like white lines in the vertical direction. This indicates a minor issue in that there was some wobbling in the beam at a given rotation angle during imaging and that the incident X-ray intensity temporarily became stronger (i.e. the image became whiter). Furthermore, this can be confirmed in Fig. 3.7b where straight white lines radiating from the origin extend.

3.2.2 Projection Theorem

The *projection theorem* is also referred to as the *central slice theorem* or *Fourier slice theorem* [13]. Figure 3.8 shows a schematic of this theorem. The projection data

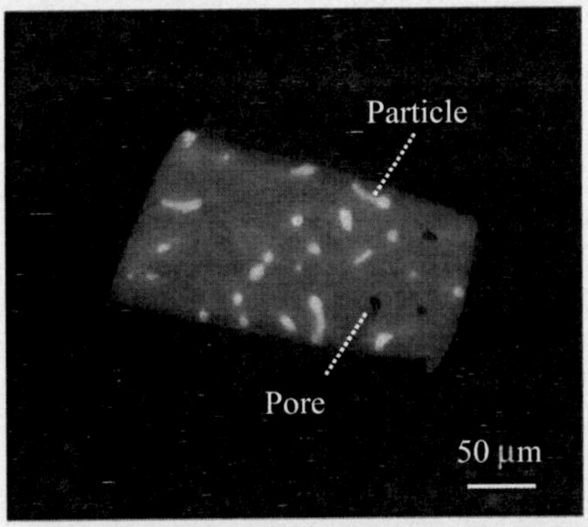

(a) Reconstructed slice

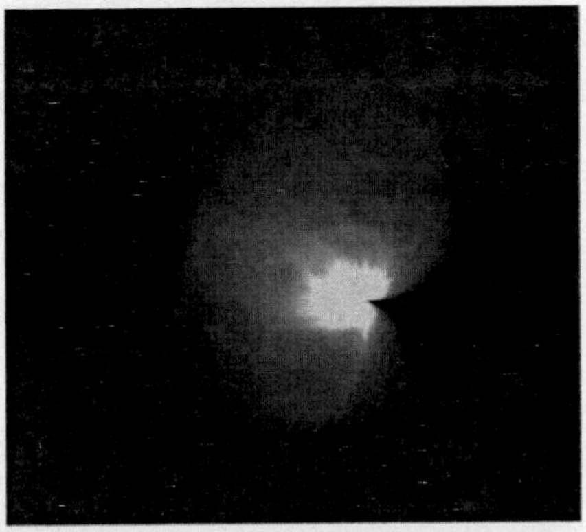

(b) Radon space (Polar coordinate)

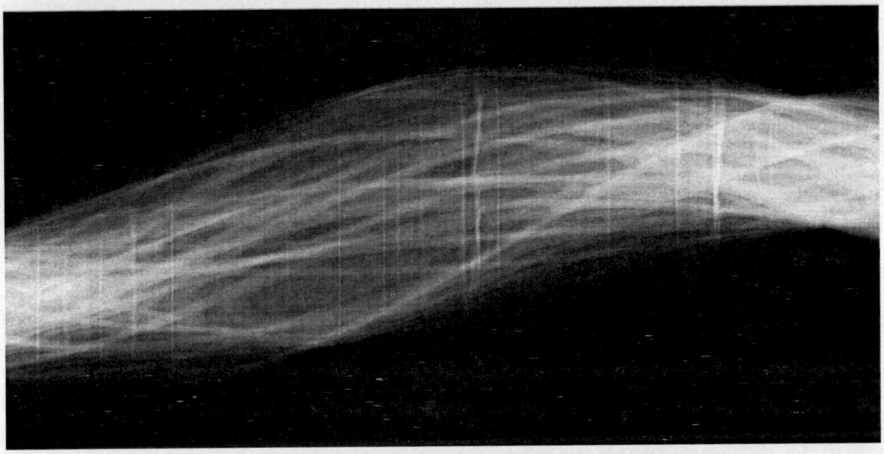

(c) Radon space (Cartesian coordinate: sinogram)

Fig. 3.7 **a** shows the reconstruction of the projection data obtained from synchrotron radiation X-ray tomography experiments; **b**, **c** show this projection data in a Radon space; **b** is in polar coordinates and **c** is in Cartesian coordinates, known as a sinogram. The sample is an aluminum-copper alloy, where a parallel beam was used to take 1800 projections

imaged when the sample rotation angle is θ is set as $p(t, \theta)$. The one-dimensional Fourier transform $F_{1D}\{p(t, \theta)\}$ for variable t in $p(t, \theta)$ is none other than the Fourier spectra in the θ direction, which passes through the origin of the two-dimensional Fourier transform $F(u, v)$ of the linear absorption coefficient distribution $\mu(x, y)$ in the object. Here, (u, v) refers to the Cartesian coordinate expression, which corresponds to the real space (x, y) of the angular frequency in the frequency space; in other words, the following is obtained when this is expressed as an equation:

$$\mu(x, y) = F^{-1}[F_{1D}\{p(t, \theta)\}] \tag{3.4}$$

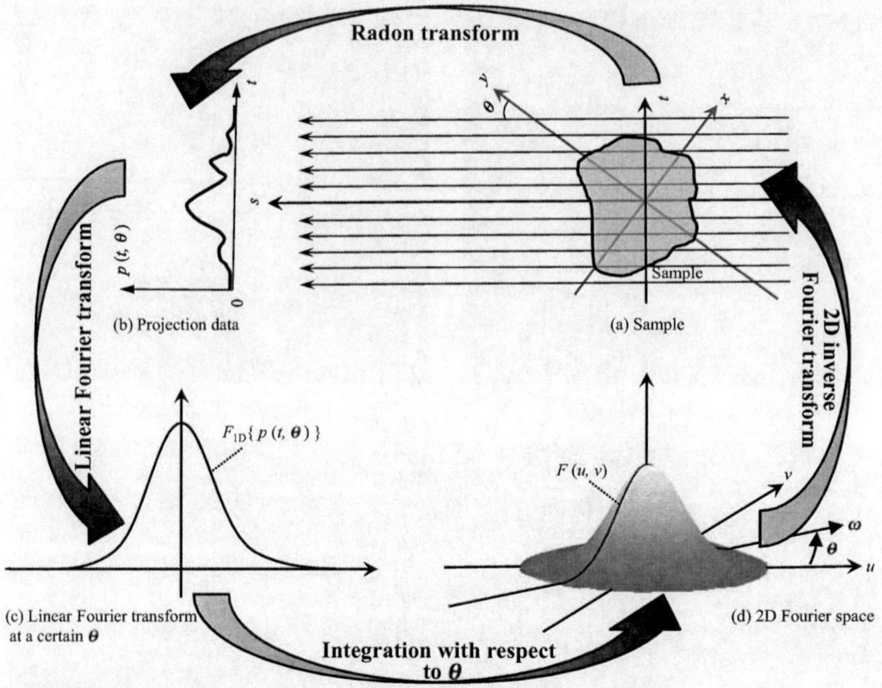

Fig. 3.8 Schematic of the Fourier transform method process. A 1D Fourier transform is applied on projection data $p\ (t, \theta\ )$. If the 2D Fourier space is filled by obtaining projections in $0 \leqq \theta < \pi$ without any defects, $F\ (u, v)$ can be completely recorded. Finally, the linear absorption coefficient distribution $\mu(x, y)$ of the original object can be obtained by applying an inverse 2D Fourier transform on $F\ (u, v)$

Thus, $F\ (u, v)$ can be recorded completely by obtaining the projection data $p\ (t, \theta)$ in the sample angle range of $0 \leqq \theta < \pi$ without any missing wedges. For this reason, a linear absorption coefficient distribution $\mu\ (x, y)$ of the imaged object can be determined and a 2D image of the object can be depicted if a 2D Fourier transform is applied on $F\ (u, v)$. This is the analytical algorithm for X-ray tomography image reconstruction, also referred to as the *Fourier transform method*.

The mathematical understanding of this method is further explored below. The 2D Fourier transform $F\ (u, v)$ of $\mu\ (x, y)$ is expressed below, based on the definition of a 2D Fourier transform:

$$F(u, v) = \int_{-\infty}^{\infty} \int_{-\infty}^{\infty} \mu(x, y) e^{-2\pi i(xu+yv)} dxdy \tag{3.5}$$

The following is obtained if the angular frequency is set as ω and transformed into polar coordinates (ω, θ).

$$F(\omega cos\theta, \omega sin\theta) = \int_{-\infty}^{\infty}\int_{-\infty}^{\infty} \mu(x, y)e^{-2\pi i\omega(xcos\theta+ysin\theta)}dxdy \tag{3.6}$$

A relationship that uses the following Jacobian matrix $|J|$ is generally established in transforms from (*x*, *y*) coordinates to (*t*, *s*) coordinates.

$$|J| = \begin{vmatrix} \frac{\partial x}{\partial t} & \frac{\partial x}{\partial s} \\ \frac{\partial y}{\partial t} & \frac{\partial y}{\partial s} \end{vmatrix} \tag{3.7}$$

A transform of the integration variables that use this matrix is possible, as shown below.

$$dtds = |J|dxdy \tag{3.8}$$

However, this was originally only for a rotation scenario, so there are no changes in differential area (i.e. $|J| = 1$) and variables can be changed to dxdy = dtds. Equation (3.6) can be transformed as follows when this is combined with Eq. (3.1) and (3.2).

$$F(\omega cos\theta, \omega sin\theta) = \int_{-\infty}^{\infty}\int_{-\infty}^{\infty} \mu(x, y)e^{-2\pi i\omega t}dtds$$

$$= \int_{-\infty}^{\infty}\left\{\int_{-\infty}^{\infty} \mu(x, y)ds\right\}e^{-2\pi i\omega t}dt$$

$$= \int_{-\infty}^{\infty} p(t, \theta)e^{-2\pi i\omega t}dt \tag{3.9}$$

The final equation takes the form of a 1D Fourier transform for a variable *t* of the projection data *p* (*t*, *θ*). Arriving at the inverse of the above process and ultimately conducting a 2D inverse Fourier transform is the image reconstruction method based on the Fourier transform method.

Figure 3.9 shows the image reconstruction obtained from imaging with X-ray tomography of the Al-Cu alloy shown in Fig. 3.7, a 1D Fourier transform of its projection data, and its re-plotting for a subsequent coordinate transformation and its 2D inverse Fourier transform. This provides a general overview of the treatment process.

In this way, the Fourier transform method is composed of a Fourier transform and its operation is fast. However, this method was not widely used when computers were undeveloped as inverse Fourier transforms required extensive computational time. Furthermore, the frequency space expressed in polar coordinates is filled with

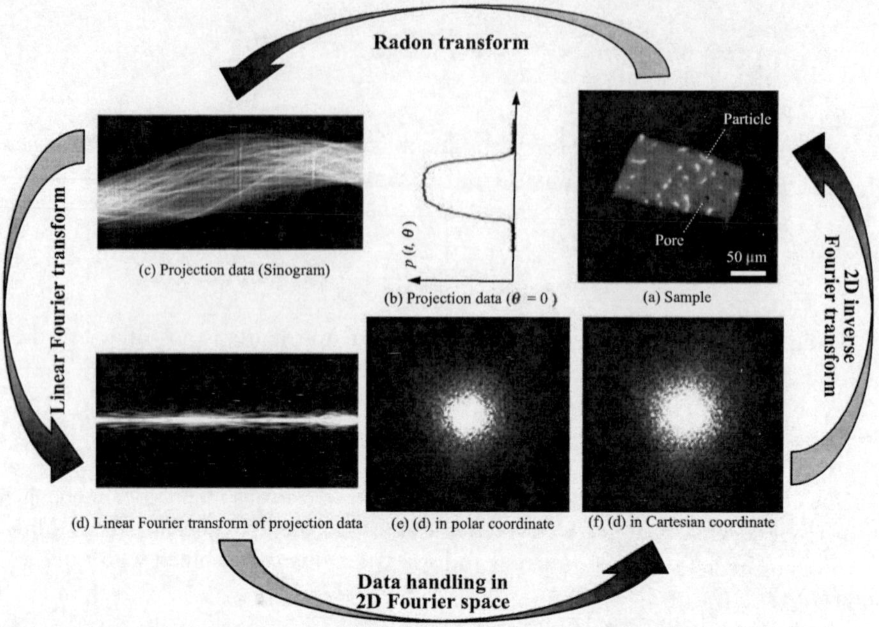

Fig. 3.9 Image reconstruction process of the projection data obtained from synchrotron radiation X-ray tomography experiments shown in Fig. 3.7; **a** is the cross-section after reconstruction, **b** is the projection data in the $\theta = 0°$ direction, **c** is the sinogram collected for all 180°, **d** is a 1D Fourier transform of the projection data, and **e** and **f** are the data after a coordinate transform was conducted prior to when 2D inverse Fourier transform was applied

projection data measured from multiple angles, as evident when looking at the series of calculation processes. As shown in Fig. 3.10, this yields a Fourier transform that follows the line section extending in the radial direction from the origin. The circle marks in Fig. 3.10 schematically represent the locations at which the Fourier transforms of the object are actually calculated. Figure 3.10 shows that the data density of the Fourier transforms greatly decreases with distance from the origin. In contrast to transforming from Cartesian to polar coordinates when moving from Eqs. (3.5) to (3.6), a transformation from polar to Cartesian coordinates is necessary for actual Fourier transform methods. At this time, a large misalignment between the lattice points of the Cartesian coordinates that express $F(u, v)$ and the lattice points of the polar coordinates which express $F(\omega\ sin\theta,\ \omega\ cos\theta)$ is formed, producing artifacts that correspond to this in the reconstructed image. Figure 3.10 shows that this error increases as the high-frequency component increases. This is difficult to identify in Fig. 3.9, but the image progressively blurs, and the morphology shifts as it approaches the edges.

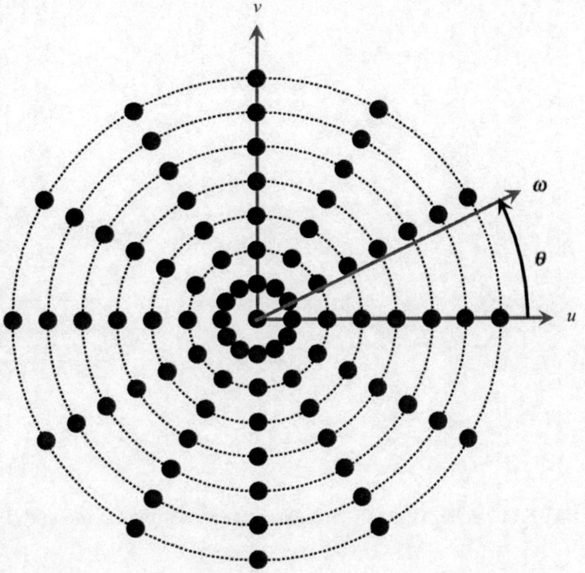

Fig. 3.10 Schematic of the frequency space filled with projection data at multiple angles

3.3 Image Reconstruction Techniques

3.3.1 Algebraic Reconstruction Technique

(1) Basic Principles

The *algebraic reconstruction technique* is often abbreviated as ART. Its computational costs are high due to its long computational time when compared to the filtered back projection discussed later, but its advantages include image noise/artifact reduction effects and its ability to conduct image reconstruction with a few projections. For these reasons, this technique is often used in transmission electron microscopy (TEM), single photon emission computed tomography (SPECT), and positron emission tomography (PET), as well as for medical X-ray tomography with the objective of low doses. A representative technique is the *iterative reconstruction method*. Furthermore, British engineer Hounsfield used this technique for his first image reconstruction [14].

Figure 3.11 shows a schematic for reconstructing the original 2D linear absorption coefficient distribution from 1D projection data; it is likely very easy to understand the fundamental thought process behind it. Figure 3.11a shows the total area including the sample divided into smaller areas of equal size in a 2 × 2 arrangement, with imaging from multiple directions using detectors with pixel number 2 and whose detection elements are aligned in the horizontal direction. The linear absorption coefficients μ_1–μ_4 of each area are set as unknowns, which can be determined by

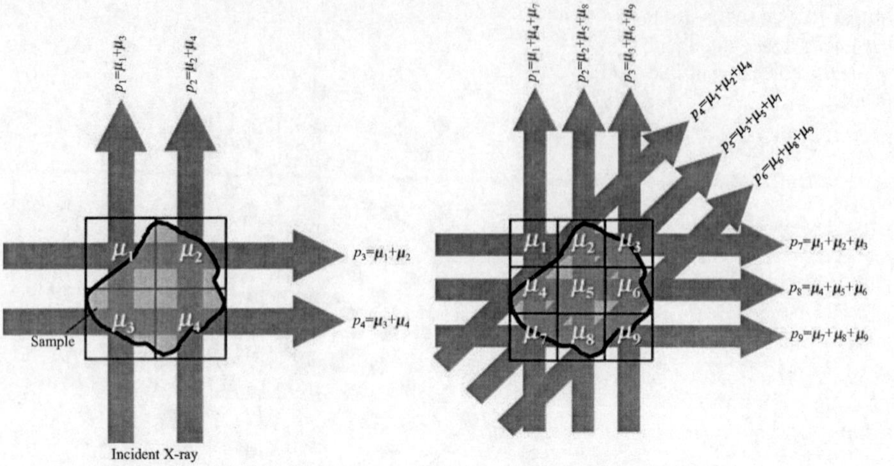

(a) Imaging using a 2 pixel detector

(b) Imaging using a 3 pixel detector

Fig. 3.11 Schematic of representative reconstruction methods

solving for the four simultaneous equations obtained by calculating the projections p_1–p_4 from two directions. There may be local distributions of the linear absorption coefficient in each area. Furthermore, external air sections may be locally present in the smaller areas near the surface. The terms μ_1–μ_4 are the result of averaging all of these aspects. Similarly, Fig. 3.11b shows a 3 × 3 division case. In this case, the nine unknowns μ_1–μ_9 can be determined by including projections at an oblique angle, calculating the projections p_1–p_9 from three directions, and setting up the nine simultaneous equations. Simultaneous equations like those shown in Fig. 3.11 can be solved using methods like Gaussian elimination [15] or lower-upper (LU) decomposition [16].

Incidentally, the traversing length across each area is different for incident X-rays in the horizontal or vertical direction relative to when the X-ray is inclined (when the vertical/horizontal area division directions are used as a standard). These need to be reflected in the simultaneous equations by introducing weighted coefficients in them. A similar consideration is necessary even when the X-ray beam width is smaller than each area. All simultaneous equations can be expressed as follows:

$$Af = p \tag{3.10}$$

Here, f refers to the unknown linear absorption coefficient set; p refers to the projection data, including all values in the Radon space; f is a K-dimensional column vector, where K is the number of areas (pixel number after 2D image reconstruction); p is an MN-dimensional column vector, where M is the number of detection elements of the 1D detector; and N is the number of projections (view number). Furthermore, the coefficient matrix A (also referred to as a system matrix) includes all the weights that express the influence of each pixel on each X-ray beam path (detection probability) and is a matrix with MN rows and K columns. The f value is considerably large in actual applications of X-ray imaging, so most coefficients are 0. The coefficient matrix A should be calculated beforehand for each device and geometric location/condition. Therefore, an advantage of this method is that actual-use conditions such as sensitivity unevenness of the detection elements can be considered by incorporating them in the coefficient matrix. The decision method of the coefficient matrix A is described in reference [17]; please refer to this there.

The column vector f can be determined from Eq. (3.10) so long as the matrix A is invertible. However, even if the matrix A inversion A^{-1} can be obtained, matrix A is extremely large and it is clear that image reconstruction calculations would consume considerable amounts of time and computer memory. For example, the device and measurement condition that the author currently uses most with synchrotron radiation-based X-ray tomography is $K = 2000 \times 2000 = 4 \times 10^6$. Furthermore, M can often be approximately 2000 pixels. N is typically around 1500–1800 projections, which can be increased up to 3600 projections when image quality must be high. Thus, $M\ N = 3.6\text{–}7.2 \times 10^6$. Therefore, the number of unknowns and equations becomes immense, ranging around the order of 10^6; solving for f by directly determining A^{-1} is not practical.

Generally, taking a high MN value by increasing the number of projections, etc. to obtain high-resolution image results in $M\ N > K$, thereby creating an overdetermined system. Meanwhile, when trying to prevent drift or prioritizing temporal resolution, N can be reduced to around 900 projections and $M\ N = 1.8 \times 10^6$. An underdetermined system is formed when there is an insufficient number of projections like in this case. There have been many reports for conducting image reconstructions in such cases. Representative methods are introduced below.

A general solution for Eq. (3.10) is to find the least-squares solution of the following equation [14]:

$$\chi^2 = |Af - p|^2 \tag{3.11}$$

This is equivalent to solving for what corresponds to an inverse rectangular matrix, also known as a Moore-Penrose inverse (pseudo-inverse matrix) [18].

One advantage of an algebraic reconstruction technique is that a degree of image reconstruction is possible even when there are missing wedges or when the rotational angle is extremely limited, such as below 60° for TEM tomography. Furthermore, image reconstruction calculations are possible for cases where the sample material or density is already known or when air in the exterior of the sample is incorporated beforehand. The area division sizes do not necessarily have to be uniform as well. For example, the number of projections necessary for image reconstruction can be efficiently reduced by dividing only the areas of interest with a higher resolution (e.g. by making coarse divisions in the external areas of the sample and fine divisions in the internal areas of the sample).

(2) Iterative Reconstruction Technique

The basic method for solving Eq. (3.10) was published by the Polish mathematician Kaczmarz in 1937 [19]. This method was later applied by Gordon et al. in the U.S. on the 3D image reconstruction of TEM tomography [20]. This method is introduced as follows.

Arranging Eq. (3.10) into simultaneous equations yields the following:

$$
\begin{aligned}
a_{11}\mu_1 + a_{12}\mu_2 + a_{13}\mu_3 + \cdots + a_{1K}\mu_K &= p_1 \\
a_{21}\mu_1 + a_{22}\mu_2 + a_{23}\mu_3 + \cdots + a_{2K}\mu_K &= p_2 \\
&\vdots \\
a_{MN1}\mu_1 + a_{MN2}\mu_2 + a_{MN3}\mu_3 + \cdots + a_{MNK}\mu_K &= p_{MN}
\end{aligned}
\tag{3.12}
$$

The 2D image expressed by (μ_1, μ_2, ..., μ_K) corresponds to a point in the K-dimensional space. Each equation in Eq. (3.12) expresses a hyperplane in this space.

Here, the simplest case is assumed to help comprehend the solution method; in other words, this is a simple projection like that shown in Fig. 3.11a, where $K = 2$ (μ_1, μ_2), and $M\,N = 2$.

$$a_{11}\mu_1 + a_{12}\mu_2 = p_1 \tag{3.13}$$

$$a_{21}\mu_1 + a_{22}\mu_2 = p_2 \tag{3.14}$$

Figure 3.12 shows a schematic of the iterative reconstruction technique, in this case, based on expressions by Rosenfeld and others [21]. The two equations shown in Eq. (3.13) on the $\mu_1 - \mu_2$ plane are two straight lines. Furthermore, their intersection expresses the solution vector. First, the linear absorption coefficients of the two pixels are set as μ_1^0 and μ_2^0, which, for the time being, can be set as 0 [22], and images obtained by filtered back projection (discussed later) or images of similar samples from the past can be included here. The first straight line (Eq. (3.13)) is drawn perpendicularly from this point. This corresponds to the acquisition of the new pair of linear absorption coefficients μ_1^1 and μ_2^1 obtained from the first projection. Next, the second straight line (Eq. (3.14)) is drawn perpendicularly from this point, and the linear absorption coefficient value is updated. Iterations of this allow for the asymptotic approach to the actual μ_1 and μ_2 values (white circles in Fig. 3.12). As seen in Fig. 3.12, cases where this does not converge are when the straight lines of Eqs. (3.13) and (3.14) are parallel or, in other words, when projections were conducted twice from the same direction. However, this solution method is generally influenced by calculation accuracy and there are thought to be cases where it does not completely converge to a single point at times.

Returning to Eq. (3.12), this iterative calculation is expressed with the following equation [17]:

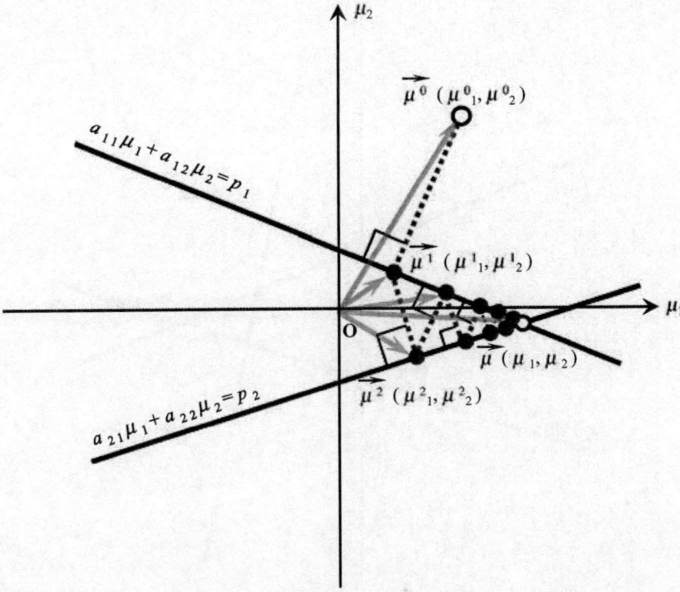

Fig. 3.12 Schematic expressing the solution method process for the iterative image reconstruction method

$$\overrightarrow{\mu^i} = \overrightarrow{\mu^{i-1}} - \frac{\left(\overrightarrow{\mu^{i-1}} \cdot \overrightarrow{a^i} - p^i\right)}{\overrightarrow{a^i} \cdot \overrightarrow{a^i}} \overrightarrow{a^i} \tag{3.15}$$

This expresses the case where $\overrightarrow{\mu^i}$ $(\mu^i_1, \mu^i_2, \ldots, \mu^i_K)$ is projected on the hyperplane based on the ith equation. Here, $\overrightarrow{a^i}$ is the coefficient matrix of the ith equation and the right-hand denominator is the scalar product between $\overrightarrow{a^i}$ and itself. The $\overrightarrow{\mu^i} \cdot \overrightarrow{a^i}$ of Eq. (3.15) expresses the projection and the remaining parts on the second term on the right-hand side correspond to the back projection. The pixel values in Eq. (3.15) are updated by adding and subtracting the correction values, but there is multiplicative feedback wherein corrections are conducted by multiplying the correction values [23]. The computational time needed to converge to a solution can be high when calculations are conducted in multiple dimensions. Furthermore, calculations with high noise can result in predicted values that are far removed from the solution. For example, Fig. 3.13 assumes a high-noise and high-artifact variation of the schematic shown in Fig. 3.12 where a third projection was added. In this case, the intersections of the three lines are not fixed at once. In this case, a solution is searched for by assuming that a pseudo-solution exists in the area enclosed by the straight lines.

Determining when to stop calculations is important in the sense of computational costs and preventing solution divergence. The decision method for stopping calculations includes simple thought processes such as stopping calculations after a fixed

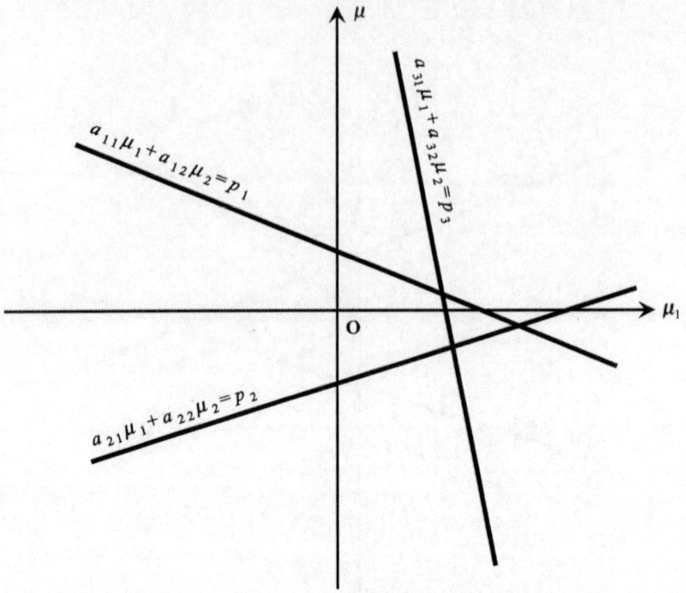

Fig. 3.13 Schematic showing issues with the iterative image reconstruction method when noise or artifacts are present

number of iterations, methods where stops are introduced after value changes due to iterations become smaller than a fixed value, and when an image quality standard that was determined beforehand is satisfied.

A relaxation parameter is generally introduced to reduce the amount of time to convergence.

$$\vec{\mu^i} = \overrightarrow{\mu^{i-1}} - \lambda_i \frac{\left(\overrightarrow{\mu^{i-1}} \cdot \vec{a^i} - p^i\right)}{\vec{a^i} \cdot \vec{a^i}} \vec{a^i} \tag{3.16}$$

Here, λ_ι is the relaxation parameter and varies by image or the number of calculation iterations. Reports have indicated that the computational time decreases when λ_ι becomes slightly larger than 1 [22].

(3) Image Reconstruction Methods for Actual Use

The development and application of algebraic reconstruction techniques and iterative reconstruction techniques have been astounding in X-ray tomography and other 3D imaging fields. However, the terms and abbreviations of the various methods are not always necessarily consistent. Here, these are summarized from the viewpoints of purely iterative reconstruction techniques, those that conduct statistical analyses, and those that use physical models.

The *simultaneous algebraic reconstruction technique (SART)* [24] simultaneously conducts pixel value updates using data from a single projection angle. This differs from the "ray-by-ray" system for updating pixel values with the above-mentioned ART. SART was developed by Andersen et al. in 1984. The introduction of this technique resulted in a reduction in computational time required and error propagation originating from noise. The *simultaneous iterative reconstruction technique (SIRT)* was first proposed by Gilbert and others in 1972 [25]. This technique conducts calculations at all projection angles, after which the correction value due to the X-ray beam in all directions is averaged and the pixel values are updated "point-by-point." Although more computational time is required relative to SART, more accurate image reconstruction is possible [17]. This method is widely used alongside ART in TEM tomography. Other techniques calculate simultaneous equations handled in ART by dividing them into numerous arbitrary blocks [26]. From this perspective, the above-mentioned SART treats data taken from a single projection angle as a block, whereas SIRT treats all equations as a single block, making them opposites.

Statistical iterative reconstruction was first published by Hurwitz in 1975 [27], after which Shepp et al. proposed an algorithm referred to as expectation maximization (EM), which introduces a maximum likelihood estimate from incomplete data [28]. On a basic level, this applies the thought process of measured photon counting statistics in the image reconstruction process to determine the expected pixel value. It is used particularly in imaging cases where there are large errors in measured data (e.g. SPECT or PET) as the *maximum likelihood-expectation maximization (ML-EM)* method [28] or the *maximum a posteriori estimation-expectation maximization (MAP-EM)* method [29]. Of these, the ML-EM method estimates images to maximize the conditional probability that a given projection data is obtained, assuming that the measured photon count follows a Poisson distribution. This has a favorable *signal-to-noise (S/N)* ratio and advantages in correcting for absorption in the subject and conducting spatial resolution corrections that factor in the dependency of the blur to the opening diameter of the collimator. However, one disadvantage is increased computational time. Meanwhile, the MAP-EM method conducts image reconstruction based on the a posteriori probability maximization estimation as well as the image reconstruction evaluation system with the a priori subject information as a constraint condition, making it advantageous for improving image quality. The *ordered subset-expectation maximization (OS-EM) method* [30], which accelerates the ML-EM method, is currently widely used in the SPECT and PET fields. The OS-EM method divides the projection data into subsets and has the advantage of increasing convergent speed and convergence when compared to using the entire projection data to update the image a single time by instead conducting forward projection, back projection, and image updating at the subset level. Furthermore, the OS-EM method has a disadvantage in which the subsets are made too small, resulting in the convergence to a periodic solution known as a limit cycle. The dynamic row-action maximum likelihood algorithm (DRAMA) method was proposed to resolve this issue [31].

Model-based iterative reconstruction (MBIR) methods can be classified into a number of different groups [32]. The first is to accurately model the X-ray imaging measurement system. This takes each *voxel* (pixel of a 3D image, which corresponds to a pixel in 2D and is formed from the terms "volume" and "pixel"), which expresses the object and considers it a cube, thereby accurately modeling their projecting effects [33] or considering that X-ray sources have a fixed size and that models of photons emitted from different positions in the X-ray source coming into a single detection element in a detector are created [34]. Furthermore, the second is to model the interactions between the X-ray photons and object. For example, for a white X-ray coming out of an X-ray tube, X-rays with longer wavelengths are rapidly attenuated near the surface whereas X-rays with shorter wavelengths easily transmit, showing a nonlinear effect of X-ray absorption. The artifacts that form from this (i.e. beam hardening) can be greatly reduced by modeling this process and factoring it in, thereby improving image quality [35]. Furthermore, there have been many reports on methods that consider information on the subject that was already known prior to experiments. For example, the reasonability of pixel values between adjacent pixels rapidly changing can be determined based on background information relating to the subject, thereby trying to both reduce noise and maintain spatial resolution [36].

3.3.2 Filtered Back Projection

(1) Back Projection

Filtered back projection (FBP) is an image reconstruction method widely used in X-ray CT. Prior to explaining this method, we will first discuss *back projection*, which is the basis of this method. Back projection is a processing technique that returns projection data to the sample position along the direction of the X-ray when conducting forward projection, though in the opposite direction.

The actual process of back projection may be difficult to understand and an overview of it even less so. With this in mind, an extremely small dataset will be handled to help understand this method, as was done in Fig. 3.11 for the algebraic reconstruction technique. As shown in Fig. 3.14, an object with a simple square shape is divided into fifths horizontally and vertically and the linear absorption coefficients of each divided area are calculated manually. The center of the 25 squares has a single square with an extremely high linear absorption coefficient (a value of 10, which is 10 times higher than its surrounding values). Qualitatively, this corresponds to an example where a square iron rod was embedded in the center of a polymer sample. Parallel-beam projections were conducted from four directions on the sample, with a total of 28 sets of projection data acquired. This projection data is also written together for Fig. 3.14. The linear absorption coefficient distribution will

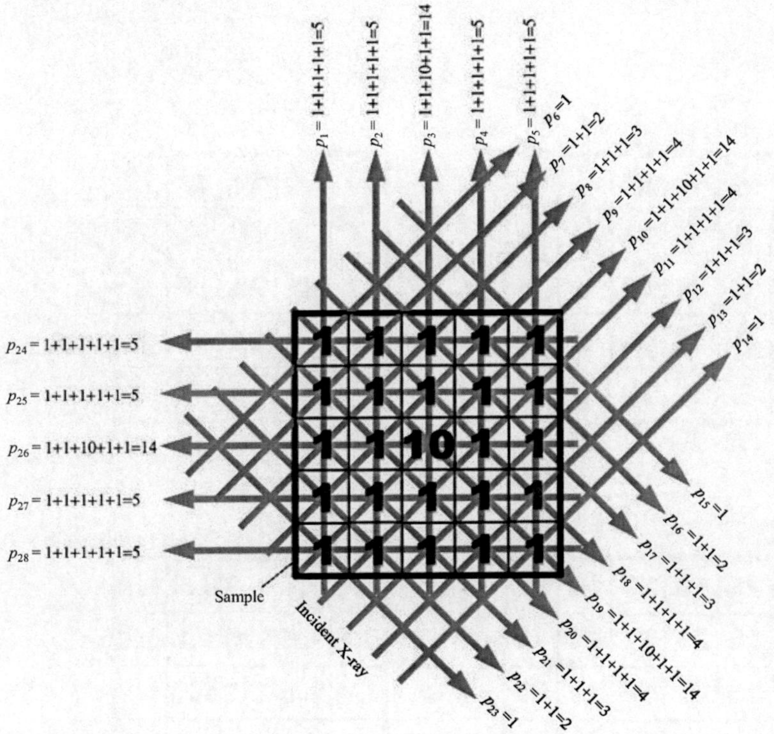

Fig. 3.14 Schematic of X-ray imaging of a square-shaped object (linear absorption coefficient of 1, with only the center at a value of 10) divided into pixels arranged in 5 × 5. Projection data acquired from four directions

be determined by solving with back projection, unlike the case with the algebraic reconstruction technique in Fig. 3.11.

First, we can see from looking at the projection data of p_1–p_5 that they accurately reflect the linear absorption coefficients in the sample interior, with p_3 having a high value, and p_1, p_2, p_3, and p_4 all having low values. Therefore, returning the projections obtained at all angles to their positions in the sample and sequentially adding them at each square highlights the large absorption of the central area with a linear absorption coefficient of 10, recreating the original linear absorption coefficient distribution. This is the thought process for back projection. Figure 3.15 shows the actual implementation of this process on the model in Fig. 3.14. The projections from four directions are back-projected sequentially on a 5 × 5 square of the object in Fig. 3.15a–c. For example, the projection p_1 at the left-most side in the vertical direction with its value of 5 is assigned to all five squares on the far-left vertical edge. The English term for conducting back projection is also referred to as "smear back" but this is indeed the essence of it. This is conducted for all projection data p_1–p_{28},

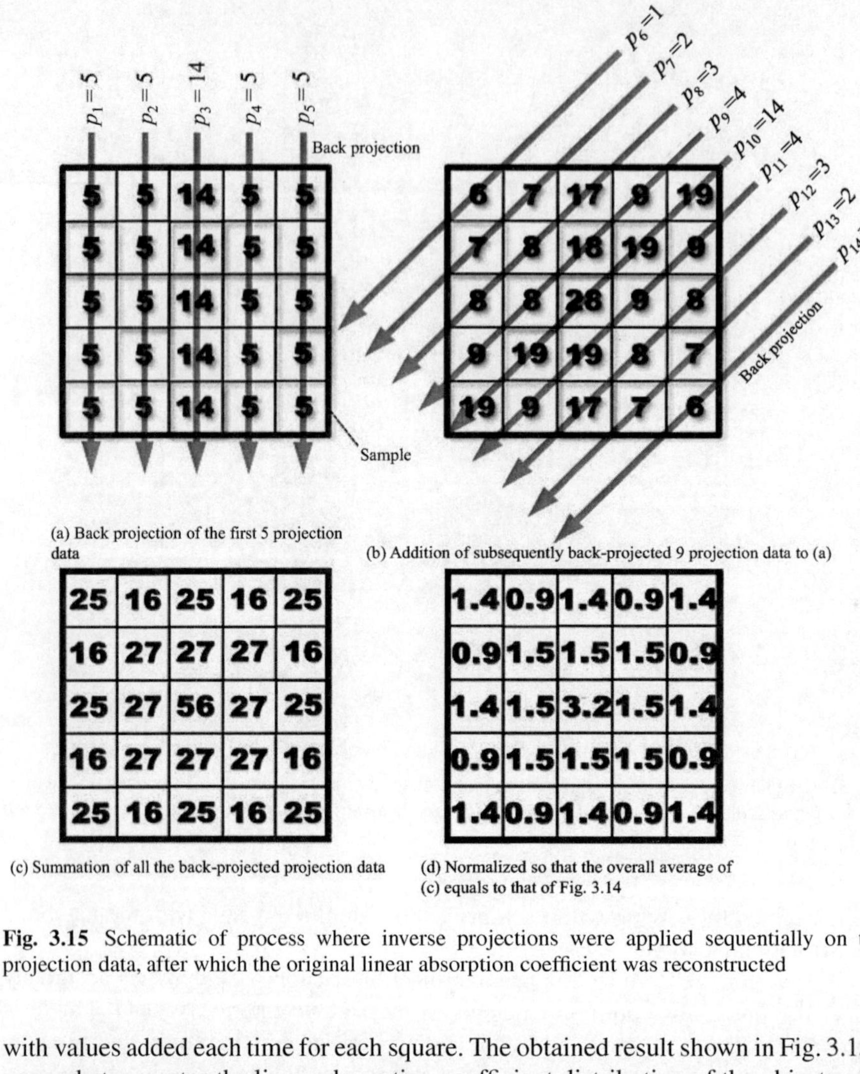

(a) Back projection of the first 5 projection data

(b) Addition of subsequently back-projected 9 projection data to (a)

(c) Summation of all the back-projected projection data

(d) Normalized so that the overall average of (c) equals to that of Fig. 3.14

Fig. 3.15 Schematic of process where inverse projections were applied sequentially on the projection data, after which the original linear absorption coefficient was reconstructed

with values added each time for each square. The obtained result shown in Fig. 3.15d somewhat recreates the linear absorption coefficient distribution of the object, with only one square in the middle of the sample having a high linear absorption coefficient value. Expressing the back projection with a mathematical formula using Eq. (3.2) yields the following equation:

$$\mu_b(x, y) = \int_{-\infty}^{\infty} p(x cos\theta + y sin\theta, \theta) d\theta \tag{3.17}$$

Here, $\mu_b(x, y)$ is the linear absorption coefficient value obtained from a simple back projection. Furthermore, Fig. 3.16 shows a schematic of a back projection

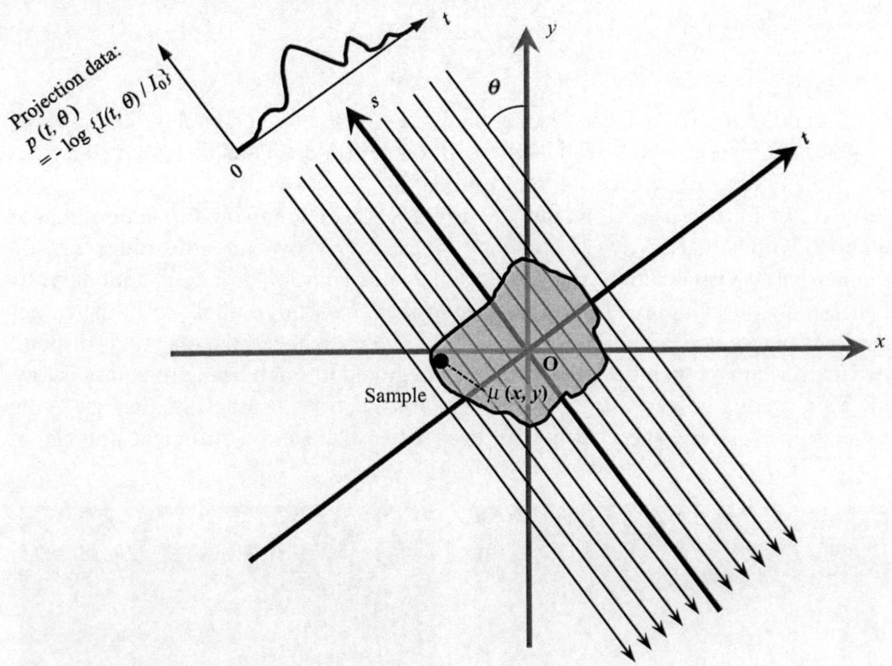

Fig. 3.16 Schematic showing the relationship between the projection data $p\ (t,\ \theta)$ at a given projection angle and given position t and its inverse projection

following a straight line expressed by the equation $t = x\ cos\theta + y\ \sin\theta$ at a given projection angle θ. The points on this line segment all have the same t value, regardless of the x and y values. Therefore, it can be observed from Fig. 3.15 that the back projection following this line segment equally contributes to the reconstructions on all the points on that line segment.

Here, care must be taken that $\mu_b(x, y)$ is considerably different with the linear absorption coefficient distribution $\mu(x, y)$ in the object. In actuality, visual comparisons between Figs. 3.15 and 3.14 show that the central peak of the linear absorption coefficient becomes lower and horizontally wider. This indicates that the peak is influenced by the surrounding areas with lower linear absorption coefficients during back projection and ultimately results in a *blurred* final image. In this case, the relationship between $\mu_b(x, y)$ and $\mu\ (x, y)$ is the following [37]:

$$\mu_b(x, y) = \mu(x, y) * h_p(r) \tag{3.18}$$

Here, $h_p\ (r)$ is the *point-spread function*, alternatively known as the *blurring function*. The "*" in the equation symbolizes a *convolution* operation. If the distance from the center of the point area were set as r, then $h_p\ (r)$ is expressed as follows [37]:

$$h_p(r) = \frac{1}{\sqrt{x^2 + y^2}} = \frac{1}{r} \tag{3.19}$$

From this equation, we can see that the image blurring when undergoing a simple back projection does not depend on the position of the object and that point-like areas like those in Fig. 3.14 have rotation symmetry.

Next, we look at image blurring due to back projection using projection images obtained with actual X-ray CT scanners. Figure 3.17 shows an industrial X-ray CT scanner-based visualization of a circular magnet used as an article of stationery. Its matrix material is a polymer with a circular magnet (iron) embedded in the center and its structure is similar to that in Fig. 3.15. The presence of the magnet is difficult to discern when the number of projections integrated through back projection is low, however, gradually increasing the number of projections enables the magnet in the center to be distinguished. Although, even when there are a sufficient number of

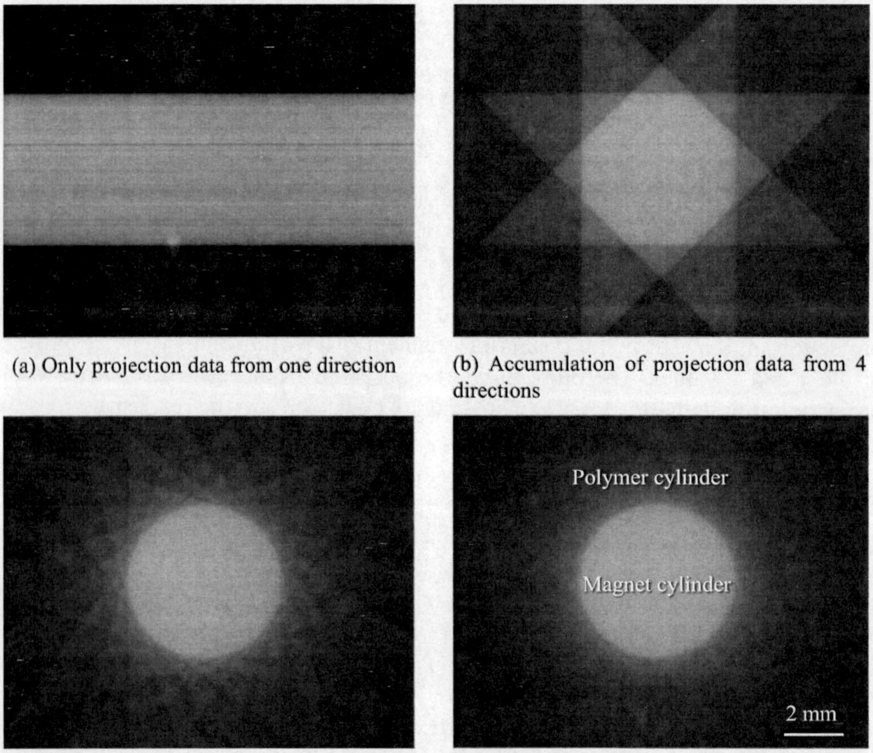

(a) Only projection data from one direction

(b) Accumulation of projection data from 4 directions

(c) Accumulation of projection data from 10 directions

(d) Accumulation of projection data from 400 directions

Fig. 3.17 Stationary magnet imaged with an industrial X-ray CT device. The number of projections is 400, the projection data had no filter correction, and the projection data from each direction underwent back projection and was accumulated projections as shown in Fig. 3.17d, the external sample shape cannot be confirmed and the magnet in the center is extremely blurred.

If even objects with more complex structures were treated as aggregates of innumerable point-like areas, then a blurred image with blurring functions convoluted throughout the object could be obtained. Projection data [12] of an aluminum-copper alloy shown in Fig. 3.7 was used as an example and data similar to Fig. 3.17 integrated after back projection without filtering are shown in Fig. 3.18. The internal structure of the sample has a large difference in the linear absorption coefficient between the high X-ray absorption area (intermetallic Al_2Cu compound particles) and matrix (aluminum); the two can be extremely clearly distinguished. However, unlike with Fig. 3.17, the number of particles is large and the overall structure seems somewhat complex. The blurring effect due to back projection becomes even more pronounced in these cases and discerning the existence, location, and shape of particles is difficult with just simple back projection.

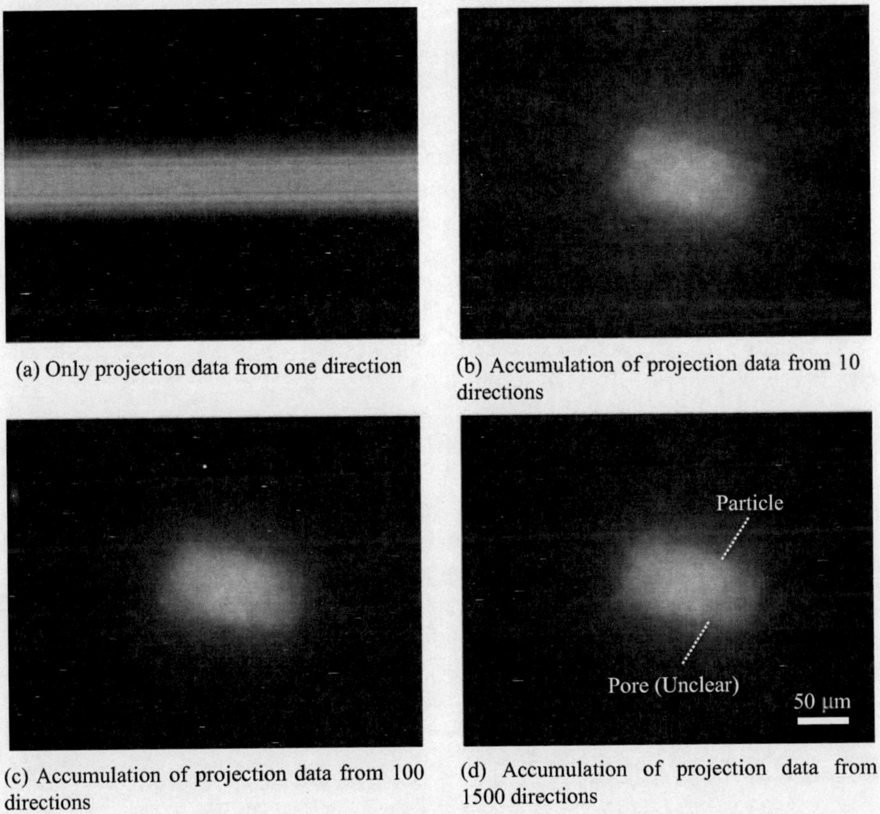

Fig. 3.18 Projection data of an aluminum-copper alloy shown in Fig. 3.7. The projection data had no filter correction and the projection data from each direction underwent back projection and was accumulated

(2) Filtered Back Projection

Filtered back projection corrects the blurring that occurs in simple back projections, making these projections shown in Figs. 3.14, 3.15, 3.16, 3.17 and 3.18 suitable for use. The projection theorem is reviewed below with an overview of its derivation.

First, returning to Eq. (3.5), this time the inverse Fourier transform is used to conversely express the linear absorption coefficient μ (x, y) of the object with F (u, v).

$$\mu(x, y) = \int_{-\infty}^{\infty}\int_{-\infty}^{\infty} F(u, v)e^{2\pi i(xu+yv)}dudv \tag{3.20}$$

As already shown in Figs. 3.8 and 3.10, (u, v) is a Cartesian coordinate expression of the frequency space corresponding to the real space (x, y). This time, we express this in polar coordinates (ω, θ). As was the case with Eqs. (3.7) and (3.8), we consider $|J|$ for the transform from (u, v) coordinates to (ω, θ) coordinates.

$$|J| = \begin{vmatrix} \frac{\partial u}{\partial \omega} & \frac{\partial u}{\partial \theta} \\ \frac{\partial v}{\partial \omega} & \frac{\partial v}{\partial \theta} \end{vmatrix} \tag{3.21}$$

Here, the relationships $u = \omega\ cos\theta$ and $v = \omega\ sin\theta$ are present between the Cartesian and polar coordinates, and, ultimately, the following relational expression can be obtained between the integration variables:

$$dudv = |J|d\omega d\theta = \omega d\omega d\theta \tag{3.22}$$

As such, Eq. (3.20) can be expressed as follows:

$$\mu(x, y) = \int_{0}^{2\pi}\int_{0}^{\infty} F(\omega cos\theta, \omega sin\theta)e^{2\pi i\omega(xcos\theta+ysin\theta)}\omega d\omega d\theta \tag{3.23}$$

The intermediate calculations are omitted below. Publications from Buzug present further details on this topic; please refer to these for additional information [37]. Schematically, the integrations relating to θ are divided between 0–π and π–2π, with the latter assumed to be the phase shift of the phase difference π. Furthermore, the symmetry of the Fourier transform $F(\omega, \theta + \pi) = F(-\omega, \theta)$ is used. Ultimately, the integration range relating to ω extends from $-\infty$ to $+\infty$, with Eq. (3.23) expressed as follows:

$$\mu(x, y) = \int_{0}^{\pi}\int_{-\infty}^{\infty} F(\omega cos\theta, \omega sin\theta)e^{2\pi i\omega(xcos\theta+ysin\theta)}|\omega|d\omega d\theta \tag{3.24}$$

As seen in Eq. (3.9), $F\ (\omega\ cos\theta,\ \omega\ sin\theta)$ is the 1D-Fourier transform for variable t in the projection data $p\ (t, \theta)$. If this is re-expressed as $P\ (\omega, \theta)$, Eq. (3.24) ultimately can be expressed as follows:

$$\mu(x, y) = \int_0^{\pi} \left\{ \int_{-\infty}^{\infty} P(\omega, \theta) e^{2\pi i \omega t} |\omega| d\omega \right\} d\theta \tag{3.25}$$

However, Eq. (3.2) is reused here and expressed using rotated coordinates. Re-examining this equation shows that the integration in the curly brackets is when an inverse Fourier transform is applied on a 1D-Fourier transform $P\ (\omega, \theta)$ of the projection data at an angle θ, considering the amount $P\ (\omega, \theta)\ |\omega\ |$. $P\ (\omega, \theta)\ |\omega\ |$ is the result of a Fourier transform being applied on the projection data, after which it was filtered with the frequency response function $|\omega\ |$. This is the reason why this method is referred to as the filtered back projection. Incidentally, Fig. 3.19 is an example of filtering applied on a 1D-Fourier transform of the projection data shown in Fig. 3.9.

Following filtering, the external integration in Eq. (3.25) can be used to add up all of the filtered projections that pass through a given point (x, y), which allows for the image reconstruction of the linear absorption coefficient $\mu\ (x, y)$ at that point, which is ultimately what Eq. (3.25) signifies. The final back projection has already been

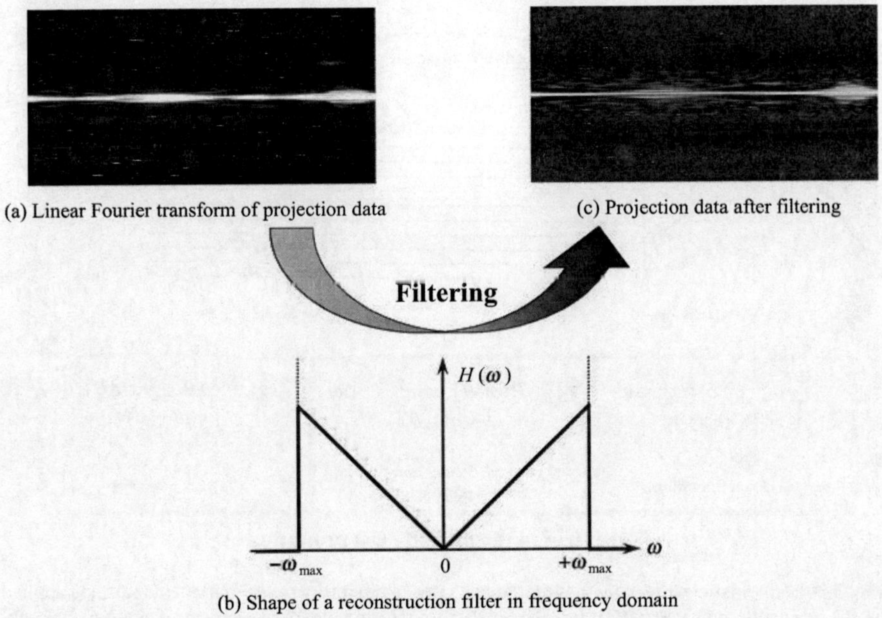

Fig. 3.19 Example where filter correction (Ramachandran-Lakshminarayanan filter) was conducted on a 1D Fourier transform of projection data shown in Fig. 3.9

confirmed as a mathematical equation in Eq. (3.17), and schematically in Figs. 3.14, 3.15, 3.16, 3.17 and 3.18.

Here, the projection data after filtering at a given angle θ is once again set as $q\ (t, \theta)$, as shown below:

$$q(t,\theta) = q(xcos\theta + ysin\theta) = \int_{-\infty}^{\infty} P(\omega,\theta)|\omega|e^{2\pi i\omega(xcos\theta+ysin\theta)}d\omega \tag{3.26}$$

The three most important processes of filtered back projection are summarized once again using this. Furthermore, Fig. 3.20 shows these in the form of a schematic; please be sure to refer to them as well.

$$P(\omega,\theta) = \int_{-\infty}^{\infty} p(t,\theta)e^{-2\pi i\omega t}dt \tag{3.27}$$

$$q(t,\theta) = \int_{-\infty}^{\infty} P(\omega,\theta)|\omega|e^{2\pi i\omega t}d\omega \tag{3.28}$$

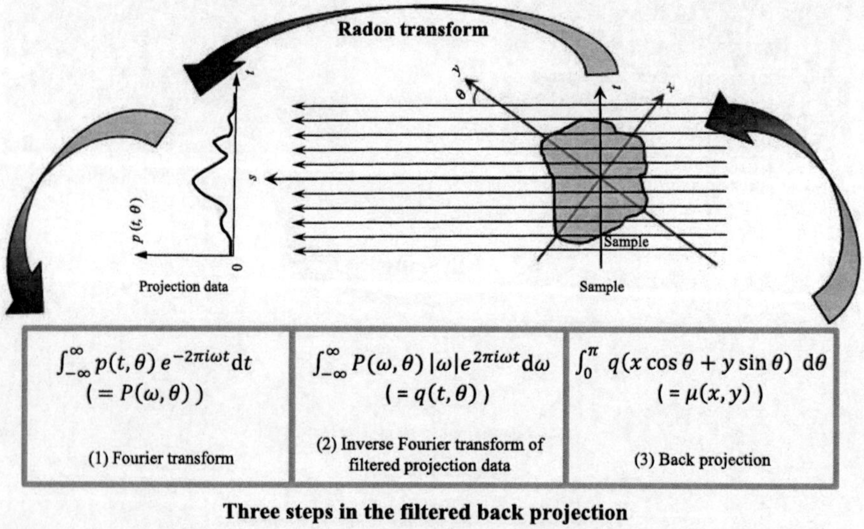

Fig. 3.20 Schematic expressing the filtered back projection process. A Fourier transform is applied on the projection data $p\ (t, \theta)$; an inverse Fourier transform is applied after this was filtered. Back-projecting this as a final step yields the original distribution of the linear absorption coefficient ($\mu\ (x, y)$)

$$\mu(x, y) = \int_{0}^{\pi} q(x\cos\theta + y\sin\theta)d\theta \tag{3.29}$$

Equations (3.27)–(3.29) are the 1D Fourier transform of the projection data, the inverse Fourier transform of the filtered data, and its back projection, respectively.

To actually confirm this effect, the filtering process is manually calculated for the model in Fig. 3.14 where image reconstruction was conducted with simple back projection. In Fig. 3.21a–c, the data are sequentially back-projected in the 5 × 5 squares of the object after filtering is applied to the projections in four directions. The filter used is straightforward to understand when examining the real spaces in

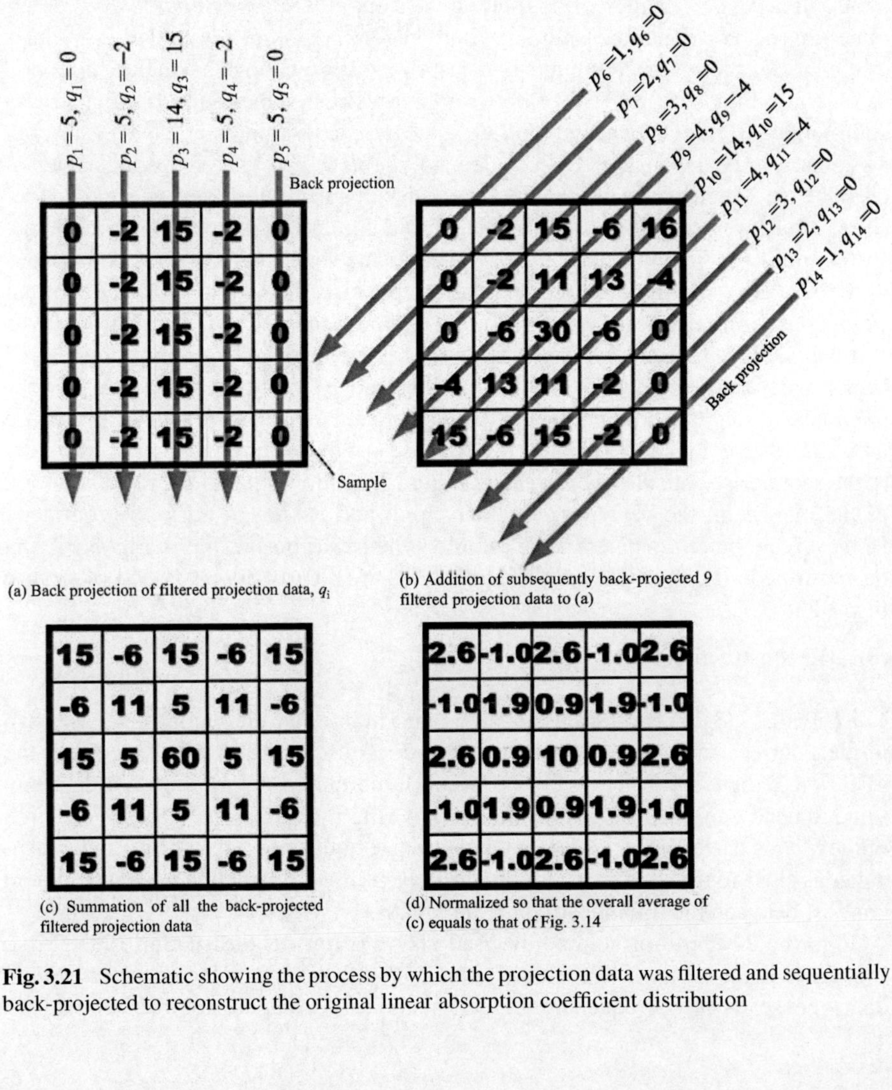

0	-2	15	-2	0
0	-2	15	-2	0
0	-2	15	-2	0
0	-2	15	-2	0
0	-2	15	-2	0

0	-2	15	-6	16
0	-2	11	13	-4
0	-6	30	-6	0
-4	13	11	-2	0
15	-6	15	-2	0

15	-6	15	-6	15
-6	11	5	11	-6
15	5	60	5	15
-6	11	5	11	-6
15	-6	15	-6	15

2.6	-1.0	2.6	-1.0	2.6
-1.0	1.9	0.9	1.9	-1.0
2.6	0.9	10	0.9	2.6
-1.0	1.9	0.9	1.9	-1.0
2.6	-1.0	2.6	-1.0	2.6

Fig. 3.21 Schematic showing the process by which the projection data was filtered and sequentially back-projected to reconstruct the original linear absorption coefficient distribution

Figs. 3.24 and 3.25 appearing later. For example, the value of the projection value $p_1 = 5$ in the furthest left in the vertical direction becomes $q_1 = 0$ after being multiplied with the reconstruction function. Here, as was the case with Eq. (3.26), the projection data q_i at the ith position can be determined after filtering. Figure 3.21d is the result of the projection data p_1–p_{28} after the back projection was conducted on its corresponding filtered data q_1–q_{28}. This figure shows that the linear absorption coefficient of the area that has a high linear absorption coefficient value in the center of the sample is properly determined and far more accurately recreates the linear absorption coefficient than the simple back projection case shown in Fig. 3.15.

Incidentally, it is not feasible to continuously conduct the measurements of Eq. (3.1) or the back projections of Eq. (3.29) at all angles. The actual process of acquiring the projection data involves acquiring the several hundred to several thousand necessary images while rotating the sample. The *sampling theorem* determines this necessary number of images and the *Nyquist frequency* actually stipulates this value. Furthermore, continuously measuring the projection data in the t-direction as shown in Figs. 3.5 and 3.16 or determining the linear absorption coefficient as a continuous variable is not actually feasible. In these cases, the sample is viewed as an aggregate of pixels with a given size, depending on the pixel number or pixel size of the detector, which can be used in measurements. For example, reconstructed images in 3D are expressed using voxels as their basic digital information. It is ultimately the sampling theorem that determines how large the voxels must be via comparisons of the spatial resolution required and the sample size. These aspects are important when conducting actual measurement; this will be discussed in Chap. 7.

Finally, Figs. 3.22 and 3.23 once again show the image reconstructions by filtered back projection on projection data that had undergone simple back projections in the previous section to better understand the filtering effect. The results in each case are different when compared to the non-filtered case and the outline and internal structure of the sample are both clearly recreated. Figure 3.22 shows that the X-ray absorption of the magnet in the center is extremely high and its image noise is prominent. However, the outer diameter of the sample, which was not visible in Fig. 3.17, can be confirmed. The multiple particles and voids are clearly distinguished as well in Fig. 3.23.

(3) Reconstruction Filter

The $|\omega|$ in Eq. (3.28) is referred to as the ramp filter due to its ramp-like waveform in the Fourier space. Although it generally does not appear to be recognized, the selection of the reconstruction filter is just as important, if not more so, for the setup selection and coordination of its conditions during measurement. Careful selection of the filter is particularly important when the size of the internal structure to be visualized is close to the spatial resolution, the noise is high, or when there is insufficient contrast between the internal structure and matrix.

Figure 3.24 shows a commonly used shape of the reconstruction filter. First, Fig. 3.24a is the $|\omega|$ itself. If the maximum spatial frequency value of the projection data were set as ω_{max}, frequencies higher than the ω_{max} that are not measured do not

(a) Only projection data from one direction

(b) Accumulation of projection data from 4 directions

(c) Accumulation of projection data from 10 directions

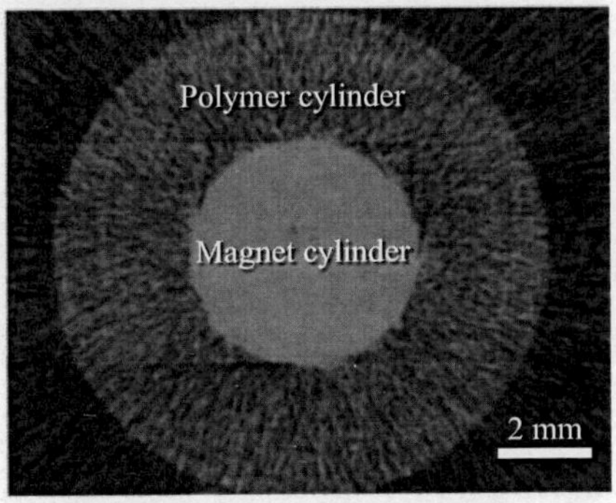

(d) Accumulation of projection data from 400 directions

Fig. 3.22 Result of the projection data of the stationary magnet shown in Fig. 3.17 undergoing filtered back projection

need to be considered in the image reconstruction process. For this reason, the reconstruction filter where high-spatial frequency domains above $\omega_{\max}$ are cut off (*cutoff frequency*) is referred to as the *Ramachandran-Lakshminarayanan filter* (often abbreviated to *Ram-Lak filter*), shown in Fig. 3.24b. The shape of this frequency domain is shown below:

$$H(\omega) = |\omega| rect\left(\frac{\omega}{2\omega_{\max}}\right) \tag{3.30}$$

Here, *rect* (x) is a rectangular function, equaling 1 at $|x| < \frac{1}{2}$ and 0 at $|x| > \frac{1}{2}$. The Ramachandran-Lakshminarayanan filter is beneficial for spatial resolution but is disadvantageous as it highlights noise. Furthermore, filters that rapidly cut away high-frequency components that exceed the cutoff frequency induce uncertainty in numerical calculations (Gibbs phenomenon) and are thought to induce artifacts in

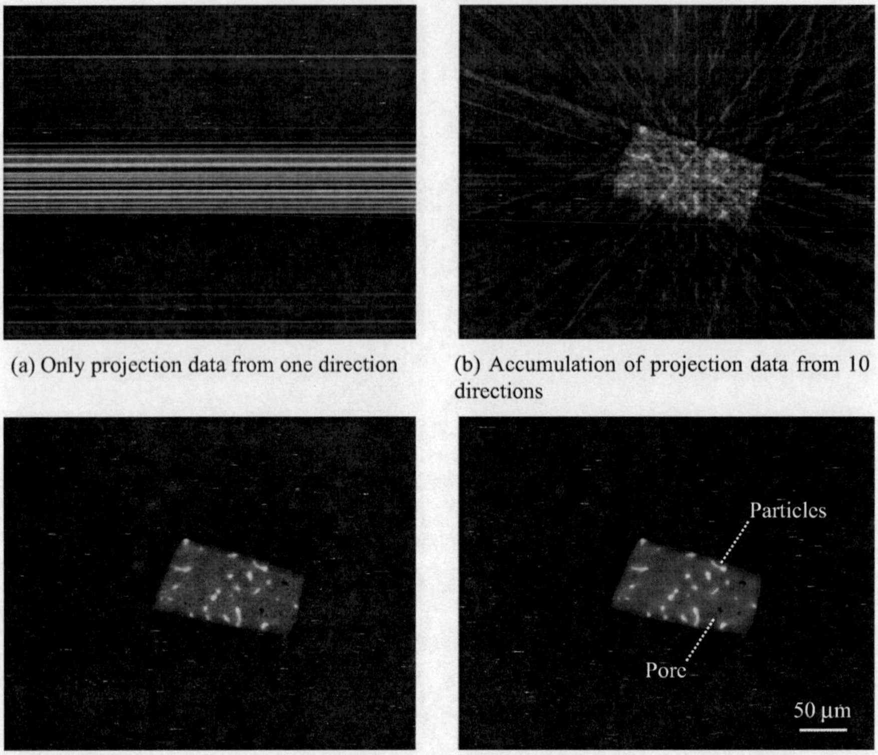

(a) Only projection data from one direction

(b) Accumulation of projection data from 10 directions

(c) Accumulation of projection data from 100 directions

(d) Accumulation of projection data from 1500 directions

Fig. 3.23 Result of the projection data of the aluminum-copper alloy shown in Fig. 3.18 undergoing filtered back projection

edge areas [38, 39]. With this in mind, reconstruction filters where a variety of *window functions* are multiplied to a ramp filter are used.

The *Shepp-Logan filter* shown in Fig. 3.24c multiplies the *sinc* (x) function to the Ramachandran-Lakshminarayanan filter in the frequency space, as shown below:

$$H(\omega) = |\omega| sinc\left(\frac{\omega}{2\omega_{\max}}\right) rect\left(\frac{\omega}{2\omega_{\max}}\right) \tag{3.31}$$

Here, *sinc* (x) is defined as $y = \frac{sin(\pi x)}{\pi x}$. The *Hann filter (Hanning filter)* shown in Fig. 3.24c can be used when noise must be further reduced.

$$H(\omega) = |\omega| \left\{ \frac{1 + cos\left(\frac{\pi\omega}{\omega_{\max}}\right)}{2} \right\} rect\left(\frac{\omega}{2\omega_{\max}}\right) \tag{3.32}$$

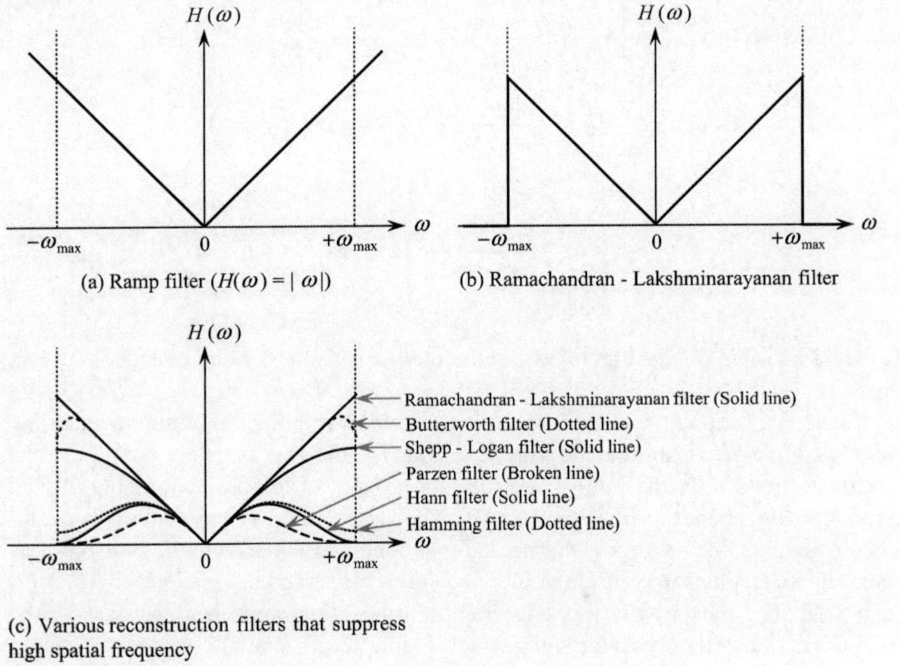

Fig. 3.24 Filters serving as the basis of filtered back projection: **a** Ramp filter; **b** Ramachandran Lakshminarayanan filter, which limits that band; and **c** various filters that further limit the amplitude of the high-frequency components

The Hann filter takes the form of 0 at $\omega_{\max}$, so high-frequency components, which have relatively high noise, are considerably filtered when compared with the Shepp-Logan filter, making the image smoother.

The Hamming filter shown in Fig. 3.24c takes the following numerical form:

$$H(\omega) = |\omega|\left\{0.54 + 0.46cos\left(\frac{\pi\omega}{\omega_{\max}}\right)\right\}rect\left(\frac{\omega}{2\omega_{\max}}\right) \tag{3.33}$$

The Hamming filter is similar to the Hann filter, but the former's amplitude value at the cutoff frequency does not drop to 0 like that in the Hann filter and instead takes the 8% value of the Ramachandran-Lakshminarayanan filter. Furthermore, the amplitude at the cutoff frequency can be adjusted by setting the 0.54 and 0.46 values of the Hamming filter as α and $1-\alpha$, respectively.

Additionally, the Parzen filter, which is separated from the ramp filter at a lower frequency than the Hann or Hamming filters and which features strong attenuation of the high-frequency component, and the Butterworth filter, which cuts closer to the cut-off frequency than the Shepp-Logan filter to maintain spatial resolution, are shown in Formulas (3.34) and (3.35) respectively.

$$H(\omega) = |\omega|\left\{1 - 6\left(\frac{\omega}{\omega_{\max}}\right)^2\left(1 - \frac{\omega}{\omega_{\max}}\right)\right\}\left(|\omega| \le \frac{\omega_{\max}}{2}\right)$$
$$= 2|\omega|\left(1 - \frac{|\omega|}{\omega_{\max}}\right)^2\left(\frac{\omega_{\max}}{2} \le |\omega| \le \omega_{\max}\right) \quad (3.34)$$

$$H(\omega) = |\omega|\left\{\frac{1}{\sqrt{1 + \left(\frac{\omega}{\omega_{\max}}\right)^{2m}}}\right\} rect\left(\frac{\omega}{2\omega_{\max}}\right)(m \ge 0) \quad (3.35)$$

Here, m is the degree and takes a value of 4 to 10; a higher degree will result in a sharper cut-off.

Figure 3.25 shows the projection data measured by rotating an aluminum quadratic prism sample 30° from the edge-on direction (i.e. the two sides are parallel to the incident direction of the X-ray) and the line profiles of the projection data filtered with the Ramachandran-Lakshminarayanan and Hann filter. The projection data after reconstruction filtering has a sharper sample edge and the surrounding air portions take negative values to control the blur and spread observed in Figs. 3.15, 3.16, 3.17, and 3.18. The reason why this edge highlighting effect appears is because of the reconstruction filter characteristics, which highlight the spatial changes, as seen in Fig. 3.24. This is a characteristic referred to as a high-pass filter. This can also be understood by looking at the shape of the reconstruction filter in real space (discussed later; see Figs. 3.29 and 3.30 for reference). Figure 3.26 shows the projection data of Fig. 3.7b with and without filtering in the form of sinograms. The filtering effect can be clearly seen with a sinogram.

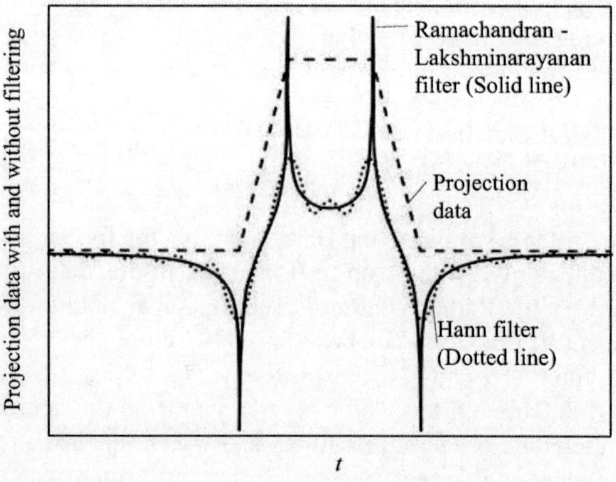

Fig. 3.25 Projection data measured when rotating an aluminum quadratic prism sample 30° in the longitudinal direction of the column and the projection data when either a Ramachandran-Lakshminarayanan filter or Hann filter was applied

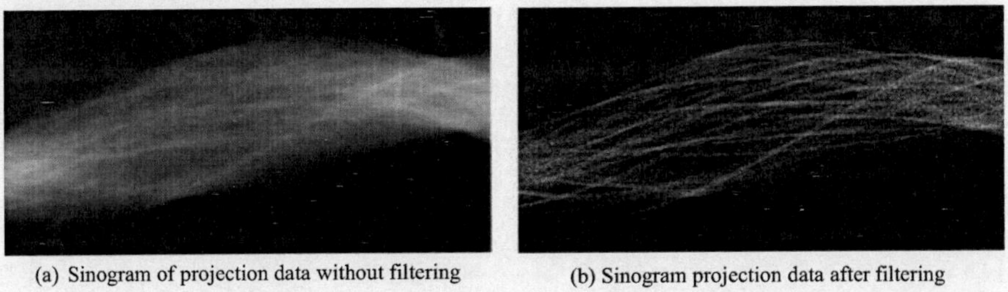

Fig. 3.26 Projection data of the aluminum-copper alloy in Fig. 3.7 shown in the form of a sinogram and a comparison of filtering and non-filtering effects

(4) Reconstruction Filter Selection

Figure 3.27 shows an image reconstruction with filtered back projection, wherein the projection data of the aluminum-copper alloy shown in Figs. 3.7 and 3.18 are used and each of the above-mentioned reconstruction filters are applied. The noise and fine structural differences can be easily observed in the magnifications in Fig. 3.28. The differences in smoothness and sharpness of images even in macro cases like Fig. 3.27 can be clearly observed when using filters that mostly preserve the high-frequency components (e.g. Ramachandran-Lakshminarayanan filter, Shepp-Logan filter, etc.) or mostly eliminate the high-frequency components (e.g. Hamming filter). The former shows the particle and pore outlines but shows noise and pixel value fluctuations beneath the structure in the matrix. As shown in Fig. 3.24c, the Hamming filter,

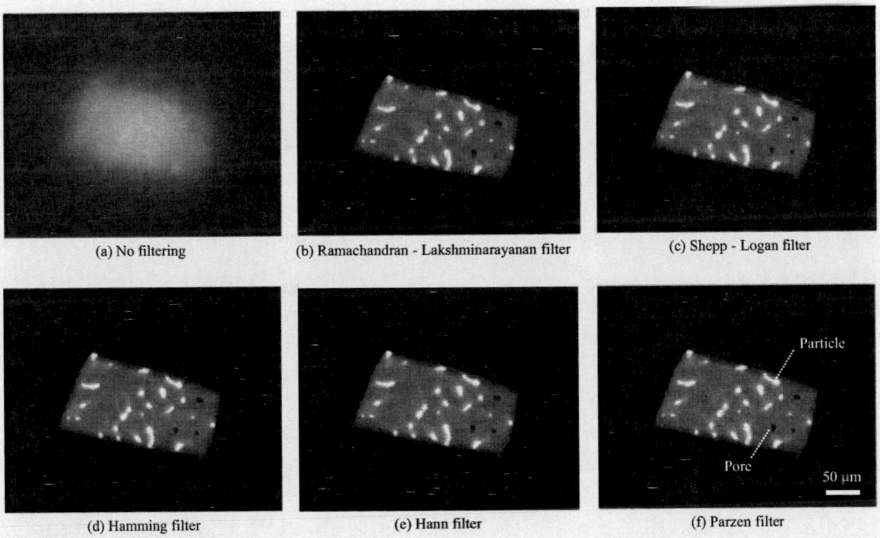

Fig. 3.27 Projection data of the aluminum-copper alloy shown in Figs. 3.7 and 3.18 reconstructed with filtered back projection using various reconstruction filters

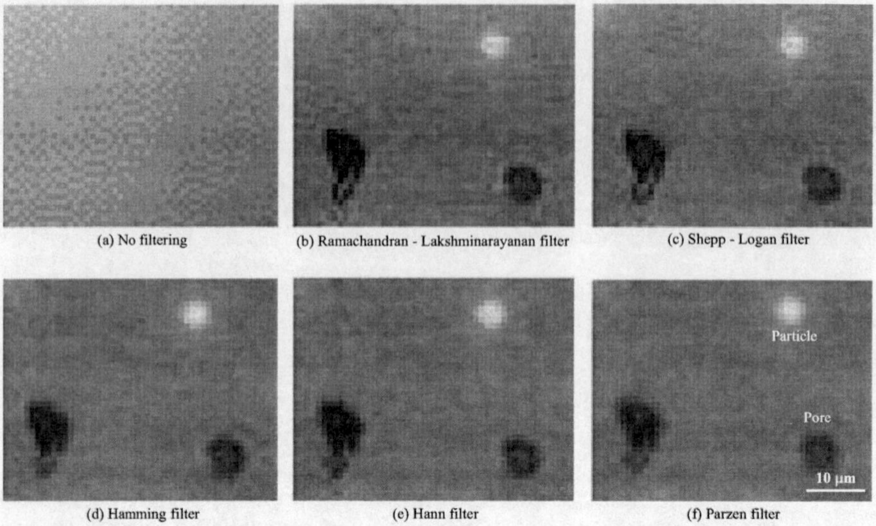

Fig. 3.28 Projection data of the aluminum-copper alloy shown in Figs. 3.7 and 3.18 reconstructed with a filtered back projection using various reconstruction filters. Magnification of Fig. 3.27

Hann filter, and Parzen filter each become further from the ramp filter with progressively lower frequency in this order. These differences are clear in Fig. 3.28d–f.

The filtered back projection method is reviewed below from a number of perspectives. The Ramachandran-Lakshminarayanan filter, Butterworth filter, and Shepp-Logan filter can be described as reconstruction filters that retain spatial resolution capabilities relatively well. Meanwhile, the Hamming filter, Hann filter, and Parzen filter effectively eliminate noise. However, with the exception of the Ramachandran-Lakshminarayanan filter, the quantitative nature of the linear absorption coefficient will be lost to some degree for each case. For example, measurement results can vary considerably depending on filter selection when trying to conduct accurate quantitative measurements of the chemical composition or density of various internal structures such as particles or inclusions by using pixel values after 3D reconstruction or measuring and assessing the size, volume fraction, or number density of regions of interest that have been segmented following the specification of a threshold for measured pixel values and binarizing them. Sufficient attention must be afforded to experimental instruments and experimental conditions in these types of experiments by using a monochromatic synchrotron radiation X-ray and avoiding X-ray refraction influences by minimizing the camera length as much as possible. However, it is somewhat of a blind spot that further attention must be paid to the appropriate selection of a reconstruction filter on top of this. Ensuring that spatial resolution, noise or quantitative characteristics and image quality are reconciled, are major issues in image reconstruction based on back projection; it should be understood from this section that there is no ideal condition in which all of these requirements are satisfied.

The spatial resolution of X-ray tomography is usually considerably inferior relative to 2D visualization devices often used in standard laboratories (e.g. scanning electron microscopes) (excluding standard light microscopes). For this reason, X-ray tomography often needs to assess the internal structures that approach the effective spatial resolution. Therefore, the first step must involve having a strong understanding of what the effective spatial resolution is for the current measurement and to measure this each time, if possible. With this in mind, examining the reconstruction filter (including its selection) also becomes important for determining the possibility of visualization. A noisy image will result in cases where sufficient exposure time cannot be assured due to the low flux of the X-ray source or when the X-ray energy cannot be properly modulated due to the chemical composition or size of the sample. The noise must be reduced in these cases, even if quantitative characteristics are somewhat sacrificed. The latter case is particularly an issue with industrial X-ray CT scanners, where the X-ray energy cannot be freely changed as with synchrotron radiation. Even if imaging was conducted after these factors were sufficiently selected, there would always be some noise (e.g. photon noise) in X-ray images and a reconstruction filter generally would need to be selected.

Incidentally, the image reconstruction program provided by SPring-8 [40] is outfitted with a Hann filter, Shepp-Logan filter, and Ramachandran-Lakshminarayanan filter. The default for this program is the Hann filter and the author typically uses the Hann filter as well. Furthermore, commercially available industrial X-ray CT scanners also include options for selecting multiple reconstruction filters.

Filtered back projection can be conducted in the mathematical analysis software MATLAB [41], which is often used for image processing, using the built-in "iradon" function. For filters, MATLAB offers the Hann filter, Hamming filter, Shepp–Logan filter, Ramachandran–Lakshminarayanan filter, and filters that use cosine functions as window functions. Furthermore, filtered back projection is available as a plugin in the free image processing software "ImageJ" [42], which is widely used for image analysis in scientific research. The filters included here are the Hann filter, Hamming filter, Shepp-Logan filter, and Parzen filter among others. It will likely be the case that reconstructed images using various reconstruction filters can be easily assessed using these programs. Incidentally, this software was developed at the National Institutes of Health in the U.S. by Rasband.

3.3.3 Convolution Back Projection

(1) Basic Principles

Generally, the *convolution integral* of the two functions $f(x)$ and $g(x)$ is expressed as follows:

$$f(x) * g(x) = \int_{-\infty}^{\infty} f(x-y)g(y)dy \tag{3.36}$$

The "*" symbol is also seen in Eq. (3.18). Equation (3.36) is referred to as a convolution. The *convolution theorem* that uses this is shown below:

$$F(f * g) = F(f)F(g) \tag{3.37}$$

Here, $F(f)$, $F(g)$, and $F(f * g)$ are the Fourier transforms of functions $f(x)$, $g(x)$, and the convolution integral of the two functions. Thus, the convolution integral of the two functions is equivalent to the inverse Fourier transform of the products of the Fourier transforms of each function.

$$f * g = F^{-1}\{F(f)F(g)\} \tag{3.38}$$

For Eqs. (3.27)–(3.29) of the filtered back projection described in the previous section, the reconstruction filter function was multiplied with the Fourier transform of the projection data, on which an inverse Fourier transform was applied, after which a back projection was finally applied. This corresponds to the right-hand side of Eq. (3.38). The convolution theorem indicates that if this filter function can be expressed in the real space, then these can be substituted with calculations in the real space using the convolution integrals in Eq. (3.36). The Fourier and inverse Fourier transform operations would no longer be necessary if this were the case. This type of image reconstruction in the real space is referred to as a *convolution back projection (CBP)*. However, the *impulse response* $h(t)$ of the reconstruction filter shown in Eq. (3.39) must be determined by applying an inverse Fourier transform on the reconstruction filter function $H(\omega)$ in the frequency space to conduct the operation in the real space.

$$h(t) = \int_{-\infty}^{\infty} H(\omega)e^{2\pi i\omega t}d\omega \tag{3.39}$$

The convolution back projection process corresponding to the filtered back projections in Eqs. (3.27)–(3.29) are summarized below:

$$q(t,\theta) = h(t) * p(t,\theta) \tag{3.40}$$

$$\mu(x,y) = \int_{0}^{\pi} q(x\cos\theta + y\sin\theta)d\theta \tag{3.41}$$

Here, $h(t)$ is the kernel of the convolution integral in Eq. (3.40).

(2) Comparisons with Filtered Back Projection

Filtered back projection and convolution back projection are numerically equivalent but the latter has the advantage that errors accompanying calculation equation approximations are reduced when conducting reconstructions of actual images [39]. On the other hand, filtered back projections are thought to be more widely used given the difficulty of handling the convolution function [39]. Furthermore, the fast Fourier transform (FFT) is generally more computationally advantageous than the convolution integral. In actual practice, it is often the case that convolution functions are converted to a different form using discrete Fourier transforms (DFTs) and then calculated using FFTs [43]. The calculated amount when the convolution integral is directly measured and when determined through DFT varies according to the pixel number but this difference can be by a factor of several dozen.

(3) Reconstruction Filter in Real Space

The $|\omega|$ seen up until now cannot be integrated, so its inverse Fourier transform does not exist. With this in mind, an integrable function $|\omega|e^{-\epsilon|\omega|}$ is defined to assess the shape of the ramp filter in real space, which does not have a cutoff frequency. This results in the following convolution function [44]. The limit $\lim_{\epsilon\to 0}|\omega|e^{-\epsilon|\omega|}$ should be taken to look at its actual shape [44].

$$F^{-1}\left(|\omega|e^{-\epsilon|\omega|}\right) = \frac{\epsilon^2 - (2\pi t)^2}{\left\{\epsilon^2 + (2\pi t)^2\right\}^2} \tag{3.42}$$

The shape of this impulse response is shown in Fig. 3.29 [44]. Here, ε is made progressively smaller from 2, 1, and finally to 0.5. The limit of Eq. (3.42) is $\frac{-1}{(2\pi t)^2}$ for $t > \varepsilon$. From Fig. 3.29, the function value around $t = 0$ sharply increases as ε becomes smaller, and the negative peak near that peak becomes equally larger and deeper.

Next, as in Eq. (3.39), an inverse Fourier transform is applied to the Ramachandran-Lakshminarayanan filter in Eq. (3.30) with values above $|\omega_{\max}|$ being cut off. The impulse response then becomes as follows [45]:

$$h(t) = \int_{-\infty}^{\infty} |\omega|e^{2\pi i\omega t}\, d\omega$$

$$= 2(\omega_{\max})^2 sinc(2\omega_{\max}\pi t) - (\omega_{\max})^2 sinc^2(\omega_{\max}\pi t) \tag{3.43}$$

Here, the shape of the impulse response when $\omega_{\max} = 1$ is shown in Fig. 3.30. Intense fluctuations can be seen when compared to Fig. 3.29, which are caused by the band limit sharply cutting off the high-frequency components that exceed the cutoff frequency [17]. These fluctuations can be reduced by using a Shepp-Logan filter or Hann filter, which have window functions that gradually drop to 0. This can also be confirmed in Fig. 3.25.

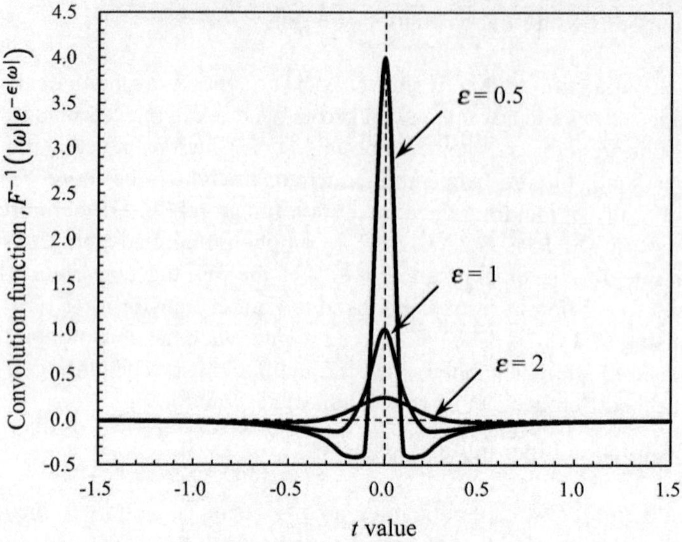

Fig. 3.29 Shape of an impulse response of an idealized reconstruction filter. Examples where three different ε values were used in Eq. (3.42)

Incidentally, the impulse response can be discretized without losing any information by invoking the sampling theorem. For this to occur, the sampling frequency needs to be double $\omega_{\max}$. Here, the discretized sampling points would be expressed by $t = n\Delta$ if the *sampling interval* was set as Δ and n was set as an integer. The reconstruction filter discretized in this way in the real space is shown in Eq. (3.44) [17].

$$h(t) = h(n\Delta) = \begin{cases} \frac{1}{4\Delta^2}, n = 0 \\ 0, n : even \\ \frac{-1}{(n\pi\Delta)^2}, n : odd \end{cases} \quad (3.44)$$

The sampling points of (Eq. 3.44) are also shown as circles in Fig. 3.30.

Other than this, the impulse responses of the Shepp-Logan filter and Hann filter are each expressed in Eqs. (3.45) and (3.46), respectively [43].

$$h(t) = h(n\Delta) = \frac{2}{\pi^2\Delta^2(1 - 4n^2)} \quad (3.45)$$

$$h(t) = h(n\Delta) = \begin{cases} \frac{1}{2\Delta^2}\left(\frac{1}{4} - \frac{1}{\pi^2}\right), n = 0 \\ \frac{1}{2\Delta^2}\left(\frac{1}{8} - \frac{1}{\pi^2}\right), n = \pm 1 \\ \frac{-n^2-1}{2\pi^2\Delta^2(n^2-1)^2}, n : even \\ \frac{-1}{2\pi^2\Delta^2 n^2}, n : odd \end{cases} \quad (3.46)$$

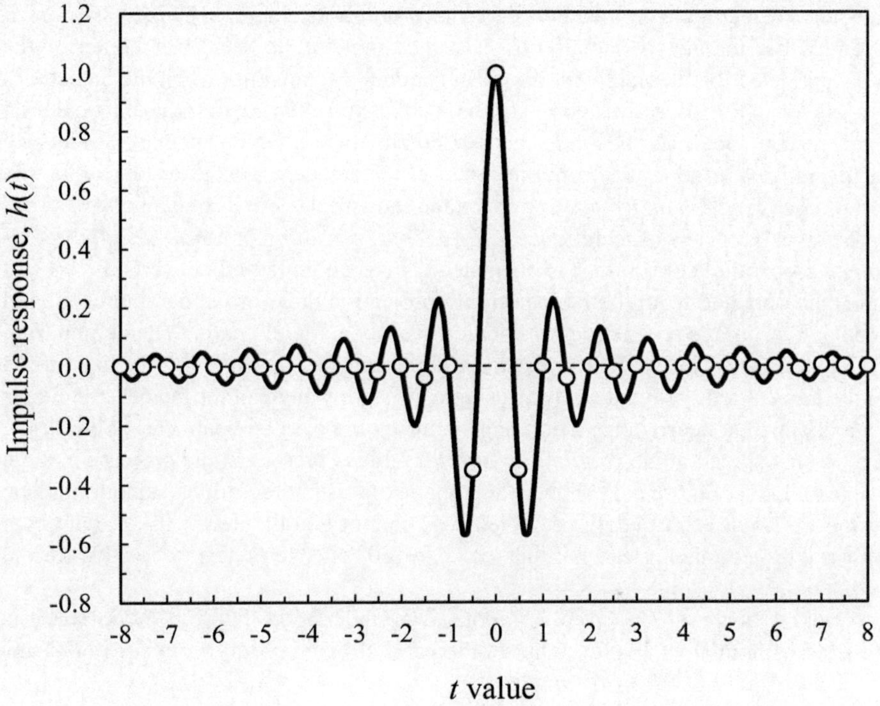

Fig. 3.30 Shape of the impulse response of the RamachandranLakshminarayanan filter with band limits at the cutoff frequencies

3.3.4 Cone Beam Reconstruction

Up until now, we have considered image reconstructions where the X-ray was treated as a parallel beam, as is frequently the case with synchrotron radiation X-ray tomography. This section provides an overview of image reconstruction where the distance between the X-ray source and detector is relatively small, as is frequently the case with industrial-use and medical X-ray CT scanners; and where the X-ray spread cannot be ignored, as seen in Figs. 3.2c, and 3.4b. The filtered back projection and convolution back projection described in detail until now will serve as the basis of the image reconstruction thought processes for these cases, as these general concepts should be sufficiently understood. Figure 3.2c is referred to as a fan beam and becomes a 2D image reconstruction in this case. Meanwhile, Fig. 3.4b is referred to as a cone beam and a 3D image reconstruction where an X-ray beam that spreads vertically and horizontally is necessary. These are discussed in order below.

(1) Fan Beam Reconstruction

A number of patterns can be discussed for detector placement when considering 2D image reconstruction using an X-ray beam that spreads in a fan-like manner. The first is when a detector curves in a bow-like shape or has its detection elements arranged in an arc-like manner, as seen in Fig. 3.1b. The angle of the X-ray beam entering the adjacent detection elements, in this case, is equal. Meanwhile, when the detector is flat as seen in Fig. 3.2c, the interval of the X-ray beam entering the adjacent detection elements are equal, but the angles between them are not, which changes nonlinearly as this moves further away from the center of the detector. This text will discuss the latter case, which is important to understand industrial X-ray CT scanners.

First, an overview of techniques referred to as *rebinning* or *fan-to-parallel rebinning* (fan-parallel conversion) is presented. This technique first converts the projection data obtained from the fan beam into projection data for the parallel beam and then applies image reconstruction methods for the parallel beam. When comparing the fan beam case in Fig. 3.2c with the parallel beam case in Fig. 3.2b, the line profile of the transmitted X-ray intensity at a single projection angle of the fan beam includes data obtained at nearby projection angles during the parallel beam case. Setting the fan beam's spread angle (*fan angle*) as $2\Psi_m$, this is present in the projection angle range of $\pm\ \Psi_m$. Figure 3.31 shows a schematic of this in the form of a sinogram. Due to the fan angle, completely projected data cannot be obtained with the fan beam unless a projection angle range that exceeds 180° is covered (i.e. $\pi + 2\Psi_m$ for the projection angle).

Looking at the entire projection data, it is understood that the Radon space is covered without any gaps for either of the parallel-beam or fan-beam cases. For this

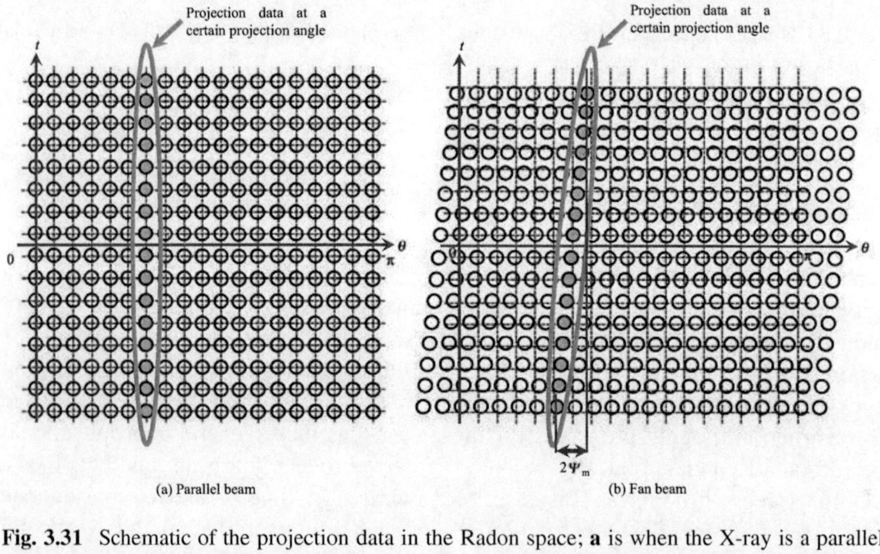

Fig. 3.31 Schematic of the projection data in the Radon space; **a** is when the X-ray is a parallel beam and **b** is when imaging was conducted with a fan beam. The projection data assigned a line profile of the transmitted X-ray intensity at a given projection angle are shown in gray for each figure. The imaging positions in the parallel beam case in (**a**) are shown by the intersections of the vertical and horizontal dashed lines in (**b**)

reason, there are always projection data acquired with the fan beam in positions that are sufficiently close to the coordinates (t, θ) of the projection data $p(t, \theta)$ of the parallel beam, as is shown in Fig. 3.31. With this in mind, the projection data experimentally acquired by the fan beam are used and interpolations are used to calculate the projection data in the hypothetical coordinates (t, θ) of the parallel beam. The image reconstruction method used for parallel beams can be directly utilized if this is conducted.

First, the parameters necessary for rebinning explanations are shown in Fig. 3.32. Here, the point source is placed at a distance D from the center of rotation of the sample and a distance D_0 from the detector. Then, we consider a scenario in which imaging is conducted where the sample is rotated β degrees from its x-y reference coordinate system. The X-ray emitted from the X-ray source enters a flat detector after transmitting through the sample. The following relational expression is obtained from this figure if the coordinates in the direction where the detection elements are arranged are set as ξ:

$$t = \frac{D\xi}{\sqrt{D_0^2 + \xi^2}} \tag{3.47}$$

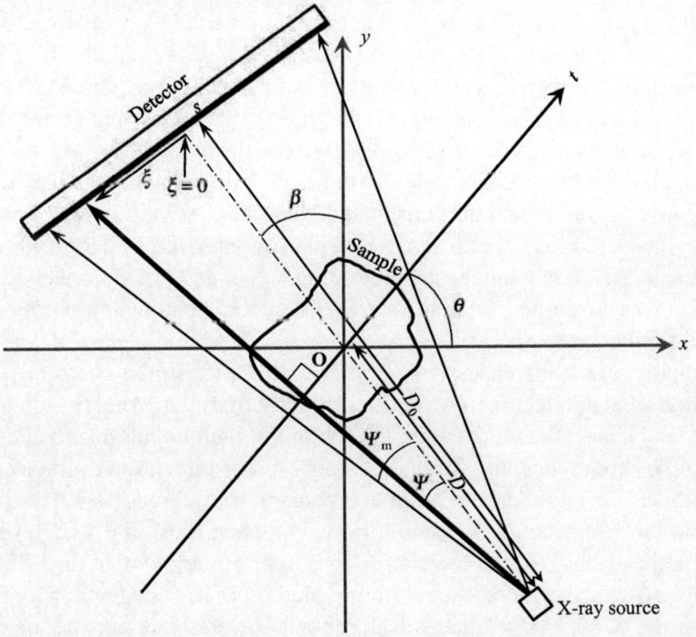

Fig. 3.32 Projection data using a fan beam are obtained for a given projection angle β. Rebinning is conducted and reconstruction is applied after converting this to projection data of the parallel beam. This figure shows a schematic of the geometric positions of the detector, sample, and X-ray in this scenario, as well as each parameter

$$\theta = \beta + tan^{-1}\frac{\xi}{D_0} \tag{3.48}$$

The following equations are obtained if these are solved for ξ and β

$$\xi = \frac{D_0 t}{\sqrt{D^2 - t^2}} \tag{3.49}$$

$$\beta = \theta - tan^{-1}\frac{t}{\sqrt{D^2 - t^2}} \tag{3.50}$$

These formulas can be used to calculate the positions at which the projected parallel beam data should be located within the projection data obtained using the fan beam. However, these are not located in the same positions in actual images due to the limited number of projection angles and the pixel number of the detector, as has already been shown in Fig. 3.31. For this reason, the projection data of the parallel beam is calculated by interpolating from the nearby measured points. This type of rebinning can be conducted within a single slice but there have been reports that simultaneously fill in multiple slices in projections where the X-ray beam is inclined relative to the slice [46].

Incidentally, rebinning can be easily conducted in the numerical software MATLAB [41] by using the built-in function "fan2para." Conversely, the function "para2fan" can be used to convert the projection data of the parallel beam into that of the fan beam. This is a digression, but the built-in function "fanbeam" calculates the projection data of a fan beam (i.e. sinogram) from a 2D image. Furthermore, the free image reconstruction program "NiftyRec" developed by University College London and Harvard University, can also conduct cone-beam and helical scan-based image reconstruction, which are introduced later [47]. A wide range of reconstruction algorithms and filters are also included in programs like ASTRA [48], provided as a toolbox in MATLAB and the programming language Python, as well as MuhRec [49], which was developed for neutron CT. These tools endow these programs with high-level utility.

Incidentally, noise increases with further distance from the center of the image with methods that conduct reconstructions directly from the projection data obtained using the fan beam [50]. Meanwhile, the rebinning method has an advantage where noise becomes homogeneous within the image. According to a simplified estimation by Xu on this issue, methods that conduct reconstructions directly from the projection data of the fan beam had a maximum noise variance ratio that was several times higher than those that conduct reconstructions after rebinning with the S/N ratio also worsening [36]. Furthermore, the rebinning method is advantageous with the ability to conduct the X-ray CT scanner misalignment correction alongside the rebinning procedure [51].

Next, we provide an overview of methods that conduct image reconstruction directly from the projection data obtained by using the fan beam. For the sake of convenience in formulating the reconstruction, this time, we place a hypothetical detector in the center of rotation of the sample, as shown in Fig. 3.33. The relational

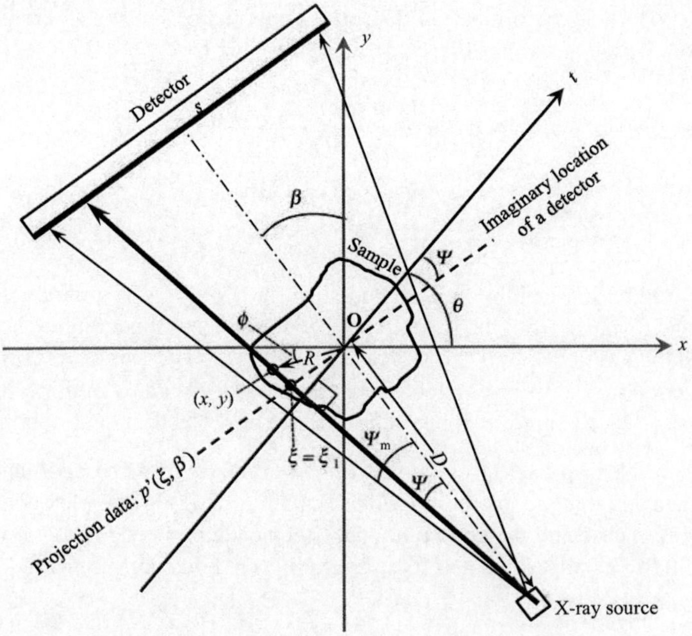

Fig. 3.33 Projection data p' (ξ, β) using the fan beam at a given projection angle β are obtained. This is directly reconstructed. This figure shows a schematic of the geometric positions of the detector, sample, and X-ray, as well as each parameter. For convenience, the sample is placed in the center of rotation, and a hypothetical detector is placed

expressions between t and θ, as well as ξ and β, are shown below:

$$t = \xi cos\Psi = \frac{D\xi}{\sqrt{D^2 + \xi^2}} \tag{3.51}$$

$$\theta = \beta + tan^{-1}\frac{\xi}{D} \tag{3.52}$$

This is substituted into the filtered back projection equation and the integral variables are converted from the dt dθ of the equation for the parallel beam to the dξ dβ of the equation for the fan beam, which is the main aspect of Eq. (3.22). The reconstructed image μ (R, ϕ) appears as the following equation when this is rearranged using the expressions of the polar coordinates (R, ϕ) in Fig. 3.33 [17]. Please see the referenced documents for its detailed derivation [17, 37, 50].

$$\mu(R, \phi) = \frac{1}{2}\int_0^{2\pi} \frac{1}{U^2}\left\{\int_{-\infty}^{\infty} p_F(\xi, \beta)h(\xi_1 - \xi)\frac{D}{\sqrt{D^2 + \xi^2}}d\xi\right\}d\beta \tag{3.53}$$

Here, $p_F(\xi, \beta)$ is the projection data of the fan beam. ξ_1 and U are expressed with the following equations [17]:

$$U = \frac{D + Rsin(\beta - \phi)}{D} \tag{3.54}$$

$$\xi_1 = \frac{D\{Rcos(\beta - \phi)\}}{D + Rsin(\beta - \phi)} \tag{3.55}$$

The multiplication weight $\frac{1}{U^2}$ is included in Eq. (3.53). U is a parameter that is related to the distance from the pixel to be reconstructed to the radiation source and varies with location.

A correction first must be conducted on the projection data by multiplying it with $\frac{D}{\sqrt{D^2+\xi^2}}$ to conduct a direct reconstruction from the projection data obtained using the fan beam. Next, a back projection with a weight $\frac{1}{U^2}$ must be conducted while convoluting a suitable reconstruction filter. Equation (3.53) is complex compared to a parallel beam case and the computational time needed for reconstruction can also become lengthy as a result. Therefore, rebinning methods are often used in actual practice.

(2) Cone Beam Reconstruction

Generally, with industrial X-ray CT scanners, a cone beam generated from an X-ray tube as in Fig. 3.4b, enters a 2D detector (e.g. a CCD camera, CMOS camera, flat-panel detector, etc.) after it transmits through the sample. In this case, the cone-beam image reconstruction method must be used to conduct an image reconstruction of a 3D image from the set of transmitted images obtained in 2D.

Figure 3.34a, b are single 2D transmission images in the Radon space, which were obtained from parallel beams and cone beams, respectively. The latter looks like a paraboloid at first glance but is actually part of a sphere. The Radon space of the single transmission image obtained with a cone-beam becomes a sphere similar to that shown in Fig. 3.35a when the detector is infinitely large. The diameter of the sphere is equivalent to the distance from the X-ray source to the center of rotation. Furthermore, the single set of projection data obtained by rotating the sample takes the shape of a donut similar to that shown in Fig. 3.35b.

A complete image reconstruction can generally be conducted for 3D image reconstruction so long as the 3D Radon space is filled in with projection data. There exists a shadow zone in Fig. 3.35b where data cannot be acquired with further distance from the origin in the z-axis direction. The sample slices corresponding to the center of the vertical direction of the detector (i.e. slices at the same height level as that of the X-ray source) can be completely reconstructed in real space. However, image reconstruction accuracy decreases as the slices become further away from the center in the vertical direction due to missing data [52]. The decrease in spatial resolution that occurs, as a result, is not an issue when the sample is small relative to the distance

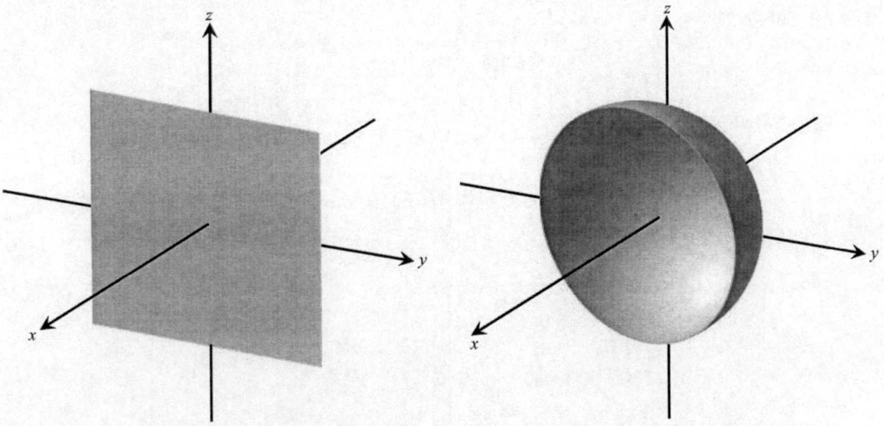

(a) Projection data obtained with a parallel beam

(b) Projection data obtained with a cone beam

Fig. 3.34 Single 2D transmission image obtained with a parallel or cone beam seen in Radon space

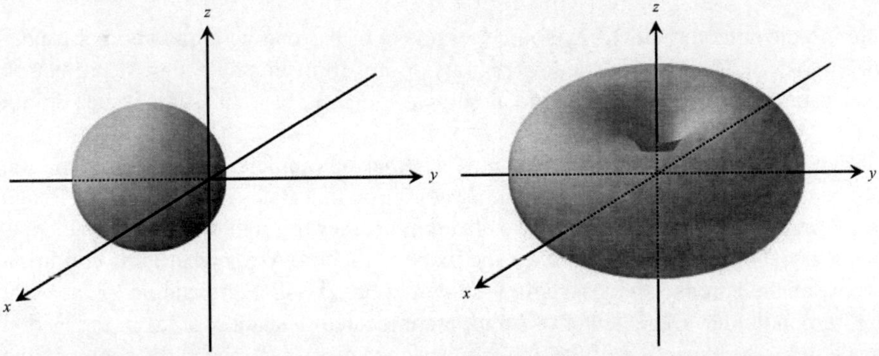

(b) One projection data obtained with a cone beam when a detector is infinitely large

(b) A set of projection data obtained with rotating a sample

Fig. 3.35 Single 2D transmission image obtained with a cone beam when the detector is infinitely large and a single set of projection data where the sample was rotated 360°. Both are seen in the Radon space

from the X-ray source; however, this becomes a progressively larger issue as the *cone angle* increases or as the cross-sections become further away from the center in the vertical direction [52]. Care must be taken with industrial X-ray CT scanners when trying to increase magnification by bringing the sample closer to the X-ray source, as this will increase the cone angle. Reports have shown methods where shadow regions were eliminated in medical-use CT scanners where the patient is fixed and where the detector and X-ray source rotate along a circular trajectory while maintaining their positional relationship by obtaining additional projection data, for example by

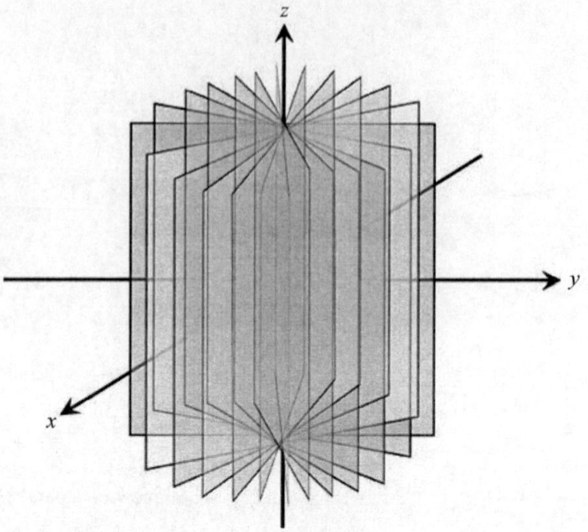

Fig. 3.36 Single set of projection data obtained by using a parallel beam, rotating the sample 360°, and showing it in Radon space

moving the detector and X-ray source vertically or by rotating along another trajectory [52–55]. Meanwhile, shadow regions do not form in single sets of projection data obtained with a parallel beam, as shown in Fig. 3.36, so the same issues do not occur.

Tuy's sufficiency condition stipulates whether a complete image reconstruction can be obtained [56]. This is a geometric condition that states that a complete image reconstruction can be obtained if the plane that passes through the sample intersects the X-ray source at least once when the plane is rotated. Areas where this condition is not satisfied are the shadow regions shown in Fig. 3.34b. For example, planes that are perpendicular to the center of rotation are located in shadow areas in the Radon space, with the exception of the above-mentioned cross-sections in the center of the detector.

An overview of the approximative image reconstruction method referred to as *Feldkamp's algorithm* is described below. Aside from those where the back projection corresponds to the cone beam, Feldkamp's algorithm is fundamentally identical to the fan beam reconstruction method. First, an imaginary detector is placed in the center of rotation of the sample for the sake of convenience when formulating the image reconstruction; this is shown in Fig. 3.37. Here, the point source is placed at a distance D from the center of rotation of the sample. We consider a scenario where imaging is conducted on the projection data $p(\xi, \eta, \beta)$ rotated with a projection angle of β degrees from the x-y reference coordinate system of the sample. The X-ray emitted from the X-ray source enters a flat detector after transmitting through the sample as shown in Fig. 3.37. Here, the coordinates in the direction where the detection elements are arranged are ξ and the coordinates in the vertical direction of the detector are η.

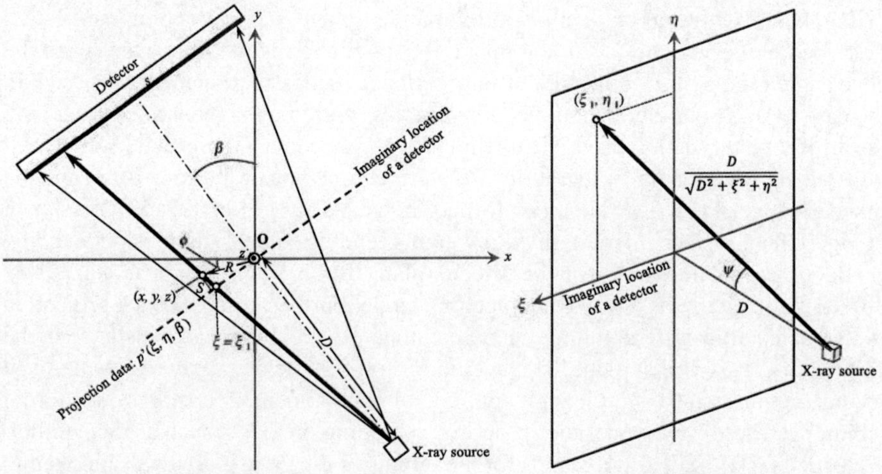

(a) An image viewed from right above in the − z direction

(b) An image viewed obliquely

Fig. 3.37 Projection data p' (ξ, η, β) using a cone beam are obtained at a given projection angle β. This is reconstructed using Feldkamp's algorithm. This figure shows a schematic of the geometric positions of the detector, sample, and X-ray, and its various parameters. An imaginary detector is placed in the center of rotation of the sample for the sake of convenience

$$\mu(R, \phi, z) = \frac{1}{2}\int_{0}^{2\pi} \frac{D^2}{(D-S)^2}\left\{\int_{-\infty}^{\infty} p(\xi, \eta, \beta)h(\xi_1 - \xi)\frac{D}{\sqrt{D^2+\xi^2+\eta^2}}d\xi\right\}d\beta \tag{3.56}$$

Here, $\frac{D}{\sqrt{D^2+\xi^2+\eta^2}}$ corresponds to the cosine of the incidence angle ψ.

$$\psi = cos^{-1}\frac{D}{\sqrt{D^2+\xi^2+\eta^2}} \tag{3.57}$$

A back projection of the projection data must be conducted by multiplying it with $\frac{D}{\sqrt{D^2+\xi^2+\eta^2}}$ to reconstruct a 3D image from the projection data obtained using the cone beam. This is a 1D filtering technique that is conducted on each horizontal row of the detector. The image reconstruction in the center of the vertical plane ($\eta = 0$) corresponds to that of the fan beam and is accurate. Meanwhile, blurring occurs with increased distance from the center of the vertical plane. Care must be taken in cases, for example, when the cone angle is over 10°.

Ring artifacts occur with Feldkamp's algorithm due to the above-mentioned weights becoming heterogeneous for each pixel [57]. Artifacts due to cross-talk between adjacent slices also occur [57]. Furthermore, some artifacts form when a part of the sample is not covered by the cone beam and sticks out due to the morphology of the cone beam [57].

Feldkamp's algorithm is an approximative image reconstruction method as described earlier. Higher-accuracy image reconstruction is possible even when the cone angle is large if a mathematically strict image reconstruction method is used. In the case of a 2D parallel beam, this corresponds with image reconstructions where the projection theorem is used. Meanwhile, for a cone beam, Grangeat derived a relationship between the line integration of the projection data of the cone beam and the first derivative of the 3D Radon transform (Grangeat's algorithm) [37, 58]. As shown in Fig. 3.38, a surface integration with weights relating to the sample is obtained by conducting a line integration on the detector plane. Rebinning is conducted on this by differentiation with the X-ray beam opening angle. Furthermore, 3D back projection is conducted after differentiating with a parameter that expresses the distance in the vertical direction of that plane. This method is too complex to apply to actual image reconstructions, and a significant magnitude of complementary errors occurs during rebinning. There have been other proposed rebinning methods, such as the parallel-FDK method (P-FDK; FDK stands for the initials of the three individuals Feldkamp,

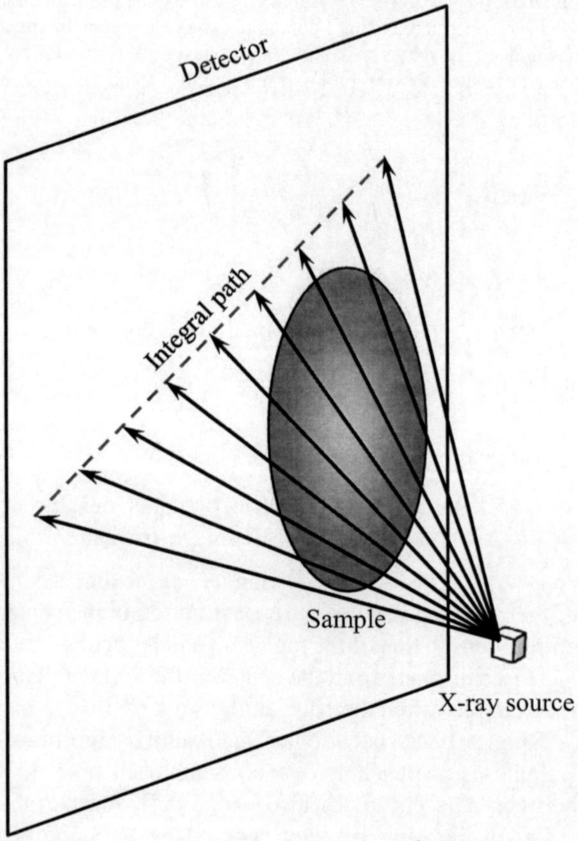

Fig. 3.38 Projection data using a cone beam at a given projection angle are obtained and this is reconstructed using Grangeat's algorithm. This figure shows the positions of the detector, sample, and X-ray in this scenario, and the line integrations on the detector plane

Davis, and Kress and refers to Feldkamp's algorithm) and tent-FDK method (T-FDK) [59]. References have been included for further information on these methods.

3.3.5 *Special Image Reconstructions*

(1) Offset Scan

A number of methods have been proposed for imaging samples that are larger than the width of the X-ray beam or detector. The observation magnification cannot be freely changed in X-ray tomography, as is the case with observations using optical microscopes or electron microscopes. If anything, it is often the case that the magnification or spatial resolution is fixed. Furthermore, increasing the magnification and conducting observations with high spatial resolution often means reducing the observable *field-of-view* (FOV), which is inversely proportional to spatial resolution. For this reason, cases that need a high spatial resolution to observe the internal structure of a large sample require that part of the sample be omitted from the FOV, as shown in Fig. 3.39a The obtainable range for the 3D image, in this case, is limited to the dotted circle shown in Fig. 3.39a.

Figure 3.39b, c schematically show imaging methods of *offset scans* [60]. The method shown in Fig. 3.39b involves offsetting the X-ray source and detector and obtaining the transmission images over multiple sessions to cover the entire sample. In this case, multiple transmission images are obtained while shifting the detector,

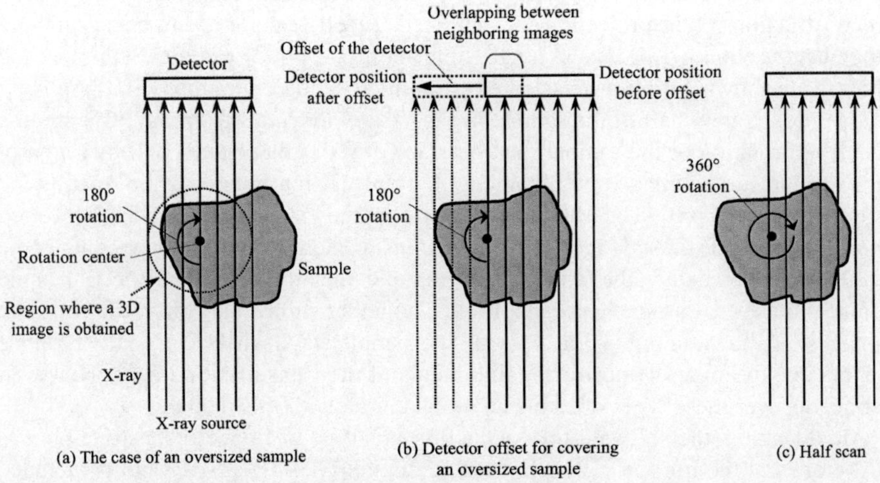

Fig. 3.39 **a** Various relationships between the sample, detector, and X-ray beam for the case where the sample does not fit within the observable field of view; **b**, **c** show representative techniques for imaging large samples with limited hardware

after which the sample is rotated and multiple transmission images are obtained again. Alternatively, a standard 180° rotation scan is conducted, after which the same scan is conducted while the detector is shifted. Often, the reality is that the X-ray source and detector are fixed, and the sample is translated/rotated parallel to the detector. Kyrieleis et al. recently reported a method that translated not the rotational axis but the sample on the *sample rotation stage*, comparing this to methods shown in Fig. 3.39b, c. Transmission images at the same projection angle obtained with the above-mentioned methods are stitched together into a single transmission image; standard methods are used for image reconstruction. In this case, the transmission image stitching accuracy determines imaging success or failure. For this reason, a number of *image registration* methods have been tested for this method [61]. The imaging beamline at SPring-8 has also published software to conduct image reconstruction on projection data obtained with the methods shown in Fig. 3.39b [62]. Furthermore, looking at Table 6.1, shown later, many representative X-ray CT scanners that are commercially available can handle offset scanning.

Incidentally, the format recently most frequently used by the author is to obtain approximately 1800 projections (about 1900 transmission images in total including the I_0 images in Eq. (2.2)) with a 2048 × 2048-pixel CMOS camera. The file size of this projection data is approximately 15.9 GB. Here, if we consider the simple case where two projection images were stitched together with no overlap, the size of the object that can be imaged and the projection data file size both increase by a factor of 2; the 3D image file size after reconstruction increases by a factor of 4. In other words, with the above example, the projection data file size becomes 31.8 GB and the 3D image data file size becomes 63.6 GB. For this reason, too many offsets are unrealistic when conducting 3D image reconstructions.

Meanwhile, although not frequently used in X-ray tomography, conducting image reconstructions of the projection data obtained at each detector position and stitching together the obtained 3D images is a feasible option. Applied examples of this method have been shown with transmission electron microscope tomography [63].

Approximately half of the sample in Fig. 3.39c protrudes from the FOV and the center of rotation of the sample rotation stage drifts to that side. The entire area of the sample can be imaged by obtaining the transmission images while rotating the sample by 360° [60]. The 180° transmission images in the latter half are inverted horizontally and stitched together with the transmission images corresponding to the 180°-projection data in the first half (for example, the 30° and 210°-projection angle data). However, this method cannot stitch together more than three transmission images, as shown in Fig. 3.39b. Moving the sample stage will occasionally produce errors due to imperfect positional alignments of the transmission image. However, concerns over these types of errors are not an issue with this method.

Incidentally, the software for conducting this type of image reconstruction was developed at the European Synchrotron Radiation Facility (ESRF) and is included in the freely available image reconstruction software "PyHST" [64].

(2) Region of Interest Reconstruction

Region of interest (ROI) reconstruction refers to image reconstruction of an ROI from incomplete projection data obtained by emitting X-rays to that particular region in a sample with a large diameter. These are also referred to as ROI CT, *local CT*, and *interior CT*. The FOV in medical X-ray tomography is often limited due to hardware drawbacks such as detector size, it is a technique to overcome these issues. However, this has recently been applied as a method to conduct 3D observations of relatively small subject areas (e.g. teeth or tumors in the heart or brain) while significantly reducing the exposure dose in non-observed areas. Meanwhile, X-ray tomography at synchrotron radiation facilities and industrial X-ray CT scanners do not need to minimize exposure, serving as an important countermeasure for the problem of narrowed FOV when increasing spatial resolution. Figure 3.41 shows an actual example where imaging was conducted under the conditions shown in Fig. 3.40 [65]. Artifacts whose *gray value* steadily becomes larger from the center of the FOV to the edge (circumference) can be observed; decreasing spatial resolution coupled with a lower S/N ratio owing to insufficient flux can be observed when the sample diameter is extremely large relative to the FOV.

The classic countermeasure most often used for this case is extrapolation using the sine curves of the *sinogram* (cosine extension) and a process known as *zero-padding* (filling a certain region with zero values) [66, 67]. A truly wide-ranging set of methods have been proposed up to now, including methods that use existing images [68], methods that utilize existing information in a small area within the ROI by employing the iterative reconstruction method [69], methods that use a wavelet as a tool for the Radon transform [70, 71], and two types of explicit solution techniques known as non-compressive and compressive sensing [72].

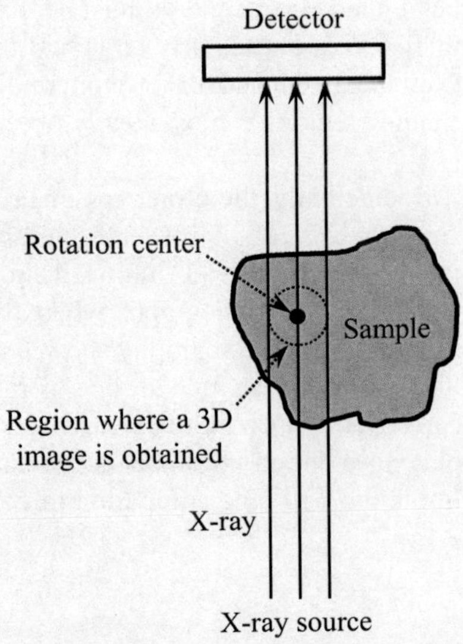

Fig. 3.40 Schematic of a scenario where the sample does not fit within the observable FOV and where the FOV is entirely within the interior of the sample

Of these, work from the author's research group (Ref. [71]) conducts an image reconstruction that takes the discrete wavelet transform, which is a time-frequency analysis method, instead of Fourier transforms and uses its localization. This research assumes that the X-ray absorption coefficient distribution outside of the ROI can be approximated (extrapolated) by the values on the edge of the ROI. This process first involves an extrapolation after creating a wavelet ramp filter. Next, a 1D discrete wavelet and 2D inverse discrete wavelet transforms are conducted instead of the Fourier/inverse Fourier transforms (corresponding to Eqs. (3.27)–(3.29)) of the filtered back-projection. Artifacts such as those in Fig. 3.41 are reduced as a result and improvements in S/N ratio and spatial resolution (2–4 times more) can be confirmed with 3D imaging of the structural material using synchrotron radiation [71].

Please refer to the references provided for further details on these methods.

(3) Laminography

Also introduced in Fig. 1.2 in Chap. 1, *laminography* refers to a classic internal observation method used from the early 20th century onward. However, the method explained here is the so-called computed laminography process, which can be said to be a variation of the X-ray tomography techniques in current use. Figure 3.42 schematically shows the configuration of a representative experimental device for laminography. The X-ray source and detector are the same as X-ray tomography and its configuration is such that the sample rotation stage is inclined by an angle Φ in the direction of the X-ray path. However, the sample must be rotated by 360° in the case of laminography. Figure 3.43 schematically shows the data distribution in Radon space for laminography. The fact that Radon space cannot be completely filled in with laminography is similar to the case with cone beams. As such, laminography is a method that should be used with sufficient understanding that spatial resolution will decrease and artifacts will occur when X-ray tomography using parallel beams cannot be applied. The angle Φ must be made large as shown in Fig. 3.43b when the plate thickness is particularly large and the shadow region in the Radon space becomes even larger. This forms even more distinct artifacts; in this way, the influence of the shadow regions is most clearly observed in the slice in the plate thickness direction [73].

Incidentally, the projection angle in X-ray tomography can be limited to a value smaller than 180° to observe flat objects like sheets or electronic substrate shown in Fig. 3.42c without cutting them. Hoshino et al. compared laminography with standard X-ray tomography where the projection angle was constrained to 150° and reported that laminography showed reconstructed images that were more faithful to the shape of the sample in slices parallel to the plate [73]. The fact that laminography can cover a much more extensive and homogeneous area with projection data in the plate face direction can be seen from imagining the rotation range in Fig. 3.36 being limited to 150° and comparing this to Fig. 3.43.

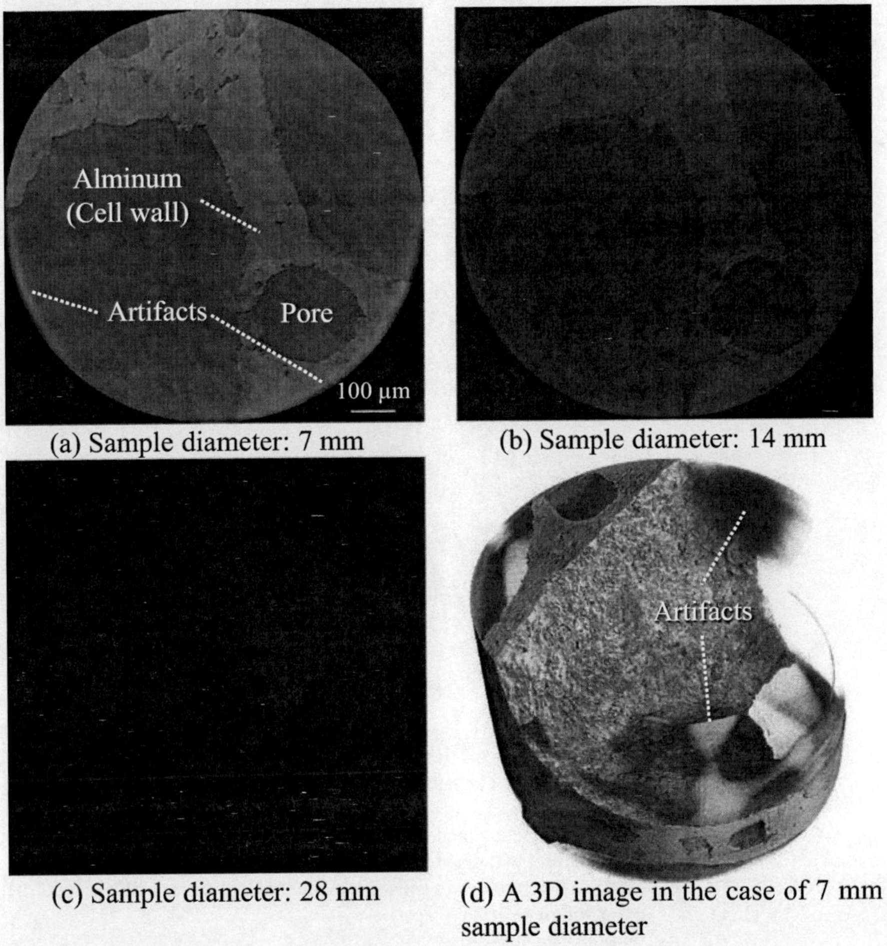

Fig. 3.41 Imaging example where the sample does not fit within the observable FOV (1 mm diameter in this case), and where the FOV is entirely within the interior of the sample. The material is foamed aluminum, with the cell wall (aluminum alloy), which surrounds the cell and where the air within the cell is shown; **a–c** show the same materials at the same locations imaged while the sample diameter was changed and **d** is a 3D image of (**a**). Artifacts when a standard reconstruction was conducted on the large-diameter sample sticking out

The image reconstruction of laminography should apply standard image reconstruction thought processes with the consideration of inclination in the sample rotation axis. If q is used to always express the inverse Fourier transform of the filtered Fourier transform of the projection data when conducting filtered back projection, its back projection is expressed as follows [73]:

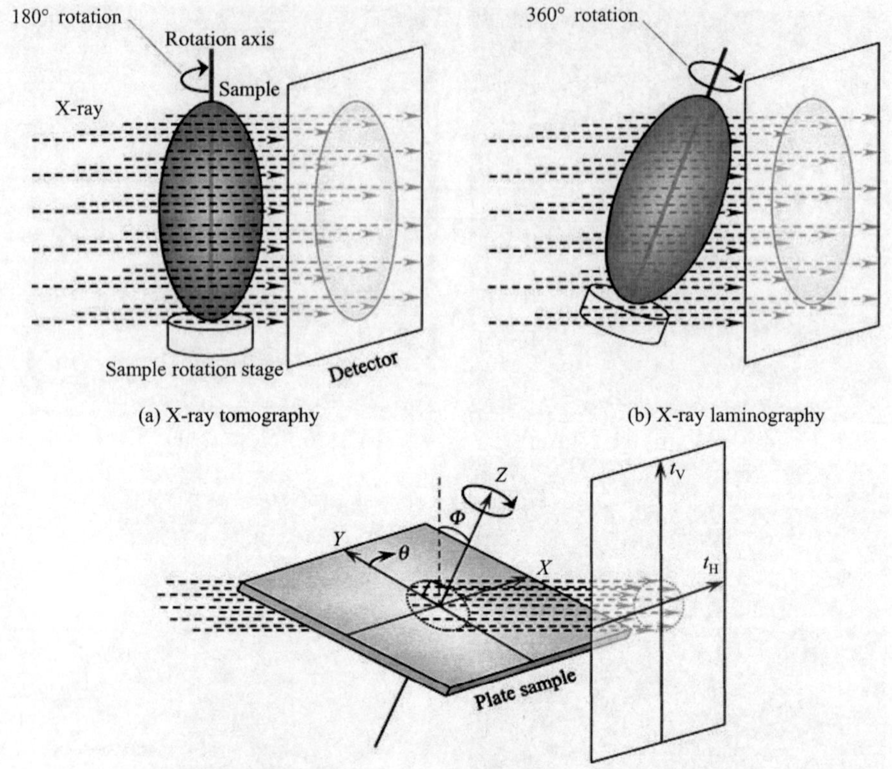

(c) Observation of a plate-like sample by means of the laminography

Fig. 3.42 **a**, **b** are the differences between standard X-ray tomography and laminography. The rotation axis of the sample is not parallel with the detector and is inclined in the direction following the X-ray beam. Thus, this is suited for observing plate-like samples like (**c**), which cannot be observed with the method in (**a**); the plate

$$\mu(X, Y, Z) = \int_{0}^{2\pi} q(t_H, t_V, \theta)d\theta \tag{3.58}$$

Here, the coordinates (t_{H}, t_{V}) on the detector plane and sample coordinates (*X, Y, Z*), shown in Fig. 3.42c, are related using the following transformation matrix:

$$T_{\theta,\Phi} = \begin{bmatrix} cos\theta & sin\theta & 0 \\ sin\theta sin\Phi & -cos\theta sin\Phi & cos\Phi \end{bmatrix} \tag{3.59}$$

Other than this, reconstructions based on iterative reconstructions that use a priori knowledge such as sample shape or material (real component δ of the complex refractive index) have been conducted with laminography as well, which can effectively reduce the influence of the aforementioned shadow areas [74].

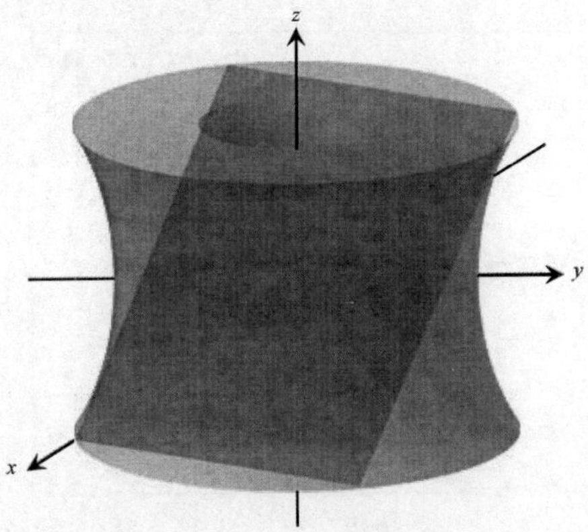

(a) Radon space for $\Phi = 45°$

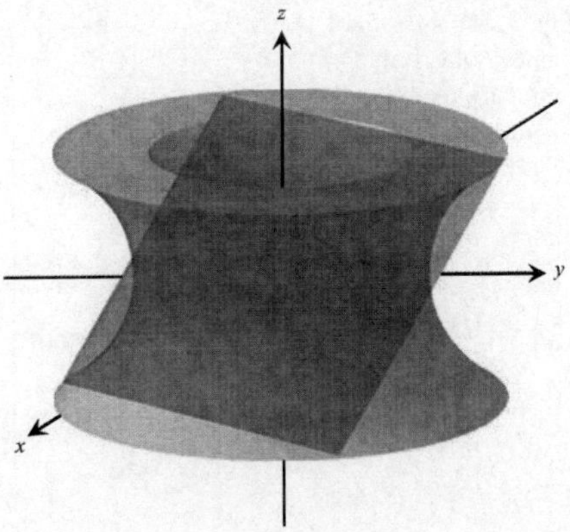

(b) Radon space for $\Phi = 60°$

Fig. 3.43 Schematic of a single set of projection data obtained using laminography and rotating the sample by 360°. Shown in Radon space. The inclination angle of the rotation axis is changed

(4) Image Reconstruction with Angle Constraints

Flat objects like plates and electronic substrates can be observed without severing them by constraining the projection angle of X-ray tomography to values below 180°, such as 120°, for example. This is a technique that is often used with transmission electron microscope tomography observations. Although this should be used when laminography can be applied, tilting the rotational axis can often be physically difficult with industrial X-ray CT scanners. The issue in this case is to how to handle the missing wedge problem, which occurs due to the fact that the Radon space cannot be filled in with projection data. Reports have indicated using simultaneous iterative reconstruction methods that control the influence of missing information in a manner similar to ROI reconstruction or laminography and the application of image processing methods that recovers to some extent missing information using information from a priori knowledge on a sample [75]. The review article on transmission electron microscope tomography should be referenced for further detail on this [75].

3.4 Realities of Image Reconstruction

As mentioned in Chap. 1, personal computers began to be used for X-ray CT scanner reconstruction and 3D imaging by the 2000s. In 1994, Cabral et al. published a pioneering article (cited over 1200 times) on image reconstruction using a *graphics processing unit (GPU)* [76]. However, the clock rate of the central processing unit (CPU) was increasing annually at this time and, perhaps due to the clear acceleration with a single core (the image reconstruction speed in the 10 years from 1994 increased by a factor of around 100 [77]), practical applications of GPU-based image

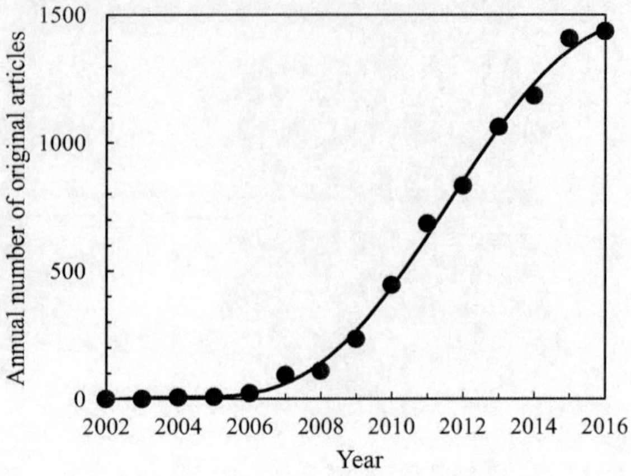

Fig. 3.44 Results of the number of original publications when the term "Tomography AND Reconstruction AND GPU" was searched in the publication database Scopus

reconstruction did not immediately occur. However, after this, the acceleration of the single-core slowed down and development of multi-cores began to increase instead, and the *parallel computing* environment was gradually established. As shown in Fig. 3.44, there has been a rapid increase in the number of publications relating to methodologies and applications of GPU-based image reconstruction since 2010. This is in part due to *application programming interfaces (API)* such as OpenGL and DirectX, which have prepared a 3D programming environment that can apply GPUs. Calculations using GPUs are referred to as general-purpose computing on CPUs (GPGPU). Although the increase in publications has slowed in the present day, GPU-based reconstructions have begun to see practical applications.

The GPU is a specialized piece of hardware that displays the results of various types of image processing conducted for graphics display. Its high parallel computing capabilities, which conduct pipeline processing of large-scale data in parallel, are particularly important. Although this originally took the form of a graphics card with a large-scale integrated (LSI) chip and video memory for image processing, forms in which it is incorporated into PCs are also currently in circulation.

Table 3.1 shows a comparison of a high-performance CPUs and GPUs. Figure 3.45 shows a photo of the external appearance of the latest GPU shown in Table 3.1 as a reference. The number of CPU cores is 30 at most, but there are over 100 times more cores in a GPU. Furthermore, the number of GPU cores has doubled approximately every 1.4 years [77]. Meanwhile, the core clock rate of the GPU is approximately half that of the CPU. The calculation core of the GPU expresses its full capabilities when multiple cores conduct the same calculations in a way similar to that of single instruction multiple data (SIMD), which is one of the processor architectures that can realize parallel processing. For this reason, GPU s can rapidly conduct parallel computing; however, its computational rate decreases significantly when parallelism is low. Furthermore, the GPU includes graphics memory but the memory capacity itself is several (to up to 100) times lower than that of the CPU. Therefore, acceleration with a GPU can be achieved only for cases where calculations, which

Table 3.1 Comparisons between high-performance CPUs and GPUs

Products	Intel Xeon Platinum 8180 M	Intel Core-i9-7980XE	AMD Radeon Pro Duo (Polaris)	NVIDIA Quadro GP100
Type	CPU (Server)	CPU (Personal computer)	GPU	GPU
Number of cores (Number of shader units)	28	18	4608	3584
Operation frequency (GHz)	2.5	2.6	1.243	1.4
Maximum memory size (GB)	1500	128	32	16
Memory type	DDR4-2666	DDR4-2666	GDDR5	HBM2
TDP (Thermal design power) (W)	205	165	250	235
Logical operation performance (TFLOPS)	2.24	1.4976	11.456	10.3

Fig. 3.45 Photograph of the external appearance of the latest GPU. NVIDIA Quadro GP100 from Table 3.1 (courtesy of Kenichi Hayashi from NVIDIA Co.)

Table 3.2 Comparison of the reconstruction time of a single cross-section when data from 900 projections obtained from a 1,000 × 1,000-pixel camera underwent convolution back projection. A single GPU (NVIDIA Tesla C1060) is compared with 4 CPUs [78]

Type	NVIDIA Tesla C1060	Xeon (E5420)	Intel Core 2 Quad	Intel Core 2 Duo	Itanium 2
	GPU	CPU	CPU	CPU	CPU
Number of cores	240	4	4	2	1
Operation frequency (GHz)	1.296	2.5	2.83	2.66	1.6
Time necessary for reconstruction (s/section)	0.11	0.82	1.02	1.94	3.72

can be conducted in this range, are divided. Other than this, GPUs characteristically consume relatively large amounts of power. Comprehensive operation capabilities are approximately 10 times faster with GPUs, according to the example in Table 3.1. Although not shown in the table, its costs are also relatively low. In summary, its merits are extremely high for specific types of computation. Comparing operation performance since 2003, that of the CPU has doubled every 1.5 years whereas that of the GPU has doubled annually [77] and the difference between them continues to increase annually.

Correspondingly, the GPU has made large contributions to the image reconstruction of X-ray tomography as well. Uesugi et al. at SPring-8 conducted GPU-based image reconstructions by obtaining data from 900 projections using synchrotron radiation and a 1000 × 1000-pixel CCD camera [78]. Their image reconstruction was based on a *compute unified device architecture (CUDA)*, which is an integrated C-language development environment provided by NVIDIA in 2006 for GPUs. Table 3.2 shows a comparison between the image reconstruction times of CPUs and GPUs [78]. Image reconstruction in these cases was conducted layer by layer. The read-out of the raw data from the next layer can be conducted simultaneously with the image reconstruction calculation, making this considerably quicker relative to CPUs. The time needed for sinogram read-out is only 83 ms if a solid-state drive (SSD) is used, which is shorter than the image reconstruction time. Meanwhile, it takes 53 ms to store the image reconstruction results as a 16-bit tiff image; along with image reconstruction, this is one of the rate-limiting factors for the total time needed.

Ino et al. used a GPU for image reconstruction based on Feldkamp's algorithm for cone-beam X-ray tomography [79]. Their data consisted of 360 projections of 512 × 512-pixel transmission images, on which CUDA-based reconstruction computations were conducted. The time required for reconstruction when an NVIDIA GeForce 8800 GTX GPU was used in a Core 2 Quad Q6700 CPU PC was 5.6 s, which is a 24-fold increase from the 135.4 s it took when a Xeon (3.06 GHz) processor was used [79]. The time required for reading out the initial-stage, reading out the raw data, filtering, back projection, and saving were 0.1, 0.1, 0.8, 4.2, and 0.4 s, respectively. The time required for back projection comprised a large fraction (75%) of the total. Furthermore, when the resolution of the projection data increased to 1024 × 1024 pixels, the data totaling 4 GB in size were divided into 0.5 GB segments, and sequential reconstruction was conducted on them. The time required for back projection increased from 4.2 s to 31.69 s, which was an increase by a factor of 7.5. The time required for raw data read-out increased from 0.14 to 6.67 s (47.6-fold increase); the time required for filtering from 0.80 to 3.28 s (4.1-fold increase); and the time required for saving from 0.37 to 2.42 s (6.5-fold increase) [79]. For this reason, data-read out acts as a bottleneck for high-resolution data.

Along with the GPU, the memory devices used for the reading and writing of data following image reconstruction greatly influence the time necessary for the task. Currently, hard disk drives (HDDs) are still mainstream, but the SSD, released by SanDisk in 2007, has facilitated rapid data access. It has been reported that the SSD has contributed to accelerated image reconstruction in X-ray tomography [78].

References

1. S.R. Deans, The Radon Transform and Some of Its Applications, Dover publications, Mineola, pp. 204–217 (1983). (Appendix A)
2. G.T. Herman, *Fundamentals of Computerized Tomography: Image Reconstruction from Projections* (Springer, London, 2009)
3. F. Natterer, F. W-bbeling, *Mathematical Methods in Image Reconstruction* (The Society for Industrial and Applied Mathematics, Philadelphia, 2001)
4. P. Grangeat, J.L. Amans (Eds.), *Three-Dimensional Image Reconstruction in Radiology and Nuclear Medicine*. Kluwer Academic Publishers (1996)
5. G.L. Zeng, *Image Reconstruction: Applications in Medical Sciences* (Walter de Gruyter GmbH, Berlin/Boston, 2017)
6. V. Palamodov, *Reconstruction from Integral Data* (Chapman and Hall/CRC, Boca Raton, 2016)
7. G. Zeng, *Medical Image Reconstruction: A Conceptual Tutorial* (Springer, Berlin, 2010)
8. G.N. Hounsfield: Nobel Lecture (1979). https://www.nobelprize.org/nobel_prizes/medicine/laureates/1979/hounsfield-lecture.pdf
9. K. Kimura, T. Katakura, K. Suzuki, Japanese Journal of Tomography **16**, 247–250 (1989)
10. K. Mori: Patent publication, Patent No. 2128454 (1990)
11. H. Hu, Med. Phys. **26**, 5–18 (1999)
12. H. Toda, K. Shimizu, K. Uesugi, Y. Suzuki, M. Kobayashi, Mater. Trans. **51**, 2045–2048 (2010)
13. H.H. Barrett, W. Swindell, *Radiological Imaging, The Theory of Image Formation, Detection, and Processing* (Academic press, New York, 1981)
14. B.M. Thorsten, *Computed Tomography, From Photon Statistics to Modern Cone-Beam CT*, Springer, Berlin, Heidelberg, Germany (2008)
15. J.E. Gentle, §3.1 gaussian elimination, in Numerical Linear Algebra for Applications in Statistics, Springer, Berlin, pp. 87–91 (1998)
16. W.H. Press, B.P. Flannery, S.A. Teukolsky, W.T. Vetterling, 2.3 LU decomposition and its applications, in Numerical Recipes in FORTRAN: The Art of Scientific Computing, 2nd ed., Cambridge University Press, Cambridge, pp. 34–42 ((1992))
17. A.C. Kak, M. Slaney, *Principles of Computerized Tomographic Imaging* (Society of Industial and Applied Mathematics, Philadelphia, Pennsylvania, 1988)

18. F. Natterer, *The Mathematics of Computerized Tomography (Classics in Applied Mathematics), Society for Industrial and Applied Mathematics* (2001)
19. S. Kaczmarz, Bulletin International de l'Académie Polonaise des Sciences et des Lettres. Classe des Sciences Mathématiques et Naturelles. Série A, Sciences Mathématiques **35**, 355–357 (1937)
20. R. Gordon, R. Bender, G.T. Herman, J. Theor. Biol. **29**, 471–481 (1970)
21. A. Rosenfeld, A.C. Kak, *Digital Picture Processing*, 2nd edn. (Academic Press, New York, 1982)
22. G.T. Herman, *1980, Fundamentals of Computerized Tomography: Image Reconstruction from Projections* (Academic Press, New York-London, 1980)
23. Y. Censor, S.A. Zenios, *Parallel Optimization: Theory, Algorithms, and Applications* (Oxford University Press, New York, NY, 1997)
24. A.H. Andersen, A.C. Kak, Ultrason. Imaging **6**, 81–94 (1984)
25. P. Gilbert, J. Theor. Biol. **36**, 105–117 (1972)
26. T. Elfving, Numer. Math. **35**, 1–12 (1980)
27. H. Hurwitz Jr., Phys. Rev. A **12**, 698–706 (1975)
28. L.A. Shepp, Y. Vardi, IEEE Trans. Med. Imaging **1**, 113–122 (1982)
29. E. Levitan, G.T. Herman, IEEE Trans. Med. Imaging **6**, 185–192 (1987)
30. H.M. Hudson, R.S. Larkin, IEEE Trans. Med. Imaging **13**, 601–609 (1994)
31. E. Tanaka, H. Kudo, Phys. Med. Biol. **48**, 1405–1422 (2003)
32. M. Beister, D. Kolditz, W.A. Kalender, Physica Med. **28**, 94–108 (2012)
33. J.-B. Thibault, K.D. Sauer, C.A. Bouman, J.A. Hsieh, Med. Phys. **34**, 4526–4544 (2007)
34. R.L. Siddon, Med. Phys. **12**, 252–255 (1985)
35. B. De Man, J. Nuyts, P. Dupont, G. Marchal, P. Suetens, IEEE Trans. Med. Imaging **20**, 999–1008 (2001)
36. W. Xu, K. Mueller, *Proceedings of the 2nd High Performance Image Reconstruction Workshop*, Beijing, China, pp. 1–4 (2009)
37. T.M. Buzug, *Computed Tomography: From Photon Statistics to Modern Cone-Beam CT* (Springer, Berlin, Germany, 2008)
38. H. Qu, F. Xu, X. Hu, H. Miao, T. Xiao, Z. Zhang, Proceedings of the 2010 symposium on security detection and information. Procedia Eng. **7**, 63–71 (2010)
39. T. Nakano, Y. Nakashima, K. Nakamura, S. Ikeda, J. Geol. Soc. Jpn **106**, 363–378 (2000)
40. JASRI/SPring-8: High-resolution X-ray CT device image reconstruction manual. http://www-bl20.spring8.or.jp/xct/manual/reconstruction_2004_07_04.pdf. accessed July 2017
41. MathWorks home page: https://jp.mathworks.com/help/images/ref/iradon.html. Accessed July 2017
42. ImageJ homepage: https://imagej.nih.gov/ij/plugins/radon-transform.html. Accessed July 2017
43. SPring-8 BL20: Reconstruction of X-ray CT images using convolution back-projection (CBP) version 2, http://www-bl20.spring8.or.jp/~sp8ct/tmp/CBP.A5.pdf. Accessed July 2017
44. A. Rosenfeld, A.C. Kak, *Digital Picture Processing* (Academic Press, New York, NY, 1976)
45. R.N. Bracewell, A.C. Riddle, Astrophys. J. **150**, 427–434 (1967)
46. E. Tanaka, S. Mori, T. Yamashita, Phys. Med. Biol. **39**, 389–400 (1994)
47. NiftyRec 2.0 home page: http://niftyrec.scienceontheweb.net/wordpress/. Accessed July 2017
48. The ASTRA Toolbox home page: https://www.astra-toolbox.com/. accessed July 2017
49. ImagingScience4Neutrons home page: http://www.imagingscience.ch/tools/. Accessed July 2017
50. J. Hsieh, *Computed Tomography Principles, Design, Artifacts, and Recent Advances*, 2nd edn, Wiley Interscience, Bellingham, WA, 101 (2015)
51. W.A. Kalender, *Computed Tomography: Fundamentals, System Technology, Image Quality, Applications*, 3rd edn. (Wiley Interscience, Bellingham, WA, 2011), p. 316
52. T. Gomi, Recent Patents Med. Imaging **1**, 1–12 (2009)
53. G.L. Geng, G.T. Gullberg, Phys. Med. Biol. **37**, 563–577 (1992)
54. X. Yan, R.M. Leahy, Phys. Med. Biol. **37**, 493–506 (1992)

55. H. Kudo, R. Clark, T.A. White, T.J. Roney, IEEE Trans. Med. Imaging **13**, 196–211 (1994)
56. H.K. Tuy, SIAM J. Appl. Math. **43**, 546–552 (1983)
57. G.L. Zeng, G.T. Gullberg, IEEE Trans. Nucl. Sci. **37**, 759–767 (1990)
58. P. Grangeat, Mathematical framework of cone beam 3d reconstruction via the first derivative of the radon transform, in Mathematical Methods in Tomography, ser. Lecture Notes in Mathematics, G. Herman, A. Louis, and F. Natterer, Eds. Springer Berlin Heidelberg, vol. 1497, pp. 66–97 (1991)
59. L. Li, Z. Chen, and G. Wang: Reconstruction algorithms. In Cone Beam Computed Tomography (Editor: C.C. Shaw). Series: Imaging in Medical Diagnosis and Therapy (W.R. Hendee, Series Editor), Taylor & Francis, (2011)
60. D. Schneberk, R. Maziuk, B. Soyfer, N. Shashishekhar, R. Alreja, AIP Conf. Proc. **1706**, 110002 (2016)
61. L.G. Brown, ACM Comput. Surv. **24**, 325–376 (1992)
62. SPring-8 BL20: Offset CT Reconstruction procedure (2008.10.11 edition, PDF format). http://www-bl20.spring8.or.jp/xct/manual/OffsetCTrec.pdf. Accessed August 2017
63. A. Kyrieleis, M. Ibison, V. Titarenko, P.J. Withers, Nucl. Instrum. Methods Phys. Res. A **07**, 677–684 (2009)
64. A. Hammersley, A. Mirone, pyHST (High Speed Tomography in python version) software, ESRF. http://www.esrf.eu/home/UsersAndScience/Experiments/StructMaterials/ID19/microtomography.html. Accessed August 2017
65. T. Ohgaki, H. Toda, K. Uesugi, T. Kobayashi, K. Makii, T. Takagi, Y. Aruga, Mater. Sci. Forum **539–543**, 287–292 (2006)
66. R.M. Lewitt, R.H.T. Bates, Optik **50**, 189–204 (1978)
67. P.S. Cho, A.D. Rudd, R.H. Johnson, Comput. Med. Imaging Graph. **20**, 49–57 (1996)
68. J. Lee, J.S. Kim, S. Cho, J. Appl. Clinical Med. Phys. **15**, 252–261 (2014)
69. P. Paleo, A. Mirone, *Advanced Structural and Chemical Imaging* 3(2017), published online 2017 Jan 19. https://doi.org/10.1186/s40679-017-0038-1
70. F. Rashid-Farrokh, K.J.R. Liu, C.A. Berenstein, D. Walnut, IEEE Trans. Image Process. **6**, 1412–1430 (1997)
71. L. Li, H. Toda, T. Ohgaki, M. Kobayashi, T. Kobayashi, J. Appl. Phys. **102**(2007), 114908, published online: https://doi.org/10.1063/1.2818374
72. H. Kudo, Medical Imaging Technol. **34**, 186–197 (2016)
73. M. Hoshino, K. Uesugi, A. Takeuchi, Y. Suzuki, N. Yagi, J. Japanese Soc. Synchrotron. Radiat. Res. **26**, 257–267 (2013)
74. S. Harasse, W. Yashiro, A. Momose, Opt. Express **19**, 16560–16573 (2011)
75. N. Baba, Kenbikyo **39**, 4–10 (2004)
76. B. Cabral, N. Cam, J. Foran, Proceedings of the Volume Visualization, ACM, New York, USA, pp. 91–98 (1994)
77. G. Pratxa, L. Xing, Med. Phys. **38**, 2685–2697 (2011)
78. K. Uesugi, M. Hoshino, A. Takeuchi, Y. Suzuki, N. Yagi, T. Nakano, AIP Conference Proceedings, **1266**, 47–50 (2010). View online: https://doi.org/10.1063/1.3478197
79. Y. Okitsu, F. Ino, K. Hagihara, Parallel Comput. Arch. **36**, 129–141 (2010)

Chapter 4
Hardware

In October 2008, a sensationalist newspaper headline was published, stating, "*Christmas could bring with it a new hazard as you wrap your gifts—X-ray-emitting sticky tape*" [1]. This was relating to the Nature publication by Camara et al. from UCLA, which stated that rapidly peeling off sticky tape would produce X-rays [2]. Based on this, the term "triboluminescence" (the phenomenon in which light is emitted when an object is rubbed, destroyed, or deformed) was coined and there were numerous reports where X-rays were actually detected. Camara et al. promptly founded a venture company named Tribogenics and developed a small-scale X-ray source based on triboluminescence. Within five years of their publication in Nature, they developed a miniature X-ray imaging device that weighed less than 2.5 kg, which was able to take X-ray images of electric calculators [3]. Progress in science and engineering devices relating to X-rays has been remarkable in recent years. We as users cannot remain ignorant to the latest technological trends as this is the era in which we currently live.

However, simply chasing after new things will not yield satisfactory measurements in X-ray tomography. This is because an X-ray CT scanner comprises various constituent devices and technical elements and, depending on its selection or setup/experimental conditions, the obtained 3D image quality may greatly decrease or phenomena that should be visible may no longer be at times. It is also attributable to the fact that the structural size of the observed subject and spatial resolution of the X-ray CT scanner are relatively similar. This aspect differs extremely from other highly generalizable observation devices that will produce some manner of observation so long as the indicated switch is turned on (e.g. light microscopes, various electron microscopes, etc).

The objective of this chapter is to assist in developing a fundamental understanding of instrument principles, types, and assessment methods, as part of a first step in selecting the constituent devices, understanding the technical elements, and correctly setting up the experimental conditions. Typical hardware that constitute an X-ray CT scanner are classified from X-ray generation to detection, are introduced here.

H. Toda, *X-Ray CT*,
https://doi.org/10.1007/978-981-16-0590-1_4

4.1 X-ray Source

4.1.1 X-ray Generation

Interactions between the electromagnetic wave and atoms were discussed in Sect. 2.1.2. An overview of the interactions between the *charged particle* beam, with a focus on electron beams, and the atoms is presented here.

(1) Bremsstrahlung

Figure 4.1 schematically shows the interactions generated at the level of a single atom when a high-speed electron is incident onto an object. Furthermore, Fig. 4.2 schematically shows the X-ray spectra emitted from the object at that time. Figure 4.1a–c shows *continuous X-ray* generation due to *bremsstrahlung* (the etymology is German. Pronounced [brɛmz ʃtraːlʊŋ]). Figure 4.2 shows that continuous X-rays are distributed across a wide energy range. These types of X-rays are referred to as *white X-rays*. Furthermore, we can see that the intensity increases as the energy becomes lower (dotted line in the figure).

In Fig. 4.1a, the electron directly collides with the nucleus. The electron loses all of its kinetic energy in a single collision, and X-rays are generated through *bremsstrahlung*. This generates high-energy X-rays. The upper limit of the spectra in Fig. 4.2 is regulated by this (so-called *Duane-Hunt law*). However, this occurrence is an extremely rare phenomenon and the X-ray intensity on the edge at the high-energy side in Fig. 4.2 is extremely small. The *tube voltage* V_a [V] of the X-ray tube and the electron charge e ($= 1.602 \times 10^{-19}$ C) are used to express the upper limit value E_max of the generated X-ray energy, which is expressed as follows:

$$E_{max} = eV_a \tag{4.1}$$

The peak value of tube voltage is expressed by kV and its units are often expressed as [kVp] but this text uses [kV] as units. As shown in Fig. 4.2, $E_\mathrm{max} = 120$ keV when tube voltage $V_\mathrm{a} = 120$ kV; $E_\mathrm{max} = 100$ keV when $V_\mathrm{a} = 100$ kV. As with the dashed line in Fig. 4.2, intensity varies while E_max remains the same when the tube voltage is the same and the object into which the electron collides is different. Incidentally, the use of solid lines in Fig. 4.2 assumes that the X-ray spectra were generated from an X-ray tube. The drop in X-ray intensity in the low-energy side in Fig. 4.2 is stipulated by the absorption of the X-ray by the target itself or in the window material of the X-ray tube. Recently, beryllium has frequently been used as a window material, and the drop in X-ray intensity in the low-energy side is in this case limited. Readers who require such data can calculate them using Eqs. (2.2) and (2.2) and the data in Table 2.10. Figure 4.2 shows extremely complex behavior, such as characteristic X-ray peaks (discussed later) and right-hand-side decay induced by the K absorption edge of the target material itself.

Figure 4.3 shows the average X-ray energy generated by *bremsstrahlung* as a function of the tube voltage of the X-ray tube [4]. Knowing the average energy is

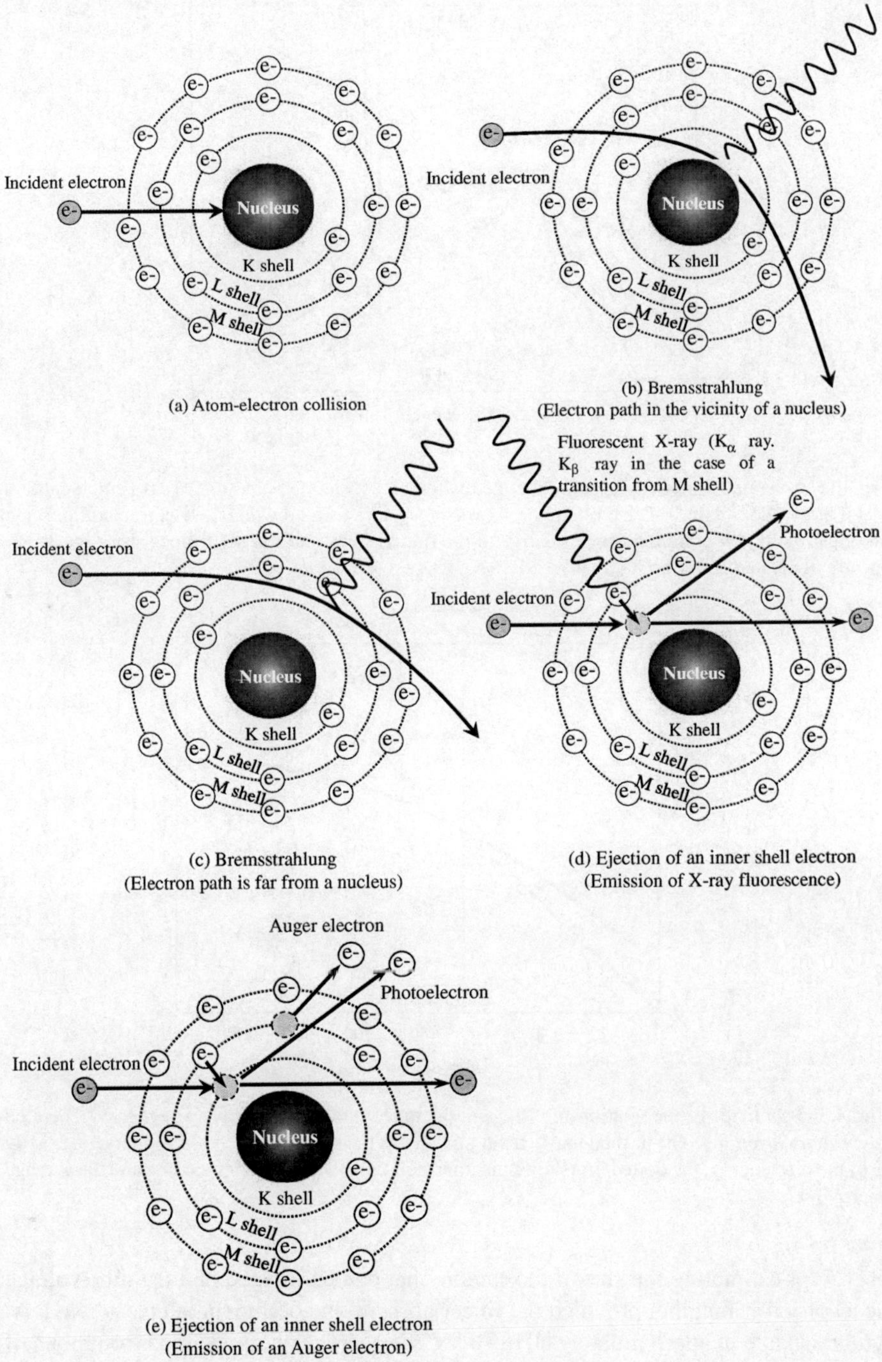

Fig. 4.1 Schematic of the various interactions between an accelerated electron and atom

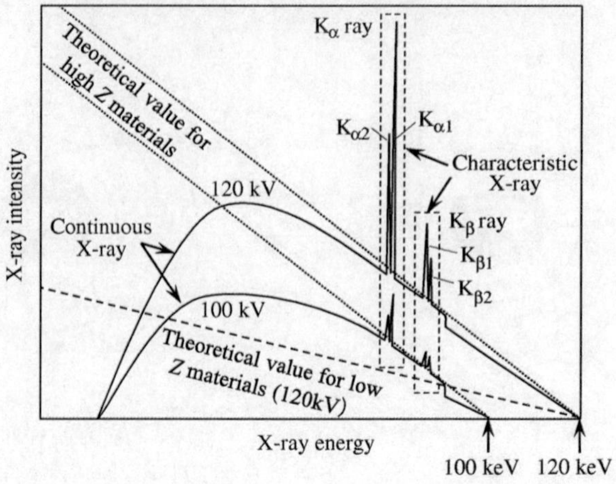

Fig. 4.2 Schematic of X-ray spectra obtained when an accelerated electron beam is incident to a metal target (dotted line). The dashed line shows the spectra formed when the electron collides into an object with low atomic number relative to the dotted lines and the solid lines show the X-ray spectra generated from the X-ray tube

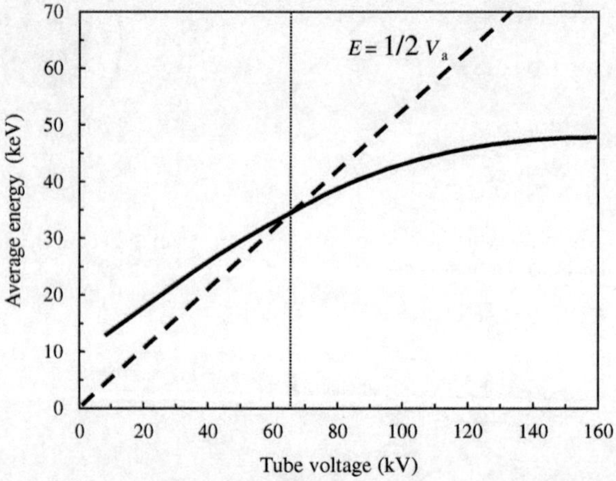

Fig. 4.3 Schematic of the relationship between the tube voltage and the average energy generated from *bremsstrahlung*. The dashed line in the figure shows the point at which X-ray energy is exactly ½ of the tube energy; the dotted line shows the intersection between that dashed line and the average energy [4]

useful for estimating the size of the sample that can be imaged and the tube voltage necessary for imaging, provided the linear absorption coefficient is investigated. At a tube voltage of approximately 60 to 70 kV, the average energy value is roughly half that of the tube voltage, a fact that should be borne in mind [4].

Incidentally, *bremsstrahlung* can occur even in charged particles other than electrons. Setting the incident charged particle mass and charge as m and ze, respectively, and the charge of the object as Ze (Z is the atomic number), then the total intensity I_b, where the X-ray intensities generated by *bremsstrahlung* are integrated across all energies and directions, is as follows [4]:

$$I_b \propto \frac{Z^2 z^4 e^6}{m^2} \tag{4.2}$$

For example, the masses of the proton and α particle (nucleus of the helium atom ^{4}He) are 1.8×10^3 and 7.3×10^3 times that of the electron, respectively. Thus, the *bremsstrahlung* due to the electron is several orders of magnitude larger than that from protons and α particles. Furthermore, Eq. (4.2) indicates that a high X-ray intensity can be obtained when metallic targets with a high atomic number are used.

If the electron does not collide with the nucleus, then it will pass through near the nucleus, as shown in Fig. 4.1b, c. In this case, the electron path is simultaneously curved and slowed in the Coulomb field of the nucleus. The accompanying X-ray generation, which occurs with *bremsstrahlung*, is the primary mechanism by which X-rays are obtained with X-ray tubes. The intensity of X-rays generated in this manner increases with the incident electron energy and atomic number of the object. When comparing Fig. 4.1b, c, the electron is curved and decelerated further in Fig. 4.1b, where the electron passes by closer to the nucleus. X-rays with higher energy are emitted as a result.

The ratio of the kinetic energy of the electron and *bremsstrahlung* energy is defined by the X-ray's *efficiency of bremsstrahlung production* η:

$$\eta = KZV_a \tag{4.3}$$

Here, $K = 9.2 \times 10^{-7}$ (kV) is a constant [5]. This equation is obtained by dividing the X-ray energy generated per unit time $KZV_a{}^2J$ (J is electric current) by the electron energy V_aJ. When η is calculated with copper and tungsten when the tube voltage is 100 kV, the values become 0.27% and 0.68%, respectively; in other words, most of the input energy is dissipated as heat. Most of the energy transmitted to the object from the incident electron serves to ionize the object and does not generate X-rays. As such, the X-ray tube target experiences a considerable temperature increase. For this reason, the usage of metallic materials that are heat-resistant and can efficiently conduct heat to its surroundings is optimal. The melting point and thermal conductivity of various metals are provided for reference in Table 4.1 [9]. In that sense, materials like tungsten, which have a high melting point and superior thermal conductivity, are suitable target materials.

(2) Characteristic X-rays

In the case of Fig. 4.1d, electrons are ejected from the K-shell due to the incident electron, and the atom enters an *excited state*. As seen in Sect. 2.1.2, fluorescent

Table 4.1 Melting point, thermal conductivity, and work function of various metallic materials and electron-emitting materials [9]

Material	Melting point (°C)	Thermal conductivity (W/mK)	Work function (eV)
Ag	961	425	4.5
Cu	1083	397	4.6
Co	1478	96	4.4
Fe	1539	78.2	4.5
Rh	1967	148	4.7
Zr	1855	22.6	4.1
Mo	2622	137	4.2
W	3382	174	4.5
LaB_6	2210	47	2.8
TiC	3067	33.5	3.5
TaC	3983	21.9	3.8
HfC	3928	30	3.4
NbC	3600	15.0	3.8

X-rays are generated owing to the transition of an electron from an outer shell. The energy of fluorescent X-rays is dependent on the fluorescing atom and this is referred to as a *characteristic X-ray* due to the emission of an X-ray with an energy that corresponds to the difference in binding energy of the trajectory before and after the transition. For this reason, there is a minimum tube voltage necessary for generating characteristic X-rays in the case of X-ray tubes, which is referred to as an *excitation voltage*. Of these, the characteristic X-ray, which occurs due to an electron transition from the L-shell, is referred to as a K_α X-ray; those due to an electron transition from the M- or N-shells are referred to as K_β X-rays. The numerous sharp peaks shown in Fig. 4.2 correspond to these characteristic X-rays. Typically, K_α and K_β X-rays can be easily distinguished. They each have extremely narrow energy ranges, so characteristic X-rays are used for experiments where monochromatic X-rays are necessary. The relationship between characteristic X-ray energy E_c and atomic number Z of the object can be easily calculated with *Moseley's law* [6].

$$\sqrt{\frac{E_c}{R_y}} = (Z - \sigma_n)\sqrt{\frac{1}{n_1^2} - \frac{1}{n_2^2}} \tag{4.4}$$

Here, $R_y = 2.18 \times 10^{-18}$ J $\approx$ 13.6 eV and corresponds to the electron binding energy of the ground-state hydrogen atom ($Z = 1$). Furthermore, n_i is 1, 2, and 3 for the K-, L-, and M-shells, respectively. The term σ_n is a constant that corresponds to the correction term from the hydrogen atom model for multi-electron systems; σ_n is dependent on n_i and Z but $\sigma_n \approx 1$ when $Z < 30$. As such, this can be simplified in the following equation when only K_α X-rays are considered:

$$\sqrt{\frac{E_c}{R_y}} = (Z-1)\sqrt{\frac{3}{4}} \tag{4.5}$$

In Table 2.3, there are slight differences in energy even for electrons that follow the same trajectory. This is typically expressed with numbered subscripts. For example, of the L_I shell (2 s trajectory of the L-shell), L_{II} shell, and L_{III} shell (2p trajectory of L-shell) in Table 2.3, only the L_{III}- and L_{II}-shells can transition into the K-shell and are each distinguished with $K_{\alpha 1}$ and $K_{\alpha 2}$ X-rays, respectively. The $K_{\alpha 1}/K_{\alpha 2}$ intensity ratio is 2:1 [7] and it is used without distinction due to its extremely close values. A weighted average of the two values is used for the wavelength value in this case [7].

$$\lambda_{K_\alpha} = \frac{2\lambda_{K_{\alpha1}} + \lambda_{K_{\alpha2}}}{3} \tag{4.6}$$

Here, λ_{K_α}, $\lambda_{K_{\alpha1}}$, and $\lambda_{K_{\alpha2}}$ are each the wavelengths of the characteristic X-ray corresponding to the average value, L_{III}-shell, and L_{II}-shell, respectively. The K_β X-ray boasts only a fraction of their strength. The characteristic X-ray energy values corresponding to representative metallic elements used in the X-ray tube target are summarized for reference in Table 4.2 [8]. When X-rays are used for analysis, it is important to not only consider the atomic number dependencies of the intensities shown in Eq. (4.2), but also that the characteristic X-rays of the target material do not overlap with those of the sample to be analyzed. Furthermore, based on Tables 2.1 and 4.2, the K absorption edge of copper, for example, is 8.98 keV and its $K_{\alpha 1}$ and $K_{\alpha 2}$ are 8.05 keV and 8.03 keV, respectively. As such, the characteristic X-ray generated when copper is used as a target, is relatively minimally absorbed in the target material itself. However, when emitting X-rays from a copper target to a sample containing iron (K absorption edge of 7.11 keV) and cobalt (K absorption edge of 7.71 keV), both of which have slightly smaller atomic numbers than copper, the absorption of characteristic X-ray due to the sample significantly increases and background levels increase due to the generation of fluorescent X-rays.

Table 4.2 Energy values (units of eV) of characteristic X-rays for representative elements used in the X-ray tube target [8]

Atomic number	Element	$K_{\alpha 1}$	$K_{\alpha 2}$	K_α (Average)	$K_{\beta 1}$
24	Cr	5414.72	5405.51	5411.65	5946.71
26	Fe	6403.84	6390.84	6399.51	7057.98
27	Co	6930.32	6915.30	6925.31	7649.43
29	Cu	8047.78	8027.83	8041.13	8905.29
42	Mo	17479.34	17374.30	17444.33	19608.30
45	Rh	20216.10	20073.70	20168.63	22723.60
47	Ag	22162.92	21990.30	22105.38	24942.40
74	W	59318.24	57981.70	58872.73	67244.30

Returning from the excited state mentioned above, Auger electrons could be emitted like the case shown in Fig. 4.1e, which is also discussed in Chap. 2. The fluorescence yield, which expresses the probability of whether fluorescent X-rays or Auger electrons being generated, has already been summarized in Fig. 2.6 for many elements.

Figure 4.2 shows two spectra for X-ray tubes when its tube voltages are changed. When a high voltage is applied, the intensities of characteristic and continuous X-rays increase. The peak positions of characteristic X-rays are unique to the material and do not change but the peaks of the continuous X-rays shift to the high-energy side.

4.1.2 X-Ray Tube

The Crookes tube used during Roentgen's era generated electrons by discharging electricity between electrodes placed in a low-pressure gas. The structure of X-ray tubes widely used today is fundamentally identical to the Coolidge tube developed by the eponymous scientist in 1913. This section provides an overview of these constituent elements in order.

X-ray tubes are important constituent devices that stipulate the imaging capability or quality based on industrial X-ray CT scanners. The corresponding physics, which serve as its basis, are described in Sect. 4.1.1. The fixed anode X-ray tube shown schematically in Fig. 4.4 comprises a *filament*, which discharges electrons (*cathode*); the target, which produces the X-ray (*anode*); and the vacuum-sealed tube container, which envelops these elements and is kept at around 10^{-6} Pa. This is referred to as a

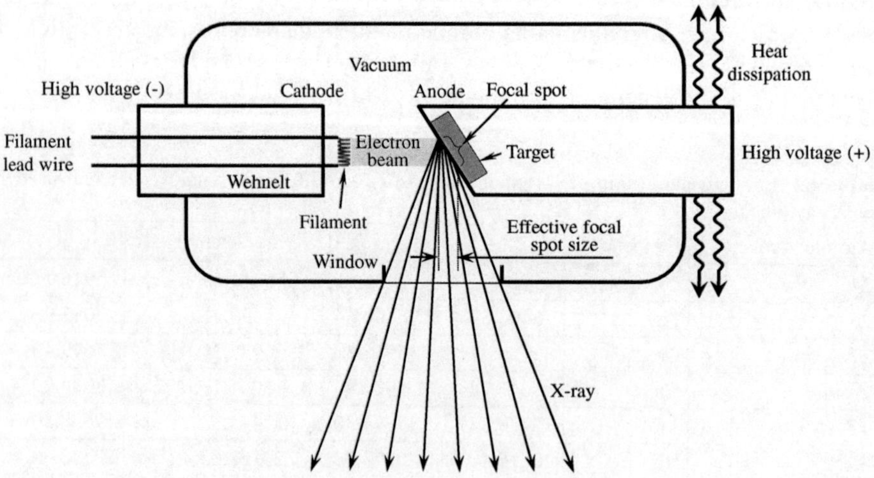

Fig. 4.4 Schematic of the fixed anode X-ray tube structure

sealed X-ray tube. Furthermore, there are also *open X-ray tubes*, which have a vacuum pump installed outside of the tube and constantly evacuate the tube interior with the pump. The target and filament can be replaced if an open system is used. A high voltage of around 500–600 kV is applied between the cathode and anode. Generally, the target is slanted in a single direction to irradiate the X-ray (reflection-type), and the generated cone-beam X-ray exits from the window portion. The window is made of either a thin glass or beryllium film (about 150–500 μm in the microfocus tube). Specialized carbon materials are also used in commercially available X-ray CT scanners shown later in Table 6.1; aluminum is also used for high-energy cases. Beryllium can be toxic and reacts with water, so care must be taken with regard to its handling. Furthermore, it is a light element at $Z = 4$, effectively transmitting X-rays. For example, a beryllium window of an approximately 500 μm thickness will transmit over 95% and 98% of X-rays with 10-keV and 20-keV energies, respectively. Some designs ensure that the thermally-conducting metals like copper are distributed and that air-cooling and even water-cooling or oil-cooling at times is installed to ensure that the high-temperature target is sufficiently cooled.

(1) Cathode

Electrons need to be obtained for X-rays to generate. Phenomena that result in the emission of electrons from metals include *thermionic emission*, photoelectron emission, field emission, and secondary electron emission. Of these, X-ray tubes primarily use *thermions*, which are obtained primarily from thermionic emission. Incidentally, this thermionic emission is also used in the electron guns of linear accelerators in synchrotron radiation facilities, vacuum tubes, and Braun tubes used in old televisions.

Thermionic emission is a phenomenon where applying energy to a free electron in the conduction band results in its overcoming the surface barrier and emission. Figure 4.5 schematically shows the relationship between the electron state density within metals as determined by the *Fermi-Dirac distribution* and potential of the cathode surface. The minimum energy needed for a single electron to jump into the vacuum is referred to as the *work function* ϕ. This corresponds to the difference between the Fermi level and vacuum level shown in Fig. 4.5. Furthermore, this ϕ value is known to vary by metal. Table 4.1 shows the work function values for representative materials used when generating thermions [9]. As indicated by the dashed line in Fig. 4.5, the area below the Fermi level is completely filled with electrons at a temperature of absolute zero and the existence probability of electrons that have energy above this is zero. Electrons at the Fermi level will be emitted into the vacuum if they obtain energy greater than that of the work function. The filament metal should be heated with Joule heat by transmitting a filament heating current to apply an energy corresponding to the work function to the electron. As indicated by the dotted line (temperature T_1) in the left half of Fig. 4.5, the existence probability of electrons gradually reaches high energy levels as the temperature increases. The top edge of the solid line in the left half of Fig. 4.5 (temperature T_2, where $T_2 > T_1$) exceeds the vacuum level and the electrons of an amount corresponding to the

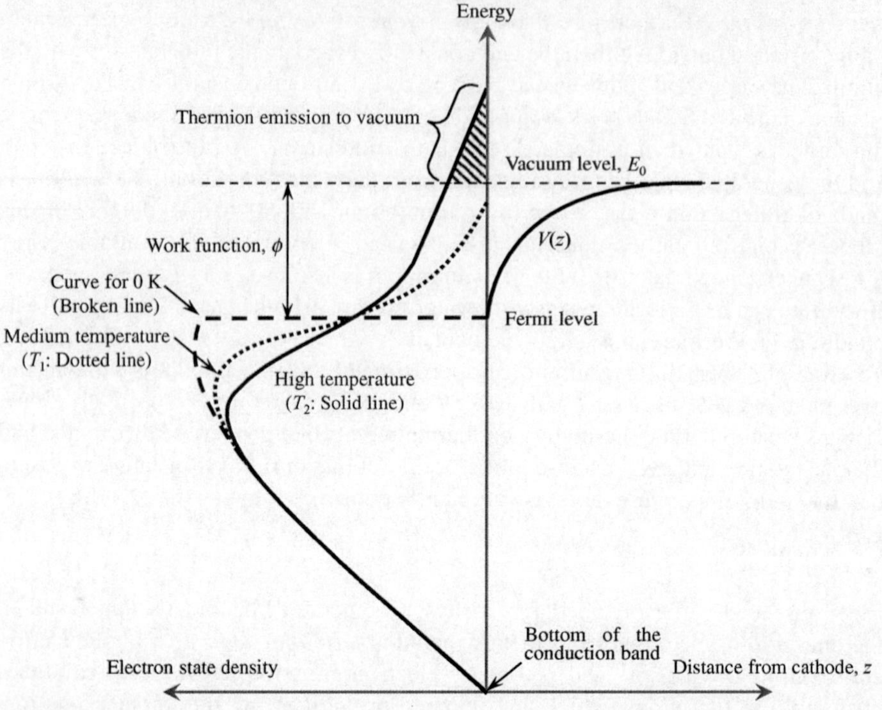

Fig. 4.5 Schematic showing an overview of thermionic emission. The right half shows the potential of the cathode surface. The left half shows the electron energy distribution at absolute zero and temperatures T_1 and T_2 ($T_1 < T_2$)

hatched area are emitted. Thermionic emission occurs all at once when the surface barrier is exceeded, so the energy distribution is biased towards the low-energy side. The potential $V(z)$ of the cathode surface shown on the right half of Fig. 4.5 is the mirror-image potential created by the electron itself on the metallic surface. This is expressed as follows, with the distance between the electron and cathode as z, and the vacuum level as E_0:

$$V(z) = E_0 - \frac{e^2}{16\pi \varepsilon_0 z} \tag{4.7}$$

Here, ε_0 is the vacuum permittivity. Also, when a relatively high electric field F is applied, an electric field potential component, which includes the slope of the electric field, is added as shown in Fig. 4.6.

$$V(z) = E_0 - \frac{e^2}{16\pi \varepsilon_0 z} - eFz \tag{4.8}$$

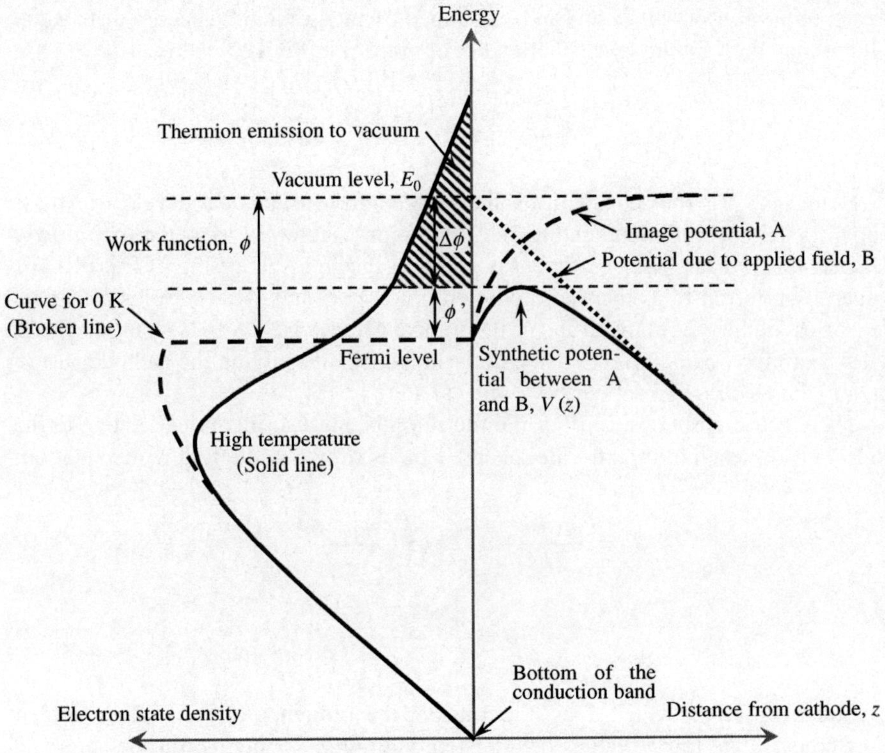

Fig. 4.6 Schematic showing an overview of thermionic emission when an electric field is applied on the cathode surface. The right half shows the potential of the cathode surface; the left half shows the electron energy distribution at high temperatures

The effective work function ϕ' in these cases is reduced by $\Delta\phi$ from its original value. This type of electric field effect is referred to as the *Shottky effect*. The decrease in the potential barrier due to the Shottky effect is at a maximum value at:

$$z = \frac{1}{4}\sqrt{\frac{e}{\pi \varepsilon_0 F}} \tag{4.9}$$

where ϕ' takes the following form:

$$\phi' = \phi - \frac{e}{2}\sqrt{\frac{eF}{\pi \varepsilon_0}} \tag{4.10}$$

So what types of materials are suited for supplying electrons when thermionic emission is utilized? If there is no electric field effect and if the amount of thermions emitted per unit surface is expressed as the electric current density J_c [A/m^2], this is expressed as follows [10] and can be obtained by integrating the energy distribution higher than the vacuum level (inclined line portion) in Fig. 4.5:

$$J_c = \frac{4\pi mek^2}{h^3} T^2 e^{-\phi/kT} \tag{4.11}$$

Here, h is the Planck constant, k is the Boltzmann constant, m is the electron mass, and T is the absolute temperature. This equation is referred to as the *Richardson-Dushman equation*. The coefficient portion on the right-hand side of Eq. (4.11) is generally referred to as the thermionic emission constant. This coefficient is fixed regardless of the type of metal and its theoretical value is 1.20×10^6 [A/(m^2·K^2)]. The ideal thermion beam current can be determined by multiplying the cathode surface area to the electric current density in Eq. (4.11).

The electric current density of the thermion beam also increases relative to that in Eq. (4.11) when the electric field is present, as shown in the following equation:

$$J_c' = \frac{4\pi mek^2}{h^3} T^2 exp\left(\frac{-\phi - \beta F^{\frac{1}{2}}}{kT}\right) \tag{4.12}$$

$$\beta = \frac{e}{2}\sqrt{\frac{e}{\pi \varepsilon_0}} \tag{4.13}$$

Equation (4.11), which does not consider the influence of the electric field, is used to calculate the J_c values of tungsten, platinum, copper, and iron. The result is Fig. 4.7. The plot is cut off near the melting point for all metals except tungsten,

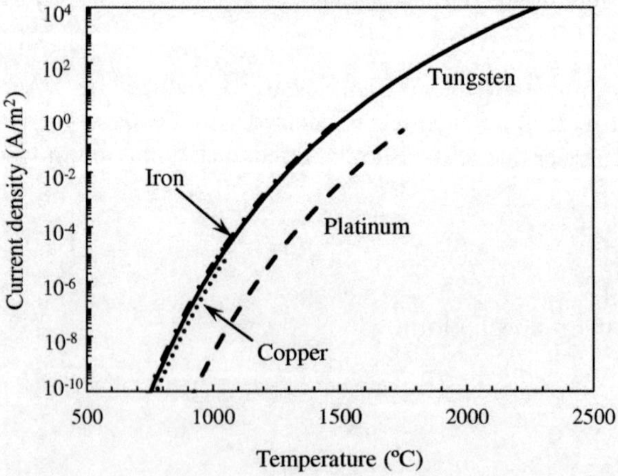

Fig. 4.7 Relationship between saturation current density of thermions, when various metals were used as an X-ray filament, and temperature. Except for tungsten (solid line), the plots of platinum (dashed line), copper (dotted line), and iron (dash-dot line) are all cut off near the melting point

which has a melting point around 900 °C higher than the upper limit of the horizontal axis. It is clear when considering heat resistance that these materials can only tolerate temperatures considerably lower than its melting point during actual use. From this figure, it is evident that tungsten, which has a high melting point and superior high-temperature strength, may not necessarily have a small ϕ but is suited as filament material. The favorable workability of the material is also one of the reasons why tungsten is used. Furthermore, other metallic materials melt before thermionic emission occurs. The working temperature of tungsten filaments is set at −2500 °C. The emission current density, in this case, is thought to be around 1000–4000 A/m^2.

A photograph of a tungsten filament from an X-ray tube is shown in Fig. 4.8. The tungsten wire is used in a coiled form. The cathode of the X-ray tube may also use *indirectly heated cathodes*, similar to those in miniature-tube-type vacuum tubes used in the past. Reported structures include a barium or strontium oxide coating on the surface of nickel cylinders, in which an insulated heating filament is placed [11]. A characteristic of oxides like oxidized barium or strontium is that their work function is significantly lower than that of the metals alone [11]. Using this enables the operation of cathodes at temperatures of 700–900 °C, which is significantly lower than that used when the tungsten filament is directly heated [11]. This also results in the longevity of the X-ray tube. Figure 4.9 shows a comparison between the thermionic emission from oxides (barium-strontium dioxide: BaO–SrO, a complex oxide of barium and strontium) and tungsten. Sufficient current density can be achieved at a temperature considerably lower than that of tungsten. Here, the work function and constant of the Richardson-Dushman equation for barium-strontium dioxide were set as 0.95 eV and 100 [A/(m^2·K^2)]. For reference, the work function of the barium used in the figure was 2.11 eV. Oxide coatings can be created by applying carbonates of barium, strontium, calcium, etc. on the substrate metal and heat treating the metal under high-temperature and high-vacuum conditions [12].

The filament is placed within a hollow referred to as a cathode cub and is set up so that the focal spot is aligned in a fixed location on the anode. An electron beam can be collimated by insulating the cathode cup from the filament and applying a negative

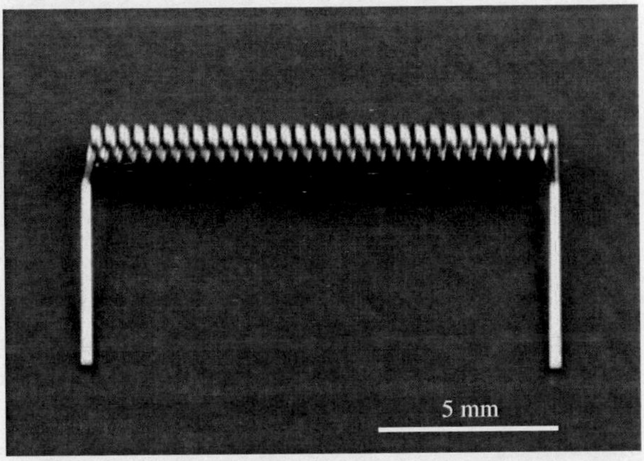

Fig. 4.8 Photograph of a cathode filament part (made from tungsten) in an X-ray tube (courtesy of Yasuaki Takeishi of Takeishi Electric, Co., Ltd.)

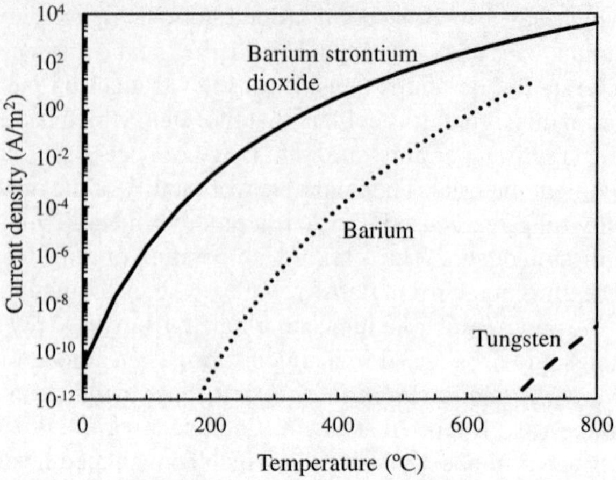

Fig. 4.9 Relationship between the saturation current density of thermions, when barium-strontium dioxide, barium, and tungsten were used as a cathode for an X-ray tube, and temperature

bias voltage. A light source that is brighter than cases where electron radiation is generated in all directions can be created using this method. For example, reports have indicated that the bias voltage application of −2000 V to the cathode cup allows for the focal spot size to be 1/4 the value of when a bias voltage is not applied [13].

As shown in Fig. 4.10a, most of the thermions generated in the filament can reach the anode when the tube voltage is high ($V_a > 40$ kV). As shown in Fig. 4.11, increasing the filament heating current and emitting more thermions results in an increased *tube current* (anode current). Meanwhile, as shown in Fig. 4.10b some of the thermions reach the anode when the tube voltage is low ($V_a < 40$ kV); the rest, however, remain as a *space charge* near the cathode. The existence of the space charge works in the direction where the emission of thermions from the cathode is suppressed. Increasing the filament heating current, in this case, does not increase the tube current as seen in Fig. 4.11 and instead saturates it. The effect of this space charge is why thermionic emission is stable and easy to control. When the space charge is predominant in this way, the following *Child-Langmuir equation* is established:

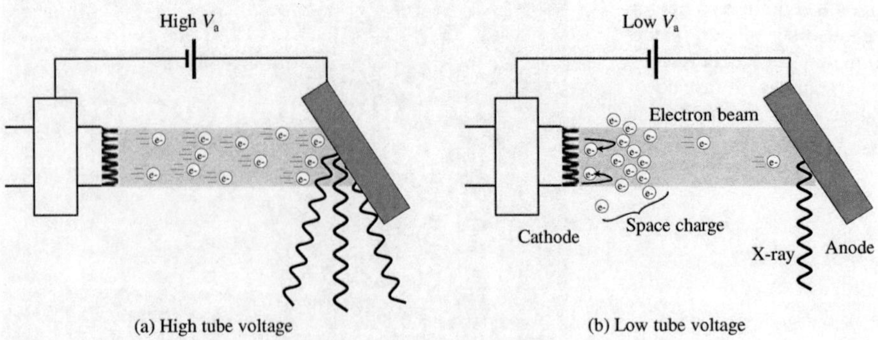

Fig. 4.10 Schematic showing the influence of space potential

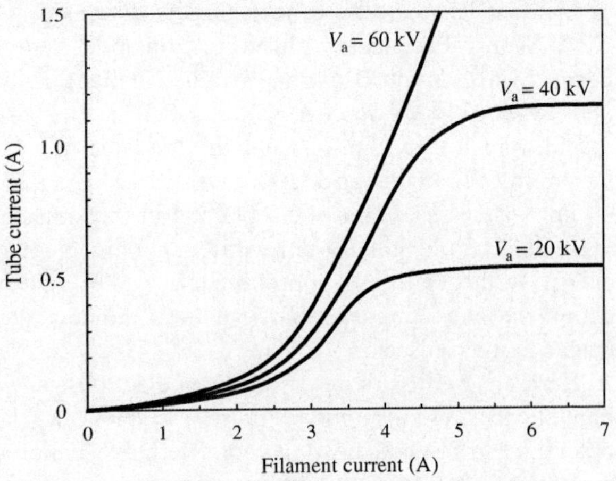

Fig. 4.11 Schematic showing the relationship between the filament heating current and tube current. Low-tube current cases where the space charge is dominant and high-tube current cases where the filament current is dominant are shown

$$J_a = K\frac{V_a^{\frac{3}{2}}}{d^2} \tag{4.14}$$

Here, K is a constant and d is the distance between the anode and cathode. Equation (4.14) indicates how the tube current increases relative to tube voltage by a power of 3/2, shown on the left-hand side of the schematic of the relationship between tube current and tube voltage in Fig. 4.12. The right-hand side of Fig. 4.12 shows that the filament heating current can be increased or decreased to regulate the filament temperature, which in turn allows for the control of the tube current. In this way, it is necessary to understand that there are cases where the tube current barely changes when the tube voltage is regulated and cases where it is acutely sensitive, depending on the focal spot size of the X-ray tube, as well as the tube current and voltage values.

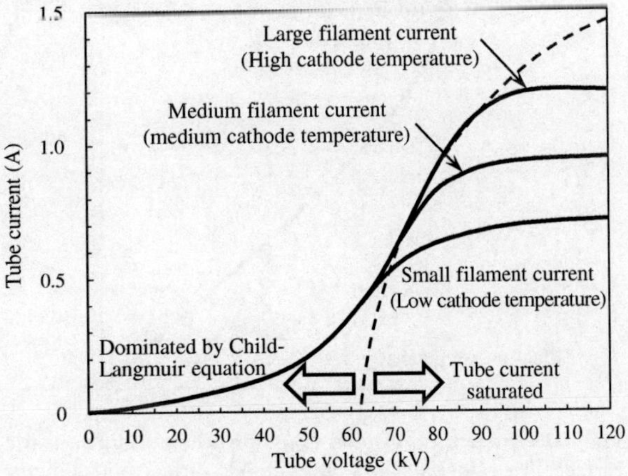

Fig. 4.12 Schematic showing the relationship between tube voltage and tube current. The figure shows three cases, ranging from where the filament current and temperature are low to cases where the filament current is high and thermionic emission is active

Incidentally, there have been reports of scanning electron microscope (SEM)-based X-ray CT scanners with the electron optics systems of an SEM that use electron beams from compounds with a ϕ value smaller than that of tungsten (e.g. LaB_6), or *field emission* guns. Recent trials by Perini et al. used targets of either bulk (i.e. several mm^2), thin-film, or nanowire tungsten in SEMs with a field emission gun, generating X-rays using a 1000 nA, 30 kV electron beam [14]. A maximum of 60 nm spatial resolution was obtained in transmission observations when using a 400-nm–2-μm length, 70-nm width nanowire as a target [14]. Although the X-ray energy, in this case, was approximately 10 keV and flux was also low, this was an extremely interesting trial that visualized the microstructures of small samples at a deep sub-micron scale.

Here, field emission refers to when electrons inside a metal are emitted when a high electric field is applied. Figure 4.13 shows a schematic of field emission. The potential barrier width becomes considerably smaller when the electric field strength significantly increases (GV/m level) compared to Fig. 4.6 and electrons are emitted through a quantum mechanical phenomenon referred to as the tunnel effect. Cathodes

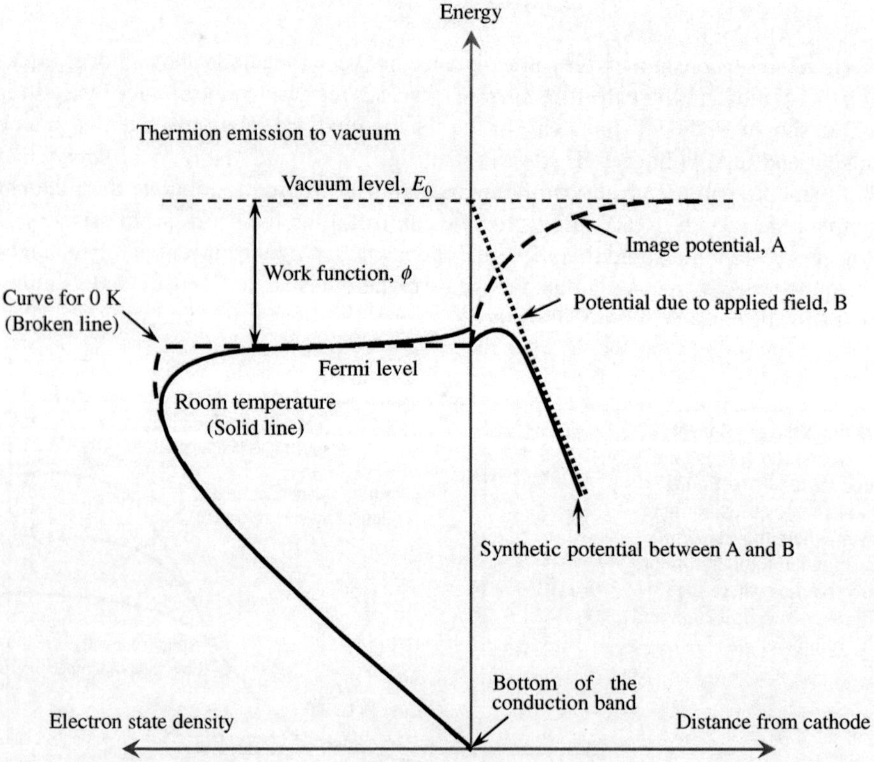

Fig. 4.13 Schematic of field emission when a large electric current is applied on the cathode surface. The right half shows the cathode surface potential; the left half shows the electron energy distribution at room temperature

do not need to be heated with field emission, so this process is also referred to as *cold emission*, in contrast to thermionic emission. This process also has the advantage of requiring less power consumption as a result. The current density for field emission cases takes the form of the Fowler-Nordheim equation, shown below:

$$J_c^F = AF^2 exp\left(\frac{-B\phi^{3/2}}{F}\right) \tag{4.15}$$

Here, A and B are constants that depend on ϕ. From the above equation, the emission current density is extremely dependent on the electric field. For this reason, even slight changes in the electric field at a level of around 1×10^9 (V/m) can rapidly change the emission current density from effectively 0 to an extremely large value.

(2) Anode

The physical phenomena of X-ray generation when the electron beam is incident to the target has been described in Sect. 4.1.1. Here, the important characteristic of X-ray tubes other than X-ray energy and intensity discussed in that section is most likely spatial resolution. Figure 4.14 shows a schematic of how focal spot size influences spatial resolution. The X-rays emitted from points A and B, which are respectively on the left- and right-hand side of the focal spot, are detected at different pixels in the detector after transmitting through the same position in the sample. This results

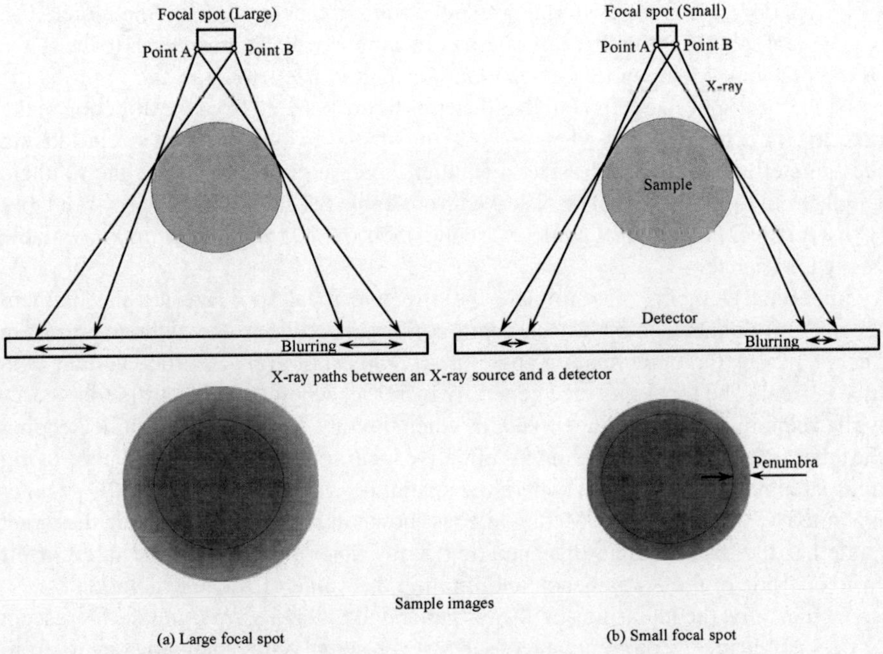

Fig. 4.14 Schematic showing the influence of focal spot size on image blur

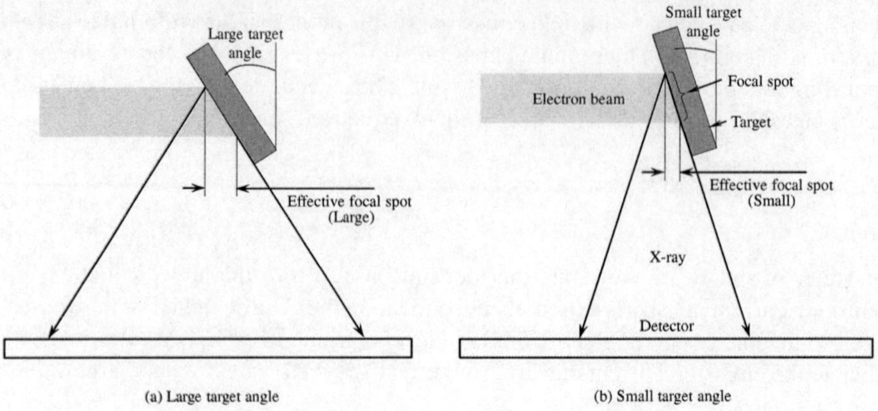

Fig. 4.15 Schematic showing the influence of the target inclination angle; a smaller target angle results in a narrower X-ray irradiation range

in image blur. Comparisons between Fig. 4.14a, b show that this blur increases when the focal spot size increases. The lower half of Fig. 4.14 shows the *penumbra* around the visualized sample. For the same reason, X-rays generated by electrons colliding into areas other than the anode target also result in decreased spatial resolution and contrast. This can be up to 10% of the overall intensity at certain times. As seen in Figs. 4.4 and 4.15, targets in the anodes of actual X-ray tubes are inclined several to several dozen degrees from the direction facing the detector from the target. The actual focal spot takes the shape of a narrow rectangle, which corresponds to the shape of the filament. Meanwhile, the *effective focal spot* size shown in the figure is the apparent focal spot size (filament length in this figure) as seen from the direction of the detector. This is important when considering spatial resolution. When reproducing the same effective focal spot size, a smaller target angle can increase the filament length, as in Fig. 4.16. It can be observed from Table 6.1 shown later that X-ray tubes with a relatively large target angle of around 45° are used for commercially-available X-ray CT scanners.

Important elements that stipulate the effective focal spot size are the filament length, coil diameter, the electron beam convergence due to the cathode cup as in Sect. 4.1.2 (1), target inclination angle (*target angle*) in Fig. 4.15, tube voltage, and tube current. The focal spot size generally increases when the tube current increases while keeping the tube voltage fixed, or when the tube voltage is low while keeping the tube current fixed. Decreasing the effective focal spot directly results in improving the spatial resolution in cases where the spatial resolution of the X-ray CT scanner is limited by the X-ray tube. Meanwhile, as shown in Fig. 4.15, decreasing the target angle has the disadvantage of generating X-ray absorption due to the target itself from one side of the X-ray beam, and limiting the range of X-ray irradiation.

Incidentally, the intensities of X-rays generated by *bremsstrahlung* or fluorescent X-rays should ideally be homogeneous. However, actual X-ray intensities are far from this ideal. This is primarily due to a phenomenon referred to as the *heel effect* shown

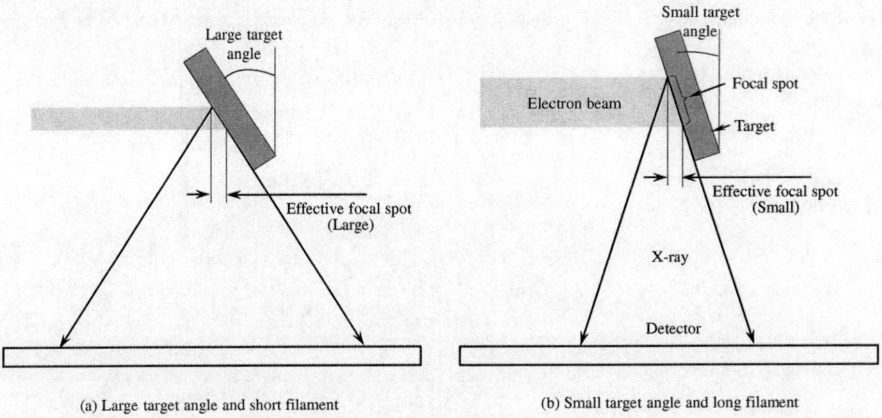

Fig. 4.16 Schematic showing the influence of the target inclination angle. The filament length needs to be decreased as shown in (**a**) when the target angle is large in order to maintain the same effective focal spot size

in Fig. 4.17. Up until now, we have only considered simplified cases in Figs. 4.15 and 4.16 where X-rays were generated on the target surface. However, in reality, electrons penetrate up to a depth stipulated by its kinetic energy and the target material, and X-rays are generated inside the target material. For example, Monte Carlo simulations conducted by Poludniowski et al. showed that electrons penetrate tungsten targets up

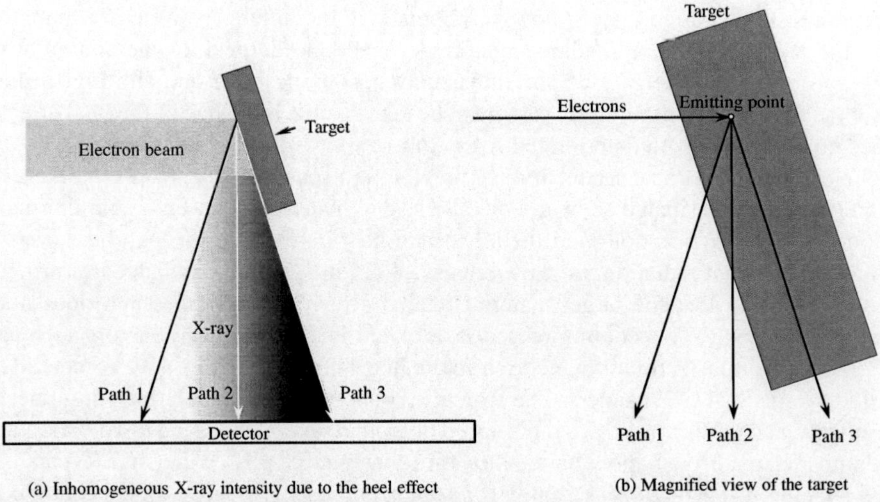

Fig. 4.17 Heel effect generated due to target inclination; **a** is a schematic showing X-ray intensity unevenness due to the heel effect and **b** is its magnified view, showing a schematic of X-ray paths 1, 2, and 3 in order from the cathode side, and the difference in transmission distances of the X-ray in the anode

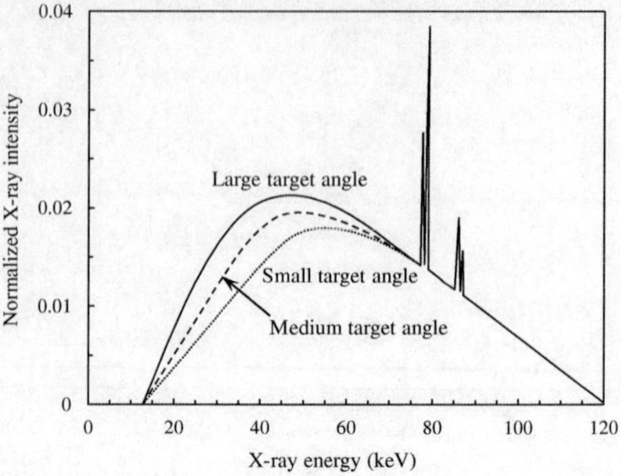

Fig. 4.18 Schematic of X-ray spectra obtained when an accelerated electron beam is incident to a metal target (solid line). The dashed lines and dotted lines are spectra for when the target angles are lower than the solid line

to a depth of around 14 μm with a tube voltage up to 150 kV [15]. For this reason, X-rays near the cathode in the generated cone beam have a relatively short distance for transmitting through the anode, while in contrast those near the anode have a longer distance. The target itself acts as a filter on the anode side, so X-rays attenuate, and their low-energy components are lost. Meanwhile, this influence is minimal on the cathode side, X-rays retain their high brilliance, and the low-energy components are not lost. Furthermore, a transition in X-ray intensity and energy distribution occurs in the intermediate position. A characteristic X-ray intensity distribution is generated in this manner. The heel effect becomes particularly strong when the target angle becomes smaller. Figure 4.18 shows a schematic of the influence of the target angle on the X-ray spectra. A smaller target angle results in the peak of the continuous X-ray on the low-energy side simultaneously becoming lower and shifting to the high-energy side. Furthermore, care must be taken as the influence of the heel effect becomes stronger with deteriorated X-ray tubes.

Next, we discuss heat production. As previously discussed, over 99% of the energy in electrons is dissipated as heat. For this reason, materials with favorable thermal conductivity are placed around the target and heat is removed through heat transmission and heat radiation, as shown in Fig. 4.4. Furthermore, coolants are further used to ensure that the target material retains a lower temperature and does not vaporize or melt. This cooling presence or extent is determined by factors such as X-ray tube output. A rotating-anode structure like that shown in Fig. 4.19 is generally used for medical CT scanners. The structure of the anode has a relatively high heat capacity, and this is rotated at a rapid speed of around several thousand to 10,000 rpm. Compared to the fixed-anode case, X-rays are generated across a wider area (typically by a factor of several dozens), and it is advantageous in that heat is not concentrated in a small area. The typically tolerable load is proportional to $\sqrt{dn_r}$, where d is the disk diameter and n_r is the number of rotations. This system has been used in medical CT scanners from the third generation (shown in Fig. 3.1b) onward. This system is used in industrial X-ray CT scanners as well for high-energy/high-output

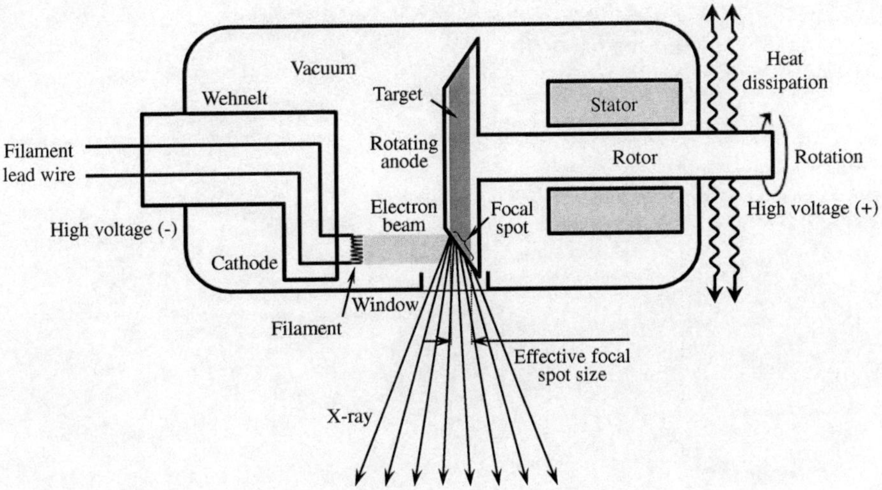

Fig. 4.19 Schematic of rotating-anode X-ray tube structure

tube cases. However, this requires bearings for the high-speed rotation of the anode in vacuum at local temperatures over 2000 °C, as well as technology for insulation from high voltages [16]. For example, gallium, which is a liquid at room temperature, has been used in liquid metal hydrodynamic bearings in order to increase cooling efficiency through liquid-metal-based conduction cooling [16]. Rotating-anode systems are ideal for medical CT scanners which conduct exposures at high output for short periods of time and which have sufficient cooldown times due to patient turnover, but have a minor role in industrial X-ray CT scanners.

Figure 4.20 shows a photograph of a tungsten target in a commercially-available X-ray tube. Fixed-anode systems like those in Fig. 4.4 typically use copper as a base and have a structure where the target is embedded in the portion where the electron hits. Rotating-anode systems like those shown in Fig. 4.19 typically have a W–10% Re alloy coated on a dispersion-hardening molybdenum alloy disk referred to as TZM (Mo–0.5Ti–0.08Zr –0.025C alloy) or MHC (Mo–1Hf –0.1C alloy). Structures whose heat capacity is assured are also used by affixing graphite, which has a high specific heat at high temperatures, to metals. Rhenium (Re) additions to tungsten increase tungsten strength and ductility at high temperatures, as well as improve its resistance to repeated thermal stress [17]. This also prevents coarsening and grain boundary embrittlement due to the recrystallization, which occurs when temperatures exceed 1500 K in pure tungsten, greatly improving its fracture toughness [18].

(3) High-Voltage Equipment

A high-voltage transformer is typically used to increase alternating current (AC) to supply tube voltages ranging from 10 to several hundred kV and tube currents ranging from 1 to several thousand mA. The high voltages obtained are applied to the X-ray tube after converting to direct current (DC) using a rectifier. Two pulses are

Fig. 4.20 Photograph of the rotating-anode portion (made from tungsten) of the X-ray tube (courtesy of Toshiba Materials, Co., Ltd.)

generated for each AC cycle when single-phase power sources are used and a voltage waveform is generated where the output peaks are continuous. Furthermore, three-phase/six-peak high-voltage X-ray devices, where a three-phase power source and six rectifiers are used to conduct full-wave rectification, produce output waveforms where six pulse peaks are superimposed per cycle. Furthermore, 3-phase/12-peak high-voltage X-ray devices, where 12 rectifiers are used, produce output waveforms that have an even smaller voltage ripple factor. Other than these, inverter-type high-voltage X-ray devices are generally widely used. In the case of the inverter-type, voltages are converted to act at high frequencies of several 100 Hz to several 10 kHz in the high-frequency inverter section, after which it is applied to the high-voltage generating section, whereupon its voltage is increased and rectified. For this reason, direct currents with even smaller ripples are output. As a result, a sufficiently smooth output can be obtained even with a single-phase power source.

(4) X-ray Tube Longevity

Most malfunctions in an X-ray tube are due to heat generation. For example, major malfunctions include melting or destruction of the anode, target material vaporization or deposition, cutting of the filament, loss of vacuum in the tube container in the sealed case, bearing malfunctions in the case of a rotating-anode type, destruction of the tube container, and decreased insulation performance of the insulating oil due to carbonization in the case of oil cooling.

For the cutting of filaments, using *thorium oxide* (ThO_2) in pure tungsten allows for a larger current density to be obtained at a lower temperature of 1900 K relative to that of pure tungsten and for longevity to be achieved [19]. For example, short-term heating of a tungsten wire with 1–2% additions of thorium oxide at temperatures around 2800 K produces a single-atom layer of thorium and reduces its work function considerably to around 2.6. For this reason, the emitted current increases by orders of magnitude at the same temperature. There have been recent investigations on carbide substitutions that do not use thorium oxide, which is a radioactive material [20]. The thermionic emission-type cathode, which uses oxides with a low work function have already been discussed. There has also been an increasing number of reports on X-ray tubes with cooled cathode-type power sources, which use a carbon nanotube and field emission to emit electrons even at low voltages. There have been over 30 publications on this since the initial 2001 publication by Sugie et al. of the Nagoya Institute of Technology [21]. There have been reports among these of X-ray tubes where the focal spot diameter was below 1 μm [22].

(5) Commercially-Available X-Ray Tubes

Specifications of commercially available X-ray tubes currently available (as of August 2017), which manufacturers have indicated are used as industrial X-ray CT scanners, are shown in Table 4.3. These are all fixed-anode X-ray tubes. The numerical specification values were directly obtained from the manufacturers' websites [23–25]. Care must be taken as the devices specified in the table have spatial resolutions of transmission images measured from test charts like the X-ray resolution chart specimens from the Japanese Inspection Instrument Manufacturers' Association [26]; their values may be considerably smaller than the focal spot dimension provided. Methods for end-users to measure the focal spot size are stipulated by ASTM, etc. (e.g. ASTM E1165-12).

Figures 4.21 and 4.22 show the external appearance of representative devices among those in Table 4.3 with large and small outputs (microfocus tube), respectively. The energy range varies widely, with a tube voltage from 5 kV to 600 kV. The tube current also ranges over four orders of magnitude, from 0.15 mA to 4900 mA. Considering the output, this can range from around 4.5 W in devices for microfocus use to a maximum of 4200 W. The focal spot size can be as small as 250 nm for microfocus tubes. However, the output, in this case, is below 10 W.

Microfocus tubes come in two forms, namely of an open and sealed type. The sealed type, which has already been considered in this book, uses a reflective target like that in Fig. 4.23a and is generally used to obtain a microfocus spot and relatively large output. An additional advantage is that it is maintenance-free. Meanwhile, the open type typically has a transmissive target as in Fig. 4.23b and has the characteristics of a focal spot size smaller than the sealed-type and a high magnification due to an even shorter focal spot to sample distance can be obtained [27]. There have also recently been products where a transmissive target is used in a sealed-type system, as shown in Table 4.3. We can see that X-ray tubes with reflective targets are primarily used in commercially-available X-ray CT scanners in Table 6.1, shown later.

In Figs. 3.32 and 3.37, we discussed projections in fan-beam or cone-beam optical systems considering a distance D between the point light source and sample center of rotation. Attention must be paid to the fact that this D does not represent the distance measured from the window in Figs. 4.4 or 4.19 but rather the location of the focal spot inside the X-ray tube. As seen in Fig. 3.32, the image magnification

Table 4.3 Representative commercially available X-ray tubes and their specifications

Manufacturer	COMET							Feinfocus	
Type	MXR-601HP/11	MXR-451HP/11	MXR-320HP/11	MXR-225HP/11	MXR-160HP/11	MXR-100HP/20	MXR-75HP/20	FXE (Transmission)	FXE (Scanning)
Output (W)	700/1500	700/1500	800/1800	800/1800	800/1800	1000	1000	–10	–40
Tube voltage (kV)	600	450	320	225	160	100	75	–225	–225
Tube current (mA)	3700/4100	4100/4200	4100/4200	4100	4100	3700	3700	0.001–1	0.001–1
Focal spot size (μm)	700/2000	400/1000	400/1000	400/1000	400/1000	1000	1000	0.3–	–10
Target	W	W	W	W	W	W	W	W	W
Target angle (°)	11	11	11	11	11	20	20	–	–
Irradiation angle (°)	40×30	40×30	40×30	40×30	40×30	40	40	170	150
Coolant	Oil	Oil	Oil	Water	Water	Water	Water	–	–
Weight (kg)	145	95	40	11	8	2.1	2.1	–	–

(continued)

Table 4.3 (continued)

Matsusada Precision

XTR-Be	XTFR	XF80-80	XRC50-50	XPR600-30	XPR1200-30	XPR600-60	XPR1200-60	XPR600-100	XPR1000-100	XPR640-160	XM6	XM10	XF40-130	XF8-90	XF18-90
−100	−60	−80	50	600	1200	600	1200	600	1000	640	6	10	−40	−8	−18
5–50	6–60	20–80	5–50	20–30	20–30	20–60	20–60	20–100	20–100	20–160	10–40	10–60	40–130	30–90	30–90
2	1	2	1	20	40	10	20	6	10	4	0.03–0.15	0.033–0.166	0–0.3	0–0.088	0–0.2
500[a]	100[a]	15–25[a]	70[a]	3000[a]	3000[a]	3000[a]	3000[a]	3000[a]	3000[a]	3000[a]	50–1000[a]	50–1000[a]	4[a]	3[a]	4[a]
W	W	W	W	W	W	W	W	W	W	W	W	W	W	W	W
–	20	–	–	–	–	–	–	–	–	–	19	19	–	–	–
50	20	33	25	40	40	40	40	40	40	40	28	20	30	30	30
Oil	Oil	Air	Oil	Water	Water	Water	Water	Water	Water	Water	Air	Air	Air	Air	Air
6	6	3.2	3	3.5	3.5	3.5	3.5	3.5	3.5	3.5	5	5	6	6	6

(continued)

Table 4.3 (continued)

Hamamatsu Photonics											X-ray WorX				
L9421-02	L10101	L10321	L9631	L9181-02	L9181-05	L12161-07	L12531	L11091	L10711-03	L10801	XWT-160CT	XWT-190CT	XWT-225CT	XWT-240CT	XWT-300CT
8	20	20	50	39	39	75	16	8	8	200	350	350	350	350	350
20–90	40–100	40–100	40–110	40–130	40–130	40–150	40–110	20–160	20–160	20–230	20–160	20–190	20–225	20–240	50–300
10–200	10–200	10–200	10–800	10–300	10–300	10–500	10–200	5–200	5–200	10–1000	50–3000	50–3000	50–3000	50–3000	50–2000
5–	5–	5–	15–	5–	15–	5–	2–[a]	1–[a]	0.25–[a]	4–[a]	2000	2000	2000	2000	3000
W	W	W	W	W	W	W	W	W	W	W	W	W	W	W	W
–	–	–	–	–		–	–	–	–	–	–	–	–	–	–
39	42	118	62	45	100	43	120	120	140	40	30	30	30	30	30
空気	Air	Air	Air	Air	Air	Air	Air	Air	Water	Water	Water	Water	Water	Water	Water
10	10	10	12	11	11	20	19	62	72	98	36	36	50	50	69

[a] Spatial resolution of a projection image measured using a resolution evaluation chart

Fig. 4.21 Example of a commercially-available X-ray tube, COMET MXR-601HP in Table 4.3 (courtesy of Mr. Matsuda from Comcraft, Co.)

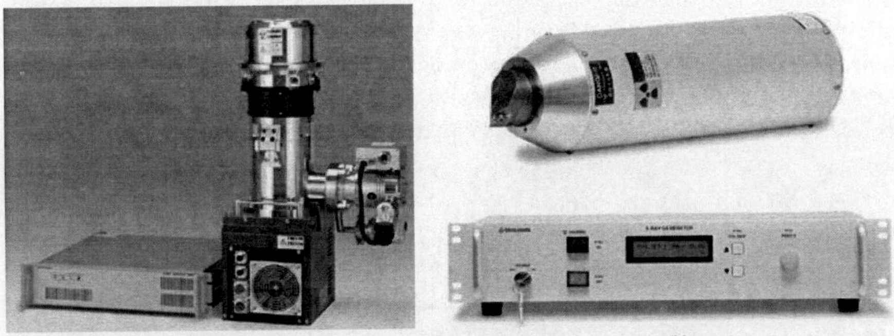

(a) (b)

Fig. 4.22 Example of a commercially-available microfocus X-ray tube from Table 4.3; **a** Hamamatsu Photonics L10711-03 (courtesy of Hamamatsu Photonics) and **b** Matsusada Precision XF40-130 (courtesy of Toyotoshi Ueyanagi from Matsusada Precision, Inc.)

M is expressed with the following equation when the distance D_0 from the point light source to the detector is considered. Making D smaller is important when it is necessary to conduct projections with a larger magnification and obtaining superior spatial resolution.

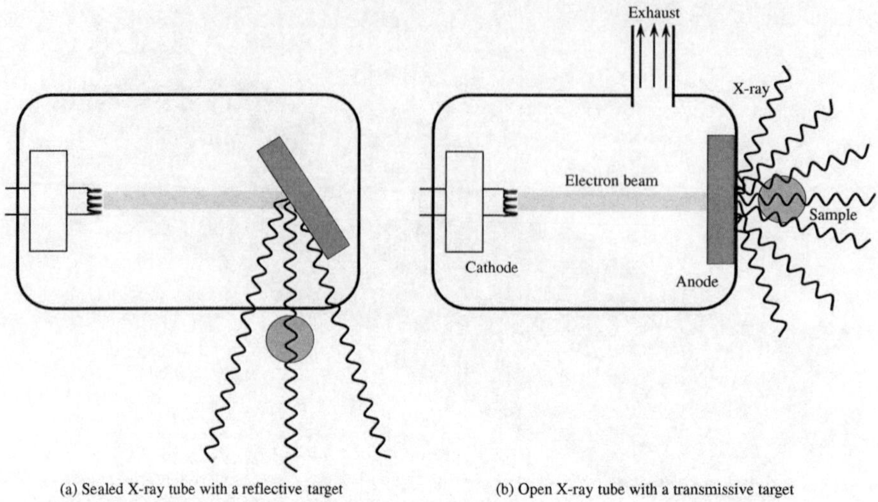

Fig. 4.23 Schematic of two microfocus tube systems

$$M = \frac{D + D_0}{D} \tag{4.16}$$

Incidentally, the minimum distance between the sample and focal spot is stipulated by the distance between the window and focal spot. Open-type microfocus X-ray tubes shown in Fig. 4.24 that use a transmissive target, described in Table 4.3, have a minimum distance of around 0.5 mm, whereas sealed-type X-ray tubes, which use a reflective target, have a relatively increased distance of between 7 to several dozen of mm. When pursuing the smallest focal spot size in open-type X-ray tubes, the electron beam should be narrowed using either a one or two-step (i.e. combination of

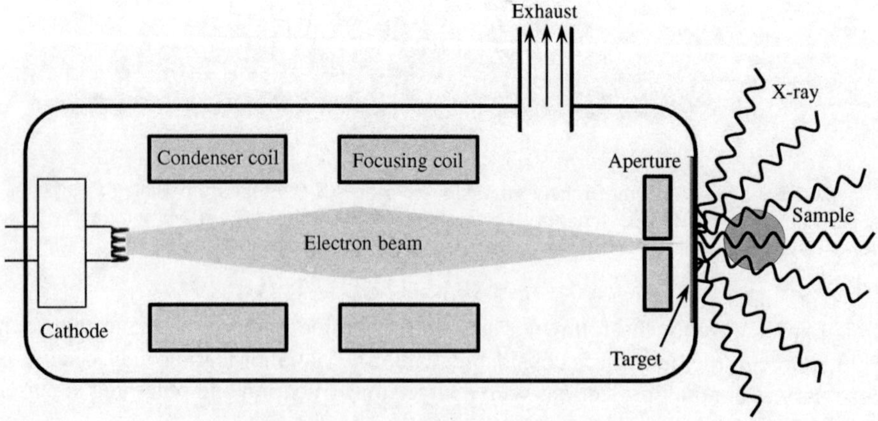

Fig. 4.24 Schematic of a microfocus tube with an extremely small effective focal spot

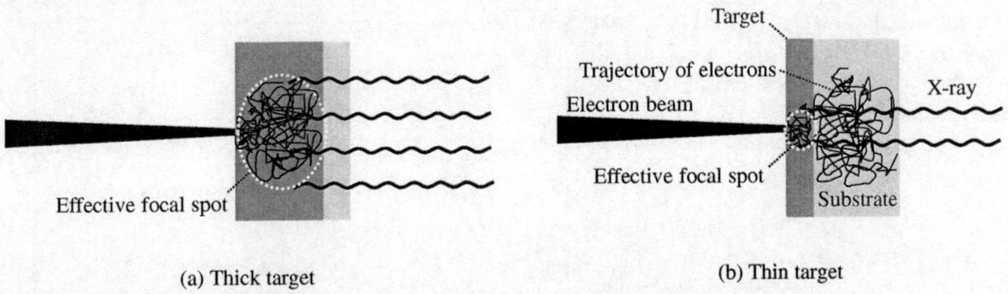

Fig. 4.25 Schematic of X-ray emission behavior from a transmissive target

condenser coil and *focusing coil*) electron beam optics system and *objective aperture*; this should be directed to the thin film of the target. A thin tungsten film coated on a substrate composed of light elements like beryllium is to be used as the target. The fact that electrons enter the interior of the target from its surface has already been discussed. As shown schematically in Fig. 4.25, the electrons spread wider due to scattering in the target and X-rays are generated from an area larger than the electron beam diameter. This is the effective focal spot size in this case. A higher tube voltage increases the scattering volume and the effective focal spot size also increases. The effective focal spot size can be reduced when a thin target is used, as in Fig. 4.25b. In contrast, a thicker target increases the intensity of X-rays generated in the target. However, the generated X-rays in transmissive targets do not come out if they do not transmit through the target interior. For this reason, thicker targets result in increased absorption by the target itself. Simulations conducted by Nasseri indicated that the maximum X-ray intensity, which considered absorption by the tungsten target itself, can be obtained when the following thickness t_{opt} (μm) is obtained [28]:

$$t_{opt} = -1.45 + 0.075E \tag{4.17}$$

Here, E (keV) is the X-ray energy; for example, the maximum transmissive target thickness value for a 30 keV X-ray is approximately 1 μm.

The *brilliance* of X-ray tubes can generally be compared by dividing the maximum output by the focal spot size (surface area). However, decreased effective focal spot sizes in microfocus X-ray tubes result in smaller X-ray outputs as well. Figure 4.26 shows the relationship between the focal spot size and maximum output of all X-ray tubes shown in Table 4.3. Although there is some variation at the smaller focal spot sizes, strong correlations are observed across four orders of magnitude for the focal spot size, and three orders of magnitude for the maximum output.

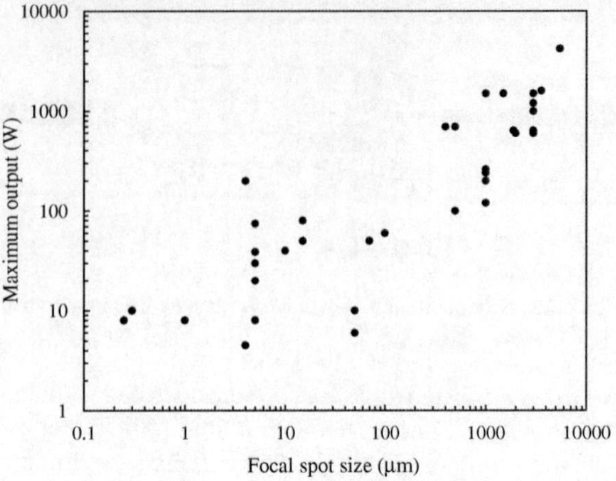

Fig. 4.26 Relationship between focal spot size and maximum output of commercially-available X-ray tubes; all products in Table 4.3 are plotted here

4.1.3 Compact Electron Accelerators

Figure 4.27 shows the thickness of iron and aluminum samples for 10% X-ray transmission calculated across a wide range of X-ray energies. In terms of image quality, this can be thought of as the maximum sample thickness under which the transmitted image can be somewhat obtained. As shown in Table 4.3, the tube voltage range of commercially-available X-ray tubes is around 5–600 kV. The maximum X-ray energies obtained in these X-ray tubes are approximately 5–600 keV and the average X-ray energies obtained from *bremsstrahlung* are thought to be about 1/2–1/3 of these values. The samples that can be imaged at the maximum X-ray energies of 200–300 keV, obtained with commercially-available X-ray tubes, are at thicknesses of at most 20–30 mm for solid iron and 70–80 mm for solid aluminum, according

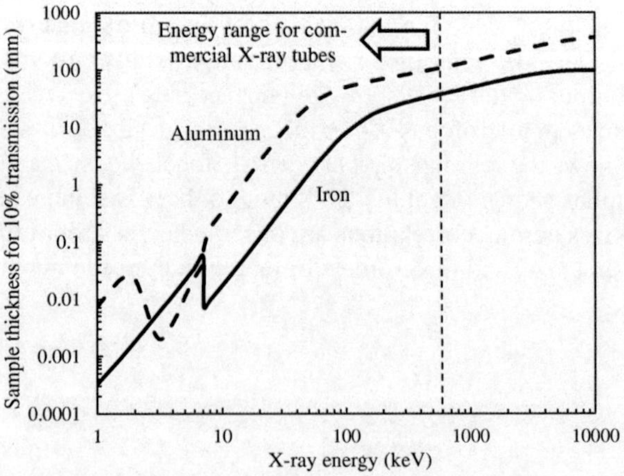

Fig. 4.27 Relationship between X-ray energy when 10% of the incident X-ray transmits through iron (solid line) and aluminum (dashed line) samples and sample thickness

to Fig. 4.27. Even higher-energy X-rays would need to be generated if imaging of thicker materials were to be conducted in the laboratory.

Van de Graaff generators, invented in 1920, and *betatrons*, invented in 1942, were used in the past to generate higher-energy X-rays. Today, we use *compact electron accelerators* and *radioisotopes* (RIs) [29]. Generally, X-ray energies at the 1-MeV level or higher are required, and RIs are used when flux is not a problem; these compact and low-cost characteristics are then effectively utilized. Meanwhile, electron accelerators are used when both X-ray energies at the 1-MeV level or higher and a sufficient flux are needed. Electron accelerators are a large-scale setup and their costs for introduction far exceed those of X-ray tubes or RIs. However, using these would enable the imaging of thicker and higher-density materials (e.g. iron with a thickness of several dozen cm); this also allows for imaging that is faster than with RI for materials with the same thickness. Electron accelerators have an additional characteristic in which there are no handling difficulties like those associated with RIs (i.e. the X-ray stops immediately when the power source is cut off). Electron accelerators first saw production in the 1950s primarily for military use [29]. Today, they are used not only for imaging of military products but for various civilian products as well. They are also used as electron injectors into synchrotrons.

Linear accelerators are referred to as *lineac* or *linac*. The former is more commonly used in Japanese, whereas the latter is more commonly used in English. As shown in Fig. 4.28, the electron linac comprises an *electron gun*, *buncher*, and *accelerating waveguide*. Figure 4.28 shows a schematic of a traveling-wave-type accelerating waveguide with a simple structure, primarily used for large-scale devices. Traveling-wave-type accelerating waveguides have disks with holes opened in them set up at equal intervals. The basic structure of the accelerating waveguide is the same for high-energy cases as well (e.g. for synchrotron injection use).

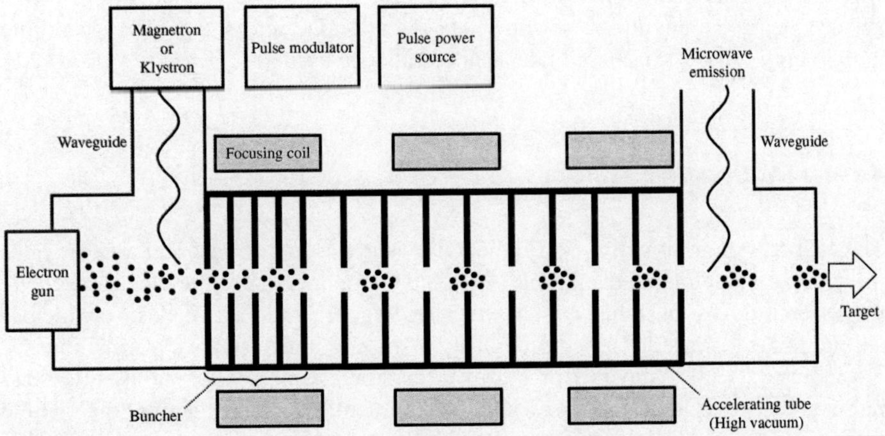

Fig. 4.28 Schematic of a compact electron accelerator with a progressive wave-type accelerating tube

First, the electrons generated with an electron gun are brought together in what is referred to as a bunch by the buncher. In a linear accelerator, an electron beam where the electrons are bunched together is obtained in the traveling direction of the electron, as well as the direction perpendicular to it. Generally, the pulse width is several μs and pulse repetition frequency is 50–500 Hz [29]. For this reason, the duty cycle (percentage of time where the pulse is on) is in the order of 0.1–0.01%.

Microwaves with a frequency of approximately 3 GHz (so-called S-band) to 9 GHz (so-called X-band) produced by a *magnetron* or *klystron* are guided through the accelerating waveguide and an electric field is generated in this cavity. The klystron is larger in scale relative to the magnetron and is more expensive, but its frequency and power control is more accurate and can be used with high-output accelerators [29]. When the microwave speed is set as equivalent to the electron speed, the electron advances under a constant drift potential within the accelerating waveguide, thereby accelerating it. Meanwhile, standing wave-type accelerating waveguides have the advantage of high acceleration efficiency per unit length, accelerating electrons over a shorter length than with progressive-wave-type accelerating waveguides. The subsequent mechanism in which the electron coming out of the accelerator collides with the target and generates X-rays is the same as that with X-ray tubes. Given the magnitude of heat inputs, the target is oftentimes water-cooled. The penetration depth of the electron into the target is approximately several hundred μm when the energy is 1 MeV and approximately several mm when the energy is 10 MeV [30]. For this reason, the target thickness would need to be 2 mm if, for example, a 9 MeV accelerator is used [30]. The range of obtained X-ray energies is around 1–15 MeV; of these, the 3–9-MeV level is widely used [29]. The focal spot size is typically approximately 2 mm and it is possible to have this below 1 mm as well [29].

A point of caution is that *neutrons*, protons, and π-mesons are also generated when the photonuclear reactions mentioned in Sect. 2.1.2 occur at the target and when X-ray energies over 6–7 MeV are used [29]. Appropriate shielding, which absorbs neutrons, should be set up in these cases. Detectors capable of handling high-energy X-rays (discussed later) should also be used.

4.1.4 Radioactive Isotopes

The half-life of commonly-used radioactive isotopes and the gamma-ray energies produced are summarized in Table 4.4 [31]. Gamma rays are the electromagnetic waves emitted by unstable nuclei when undergoing radioactive decay, which is similar to the X-rays discussed previously. Atoms in a metastable state in Table 4.4 are indicated with a superscript of "m" (for "metastable") after the atomic mass number. Atoms from Table 4.4, which are frequently used in imaging, include the following, in ascending order of gamma-ray energy: ^{241}Am (energy of 59.5 keV), ^{75}Se (270 keV), ^{192}Ir (310.5, 469.1 keV), ^{137}Cs (662 keV), and ^{60}Co (1173, 1332 keV). A characteristic of RIs is that they can be used to obtain either monochromatic or bichromatic gamma rays (gamma rays where energies are concentrated at two

Table 4.4 Half-life and gamma-ray energy of various radioactive isotopes [31]

Radioactive isotope	Half-life (s)	γ ray energy (MeV)
^{3}H	3.9×10^{8}	–
^{7}Be	4.6×10^{6}	0.48
^{10}Be	4.7×10^{13}	–
^{14}C	1.8×10^{11}	–
^{22}Na	8.2×10^{7}	1.28
^{24}Na	5.4×10^{4}	1.37
^{26}A1	2.3×10^{13}	1.84
^{32}Si	5.4×10^{9}	–
^{32}P	1.2×10^{6}	–
^{37}Ar	3.0×10^{6}	–
^{40}K	4.0×10^{16}	1.46
^{51}Cr	2.4×10^{6}	0.325
^{54}Mn	2.7×10^{7}	0.84
^{55}Fe	8.6×10^{7}	0.006
^{57}Co	2.4×10^{7}	0.122
^{60}Co	1.7×10^{8}	1.17 & 1.33
^{66} Ga	3.4×10^{4}	1.04
^{68}Ga	4.1×10^{3}	1.07
^{85}Kr	3.4×10^{8}	0.52
^{89}Sr	4.4×10^{6}	–
^{90}Sr	9.1×10^{5}	–
^{90}Y	2.3×10^{5}	–
^{99}mTc	2.2×10^{4}	0.14
^{106}Ru	3.2×10^{7}	–
^{106}Rh	3.0×10^{1}	0.51
^{112}Ag	1.1×10^{4}	0.62
^{109}Cd	4.0×10^{7}	–
^{109m}Ag	4.0×10^{1}	0.088
^{113}Sn	9.9×10^{6}	0.392
^{132}Te	2.8×10^{5}	0.23
^{125}1	5.2×10^{6}	0.035
^{129}I	5.0×10^{14}	0.038
^{131}I	7.0×10^{5}	0.36
^{133}Xe	4.5×10^{5}	0.08
^{134}Cs	6.5×10^{7}	0.61
^{137}Cs	9.5×10^{8}	0.66
^{137m}Ba	1.6×10^{2}	0.66

(continued)

Table 4.4 (continued)

Radioactive isotope	Half-life (s)	γ ray energy (MeV)
^{133}Ba	3.3×10^{8}	0.36
^{140}La	1.4×10^{5} s	1.60
^{144}Ce	2.5×10^{7}	0.13
^{144}Pr	1.1×10^{3}	0.69
^{144}Nd	7.3×10^{22}	–
^{152}Eu	4.3×10^{8}	0.122
^{192}Ir	6.4×10^{6}	0.32
^{198}Au	2.3×10^{5}	0.41
^{204}Tl	1.2×10^{8}	–
^{207}Bi	1.0×10^{9}	0.57
^{222}Rn	3.3×10^{5}	0.51
^{218}Po	1.9×10^{2}	–
^{214}Pb	1.6×10^{3}	0.35
^{214}Bi	1.2×10^{3}	0.61
^{226}Ra	5.0×10^{10}	0.19
^{228}Th	6.0×10^{7}	0.24
^{234}U	7.9×10^{12}	0.05
^{235}U	2.2×10^{16}	0.19
^{238}U	1.4×10^{17}	0.05
^{239}Pu	7.6×10^{11}	0.05
^{240}Pu	7.6×10^{11}	0.05
^{241}Am	1.4×10^{10}	0.06
^{252}Cf	8.2×10^{7}	0.04
^{252}Fm	9.0×10^{4}	0.096
^{268}Mt	7.0×10^{-2}	–

energy levels). The usage of monochromatic light does not result in beam hardening issues (discussed later), which occurs when using an X-ray tube or compact electron accelerator. For example, the low gamma-ray energy ^{241}Am is used for imaging of wooden materials [32]. Meanwhile, ^{137}Cs or ^{60}Co is used for large-scale containers made of iron [33, 34]. Reference [33] reports imaging of aluminum structures within a plexiglass container with a diameter of 440 mm using a 3.7 GBq (giga-becquerel, 10^9 becquerels; which are the quantity of radioactive decay of a radioactive isotope element for a unit period of time) ^{137}Cs. Reference [34] reports the imaging of a stainless-steel disk with a diameter of 60 mm using a 74 GBq ^{137}Cs. The obtained flux in each case is in the order of 1/100 compared to using X-ray tubes [35]. This results in long imaging times, as well as lower image quality.

RI-based imaging can also quantitatively determine local density or average atomic number from images utilizing gamma rays with two different energy levels from bichromatic sources, such as ^{192}Ir [36]. As seen in Sect. 2.1.2 (1) and (2), the extent of photoelectric absorption decreases with increased X-ray energy, whereas Compton scattering becomes more active with increased X-ray energy, up to a level of several MeV. Furthermore, photoelectric absorption has a large atomic number dependency, whereas Compton scattering does not depend on the atomic number. These trends can also be re-confirmed in Figs. 2.10, 2.11, 2.12 and 2.13. In this manner, quantitative assessments can be conducted with bichromatic gamma rays as two absorption/scattering processes with completely different energy and atomic number dependencies compete with one another.

4.1.5 Synchrotron Radiation

Synchrotron radiation refers to electromagnetic waves generated in the tangential direction of the synchrotron when the advancing direction of the electrons orbiting the synchrotron is changed using magnets. The advancing and accelerating directions of the electron are parallel in the case of *bremsstrahlung*, whereas both intersect in the case of synchrotron radiation. Synchrotron radiation has the advantages of a small radiation source size, high X-ray beam interference due to a long beamline, and high brilliance; additionally, its monochromatic capabilities indicate that the X-ray energy can be changed relatively freely. Furthermore, it has the characteristics of polarizability and pulsed light. A higher resolution, superior image quality, and S/N ratio can be obtained using these characteristics. In that sense, it would not be an exaggeration to say that synchrotron radiation X-ray tomography is the flagship of X-ray tomography technology.

The generation of synchrotron radiation was first experimentally confirmed by Langmuir et al. in 1947. The electron energy of the synchrotron at this time was 70 MeV. Afterward, first-generation radiation facilities, where accelerators subsequently created for high-energy physics research were diverted to conduct applied radiation research; second-generation radiation facilities from the 1970s onwards such as SOR-RING (1975) and the Photon Factory (1982), which used radiation generated from bending magnets; and third-generation radiation facilities like SPring-8, APS (1996, U.S.), and ESRF (1994, 19 countries of the EU), which used undulators and wigglers to generate high-brilliance radiation, were developed and used in order. Figure 4.29 shows synchrotron radiation obtained from SPring-8 compared with X-ray tubes [37]. X-ray brilliance is expressed in units of [photons/s/mm^2/mrad^2/0.1% bandwidth]; where rotating-anode X-ray tubes are approximately 10^9 in this unit, this is about 10^{12} in first-generation radiation facilities, 10^{14} in second-generation facilities, and up to 10^{19}–10^{20} in standard undulator light sources in SPring-8. It is evident that radiation facility performance has improved rapidly at each order of magnitude. Incidentally, fourth-generation radiation refers not to X-ray free-electron lasers (XFEL) based on linac principles, but radiation sources with lower emittance than third-generation radiation (e.g. MAX-IV, Sweden).

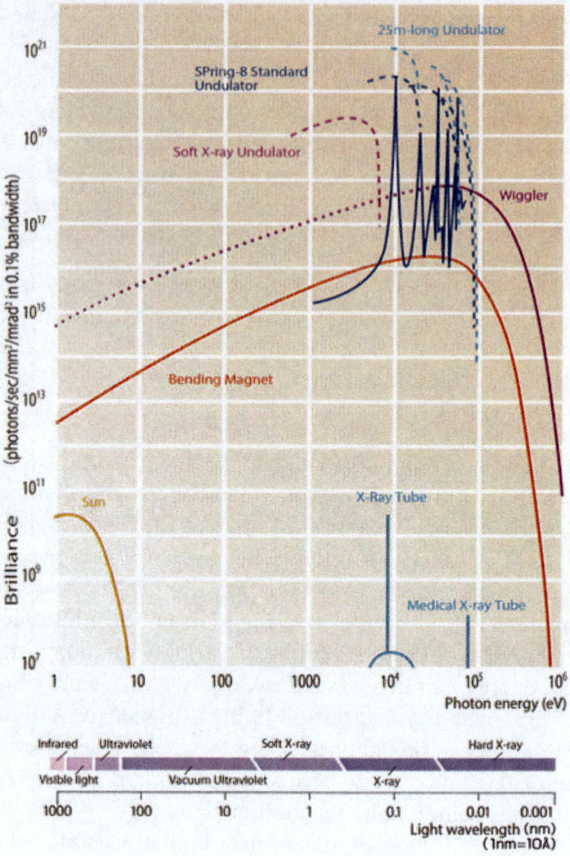

Fig. 4.29 Brilliance of radiation at SPring-8 (provided by Riken) [37]

Focusing on the wavelength in Fig. 4.29, it can be observed that synchrotron radiation covers a wide range from infrared to visible light to hard X-rays. Furthermore, synchrotron radiation is a row of pulsed light, where bunches lasting for 20–500 ps are repeated at 2 ns–1 μs -interval frequencies; for example, the length of a bunch with a width 100 ps is 3 cm.

Incidentally, the synchrotron itself cannot accelerate low-energy electrons. For this reason, electrons must be accelerated to an energy of at least several dozens of MeV before being injected into the synchrotron; in other words, the general structure of the synchrotron comprises sections that accelerate the electron beforehand (e.g. a linac) and *storage rings*, as shown schematically in Fig. 4.30. The University of Hyogo NewSUBARU adjacent to SPring-8 primarily uses electrons in the extreme ultraviolet ray to soft X-ray region, so they are directly injected from the linac to the storage ring as shown in Fig. 4.30a. Meanwhile, SPring-8 injects at the GeV energy level and combines the same electron linac as with NewSUBARU with an

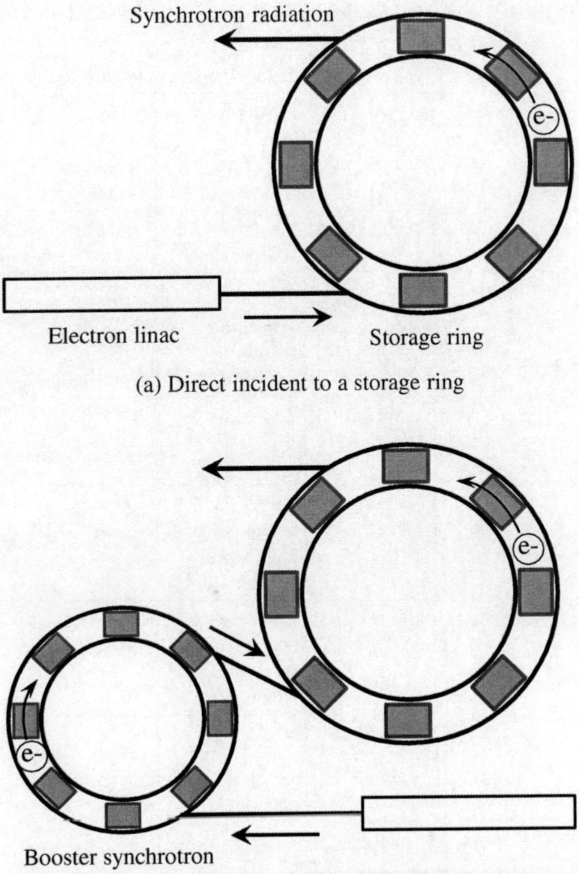

(a) Direct incident to a storage ring

(b) Booster synchrotron as an electron accelerator

Fig. 4.30 Schematic of a synchrotron structure

electron synchrotron, as shown in Fig. 4.30b. Some microtrons combine a high-frequency electric field for accelerating electrons and a DC magnetic field for creating a circular trajectory but these are omitted here as they are not widely used in Japan. The structures in Fig. 4.30 are discussed in order below.

(1) Electron Linac

The basic principles of an electron linac in a synchrotron radiation facility are the same as those of a compact electron accelerator introduced in Sect. 4.1.3; what varies is their size and energy. Here, we primarily take SPring-8 as an example and provide an overview with a focus on its specifications.

Table 4.5 shows an overview of the electron linacs and electron synchrotrons in the major synchrotron radiation facilities in Japan [37–40]. Furthermore, Fig. 4.31 shows a photograph of the external appearance of a SPring-8 electron linac [37]. The SPring-8 electron linac is 140 m in length and accelerates an electron beam

Table 4.5 Electron accelerators of representative synchrotron radiation facilities used for X-ray microtomography [37–40]

	Type	Linac + booster synchrotron		Linac	
	Facility	SPring-8	Aichi Synchrotron Radiation Center	Photon Factory (KEKB)	Kyushu Synchrotron Light Research Center
	Location	Sayo town, Hyogo, Japan	Seto city, Aichi, Japan	Tsukuba city, Ibaraki, Japan	Tosu city, Saga, Japan
Electron linac	Energy (GeV)	1	0.05	2.5	0.26
	Type	Travelling wave accelerator	Travelling wave accelerator	Travelling wave accelerator	Travelling wave accelerator
	Number of accelerating tubes	25	2	240	6
	Peak current (mA)	2000 (Single bunch mode) 350 (Multi bunch mode)	100	50	–
	Pulse width (ns)	2 (Single bunch mode) 40 (Multi bunch mode)	1	3	200
	Repetition rate (Hz)	60	1	25	1
	High frequency acceleration frequency (MHz)	2856	2856	2856	2856
	Length (m)	140	10	400	30
Booster synchrotron	Energy (GeV)	8	1.2	–	–
	Emittance (nmrad)	230	200	–	–
	Beam current (mA)	10	5	–	–
	Repetition rate (Hz)	1	1	–	–
	High frequency acceleration frequency (MHz)	508.6	499.654	–	–
	Perimeter (m)	396	48	–	–

(a) First stage of the linac (b) Whole linac

Fig. 4.31 Photograph of the electron linac in SPring-8 (provided by Riken) [37]

emitted from an electron gun operating with a voltage of approximately 180 kV up to 1.0 GeV [37]. The electron gun used is the thermionic type introduced in Sect. 4.1.3, it uses porous tungsten with impregnated barium as a filament material [37]. High-temperature activation following barium impregnation can reduce its work function to approximately 1.6 eV due to the presence of free barium; owing to this, electrons can be emitted at a relatively low temperature. The process of forming bunches of the electron beam using the buncher is the same as in Fig. 4.28. As in Sect. 4.1.3, a high-frequency electric field was used to accelerate the electron beam. The accelerating waveguide portion of the SPring-8 electron linac is composed of 25 individual accelerating waveguides, each measuring 3 m in length, all joined [37]. A single accelerating waveguide is composed of 81 accelerating cavities [41] and the electron acceleration from a single accelerating waveguide is approximately 40 MeV [41]. High-frequency power is applied from a klystron every two accelerating waveguides [41]. The high-energy (1.0 GeV) electron beam obtained from this process undergoes energy width reduction in the *energy compression system* section, after which it is supplied directly to the electron storage ring of the adjacent University of Hyogo NewSUBARU through the beam transfer system. For SPring-8, it is first injected into the booster synchrotron. The energy compression system is set up at the end of the linac and is comprised of a chicane electromagnet section, where the bunch length is extended according to the energy distribution of the electron beam and an accelerating waveguide section, where energy modulation is applied to compress the energy spread [42]. Energy stability also improves due to the energy compression system and improvements in large current injection and energy stability (rms of less than 0.01% for small current) are achieved [42].

Incidentally, for electron beam obtained at the synchrotron radiation facility, high electric current, high energy, and high quality are required. *Emittance* is the primary factor that stipulates these qualities. A major objective at the currently-planned SPring-II is also reducing the emittance of the storage ring (approximately 2.4 nm rad at present to approximately 150 pm rad). The emittance, which is used to show beam performance at the accelerator, primarily expresses the volume occupied by particles in the phase space. When considering an electron beam, the horizontal direction of emittance relative to the advancing direction of the beam is often expressed. Beams with low emittance are essentially parallel beams with a small size and whose directions are aligned. Here, the emittance of the electron beam emitted from the electron gun (thermal emittance) $\varepsilon_{\mathrm{th}}$ is as shown below:

$$\varepsilon_{th} = \sigma \sqrt{\frac{kT}{mc^2}} \tag{4.18}$$

Here, σ is the cathode spot size and c is the speed of light. The emission current from the cathode is determined from the work function and temperature, but the emittance is not dependent on the work function; moreover, if the temperature is equivalent, it is only dependent on the cathode spot size. As such, it is important to make the cathode radius as small as possible. Incidentally, as can be seen in Eq. (4.18), emittance usually has the dimensions of length. However, it can also be related to the beam inclination and expressed in units of mm·rad or multiplied by π and expressed in π mm·rad. Hence, care must be taken for its definition.

The emittance of electron beams in synchrotron radiation facilities can vary widely by the performance of the upstream electron gun. However, emittance worsens even during acceleration in a linac. This is because the electron is influenced by the electromagnetic field generated by itself.

(2) Electron Synchrotron

The electron synchrotron mentioned here is the so-called booster synchrotron, which is used to increase the energy of the electron beam injected into the storage ring. This enables top-up operations, where the electron is not accelerated in the storage ring. Synchrotron radiation with a fixed intensity can be constantly generated and makes large contributions to X-ray imaging. Circular accelerators that change the magnetic field with increases in acceleration energy and where the electron orbits a fixed trajectory are referred to as synchrotrons. *Bending magnets* are arranged in a circle, electrons are moved in a pseudo-circular manner, and they are accelerated up to high energy through high-frequency cavities (Fig. 4.32 [43]) placed in the electron trajectory. In the case of synchrotrons, the magnetic field is strengthened with increased energy and the electron is extracted once it reaches a given energy level. Much content overlaps with the "(3) Storage Ring" (discussed later). When compared to a storage ring, the electron synchrotron has a relatively small number of *quadrupole electromagnets* and *six-pole electromagnets* relative to bending magnets. This is due to the different usage applications of the synchrotron, such as not needing to store the electron beam [41]. Furthermore, this typically has a structure referred to as FODO, where bending magnets and quadrupole magnets are arranged in alternate order.

In the case of SPring-8, an electron beam accelerated up to 1 GeV with an electron linac is injected into the electron synchrotron at an angle of 15° using an on-axis injection method. This is accelerated up to 8 GeV and then outputted to the storage ring, which is repeated in 1 s intervals [43]. This can be broken down as follows: 1 GeV injection, 150 ms; electron acceleration, 400 ms; 8 GeV electron beam ejection, 150 ms; and deceleration period, 300 ms [43]. Electron synchrotrons have two klystron units set up in the klystron facility; high-frequency power for acceleration is supplied through a waveguide [41]. Progressively higher energies can be obtained if the synchrotron trajectory radius is increased and the magnetic field is strengthened. Electron beams accelerated by a synchrotron are transferred from a junction, such as in Fig. 4.32b, to a storage ring using a beam transfer setup as in Fig. 4.33. The synchrotron building, which houses the electron synchrotron, is placed in a position approximately 10 m below that of the storage ring, enters the interior of the storage ring by going underground as seen in Fig. 4.33, and is connected to the storage ring from the inside.

(3) Storage Ring

Figure 4.34 shows the storage ring size and storage electron energy of the major synchrotron radiation facilities in Japan and overseas. There are a number of synchrotron radiation facilities of various scales throughout the world, with storage rings up to 1 km or more in perimeter and storage electron energies ranging from around 0.5 GeV to 8 GeV in SPring-8. In Japan, facilities from Kyushu and the Kanto region, particularly the relatively low-energy facilities and SPring-8, which has the world's largest storage electron energy, serve as opposite ends, with a total

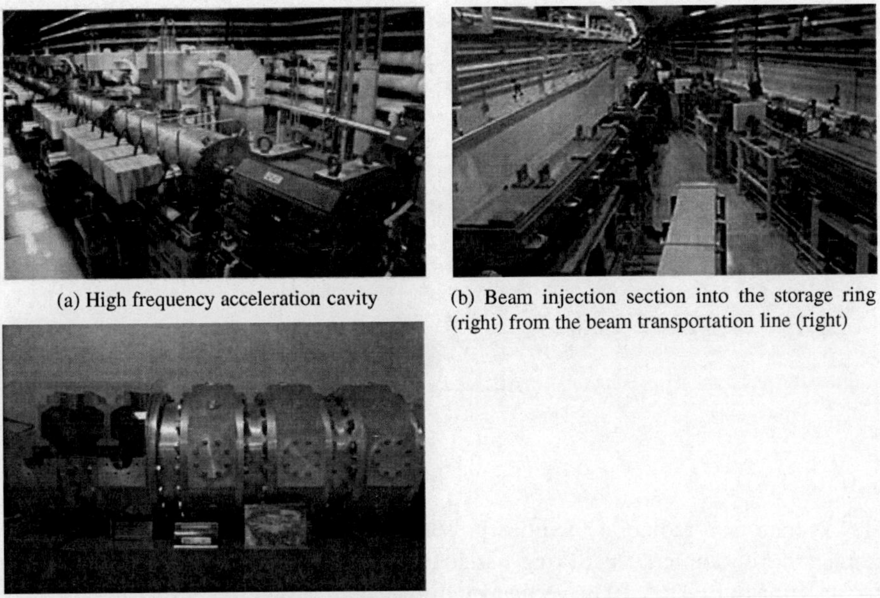

(a) High frequency acceleration cavity

(b) Beam injection section into the storage ring (right) from the beam transportation line (right)

(c) 5-cell acceleration cavity for synchrotron

Fig. 4.32 Photograph of electron synchrotron in SPring-8 [37, 43] (**a**: courtesy of Hiroto Yonehara, **b** provided by Riken, and **c** is a photograph of a SPring-8 display)

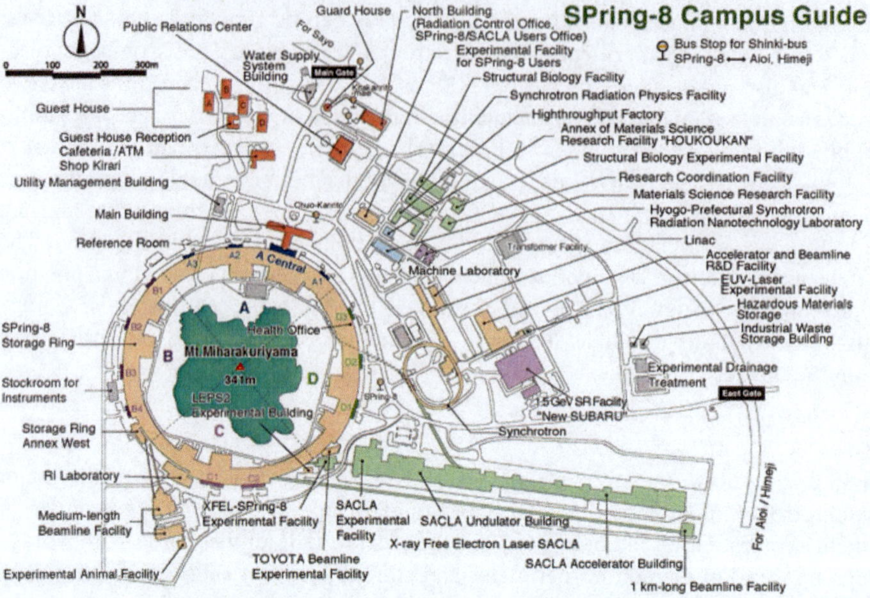

Fig. 4.33 Entire structure of SPring-8 (provided by Riken) [37]

Fig. 4.34 Storage ring size (perimeter) and storage electron energy of various international synchrotron radiation facilities

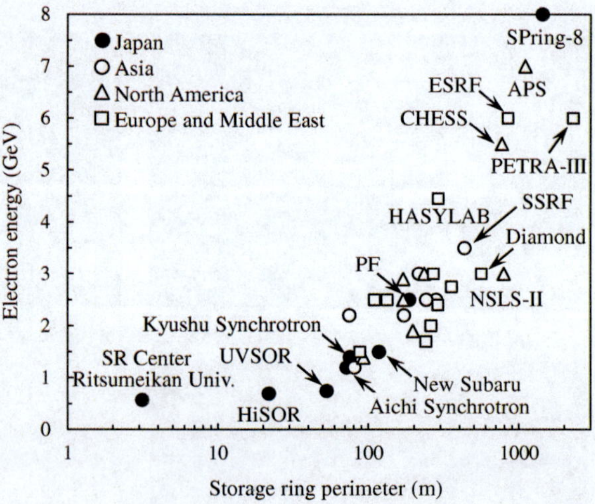

of 9 synchrotron radiation facilities present in the country. Considering only the storage ring circumference, SPring-8 is no longer the world's largest facility after the construction of PETRA-III in Germany in July 2009. Considering global trends, there have been a number of smaller 3 GeV-class, high-brilliance/moderate-scale radiation facilities, with emittance values lower than 4 nm/rad, built since 2000. These include

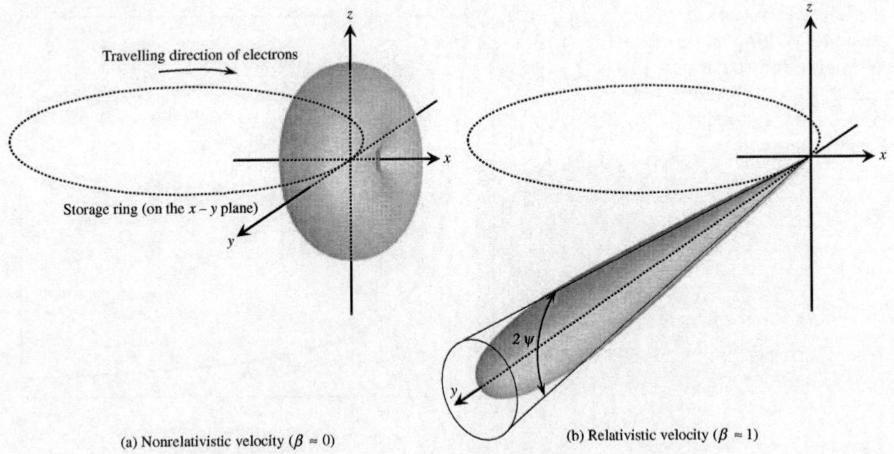

Fig. 4.35 Schematic showing the influence of electron acceleration in the horizontal direction (x-direction) on patterns of synchrotron radiation due to the bending magnet

SOLEIL in France, Diamond in the U.K., SSRF in China/Shanghai, and ALBA in Spain. It is anticipated that there will be more high-brilliance/moderate-scale radiation facilities with an emittance of 1 nm/rad or lower in the future.

As shown in Fig. 4.35a, electric *dipole radiation* is generated when applying acceleration by using bending magnets perpendicular to the direction of travel of the electron. In this case, acceleration is in the y-direction of the figure. Figure 4.35a shows the case in which β $(= v/c)$, where the electron drift speed v is normalized by the speed of light c, is sufficiently small. The radiation, in this case, is 0 in the direction in which acceleration is applied and the opposite direction, and maximum in the plane perpendicular to those directions. Meanwhile, when looking at synchrotron radiation emitted from electrons at relativistic speeds ($\beta \approx 1$) from laboratory systems, radiation becomes concentrated in a narrow area referred to as the radiation cone as shown in Fig. 4.35b. When the diverging angle, in this case, is set as 2 ψ, the following equation is obtained:

$$\psi \approx \frac{mc^2}{E_e} \tag{4.19}$$

Here, E_e is the storage electron energy of the storage ring. Incidentally, $\beta =$ 0.999999998 in the case of SPring-8 and 0.999999869 when storage electron energy is 1 GeV, effectively equal to the speed of light. If calculated using the range of E_e currently capable in synchrotron radiation facilities, the synchrotron radiation divergence is in the order of [mrad], as shown in Fig. 4.36. Based on this, synchrotron radiation can be said to have high directionality, wherein extremely high fluxes are sharply concentrated in a small area. In the case of bending magnets, electrons orbit a trajectory while generating synchrotron radiation in the direction tangent to it,

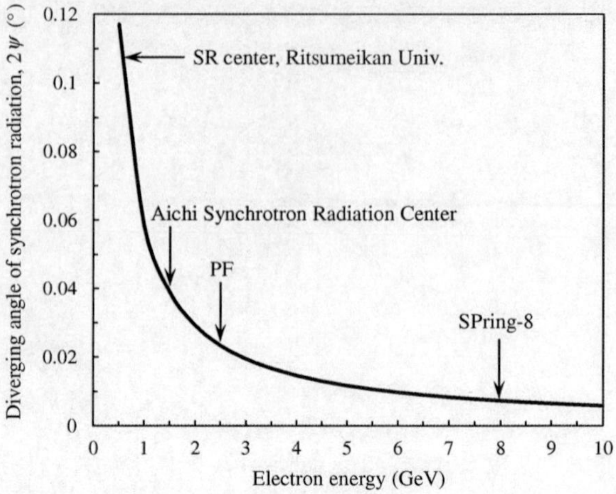

Fig. 4.36 Relationship between diverging angle of the synchrotron radiation emitted from the bending magnet and the storage electron energy of the storage ring

so synchrotron radiation diverges in the horizontal plane that includes the storage ring. This process is shown in Fig. 4.37. The divergence angle φ_B is dependent on the length of the bending magnet and is equal to 4.04° in the case of the bending magnets in SPring-8 [37]. Meanwhile, the *collimation* set by Eq. (4.19) is mostly

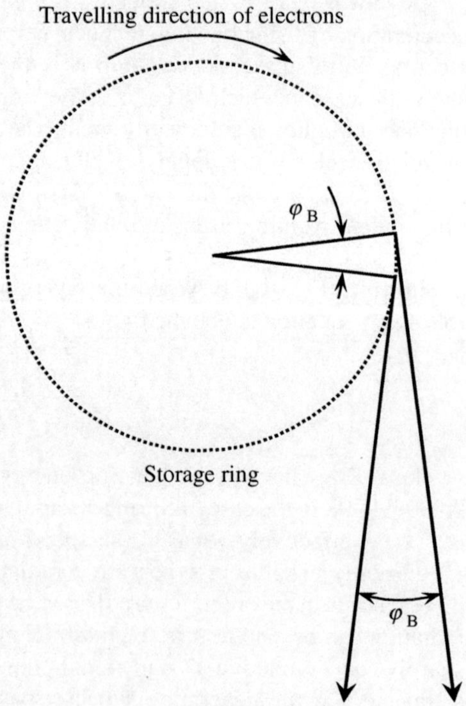

Fig. 4.37 Schematic showing the electromagnetic wave superposition and divergence in the horizontal direction of the synchrotron radiation due to the bending magnet

maintained in the direction perpendicular to the electron trajectory. The diverging angle of synchrotron radiation is 0.0036° for a 28.9 keV energy X-ray in the case of bending magnets in SPring-8 [37].

Incidentally, integrating the instantaneous *radiation power* P_e of synchrotron radiation emitted by a single electron orbiting a circular trajectory across all angles and wavelengths results in the following, where the radius of trajectory curvature (not the radius of the storage ring) is set as R [44].

$$P_e = \frac{e^2 c E_e^4}{6\pi \varepsilon_0 \left(mc^2\right)^4 R^2} \tag{4.20}$$

The total power of the synchrotron radiation is strongly dependent on the storage electron energy. Furthermore, integrating the amount of energy lost by a single electron as it circles a storage ring a single time (energy loss per turn) for the entire turn around the storage ring results in the following equation [44]:

$$\Delta E_e = \frac{e^2 E_e^4}{3\varepsilon_0 \left(mc^2\right)^4 R} \tag{4.21}$$

Table 4.6 shows the calculated ΔE_e values for the storage rings of major synchrotron radiation facilities. When E_e and R are expressed in units of [GeV] and [m], ΔE_e [keV] can be calculated as $\Delta E_e = 88.5\ E_e^4/R$. The storage electron energy in large-scale radiation facilities is high, therefore, energy loss is also extremely significant. Conversely, this amount of energy must be supplemented during that turn. For this reason, cavity spaces for conducting the high-frequency acceleration

Table 4.6 Energy loss, total radiation power, and critical energy of storage rings in representative synchrotron radiation facilities that can conduct X-ray microtomography, calculated by Eqs. (4.21), (4.22), and (4.24) [37–40]

	E_e (GeV)	R (m)	I_{sc} (mA)	ΔE_e (keV)	P_{SR} (kW)	E_c (keV)
SPring-8	8.0	39.27	100	9230	923	28.9
Aichi Synchrotron Radiation Center	1.2	2.7	300	68.0	2.04	1.4
Photon Factory	2.5	8.66	300	399	120	4.0
Kyushu Synchrotron Light Research Center	1.4	3.2	300	106	31.8	1.9
New Subaru	1.5	3.2168	300	139	41.7	2.3
ESRF	6.0	23.4	200	4900	980	20.5
APS	7.0	38.9611	200	5450	1090	19.6
ALS	1.5	4.89	400	91.6	36.6	1.5
PETRA-III	6.0	22.92	100	4660	351	20.907

of the electron and bending magnets, which curve the electron trajectory, are placed alternative in the storage ring.

In actual practice, the average stored current value (also shown in Table 4.6) should be set as I_{sc}, the electron number should be multiplied by Eq. (4.21), and this should be understood as the *total radiation power* P_{SR} from the storage ring [44].

$$P_{SR} = \Delta E_e \frac{I_{sc}}{e} \tag{4.22}$$

The P_{SR} values of representative synchrotron radiation facilities are summarized in Table 4.6. Third-generation large-scale radiation facilities like SPring-8 and ESRF are known to have a total radiation power of approximately 1 MW. Meanwhile, using Eq. (4.3) and calculating the radiation power from the various X-ray tubes from Table 4.3 shows that those from synchrotron radiation facilities have values higher by up to 8 orders of magnitude.

The *flux density* of synchrotron radiation can be determined from the following equation [45]:

$$\frac{d^3N}{dtd\Omega d\lambda/\lambda} = 3.46 \times 10^3 \gamma^2 \left(\frac{\lambda_c}{\lambda}\right)^2 \{1 + (\gamma\Psi)^2\} \left\{ K_{2/3}^2(\xi) \frac{+(\gamma\Psi)^2}{1 + (\gamma\Psi)^2} K_{1/3}^2(\xi) \right\} \tag{4.23}$$

Here, N is the photon number, λ is the wavelength of the X-ray, λ_c is the critical wavelength of the X-ray, Ω is the solid angle of radiation from the light source, $\frac{1}{\gamma} = \sqrt{1 - \beta^2}$, and $\xi = \lambda_c\{1 + (\gamma\Psi)^2\}^{2/3}$. $K_{2/3}$ and $K_{1/3}$ are the modified Bessel functions; there is a useful graph regarding these functions in Wiedemann [46]. The first and second terms within the second set of parentheses on the right edge in Eq. (4.23) express the polarized components parallel and perpendicular to the trajectory plane, respectively. Furthermore, λ_c is obtained from the following equation and defined so that the total radiation power is divided into two at λ_c [45]. The critical frequency ω_c is expressed as follows.

$$\omega_c = \frac{3}{2} \frac{\gamma^3 c}{R} \tag{4.24}$$

It is clear from Eq. (4.23) that synchrotron radiation is a continuous spectrum across a wide range of wavelengths; in other words, X-rays obtained from bending magnets are white if they do not undergo spectral diffraction with a monochromator. This is depicted as a graph in Fig. 4.38, which shows that electromagnetic waves across a wide wavelength range from infrared to hard X-rays are emitted. The position of critical energy E_c, corresponding to the critical wavelength, is also illustrated in the figure. Furthermore, the critical energy values in representative synchrotron radiation facilities are shown in Table 4.6. When using X-rays, a higher storage electron energy has minor influence on the low-energy side, whereas the energy range in the high-energy side greatly increases. It is clear that large-scale synchrotron

Fig. 4.38 Schematic comparing the flux density and critical energy of synchrotron radiation emitted from bending magnets in storage rings at SPring-8 and the Photon Factory

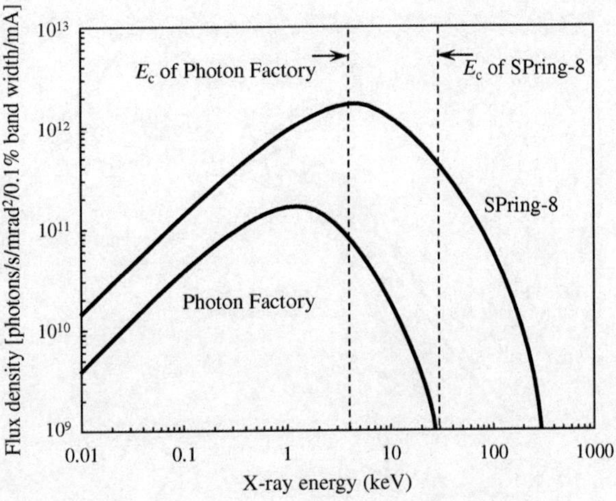

radiation facilities with high storage electron energies at a SPring-8 level are required when considering X-ray imaging of metallic materials.

Incidentally, the units of Eq. (4.23) are [photons/s/mrad^2/0.1% band width/mA], which are the number of photons emitted for a unit period of time, unit relative bandwidth, divergence angle in both the horizontal and vertical directions, and a storage current of 1 mA. Meanwhile, integrating by angle yields the *flux*. Furthermore, brilliance is dividing the flux density by the effective light source size in both the horizontal and vertical directions, and converting it to a unit surface area. The brilliance of synchrotron radiation is an important index that determines the success or failure of many experiments. Radiation facilities with low emittance are advantageous for obtaining high brilliance. Furthermore, the maximum flux value $F_{\max}$ is obtained when $\lambda = 3.4\,\lambda_c$ and can be approximated using the following equation [47]:

$$F_{max} \approx 1.29 \times 10^7 \gamma \tag{4.25}$$

Next, Fig. 4.39 shows a schematic of a basic storage ring structure. The role of the storage ring is to store electrons on a pseudo-circular trajectory for long periods while retaining a fixed amount of energy. Multiple bending magnets are set up on the trajectory to achieve this. As there is no need to increase the electron energy in storage rings, direct current is used in bending magnets. With the example of SPring-8, a total of 88 bending magnets with 0.68 T (Tesla) are set up (Fig. 4.40a) and the radius of curvature due to the bending magnet is set as 39.27 m [37]. These bending magnets are dipoles, however, as in the schematic in Fig. 4.39, there are also quadrupole (Fig. 4.40b) and six-pole (Fig. 4.40c) magnets set up on the trajectory; the former is for the convergence of the electron beam and the latter for correcting the deviations in the closed orbit or focal spot location. Furthermore, there are forty 7-m linear sections and four 30-m linear sections set up for a single orbit in SPring-8 [37]. High-frequency accelerator cavities for supplementing the radiation loss shown

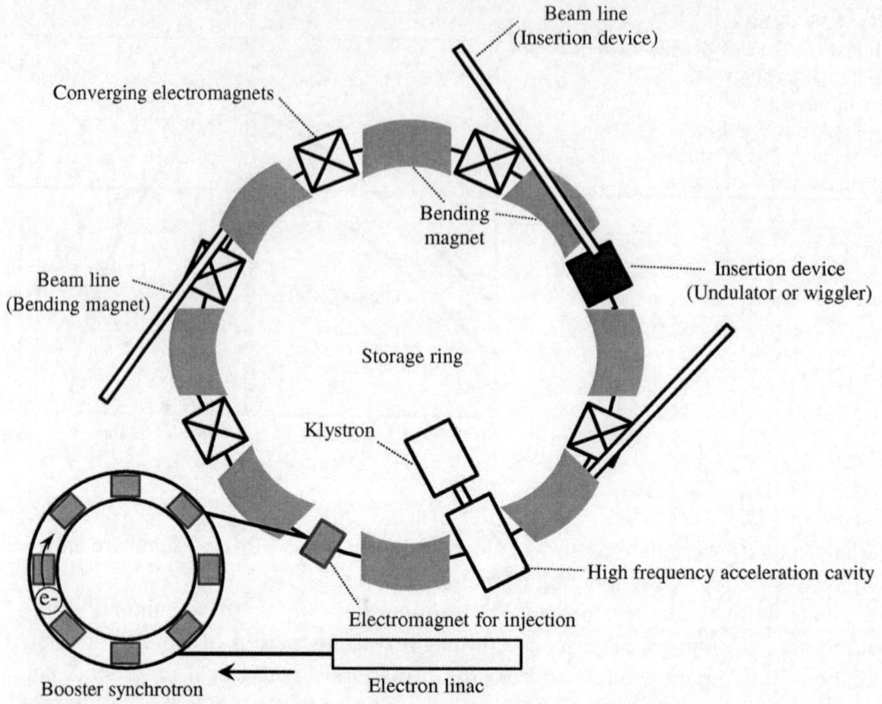

Fig. 4.39 Simplified schematic showing a basic structure of a storage ring in a synchrotron radiation facility

in Eq. (4.21) are connected to klystrons through waveguides and set up in the linear section. This basic principle is the same as a linac. Furthermore, the insertion device discussed in (4) is also set up in the linear section. The electron beam itself propagates through the vacuum chamber with an elliptical cross-section, as shown in Fig. 4.40d. The origin of the elliptical shape was made to resemble the cross-sectional shape of the electron beam. Furthermore, an extremely high vacuum of 10^{-8}–10^{-6} Pa is maintained in the chamber interior in the case of SPring-8 [37]. The vacuum chamber is made of an aluminum alloy extrusion that is non-magnetic and has favorable vacuum characteristics, as well as limited radioactivation.

Incidentally, top-up operations have been conducted at SPring-8 since 2004. Typically, the storage current gradually decreases due to the scattering by residual gas in the vacuum chamber or scattering among electrons within bunches [48]. To address this, top-up operations involve systematically yet frequently repeating electron beam injections from the injector to macroscopically maintain the storage current. As shown in Fig. 4.41, past methods have involved increasing the storage current value to around 100 mA through electron beam injections from the electron synchrotron twice a day and subsequently decreasing it to 75 mA over the course of half a day [37]. In response, image quality obtained with X-ray imaging varied considerably

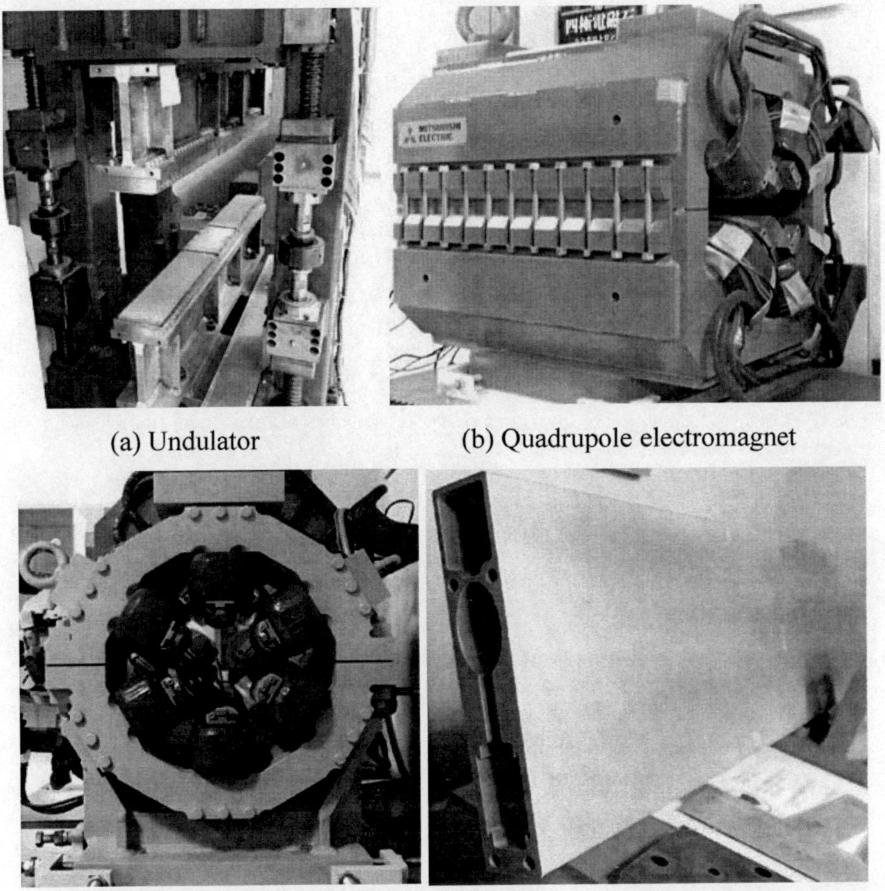

(a) Undulator (b) Quadrupole electromagnet

(c) Sextupole electromagnet (d) Vacuum chamber of the storage ring

Fig. 4.40 Constituent machines of the storage ring in SPring-8; photographs of SPring-8 displays are shown

before and after injection; currently, 1 to 3 injections every 1 to 5 min for a 100-mA storage current around 100 mA [37] control variations in the storage current to less than 0.1% [48]. For this reason, it can be guaranteed that there will be no changes in image quality or spatial resolution due to variations in storage current across the course of a single day of beam time when conducting X-ray imaging. Currently, top-up operations are a global standard used in the operation of high-brilliance light sources [48].

(4) Insertion Device

An *insertion device* (ID) is a device that is inserted in the linear portion between bending magnets to obtain higher quality synchrotron radiation than that from

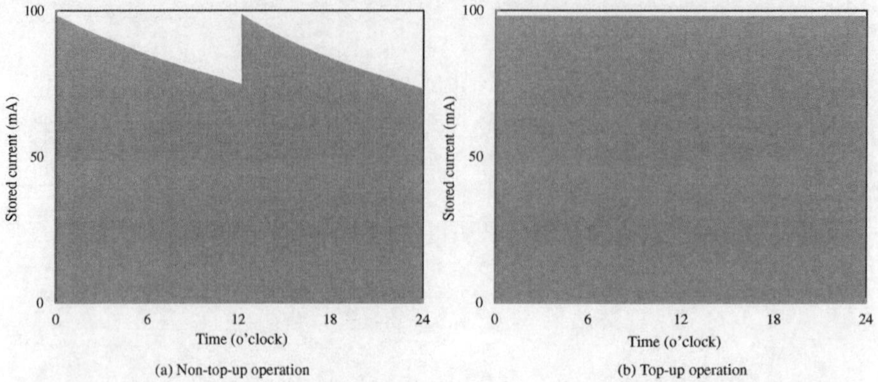

Fig. 4.41 Schematic showing the storage current value changes of the storage ring when top-up operations are and are not conducted [37]

bending magnets in terms of brilliance or directionality. There are two types of insertion devices: an *undulator* and a *wiggler*. Third-generation radiation facilities like the SPring-8 often use undulators in particular. For beamlines in SPring-8, such as BL20B2, the "B" that comes after the beamline number refers to the bending magnet, the "XU" in beamlines such as BL20XU refers to undulators, and the "W" in beamlines such as BL08W refers to wigglers. An undulator displayed in SPring-8 is shown as an example of an insertion device in Fig. 4.42. A schematic of an insertion device is also shown in Fig. 4.43. A periodic magnetic field as in Fig. 4.43 can be created by alternating the N- and S-poles of multiple permanent magnets as well as placing them above and below the electron trajectory. Electrons flying through this environment will follow a meandering path. The radiation generated at each

(a) Inside of an undulator (b) Magnified view of a magnet

Fig. 4.42 Undulator of a storage ring in SPring-8 (courtesy of Yoshio Suzuki, formerly of SPring-8)

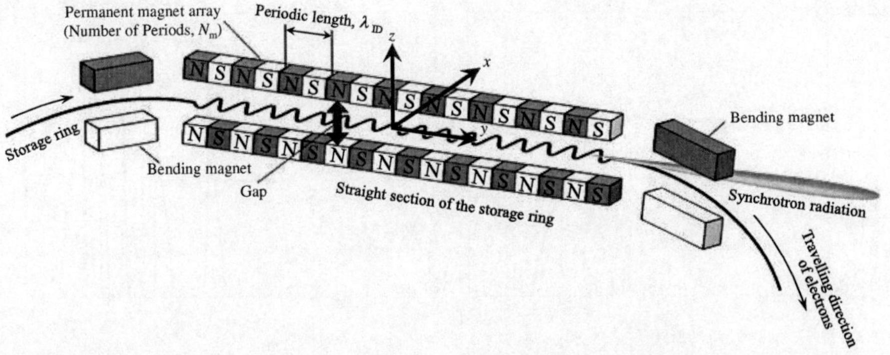

Fig. 4.43 Schematic showing radiation generation from an insertion device placed within a storage ring in a synchrotron radiation facility

magnet will be superimposed due to this, greatly increasing intensity and directionality. Changing the *gaps* of the magnets above and below the trajectory alters the magnetic field intensity as well as the energy of the synchrotron radiation. Permanent magnets are generally used to reduce the periodic length while creating a strong magnetic field. Neodymium magnets (10-mm gap, 0.78 T maximum magnetic field), where neodymium, iron, and boron are sintered together as the primary compositions, are used at SPring-8. Magnets need to be placed and coordinated to fulfill the following condition to ensure the electron beam does not deflect as a result of placing an insertion device:

$$\int B_z dy = 0 \tag{4.26}$$

The deflection angle of the electron trajectory from the y-axis in Fig. 4.43 is expressed using the constant K parameter, as shown below:

$$\Phi_{max} = \frac{K}{\gamma} \tag{4.27}$$

Here, K is expressed as follows:

$$K = \frac{eB_0\lambda_{ID}}{2\pi mc} \tag{4.28}$$

λ_{ID} is the periodic length of the insertion device magnet (see Fig. 4.43 for reference) and B_0 is the maximum value of the magnetic field. General rules-of-thumb for undulators and wigglers are $K \leqq 1$ and $K \gg 1$, respectively. However, undulators with $K > 1$ are also used, therefore, in actual practice, the distinction between the two is based on the coherency of the obtained X-rays. Synchrotron radiation characteristics from undulators and wigglers are entirely different. We can see from Eq. (4.28)

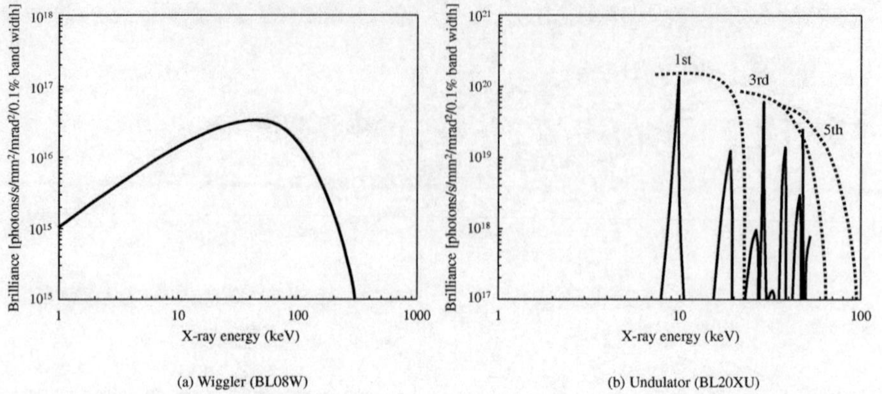

Fig. 4.44 Schematic comparing the performances of an undulator and wiggler in a SPring-8 storage ring [37]

that the $K \gg 1$ condition is satisfied when the magnetic field is relatively strong. The spectral shape is as shown in Fig. 4.44a in the case of the wiggler and is similar to the bending magnet in Fig. 4.38 [37]. However, bending magnets have higher brilliance and are shifted to the high-energy side. This is because with wigglers, there is insufficient interference from the radiation emitted from each magnet, and they are simply added based on photon numbers. For this reason, the brilliance becomes greater by a factor of $2N_m$ when $2N_m$ magnets are arranged. In contrast, this value is the sum of amplitudes as discussed later in undulators, so the intensity becomes proportional to $N_m{}^2$ under conditions where it is mutually strengthened through interference. Furthermore, the critical energy, in the case of the wiggler, becomes considerably higher than was the case with bending magnets. Meanwhile, the peak can be shifted to the low-energy side if the magnetic field is weakened. For this reason, the wiggler is also referred to as a *wavelength shifter*. Incidentally, the BL08W at SPring-8, which is a beamline with a wiggler installed, has values of periodic length λ_{ID} of 12 cm, periodicity N_m of 37, a minimum gap of 30 mm, and a maximum K-value of 9.89.

Meanwhile, the undulator magnet interval is shorter than a wiggler and its magnetic field is also weak. In the case of an undulator, the obtained synchrotron radiation becomes a *quasi-monochromatic beam* (*pink beam*), as shown in Fig. 4.44b [37]. The wavelength bandwidth, in this case, is shown in the following equation:

$$\frac{\Delta\lambda}{\lambda} = \frac{1}{nN_m} \tag{4.29}$$

For example, the BL20XU at SPring-8, which is a high-resolution imaging beamline with a vacuum-sealed undulator, has values of $\lambda_{ID} = 2.6$ cm, periodicity $N_m = 173$, a minimum gap of 7 mm, and a maximum K value of 2.12. Thus, $\Delta\lambda/\lambda$ is below 5.8×10^{-3}. However, this is an ideal scenario, where the electron energy, position, and direction are entirely uniform. Current storage rings have a limit of $\Delta\lambda/\lambda \lesssim 10^{-2}$ depending on the electron energy distribution. As in Eq. (4.29) and Fig. 4.44b, the bandwidth becomes even narrower with *harmonics* (X-rays that have an integral multiple of energy of the fundamental wave). Figure 4.44b has a spectrum in which the fundamental wave and multiple harmonics, which have higher energy (generally, higher-order waves at an odd order), are superimposed. The wavelength λ_n of the n-order harmonics is expressed with the following equation:

$$\lambda_n = \frac{\lambda_{ID}}{2n\gamma^2}\left(1 + \frac{K^2}{2} + \gamma^2\theta^2\right) \tag{4.30}$$

Here, θ is the angle between the y-axis (beam direction) and the observation axis of Fig. 4.43; the wavelength becomes shortest in the y-axis and that the wavelength increases with increased θ. Although the fundamental wave is strongest, higher-order harmonics also strengthen with $K > 1$. The undulator, in which $K > 1$ is referred to as a strong magnetic field undulator, promotes the generation of harmonics [49].

Incidentally, as λ_n is proportional to K^2, λ_n can be controlled by Eq. (4.28) through magnetic field adjustment. Undulators mainly use permanent magnets, so this refers to the adjustment of the gaps between the rows of overlying and underlying magnets. Permanent magnets are vacuum-sealed (degree of vacuum 3×10^{-8} Pa [37]) inside chambers at SPring-8 and the minimum gap can be reduced to 8 mm. For this reason, an X-ray energy range of 5–80 keV can be covered by using the 1st- through 5th-order waves of the undulator [37]. Meanwhile, there is no point in adjusting the gaps in wigglers that have a wide bandwidth.

In the case of the undulator, the brilliance becomes larger by a factor of N_{m}^2 when $2N_{\mathrm{m}}$ magnets are arranged. This is because the radiation generated by each magnet strengthens each other through interference; in other words, radiation that is brighter than bending magnets or wigglers by orders of magnitude can be generated. Furthermore, the directionality of the obtained synchrotron radiation also becomes significantly higher. The diverging angle φ_u of the synchrotron radiation is as expressed below:

$$\varphi_u \approx \sqrt{\frac{\lambda}{N_m\lambda_{ID}}} = \frac{1}{\gamma}\sqrt{\frac{1 + K^2/2}{2nN_m}} \tag{4.31}$$

$N_m\,\lambda_{ID}$ is the length of the row of magnets; for example, φ_{u} is approximately 3.7 μ rad for the leftmost peak in Fig. 4.44b. Furthermore, the diverging angle becomes approximately $1/\sqrt{N_m}$ compared to bending magnets when considering the fundamental waves of undulators. The reason why brilliance is high in undulators can be attributed to the perfectly mutually strengthened synchrotron radiation from each magnet due to interference in the optical axis only for a specific wavelength. However, the total radiation power even in undulators is approximately the same as that in bending magnets. In actuality, and as shown in Fig. 4.29, only X-rays that are weaker than those from wigglers or bending magnets can be observed at wavelengths removed from the undulator peaks.

Finally, if we assume that the spatial coherence discussed in Sect. 2.2.2 (3) can be approximated with a Gaussian function, the beam size σ of synchrotron radiation emitted from an undulator can be expressed as follows. However, the Gaussian function, in this case, cannot be considered a good approximation, so care must be taken to note that the following equation does not accurately reflect the effective beam size.

$$\sigma = \frac{\sqrt{\lambda N_m \lambda_{ID}}}{4\pi} \tag{4.32}$$

Meanwhile, the temporal coherence similarly discussed in Sect. 2.2.2 (3) is expressed as follows for an undulator:

$$l_t = \frac{\lambda^2}{2\Delta\lambda} = \frac{N_m \lambda_{ID}\left(1 + K^2/2\right)}{4\gamma^2} \tag{4.33}$$

In this manner, temporal coherence is improved by increasing the length of the row of magnets in the undulator. This effect is proportional to the length of the row of magnets. Spatial coherence is not dependent on undulator length in the case of a completely coherent light source.

(5) Beamline

Using SPring-8 as an example, there are a total of 57 beamlines (62 are installable), as shown in Fig. 4.45 [37]. Of these, the BL20XU, BL20B2, BL47XU, BL37XU, and BL19B2 are used for X-ray imaging. Beamlines are composed of a *front end* (*FE*), *optics hutch*, and *experimental hutch*. In SPring-8, the distance from the X-ray source to the entrance of the front end is 10–20 m and the distance from there to the front-end terminal is another 10–20 m, with a total distance from the X-ray source when considering the experimental hall terminal of approximately 80 m (Fig. 4.46a [37]). The beamlines BL20XU and BL20B2 extend out of the experimental hall, with a total extension of 250 m (Fig. 4.46b [37]). A detector and sample rotation stage (discussed later), are set up in the experimental hutch.

Synchrotron radiation generated at the storage ring is introduced in the front end, composed of a gate valve for vacuum shielding, *main beam shutter* (*MBS*), *slit*, and a *beryllium window* (refer to Fig. 4.47 [50]). The front end is maintained in an ultra-high vacuum with an ion pump and titanium getter pump. Furthermore, this is stored in what is referred to as a storage unit and general users are not permitted to enter. This storage unit is covered by a thick concrete shielding wall to ensure that gamma rays, neutrons, and excess synchrotron radiation generated at the storage ring does not enter the experimental hall. The function of the front end is to guide only high-quality synchrotron radiation near the optical axis (y-axis of Fig. 4.43) of the undulator, shielding synchrotron radiation. Unneeded synchrotron radiation, which is distant from the optical axis and removed by slits, is cut by XY slits and released

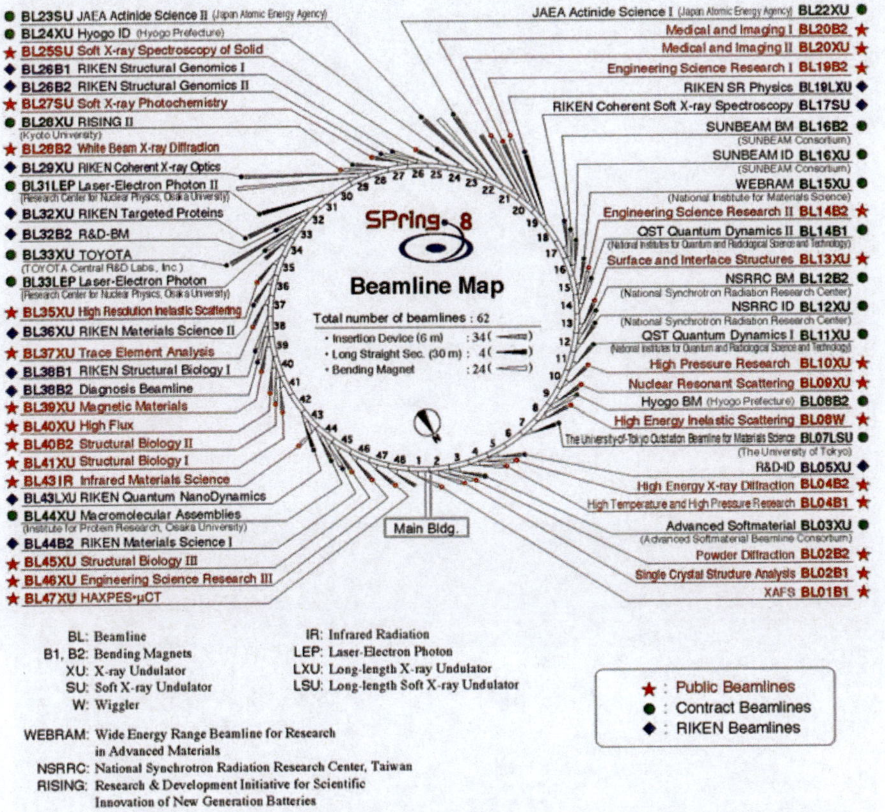

Fig. 4.45 Examples of beamlines in synchrotron radiation facilities (SPring-8 beamline map provided by Riken. Current as of April 1st, 2017) [37]

as heat. Therefore, it is important to ensure that the front end is properly treated for high heat loads; for example, the total power for BL20XU in SPring-8 would be 13.83 kW when the minimum gap is 7 mm [51]. Most of this enters the front end slit and the output is approximately 290 W when the slit opening is 0.5 mm (vertical) × 0.8 mm (horizontal) [51]. This difference of approximately 13.5 kW is processed as heat. The main beam shutter is the general term for the *absorber* that shields the heat component and the *beam shutter* that shields the radiation component; both work together during operation [52]. For example, closing the main beam shutter involves first closing the absorber and then automatically closing the beam shutter [52]. This prevents heat loads on the beam shutter, which is not cooled. In the case of SPring-8, a 38-cm-thick tungsten block is used for the beam shutter and a water-cooled copper block with high heat conduction is used for the absorber. Both are raised and lowered using a compressed-air cylinder [52].

Figure 4.48 shows the workstation display of an experimental hutch beamline where a front end is controlled. Some of the machinery for a front end can be easily

(a) Appearance of beam lines in the experimental hall

(b) Appearance of beam lines outside the experimental hall

Fig. 4.46 Photograph of beamlines in synchrotron radiation facilities (courtesy of Kentaro Uesugi from SPring-8 for BL20B2 at SPring-8) [37]

controlled by a general user from the experimental hutch. However, troubles with the front-end machinery can damage the monochromator of the beamline in use and can at times have serious effects on other beamline experiments through a beam abort; hence, sufficient care must be taken. Please refer to specialized texts for further detail on this [52].

The beryllium window is an important component with regard to the front end when the user conducts X-ray imaging. As shown in Fig. 4.49, the beryllium window shields both the ultra-high front end vacuum (10^{-8}–10^{-7} Pa) and the high optics hutch vacuum (10^{-6}–10^{-3} Pa), as well as enabling the transmission of only hard X-rays above approximately 3 keV [52]. The beryllium window of SPring-8 is comprised of a high-purity beryllium foil affixed on a water-cooled copper mask and a structure to withstand heat loads due to the white X-ray. For example, there are two beryllium

Fig. 4.47 Example of a beamline in a synchrotron radiation (courtesy of Yoshiharu Sakurai of SPring-8 for BL47XU at SPring-8) [50]

windows with thicknesses of 250 μm in the upstream side of the optics hutch of BL20XU of SPring-8. The downstream side of the optics hutch is shared with the experimental hutch and there are two PVD-based beryllium windows with a 60 μm thickness (two 30 μm -thick windows affixed together). The surface roughness of the beryllium foil, inclusion particles due to trace impurities, and material defects like pores all produce fringe patterns or intensity irregularities referred to as *speckle noise* in the image obtained with X-ray imaging, therefore, care must be taken.

As can be observed in Fig. 4.50 [37], the X-ray coming out of the front end is not immediately released into the atmosphere, instead, passing through a vacuum pipe where both ends are sealed with beryllium and guided into a *monochromator*. The monochromators for BL20XU and BL20B2 in SPring-8 are set up 46 and 36.8 m, respectively, from the beam source. There are two experimental hutches in BL20XU, with the first experimental hutch placed 80 m from the beam source and capable of using a beam size of 0.7 mm (vertical) × 1.4 mm (horizontal). The second experimental hutch is placed 245 m from the beam source and is capable of using a beam size of approximately 2 mm (vertical) × 4 mm (horizontal) [37]. A flux of 10^{13} [photons/s/mm^2] can be obtained when a Si (111) plane is used on the monochromator in the first experimental hutch [37]. Meanwhile, BL20B2 has three experimental

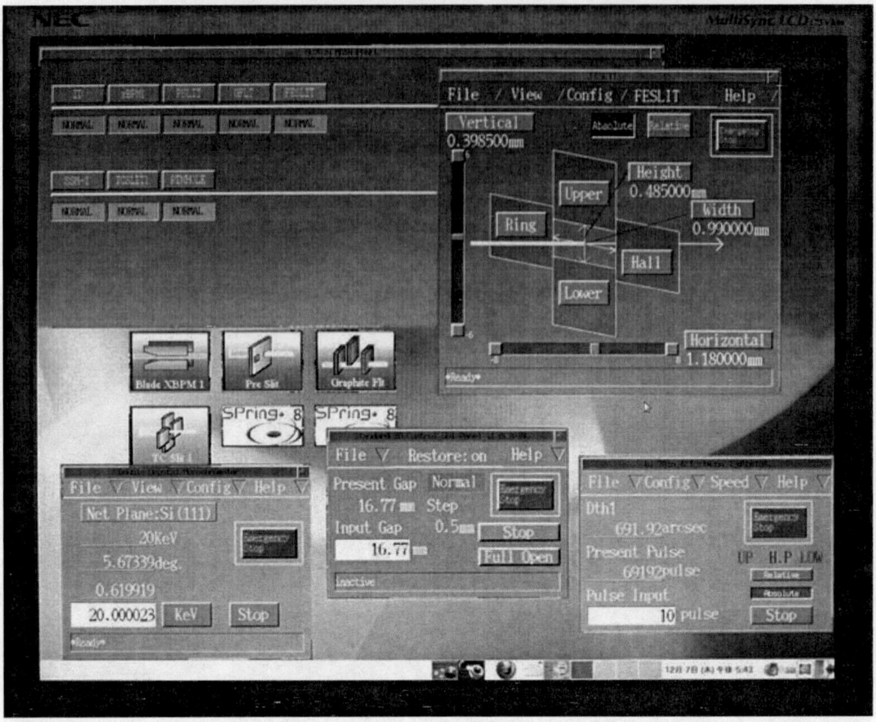

Fig. 4.48 Workstation display of the front-end slit and monochromatorcontrolling beamline in the experimental hutch of the synchrotron radiation facility (BL20XU at SPring-8)

hutches, with the first experimental hutch being located 42 m from the light source. The useable beam size is 75 mm (vertical) × 5 mm (horizontal). The second and third experimental hutches are located 200 m and 206 m, respectively, from the beam source and are capable of using a beam size of approximately 300 mm (vertical) × 20 mm (horizontal) [37]. A flux of 6.5×10^9 [photons/s/mm^2] can be obtained when a Si (111) plane is used in the monochromator in the first experimental hutch [51].

Monochromators include *diffraction grating*-, *multilayer*-, and *perfect crystal*-types. The diffraction grating-type has multiple grooves etched at equal intervals and parallel on a substrate. They are used for soft X-rays. The multilayer-type has both light and heavy thin films periodically layered on top of each other. The beam becomes approximately monochromatic with an *energy resolution* $\Delta E/E$ of around 10^{-2}, whereas the *bandwidth* expands about 100 times more than that in a perfect crystal-type. These specifications are considered to be sufficient for small-angle scattering or X-ray imaging based on absorption contrast, which does not have any issues with regard to quantitative characteristics. Multilayer monochromators are used at approximately half of the beamlines in ESRF. Meanwhile, perfect crystal-type monochromators can achieve an energy resolution of around 10^{-4}. For example, X-rays with an energy of 20 keV have a wavelength of 0.062 nm and wavelength

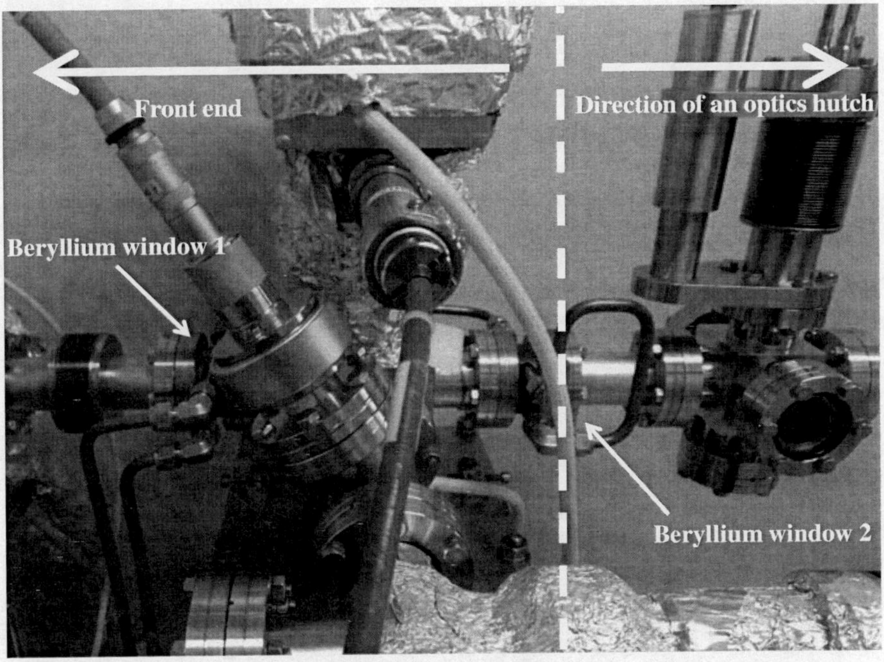

Fig. 4.49 Beryllium window in a beamline in a synchrotron radiation facility (courtesy of Kentaro Uesugi of SPring-8 for BL20XU at SPring-8)

dispersion of around 0.000006 nm. Except for soft X-ray beamlines, almost all of the beamlines in SPring-8 use *double crystal monochromators*, which use perfect crystals. Double crystal monochromators are briefly discussed below.

The Bragg condition shown below is the basis for X-ray spectroscopy using a monochromator (Fig. 4.51a):

$$2dsin\theta_B = n\lambda \tag{4.34}$$

Here, d is the lattice spacing and θ_B is the Bragg angle. However, the X-ray beam leaves the beamline axis in diffraction based on a single crystal (Fig. 4.51b), so double crystal monochromators use two crystals to guide X-ray beams parallel to the incident beam and in the direction of the experimental hutch. As seen in Fig. 4.51c, translating/rotating stepping motors are attached to each crystal and the first crystal translates up and down as well as front to back to maintain a constant outputted beam position even if the energy is changed and the Bragg angle alters. The user has to make fine adjustments in the energy and first crystal angle ($\Delta\theta_1$) from the experimental hutch shown in the display in Fig. 4.48, ensuring that both crystals are parallel and the beam hits the sample.

Other than the fundamental wave, n-order waves with a wavelength of $1/n$ transmit through the monochromator with the Bragg condition using the same set up. This

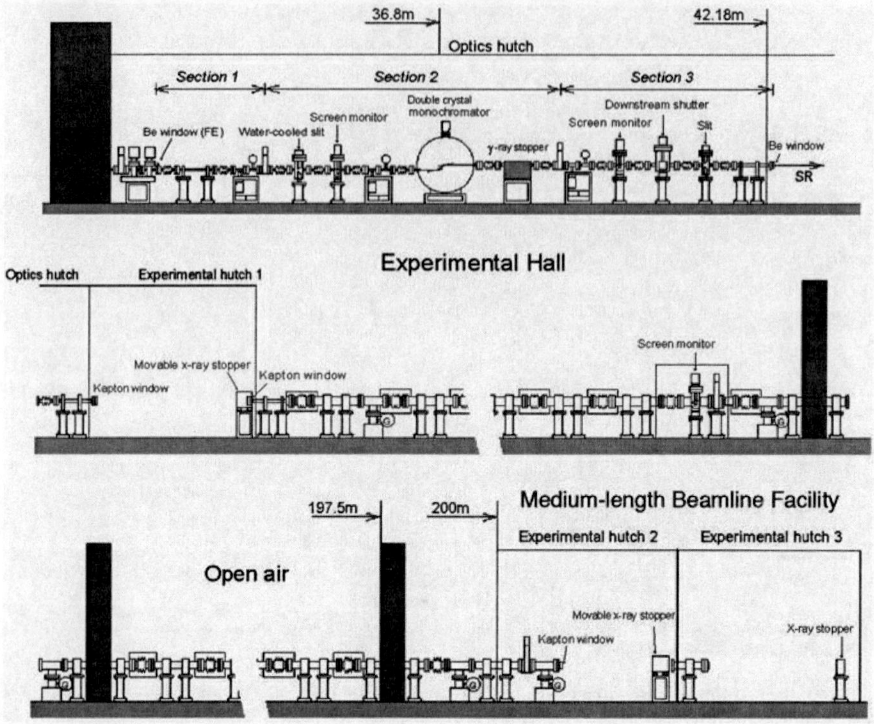

Fig. 4.50 Example of a beamline in a synchrotron radiation facility (courtesy of Kentaro Uesugi of SPring-8 for BL20B2 at SPring-8)

has the possibility of inducing errors in the quantitative analyses conducted using monochromaticity or by creating artifacts. The X-ray diffraction angle range due to a plate crystal such as in Fig. 4.51b can be assessed with the *Darwin width* ω; in other words, the monochromator can diffract X-rays by an angle equivalent to the Darwin width, with the Bragg angle at the center [53].

$$\omega = \frac{2.12}{sin2\theta_B}\frac{r_e\lambda^2}{\pi V}C|F_{hr}|e^{-B\left(\frac{sin\theta_B}{\lambda}\right)^2} \tag{4.35}$$

Here, $r_e = e^2/(m_e\ c^{\ 2})$ is the classical electron radius, V is the volume of the monochromator material per unit lattice, C is the polarization factor, F_{hr} is the crystal structure factor determined from the real part of the atomic scattering factor, and $e^{-B\left(\frac{sin\theta_B}{\lambda}\right)^2}$ is the Debye-Waller factor [53]; F_{hr} and $e^{-B\left(\frac{sin\theta_B}{\lambda}\right)^2}$ are shown in reference [53]. The Darwin width when the X-ray energy is 10 keV is 5.9" and 1.4" for Si (111) and Si (333), respectively, being roughly proportional to $1/n$. At an energy of 100 keV, they each become approximately 1/10 at 10 keV [53]. Meanwhile, higher-order harmonics are proportional to $(1/n)^2$ and n-order harmonics can be cut without

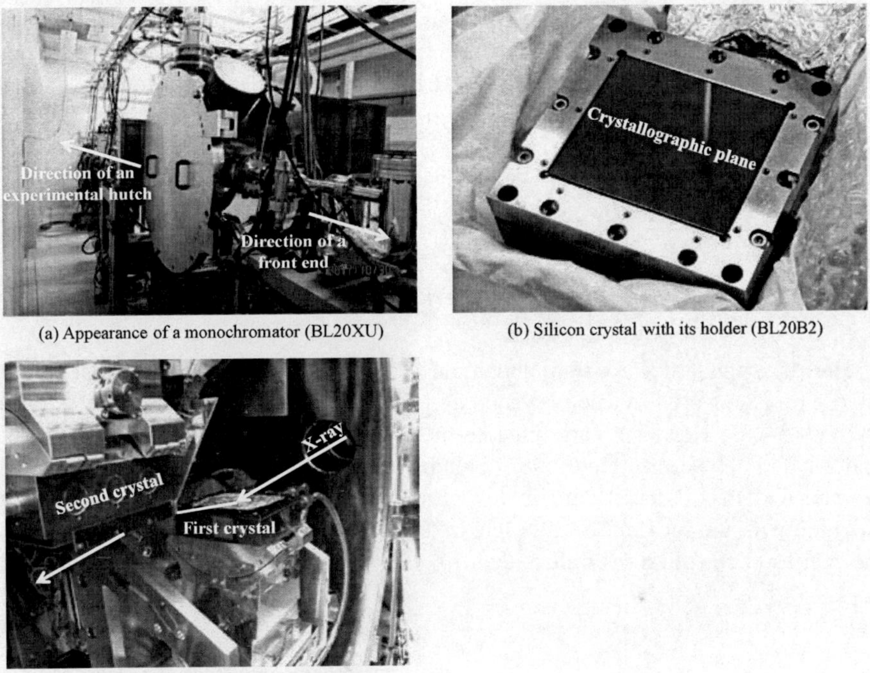

(a) Appearance of a monochromator (BL20XU)

(b) Silicon crystal with its holder (BL20B2)

(c) Inside of a monochromator (BL20B2)

Fig. 4.51 Photograph of a monochromator set up in a beamline of a synchrotron radiation facility (courtesy of Kentaro Uesugi of SPring-8 for BL20XU and BL20B2 at SPring-8) [37]

utilizing a harmonic cutting mirror using this and shifting it by an angle equivalent to the Darwin width of the *n*-order harmonic from the Bragg angle. Taking Si (111) reflection as an example, shifting the angle so that the fundamental wave intensity is 90–80% of the peak as a result, orders higher than (333) can be cut without losing much of the fundamental wave intensity ((222) reflection is prohibited and thus does not exist). This type of diffraction angle adjustment is referred to as *detuning*.

Silicon is primarily used as the crystal for diffraction. Other X-ray spectroscopy crystals include indium antimonide (InSb), crystallized quartz (SiO_2), diamonds (C), alumina (Al_2O_3), beryl, and yttrium boride (YB_{66}). The lattice constants of each are greater than silicon and are suited for low energies such as soft X-rays.

Incidentally, the lattice spacing of silicon crystals needs to be made completely homogenous to prevent X-ray intensity variations in synchrotron radiation facilities that use X-rays in areas between several m to 200 m away from the monochromator. Polycrystal silicon is not suited for this and lattice defects or impurity atoms in the solution must be removed insofar as possible even in single crystals. For this reason, SPring-8 uses artificial silicon grown unidirectionally using the floating zone melting method, which has a purity of 99.999999999% [52]. Furthermore, any strains introduced due to crystal cutting are removed beforehand using machine polishing and etching [52]. Crystals whose Si (111) plane has shown in its surface can be exchanged with those whose Si (220) plane has shown in its surface for the analyzer crystals in SPring-8 BL20XU [37]. These two crystals can enable the use of X-rays with energies of 7.62–37.7 keV and 37.7–61 keV, respectively [37]. Both are cooled using an indirect cooling method based on liquid nitrogen [37]. The energy values at BL20B2 are as follows: Si (111), 5.0–37.5 keV; Si (311), 8.4–72.5 keV; and Si (511), 13.5–113.3 keV [37]. The lattice spacing $2d(h\ k\ l)$ in the $(h\ k\ l)$ plane can be determined using the following equation:

$$2d(hkl) = \frac{2a}{\sqrt{h^2 + k^2 + l^2}} \tag{4.36}$$

Here, a is the lattice constant and equal to 5.4310623 Å (25 °C) for silicon. The energy range covered by each crystal can be calculated if we assume an incidence angle of 3–27°. However, care must be taken as Si (311) has a flux that is smaller than Si (111) by several factors; Si (511) has a flux that is even smaller by several more factors. In contrast, the energy resolution is expressed as shown in the following equation with the incident beam angle divergence φ and ω being determined from the beam source emission angle and slit:

$$\frac{\Delta\lambda}{\lambda} = \sqrt{\varphi^2 + \omega^2} cot\theta_B \tag{4.37}$$

Thus, energy resolution increases when a higher-index plane is used.

A *gamma-ray stopper* is set up behind the monochromator, as shown in Fig. 4.52.

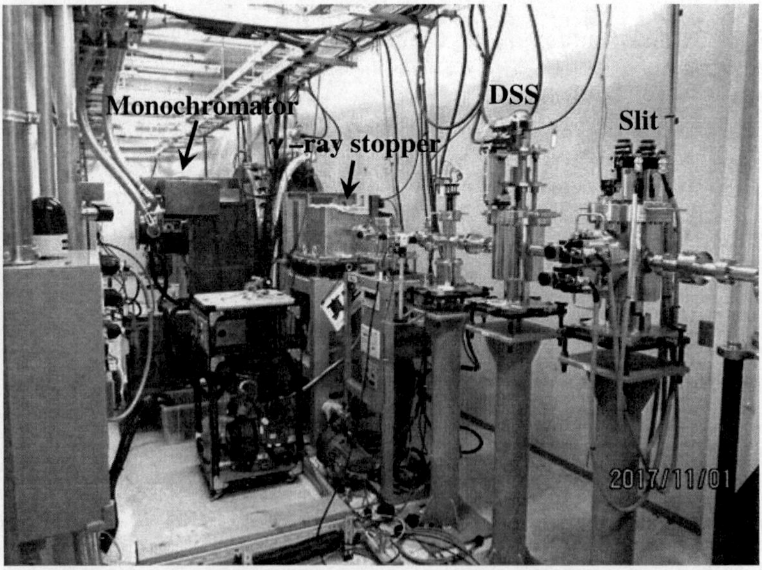

Fig. 4.52 Optics hutch of a beamline in a synchrotron radiation facility (courtesy of Kentaro Uesugi of SPring-8 for BL20XU at SPring-8)

Gamma rays at a level of several GeV are produced from *bremsstrahlung* when the electron beam passes through trace amounts of residual gas in the storage ring; the gamma-ray stopper is set up to stop these gamma rays. Gamma-ray generation is the same *bremsstrahlung* from X-ray tubes but with a frontward directionality due to its extremely high electron energy similar to synchrotron radiation. For this reason, only a narrow area near the optical axis must be shielded. Lead blocks with 30- to 35-cm thickness are used in SPring-8 [52] and monochromatic X-rays pass through a pathway that is offset by approximately 30 mm. Behind this is the *downstream shutter* (*DSS*), as shown in Fig. 4.52. This is for shielding the monochromatic X-rays after spectral diffraction. Closing the DSS enables synchrotron radiation to irradiate the monochromator even when the user is working at the experimental hutch, hence, the thermal stability of the monochromator can be assured. Typically, a sealed lead block with a 10-cm thickness is brought up and down using an air cylinder. A 9.4 cm-thick lead block is used as a DSS in SPring-8 [52]. There is also a slit downstream of the monochromator (transport channel slit, TC slit); however, with the exception of special experiments, only the front-end slit is used.

4.2 Filter

A filter is used for the absorption of endogenous X-rays generated in the window material or target of X-ray tubes, as well as for the systematic lowering of incident X-ray intensity and selective attenuation of low-energy components in incident X-rays by inserting thin metal plates between an X-ray source and sample. They are a particularly important component in industrial X-ray CT scanners and, with regard to the latter, they are used for reducing artifacts referred to as *beam hardening*, discussed later in Sect. 7.6. The filter can be adjusted to the imaging conditions corresponding to the sample and scanner by changing the inserted metal type and plate thickness. Industrial X-ray CT scanners include those with multiple filters installed, while some automatically select the optimal filter out of several. An example of a filter changer in commercially-available X-ray CT scanners for industrial use is shown in Fig. 4.53.

Figure 4.54 shows a schematic of the effects of filter insertion. X-ray intensity in the low-energy side greatly decreases with the insertion of a filter and the X-ray energy peak shifts to higher energies. The latter is referred to as beam hardening, where this becomes an artifact that appears in the sample. For this reason, removing low-energy components beforehand with a filter is a useful strategy for preventing artifacts. Filter 2 in the figure has elements with an absorption edge at an energy lower than K_α X-rays, where virtually all high-energy components above the absorption edge have been cut. This is used in X-ray diffraction, for example, by combining a tube with a copper target and a nickel filter, K_β X-rays and continuous X-rays in the short-wavelength side are eliminated, and almost all of the spectrum is composed of close to monochromatic X-rays derived from K_α X-rays. This is because the nickel absorption edge is located between the K_α and K_β X-rays. The absorption edge effect in Fig. 4.54 is considered problematic when conducting 3D imaging. For this reason,

Fig. 4.53 Example of a filter changer used in an X-ray CT scanner. A filter changer was incorporated in a Karl Zeiss AG ZEISS Xradia 520 Versa (Fig. 6.18) (courtesy of Nobuhiro Hayamizu of Karl Zeiss AG)

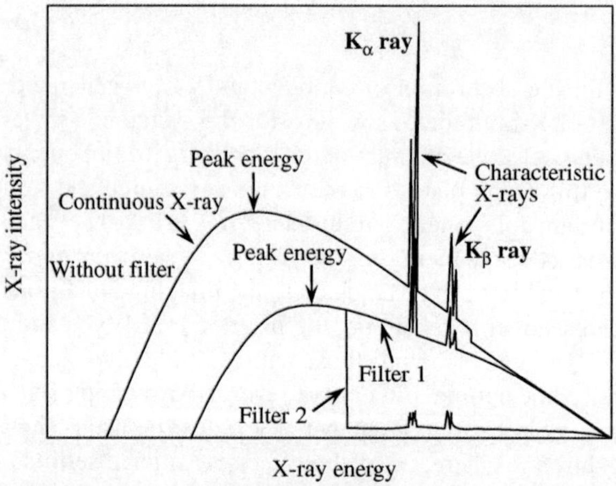

Fig. 4.54 Schematic showing the effect of a filter on X-ray intensity when using an X-ray tube. The figure shows the effects of Filter 1, composed of an element that has no absorption edge in the X-ray energy shown in the figure and Filter 2, composed of elements where the absorption edge is virtually in the center of the horizontal axis

elements like silver or tin, which have an atomic number of approximately 40 or above and an absorption edge at an X-ray energy of several dozens of keV, are not suitable as a filter, depending on the experimental conditions.

As shown in Figs. 4.2 and 4.18, X-ray tubes have characteristic X-ray spectra depending on the target material, tube voltage, and target angle. The filter material and thickness are selected by considering the energy range of the X-rays to be attenuated. For example, suppose that aluminum, titanium, copper, and niobium were four metal types used as a filter. Their atomic numbers vary widely from 13 to 41 and, as seen in Chap. 2, their X-ray attenuation behavior can vary widely. Nickoloff et al. reported that when tungsten was used as a target at an 80-kV tube voltage (target angle was 12°) and when the filter thickness was adjusted so that these all took the same peak X-ray energy (48.0 keV), the thickness values were 4.1, 0.50, 0.11, and 0.05 mm, respectively [54]. Assessments on beam intensities in these cases showed results of 58.0, 65.6, 66.8, and 65.3% of the non-filtered cases, respectively [54]. Virtually identical X-ray spectra were obtained regardless of the filter when these were adjusted with tube current [54].

An optimal filter thickness can be calculated for medical-use X-ray CT scanners when considering patient exposure (particularly X-ray absorption near the skin) and tube load by considering the X-ray spectra obtained in this way [54]. However, sample exposure is typically not an issue with industrial-use CT scanners, so these types of discussions are not applicable. Thus, for cases where the X-ray tube spectra used in the X-ray CT scanners are determined by the user and the sample cross-section morphology is simple, filters can be selected simply by considering the X-ray energy dependencies of the linear absorption coefficients, as shown in Chap. 2. Figure 4.55 shows the X-ray spectra adjusted by X-ray tube voltage and the filter so that a polymer, aluminum, and steel disk sample with 1-mm diameter can be effectively imaged using a commercially available microfocus tube. Suitable filters are often selected by actually imaging a sample and confirming the quality of 3D images in cases where the X-ray spectra specifications are unknown for the X-ray CT scanner used.

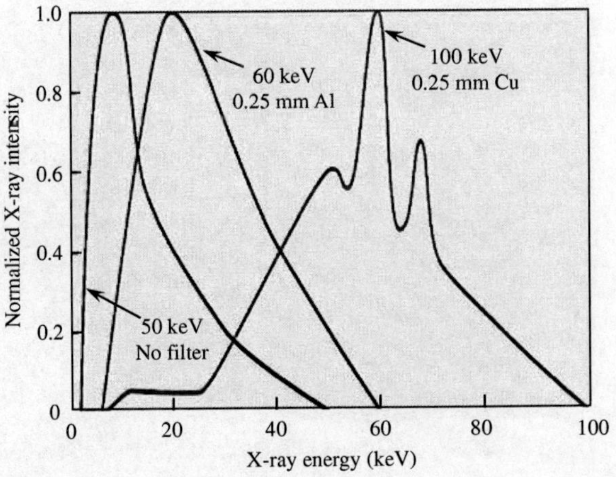

Fig. 4.55 Schematic of X-ray spectra where a single microfocus tube was used and where adjustments in the tube voltage and filter were made to adjust to the X-ray spectra matching a polymer, aluminum alloy, and steel. The vertical axis shows normalized X-ray intensity

4.3 Positioning Stage

At the very least, an X-ray CT scanner must fundamentally have its X-ray source, sample rotation stage, and detector perfectly aligned in the vertical and horizontal directions. Furthermore, when using various slits, zone plates, condenser plates, phase plates, etc., these must be accurately positioned with the detector and sample rotation stage as well. To achieve this, a sample rotation stage, *XY linear stage*, *slider module*, and a *goniometric stage* must be used as required. Examples of these are shown in Fig. 4.56. Synchrotron radiation X-ray tomography in particular requires many combinations of these due to the multiple uses of the experimental hutch and diversity of user experiment. Of these, the sample rotation stage directly affects the spatial resolution of the obtained 3D images and is related to artifact generation.

As discussed later in Eqs. (5.29) and (7.30), among the various factors in X-ray tomography—the effective focal spot size of the radiation source, the Nyquist frequency stipulated by the rotation step and the pixel size of the detector, the positional accuracy of the sample rotation stage, sample drift, and the detector—the single factor with the lowest accuracy acts as the rate-determining step that stipulates the spatial resolution. Of these, Sect. 4.1 discusses the X-ray source and Chap. 7

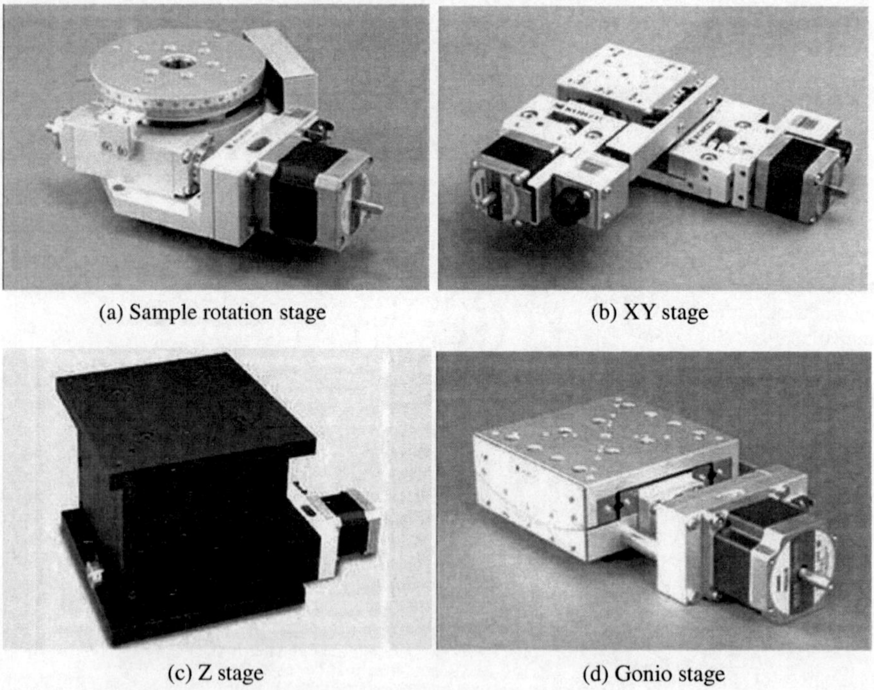

(a) Sample rotation stage (b) XY stage

(c) Z stage (d) Gonio stage

Fig. 4.56 Various stages used in X-ray CT scanners (courtesy of Masamichi Kojima of Kohzu Precision Co., Ltd.)

discusses the Nyquist frequency. The remaining factors of positional accuracy of the sample rotation stage and sample drift are particularly important for high-resolution imaging and synchrotron radiation X-ray tomography. Furthermore, sample rotation stage accuracy is presumed to lower the spatial resolution even in the case of standard industrial X-ray CT scanners if the spatial resolution measured in 3D images cannot be explained by the effective focal spot size of the X-ray source or the Nyquist frequency.

4.3.1 Sample Rotation Stage

Although there are components without which an X-ray CT scanner cannot function, it is perhaps the sample rotation stage whose importance remains most unrecognized. An X-ray CT scanner requires a *thrust bearing* that can rotate a sample while withstanding the sample weight; *ball bearings* are primarily used. *Air bearings* and *slide guides*, the latter of which operates on a sliding bearing, may be used depending on the desired accuracy if a higher spatial resolution is desired. Figure 4.57 shows a schematic of each mechanism.

A general-use rotation stage with a ball bearing is typically used for industrial X-ray CT scanners, even for microtomography setups. These cases usually use a 5-cm diameter sample rotation stage with a *load capacity* of 2–3 kgf, *eccentricity* of 5 μm, and *surface runout* of approximately 20 μm; alternatively using a 20-cm diameter sample rotation stage with a load capacity of 30 kgf, eccentricity of 10 μm, and surface runout of approximately 20 μm. Furthermore, diameters of 7–10 cm are presumed to yield the highest accuracies, where eccentricities of up to 0.5 μm can be used. From Table 6.1 shown later, commercially-available high-resolution X-ray CT scanners use rotation stages with a load capacity of 2–12 kgf. Discussions with the manufacturer are particularly important for high-precision sample rotation stages, as performance varies for each individual rotation stage. Here, eccentricity is the runout in the rotation radius direction and surface runout is the angle runout in the plane normal to the rotation axis. Photographs depicting the actual measurement of eccentricity and surface runout are shown in Fig. 4.58. Methods stipulated by JIS involve contact- or non-contact-type measurement methods where a true sphere is placed on the stage. The measured rotation errors such as for eccentricity and surface runout are the result of superpositions of displacements in the radius direction and axial direction (i.e. vertical direction), shifts in centering, and angle fluctuations in the rotation axis due to the *motion of precession*. What is important for X-ray tomography is not the upper surface of the stage but the vertical and horizontal fluctuations at sample positions located several dozen cm at maximum in the vertical direction from the upper stage surface. Thus, care must be taken so that effects on spatial resolution cannot be directly determined simply from the specifications of sample rotation stages like those shown above. Furthermore, accuracy is thought to further worsen in cases of laminography where the rotation stage is inclined.

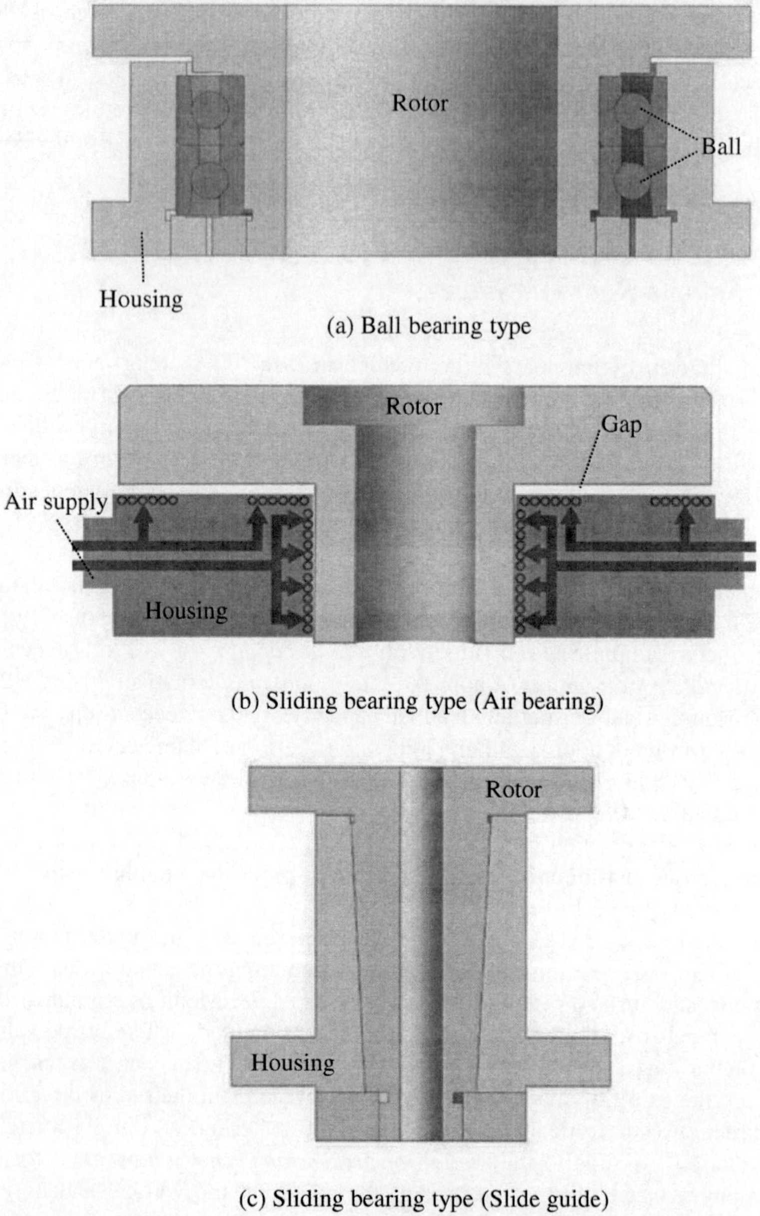

Fig. 4.57 Various rotation stages used in X-ray CT scanners (courtesy of Masamichi Kojima of Kohzu Precision Co., Ltd.)

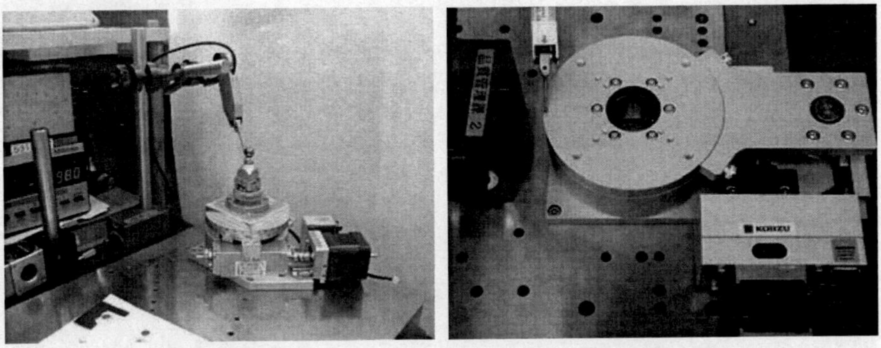

(a) Eccentricity (b) Surface runout

Fig. 4.58 Measurement examples of eccentricity and surface runout of the rotation stage used in the X-ray CT scanner (courtesy of Masamichi Kojima of Kohzu Precision Co., Ltd.)

A stage rotation speed of 0.2 rotations/s is considered the limit for a standard *worm-gear type* (combination of a worm shaft and worm gear) shown in Fig. 4.59a. A sample rotation stage may at times need to be rotated at faster speeds in cases of fast tomography using synchrotron radiation. For these types of cases, rotation speeds of 2 rotations/s and 5 rotations/s can be achieved with a *belt-drive type* (Fig. 4.59b) and *direct-drive type* (Fig. 4.59c), respectively. Equipment selection alone is insufficient for high-accuracy rapid rotation and ensuring that the sample's center of gravity is accurately placed on the rotational axis.

The eccentricity can be controlled to 100–200 nm when the sample rotation stage is an air-bearing type (referred to as an *air spindle* when seen as a rotation unit) such as shown in Fig. 4.57b, with a maximum load capacity of approximately 10 kg and a maximum diameter of approximately 20 cm. Generally, when using fluid lubrication (e.g. sliding bearing), the fluid viscosity is proportional to the friction coefficient. The air viscosity is three orders of magnitude smaller than the lubricant oil, so the air bearing type has less friction loss and heat generation, making it suitable for faster speeds. It also has the advantage of working well in clean environments or high temperatures. Meanwhile, a clean dry air must be stably supplied at a fixed pressure of approximately 0.5 MPa. Care must be taken, as the actual eccentricity accuracy of the air spindle is strongly dependent on the stability of the supplied air pressure and temperature. Meanwhile, a slide guide type as shown in Fig. 4.57c has its load capacity-controlled to approximately 2 kgf (20 cm in diameter) but a relatively low-cost and compact yet high-accuracy sample rotation can be obtained as a result. A maximum eccentricity of approximately 70 nm (up to 200 nm) can be achieved. However, it is essential that lubricant oil enters the gaps and forms an oil film. The rotation stage will be scorched without a gap, so maintenance is necessary to ensure that the lubricant oil does not run out. Care must be taken particularly in synchrotron radiation experiments where non-use periods are extensive.

Active eccentricity/surface runout control techniques are necessary when further rotation accuracy is needed; for example, when a sub-micron spatial accuracy is

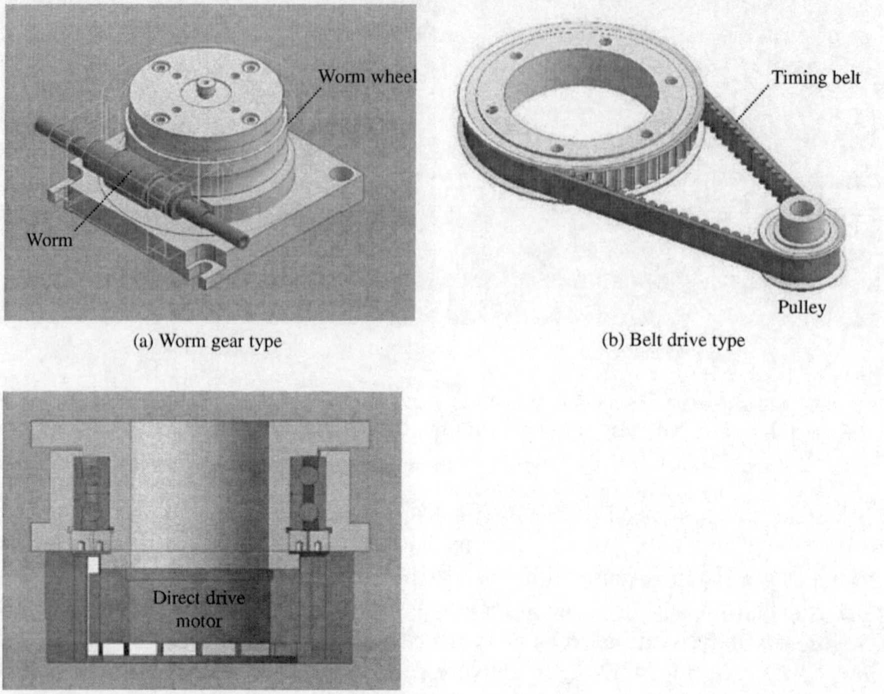

(a) Worm gear type (b) Belt drive type

(c) Direct drive type

Fig. 4.59 Various rotation mechanisms used in X-ray CT scanner (courtesy of Masamichi Kojima of Kohzu Precision Co., Ltd.)

needed for synchrotron radiation X-ray tomography [55]. For example, there are application reports at the synchrotron radiation facility SOLEIL, near the suburbs of Paris, in France [56]. Here, the eccentricity and surface runout characteristics unique to the sample rotation stage used were measured beforehand to conduct corrections with feed-forward controls with an XY stage. Interferometers were used during imaging as well to measure eccentricity and surface runout and additional corrections were conducted using feedback controls [55].

The helical-scan type X-ray CT scanners of the sixth-generation CT introduced in Chap. 3 has a *slip ring* mechanism installed, which allows for continuous rotation in a single direction by supplying power and transferring data to the X-ray tube and detector attached to the gantry. A slip ring is also used for high-speed tomography discussed in Sect. 5.3, because the rotation stage is continuously rotated multiple times while keeping its test rigs for in situ observation (discussed later in Sect. 4.5) and heating furnace on the rotation stage. Conducting metal disks made of copper are stacked and fixed by the number of necessary channels in the slip ring, so each disk is in constant contact with separate brushes even during rotation. This can be used to conduct power voltage and data signal exchange without connecting the top and bottom of the rotation stage with cables.

4.3.2 Other Positioning Stages

The number of strokes is the only issue when using an XY-stage as shown in Fig. 4.56b or *Z-stage* as shown in Fig. 4.56c for industrial X-ray CT scanners. However, the main rotation stage body rotates at an inclined position relative to the incident X-ray in laminography, so care must be taken to ensure that the XY-stage is not detached. Magnets are occasionally used to keep this fixed. The sample is sometimes removed from the field-of-view and an I_0 image (Eqs. (2.2) and (3.1)) is acquired in synchrotron radiation X-ray tomography. An issue that arises in these cases is the reproducibility of the positioning when the XY stage returns. This has been stipulated by JIS as repeated positioning accuracy and its specifications are indicated in their catalog. Reproducibility is primarily influenced by the heat generation of the motor. For this reason, countermeasures such as including heat insulation in the rotation stage or sufficiently moving it beforehand and using it when temperature fluctuations are low, have been used.

A gonio stage, as shown in Fig. 4.56d, must be used to carefully adjust the alignment of the detector and rotation axis of the sample rotation stage when conducting synchrotron radiation X-ray tomography, where the detector is frequently taken on and off. The gonio stage can conduct micro-adjustments in the tilt angle of the detector in a range of around $\pm 10°$ by moving the upper gonio stage surface along an arc in which the center of rotation is a point well above the stage. The detector shifts significantly in the vertical and horizontal directions simultaneously with this adjustment, so coordination with the XY stage is essential.

4.4 Detector

Within the history of X-ray tomography, detectors have shown more rapid progress than the introduced X-ray sources and positioning stages, with many existing variations. With this in mind, an overview will be presented on a limited number of detectors that are in frequent use today. For example, *imaging plates*, which use ion crystals that exhibit photostimulated luminescence, have been omitted from the following discussion. Figure 4.60 shows the types of detectors used in the primary commercially-available industrial X-ray CT scanners and their relationship with maximum tube voltage. Detectors can be classified between a *charge-coupled device* (*CCD*), *flat panel detector* (*FPD*), *image intensifier* (*II*), and *line sensor camera*. Except for line sensor cameras, all of these are two-dimensional detectors that can directly capture two-dimensional information as is shown in Fig. 3.4. Meanwhile, the line sensor camera is a linear or curved one-dimensional detector similar to those shown in Figs. 3.1 or 3.2c. As shown in Fig. 4.60, line sensors are frequently used when the tube voltage is over 450 kV. Furthermore, the FPD or image intensifier is used up to a tube voltage of approximately 300 kV. Furthermore, a CCD camera is

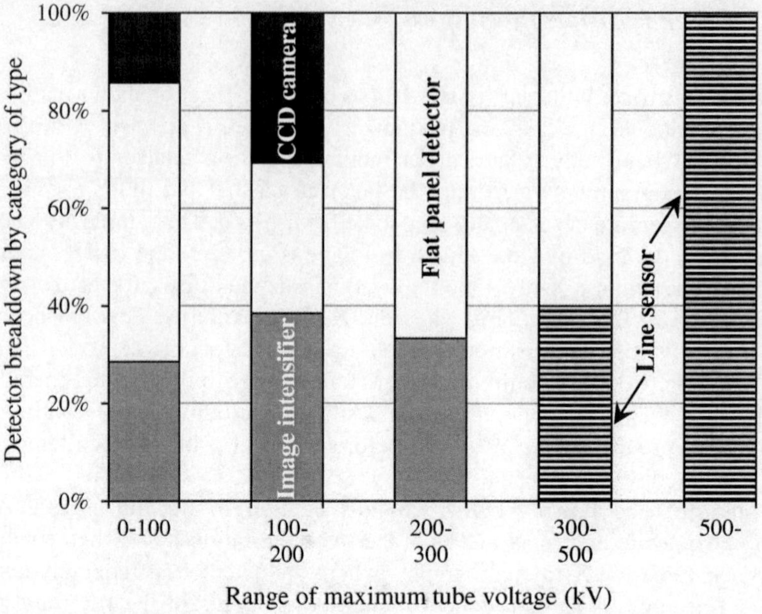

Fig. 4.60 Maximum tube voltage and type of detector used in commercially available industrial-use X-ray CT scanners as of the fall 2017

used when observing small samples with a low tube voltage. *Complementary metal-oxide-semiconductor* (*CMOS*) cameras have recently been used in place of CCD cameras in synchrotron radiation facilities. The use of CMOS cameras, in industrial X-ray CT scanners and elsewhere, is expected to increase in the future.

This section first provides an overview of the important indicators required for assessing the performance of each detector, after which each detector is investigated in further detail.

4.4.1 Characteristic Assessment of Detectors

X-ray intensity measurements involve factors like *gain*, *quantum efficiency* (*QE*), *noise*, *dynamic range*, and *linearity*. The detectable X-ray energy range is then determined from combinations with a scintillator, which determines the overall efficiency of the detector system. This is described in the scintillator section. Furthermore, the pixel number and size of the detector influence the sample size that can be covered by X-ray imaging and the achievable effective spatial resolution through a combination with the glass fibers and the optical lens, which magnify or demagnify the image after converting it to visible light. Meanwhile, *temporal resolution* is determined from the *frame rate* of the detector and the *decay time* of the scintillator.

(1) Gain

Gain is also referred to as the *conversion ratio* or *sensitivity*. The photons, which enter the detector are accumulated within the detector after being converted to a charge, are further converted to voltage levels through the readout pathway and ultimately outputted after AD conversion. The detector gain is stipulated by the ratio between the charge number per pixel and the pixel output signal, expressed by the *digital number* (*DN*) or *analog-to-digital unit* (*ADU*). A single photon corresponds to a single charge in the former for standard CCD cameras, with the exception of EM-CCD cameras. The latter is in units of count number. Both have a linear relationship and multiplying the gain to ADU yields the measured photon number. The gain is dependent on both the quantum efficiency of the element and the electron-to-voltage conversion efficiency. The latter is dependent on the capacitance and gain of the output amplifier. A higher gain generally increases the dynamic range as well as noise, so the gain must be set to balance the two.

Gain heterogeneity is referred to as *shading*, with areas where the image brightness is not uniform and dark areas appear across wide ranges. This can be corrected using calibration data prepared in advance.

(2) Quantum Efficiency

Quantum efficiency is defined as the number of photon charges accumulated at the detector divided by incident photons. This naturally takes a value less than one and values closer to one are considered superior. Quantum efficiency can be estimated by considering the material used for detection, but this cannot be directly measured for individual detectors. Furthermore, quantum efficiency does not consider image information deterioration such as increased noise based on signal processing within the detector interior. Figure 4.61 shows the quantum efficiency of two CCD cameras: the Hamamatsu Photonics C4880-41S camera (4000 × 2624 pixels, pixel size 5.9 μm, back-side illuminated type) and the ESRF FreLoN 2000 camera [57]. The former has a maximum quantum efficiency of approximately 50% when the incident ray is 460 nm (blue) and over 40% when between 420–550 nm. The latter has high quantum efficiency in the 500–800-nm range (green–red). In this manner, the quantum efficiency of the detector has a visible light wavelength dependency and the scintillator is selected with this in mind. Quantum efficiency in front-side illuminated type CCD cameras decreases when the transparent electron acts as an optical filter, whereas a back-side illuminated type exhibits high quantum efficiency across a wide wavelength range. The detector mechanism thus affects quantum efficiency in this manner.

Incidentally, a *detective quantum efficiency* (*DQE*) is used as a practical indicator for assessing the differences in image quality for each detector, which is expressed through directly measurable quantities. The DQE is expressed as follows [58]:

$$DQE = \frac{(SNR_{\text{out}})^2}{(SNR_{\text{in}})^2} \tag{4.38}$$

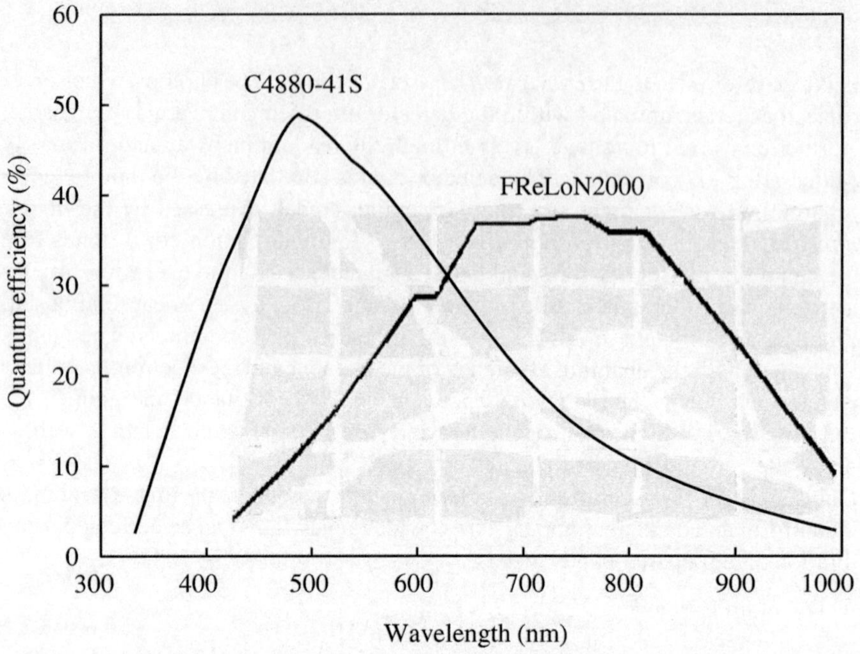

Fig. 4.61 Example of quantum efficiencies in a CCD camera [57]

Here, SNR_{in} and SNR_{out} refer to the S/N ratios of the input and output image, respectively. Setting the input/output signals as S_{in} and S_{out} and input/output noise as N_{in} and N_{out} yields $SNR_{\text{in}} = S_{\text{in}}/N_{\text{in}}$ and $SNR_{\text{out}} = S_{\text{out}}/N_{\text{out}}$. The DQE, in an ideal detector without any noise, is equal to the quantum efficiency. SNR_{in}^2 increases linearly with incident photon flux. The input noise relating to this is the *photon noise* (also referred to as *quantum noise* or *shot noise*). This is due to the random fluctuations centering on the average value seen in the photon flux. SNR_{in} can be improved by adjusting the imaging conditions, such as increasing the exposure time. However, N_{in} cannot be eliminated; in other words, photon noise is an unavoidable form of noise. In general, X-ray intensity fluctuations follow *Poisson statistics*. The square root of the average parameter value is equal to the standard deviation when a given data set follows Poisson statistics. For this reason, the photon noise corresponding to the standard deviation of image signals can be expressed as proportional to the square root of the X-ray photon number.

$$\sqrt{S_{in}} = N_{in} \tag{4.39}$$

Figure 4.62 shows the relationship between the input signal and photon noise. *Secondary quantum noise*, which is generated through a fluorescence emission process, and *excess noise*, which is produced through scattering in X-ray photon absorption processes, similarly satisfy Eq. (4.39) in addition to photon noise.

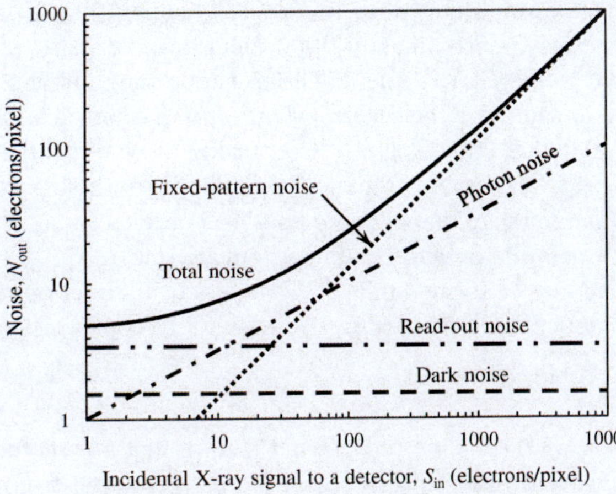

Fig. 4.62 Schematic showing the relationship between different types of noise and the intensity of an X-ray incident to the detector

If gain is defined as $G = S_{out}/S_{in}$, Eq. (4.38) can be re-written as follows [59]:

$$DQE = \frac{G\,S_{out}}{(N_{out})^2} \tag{4.40}$$

The right-hand side of Eq. (4.39) is the output signal, output image noise, and detector gain, where each is a measurable amount. These can be used to quantitatively compare the performance of detectors of otherwise different types. The ideal detector is $DQE = 1$.

(3) Full Well Capacity and Dynamic Range

Full well capacity is the amount of charge that saturates each pixel. Dynamic range is defined as the ratio between the maximum image signal level where linearity can be maintained without saturation and the minimum detectable signal level; the former is stipulated by the full well capacity and the latter is substantially defined by background noise levels. The dynamic range is generally expressed in units of bits. For example, a dynamic range of 10,000 is expressed as $\log_2(10{,}000) = 13.3$ (bits).

The full well capacity varies according to detector pixel size or operating voltage. A saturated charge in a pixel results in a loss of linearity and quantitative characteristics, as well as a phenomenon referred to as *blooming*, where the charge leaks out into adjacent pixels and influences the measured values in those pixels. A larger pixel size reduces spatial resolution but also increases full well capacity and dynamic range. Even extremely large differences in local X-ray absorption within a sample can be effectively distinguished when using a high-dynamic-range detector. Furthermore, quantitative measurements can be assured. As such, dynamic range is an important detector characteristic alongside sensitivity and spatial resolution.

Using a CCD camera as an example, converting X-rays into visible light and emitting them toward a camera is more effective in obtaining a high dynamic range than directly emitting X-rays to the CCD camera. An X-ray with an energy of 10 keV generates approximately 3000 electron-hole pairs when X-rays are directly incident to the CCD camera. The dynamic range is approximately 30–300, based on the capacity of the *electron-hole pairs*, which can be accumulated in each pixel (100,000–1,000,000) [60]. Converting these to visible light and then emitting them to the CCD camera produces several electron-hole pairs for each X-ray photon and a dynamic range that encompasses 4–5 orders of magnitude can be obtained if a suitable scintillator and transfer system are selected [60]. For this reason, images should generally be formed in a CCD camera using fiber optic taper for demagnification or an optical lens after using a scintillator to convert X-rays to visible light [60].

(4) Noise

Photon noise is the only form of noise that affects the S/N ratio on the input side. However, other noise types in the image output from the detector include *readout noise*, *dark noise*, and *fixed-pattern noise*.

Fixed-pattern noise is caused by the scintillator microstructure, microdefects or thickness irregularities, variations in transmission capability of each optical fiber, and sensitivity irregularities in each pixel. The extent of fixed-pattern noise is presumed to be proportional to the input signal level, as shown in Fig. 4.62. Fixed-pattern noise has a spatial distribution but no temporal variability, therefore, it can be removed with *flat-field image correction*. In this case, a high-accuracy pixel-based correction is possible by taking multiple images based on homogeneous incident radiation without the sample and averaging these images.

Furthermore, *calculation noise*, described earlier in Sect. 3.3.4 (1), also occurs during the reconstruction process. Furthermore, *aliasing noise*, which occurs in images with frequency components exceeding the Nyquist frequency that stipulates the sampling theorem described in Chap. 7, can be an issue under certain conditions. These types of noise appear after the data output from the camera is reconstructed.

Incidentally, N_r and N_d are each defined as the readout noise and dark noise, F (photons/pixel/s) as the photon flux, τ (s) as the exposure time, η_q as the quantum efficiency of the detector, and i_d as the *dark current*. The photon number N is expressed as $F_{\tau\eta_\mathrm{q}}$. The total noise N_t is expressed by Eq. (4.41) [57].

$$N_t = \sqrt{N + N_d^2 + N_r^2} = \sqrt{F\eta_q\tau + i_d\tau + N_r^2} \tag{4.41}$$

From Eq. (4.41), contributions of the readout noise become relatively large under insufficient quantities of light. This is clear even with the relationship between noise and signal, shown in Fig. 4.62. Furthermore, the S/N ratio of the image is expressed with Eq. (4.42) [57].

$$SNR_{out} = \frac{F\eta\tau}{\sqrt{F\eta_q\tau + i_d\tau + N_r^2}} \tag{4.42}$$

The S/N ratio greatly decreases with larger readout noise under insufficient quantities of light. Equation (4.42) demonstrates that increases in photon flux and exposure time in these cases can have an equivalent effect. Detectors with superior S/N ratios have low readout noise and dark current.

The readout noise is electric noise generated by signal transmission, amplifier circuits, and electronic circuits such as switches. This is observed even under short exposure times where dark noise is not prominent. For this reason, its influence is particularly large under low-exposure conditions. This amount varies depending on the detector output type. Furthermore, although this does not vary with exposure amount, as in Fig. 4.62, there is a frequency dependency. With a CCD camera, for example, a higher readout frequency results in rapidly increased readout noise.

Meanwhile, dark noise occurs when valence band electrons are thermally excited within the interior of materials with a narrow energy band gap such as silicon. The abundance of dark currents is heterogeneous and induces fixed pattern noise. Furthermore, they can appear as random noise due to temporal variations. Dark noise is proportional to the square root of dark current. Furthermore, dark current is dependent on temperature as shown in the following equation:

$$i_d = CT^{3/2}e^{-E_g/2kT} \tag{4.43}$$

Here, C is a constant relating to the light-receiving area of the pixel and E_g is the energy band gap. We can see from this equation that the dark current decreases by approximately 1/2 for every 7° drop in temperature. For this reason, elements are sometimes cooled with a *Peltier device* to reduce the dark current. In practice, the Hamamatsu Photonics C4880-41S camera used in SPring-8 and shown in Fig. 4.61 controls its dark noise to approximately 1 e^-/s by cooling it to approximately −50 °C with a Peltier device. However, condensation countermeasures are also necessary for detector cooling. Although this is also dependent on exposure time, dark noise at about room temperature is typically predominant with regard to the minimum detectable signal level, which stipulates the lower limit of dynamic range. Readout noise becomes predominant when the element is cooled to a point where dark noise is negligible.

Figure 4.63 shows a noisy image undergoing *averaging* across 20 continuous images. This is a transmission image of an aluminum alloy microstructure that was obtained with a deliberately poor image quality by reducing the exposure time to a level considerably below normal. The specimen thickness and X-ray energy, in this case, were 600 μm and 20 keV, respectively. Averaging a noisy image can significantly improve the image quality from the perspective of the noise's statistical nature.

Incidentally, *binning treatment* (additive treatment of multiple pixels) can improve the photon count per pixel and sensitivity. At the same time, it can also decrease the readout frequency, which decreases the readout noise and improves the frame rate. This treatment is commonly expressed by labels such as 2 × 2; this indicates the addition of four pixels in the horizontal and vertical directions.

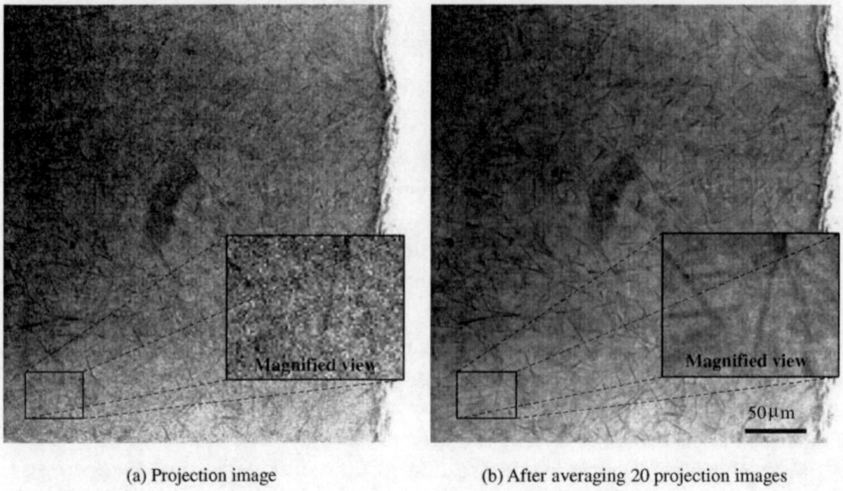

Fig. 4.63 Transmission image of an Al-7% Si cast alloy obtained at SPring-8. The exposure time was minimized, and an image with a deliberately poor S/N ratio is shown in (**a**). This was imaged 20 consecutive times and the averaged result is shown in (**b**). The silicon and intermetallic compounds including iron are clearly visualized in (**b**)

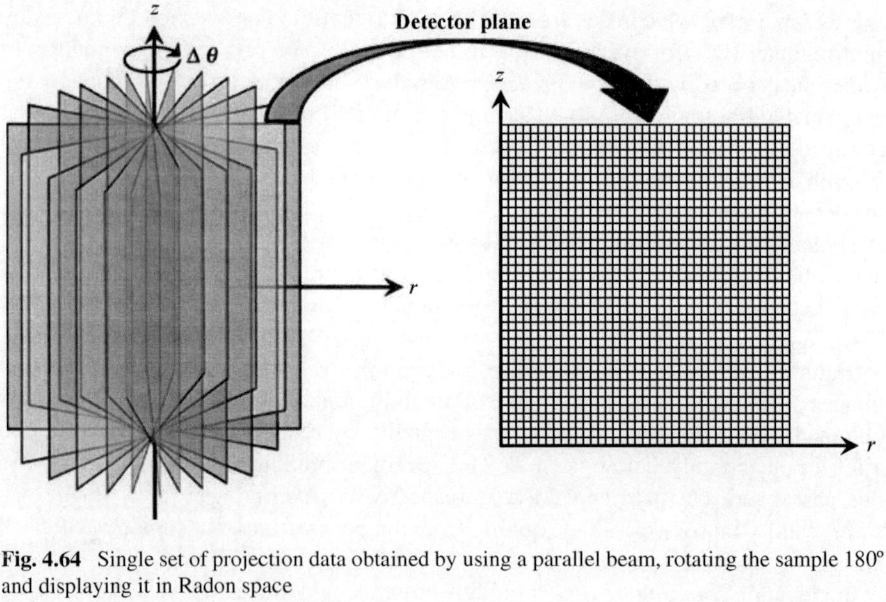

Fig. 4.64 Single set of projection data obtained by using a parallel beam, rotating the sample 180° and displaying it in Radon space

(5) Pixel

As shown in Fig. 4.64, the vertical and horizontal pixel sizes within a detector plane when a two-dimensional detector is used stipulate the sampling pitch, which in turn determines the spatial resolution. The images obtained from two-dimensional detectors are 3D, hence, the last remaining direction is determined from the *rotation pitch* ($\Delta\theta$) of the rotation stage. The effective element size is determined by multiplying the pixel size with the *pixel number*. The detector combines glass fibers and an optical lens, which magnify or demagnify an image after visible light conversion; hence, the field of view in the detection system considers the magnification or demagnification.

Meanwhile, as shown in Fig. 4.65, the percentage of effective-use area, where areas for charge readout wires in the light-receiving area are removed, in a single-pixel area is referred to as the *fill factor*. Even in CMOS cameras, the fill factor can go as high as 90%. Although CCD cameras vary widely according to their mechanism, those with a 100% fill factor are commercially available. The fill factor is important as it determines X-ray detection efficiency.

(6) Frame Rate and Dead Time

As shown in Fig. 4.66, the frame rate is the inverse of a single cycle time T_c, which includes the *exposure time* T_e, read out time T_r, and *dead time*. Naturally, the single

(a) Detector of 70% fill factor

(b) Detector of 90% fill factor

Fig. 4.65 Fill factors of detectors

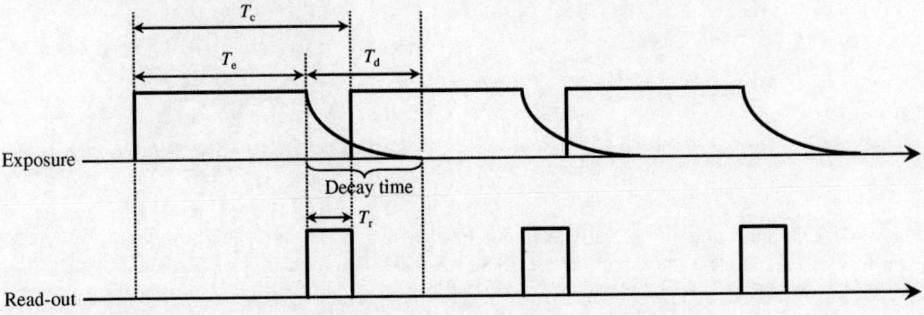

Fig. 4.66 Cycle time and frame rate of the detector

cycle time is not a simple addition in cases where the exposure and readout can be conducted in parallel. The frame rate is expressed in units of frames per second (fps). Furthermore, the T_d in the figure is the decay time of the scintillator. Other than this, a single cycle time in actual practice depends on the detector used but also includes the time required for clearing the accumulated charge, as well as the time required for opening and closing the shutter. It also significantly depends on the imaging mode, such as with binning treatment or imaging regions of interest. The frame rate, which is the specification of the detector, is the value under standard exposure conditions and care must be taken for the frame rate value when the exposure time is zero as it may at times be used.

4.4.2 Various Detectors

(1) Line Sensor Camera

The line sensor camera is an important detector, which has supported the development of medical-use X-ray CT scanners. Even today, it is widely used in non-destructive observations and the detection of foreign objects (e.g. detection of foreign bodies in food products, medical products, and electronic components transported on a belt conveyor, as well as hand-luggage detection at airports). Figure 4.67 shows an example of a commercially-available line sensor camera.

As shown in Fig. 4.68, combinations of scintillators (e.g. GOS:Tb, CsI:Tl, CWO), and *photodiodes* (Fig. 4.68a) or semiconducting detectors (Fig. 4.68b) have been used as line sensor cameras. Visible light generated in the scintillator of the former is radiated in all directions. For this reason, even if the visible light was effectively guided with light-reflective materials or *structured scintillators*, only a tiny portion of the generated light would be detected at the photodiode. Silicon, with high sensitivity

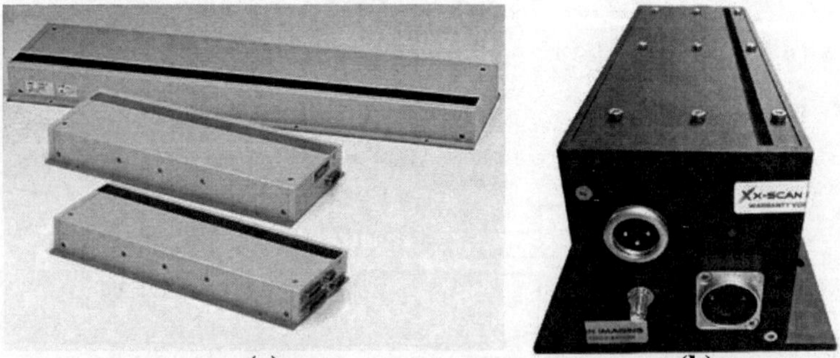

(a) (b)

Fig. 4.67 Examples of commercially available line sensors; **a** Hamamatsu Photonics C9750 series (courtesy of Hamamatsu Photonics) and **b** X-SCAN XIH8800 (courtesy of Mr. Tsuchiya from Ad Science)

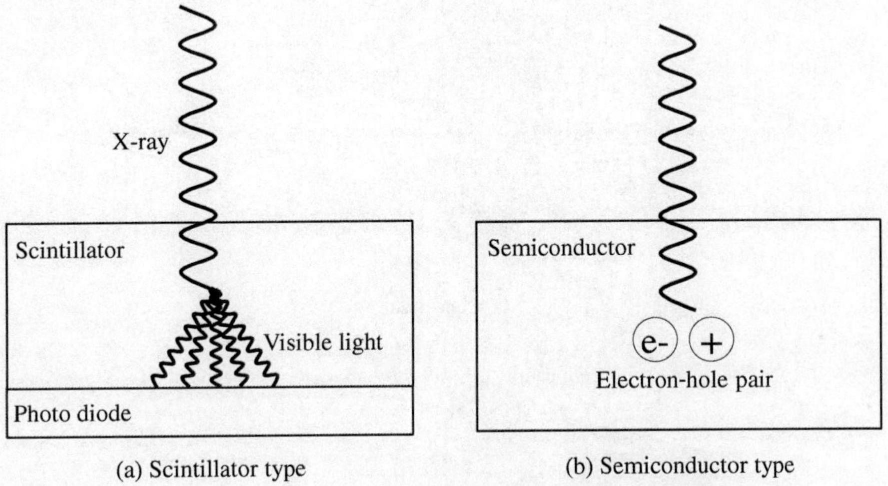

Fig. 4.68 Line sensor camera unit. A line sensor camera is composed of these units arranged in a linear fashion

for high-energy X-rays, has been used for the latter. CdTe and CdZnTe have also been used in recent years for *photon counting* purposes (discussed later). Approximately 500 to 1000 detection units such as those in Fig. 4.68 are generally arranged one-dimensionally at a fixed interval in line sensor cameras. Two-dimensional images must be obtained from one-dimensional images taken continuously while moving the sample in the rotational axis direction of the sample rotation stage to obtain 3D images with a line sensor camera. The X-ray is narrowed into a fan-beam shape using a collimator as shown in Fig. 4.69b to make it incident to the sample such that the field-of-view alone of the line sensor camera is incident to the sample. As shown in Fig. 4.69a, scattered X-rays are incident to the pixels in the detector from all locations on the sample because of forward-scattering of high-energy X-rays in the case of a parallel or cone beam. This can be limited to scattering from the width direction in the case of a fan beam in Fig. 4.69b. As a result, forward-scattering-induced decreases in both contrast and spatial resolution can be effectively prevented.

Time delay integration- (*TDI*) and *dual energy*-types exist as variations of X-ray line sensor cameras. There are multiple (typically 128) rows unlike the single detector element row for the former. Arithmetic means from multiple rows enable images to be obtained with a high S/N ratio [61]. Detector elements are in two stages in the X-ray propagation direction for the latter, with the first and second stages detecting X-rays with low and high energies, respectively [61]. The constituent elements of materials can be identified with this type.

(2) Image Intensifier

An image intensifier has high sensitivity due to its electron multiplication mechanism. It was historically developed for military purposes with regard to night vision under

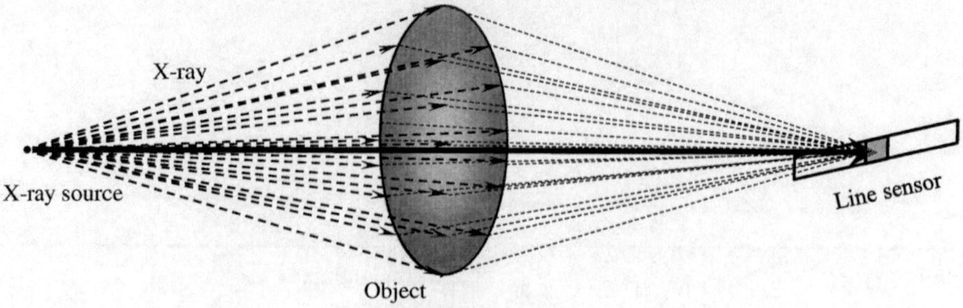

(a) Cone beam

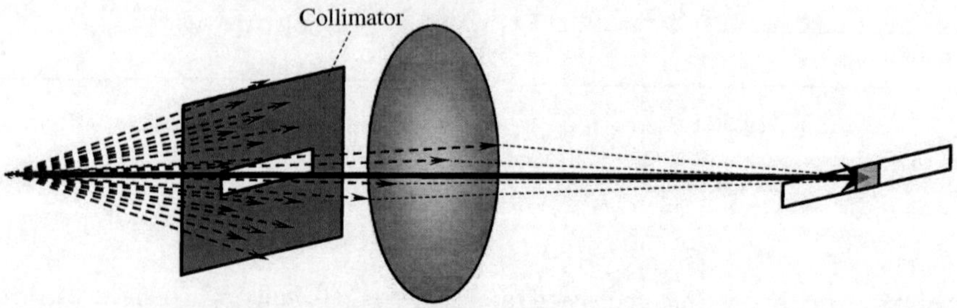

(b) Fan beam

Fig. 4.69 Schematic comparing the extent of forward-scattering due to high-energy X-rays when imaging with a line sensor and either a cone beam or fan beam. The number of scattered X-rays incident to specific pixels of the line sensor is evident

starlight or moonlight. It is typically classified between first–third generation devices, depending on its structure or the material used.

First-generation image intensifiers are a *photomultiplier tube*-type, which has a structure as is shown in Fig. 4.70a. In order from the input side, they are composed of an incidence window, a fluorescent screen on the input side, a *photocathode*, a focusing electrode, anode, a fluorescent screen on the output side, a coupling lens, and a visible light detector (a CCD camera is primarily used). Materials like aluminum are used for casing and its interior is maintained in a vacuum. X-rays are first converted to visible light with fluorescent screens made of CsI columnar crystals, which exhibit an optical fiber-like structure. Next, the visible light is converted to electrons with a *photoelectric surface* (e.g. Sb_2Cs_3) in the vacuum side. The spectra of the generated visible light, in this case, must be matched with the *spectral sensitivity* (the relationship between quantum efficiency and wavelength) of the photoelectric surface. The electron is focused and accelerated through an electron lens due to an electric field generated by voltage differences of several dozens of kV; the electron then collides with a thin and fine fluorescent screen (e.g. ZnCdS:Ag or ZnS:Cu,Al) on the output side. The accelerated electron generates multiple photons as a result. The material on the incident window is aluminum, but a beryllium window is preferred instead when a low-energy X-ray is used.

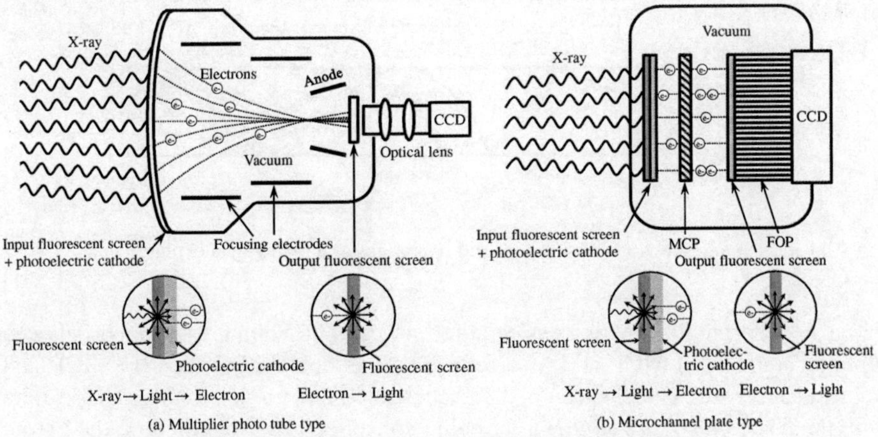

Fig. 4.70 Schematic showing a representative image intensifier structure

The incident window is spherical as the interior is in a vacuum, which results in shading wherein the center is bright and the periphery is dark [61]. Additional disadvantages include image distortion with a closer distance to the outer edge and *halation* caused by charge saturation of the CCD camera [61]. Flat-panel sensitivity has increased in recent years, and devices with larger diameters can be used; therefore, flat panel detectors have been exchanged with image intensifiers.

The gain of first-generation image intensifiers is determined from the *flux gain* due to electron acceleration and *minification gain* due to focusing. The former is defined as the number of photons generated in a fluorescent screen on the output side per unit photon generated in the fluorescent screen on the input side. This has a typical value of approximately 100. The latter is expressed as the percentage of input surface area to output surface area. As such, the gain can be reduced if the applied voltage is reduced. For example, an image intensifier with an incident window size of 25 cm and an output window size of 2.5 cm has a minification gain of approximately 100. The total gain, in this case, is 10,000, which is the product of the flux gain and minification gain. An incident window size of 10–40 cm can be used. Image intensifiers with a wide field-of-view have a larger gain due to minification gain but are also more likely to generate image distortion.

Second-generation image intensifiers use a *microchannel plate* (*MCP*) instead of an electron lens due to electron multiplication, as shown in Fig. 4.70b. An MCP has a structure comprised of multiple glass pipes with a diameter of 10 μm and length of 1 mm clustered together in a slight inclination. This acts as an electron multiplier by having electrons repeatedly collide into channel walls due to the voltages applied on both ends, which in turn emit secondary electrons each time this occurs. The photocathode, MCP, and output fluorescent screen are placed immediately adjacent to one another; the MCP interior is as shown in Fig. 4.71. Moreover, accelerating an electron in each channel interior increases these numbers and MCPs are used

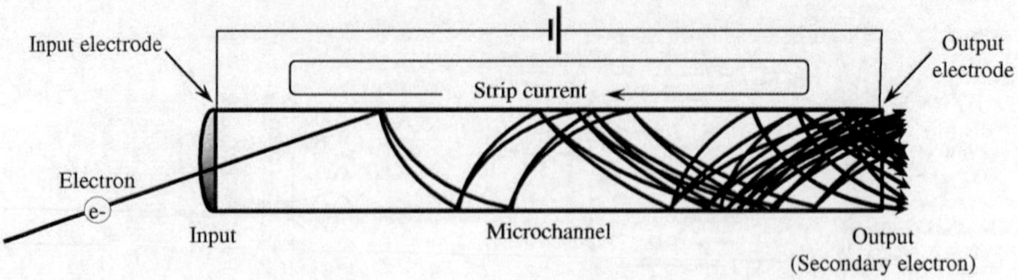

Fig. 4.71 Schematic showing the structure and operations of a microchannel plate

in stacks when high sensitivity is required. A cathode voltage is applied between a photocathode and MCP and a screen voltage is applied between the MCP and fluorescent screen. The electron is then accelerated and guided to the CCD camera with the *fiber optic plate* (*FOP*) after being converted to visible light on the output side.

Similarly to the second-generation image intensifiers, third-generation image intensifiers use an MCP but also increase their sensitivity by using semiconducting crystals such as gallium arsenide (GaAs) or gallium arsenide phosphide (GaAsP) on the photoelectric surface.

Others in use include color image intensifiers, which use fluorescent screens emitting multiple colors to conduct imaging with a color camera and high-speed image intensifiers, which use high-speed response fluorescent screens to image with high frame rate cameras. Aluminum is used as the material in the incident window but beryllium can also be used. This enables imaging with low-energy X-rays.

Finally, representative image intensifier products that are currently commercially available are shown in Fig. 4.72.

(3) CCD Camera

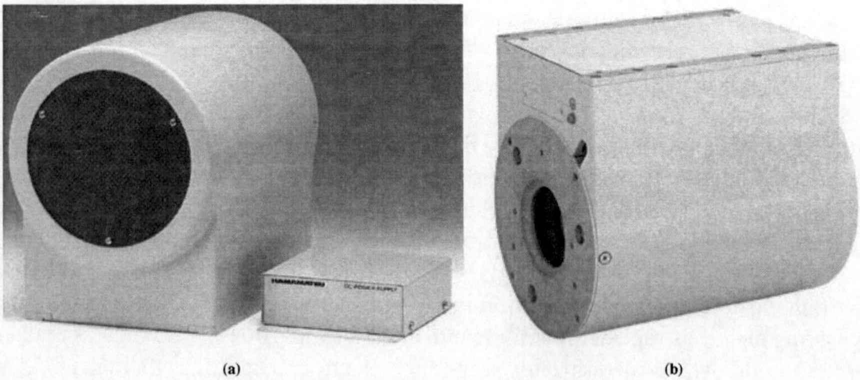

Fig. 4.72 Example of a commercially available image intensifier; **a** Hamamatsu Photonics V10709P (courtesy of Hamamatsu Photonics) and **b** Canon Electron Tubes & Devices E5877J-P1K (courtesy of Hitoshi Chiyoma of Canon Electron Tubes & Devices)

The CCD camera primarily involves conducting impurity doping or surface treatment (e.g. oxide film production) on a silicon wafer, creating a chip by transcribing a detailed pattern like an IC on the wafer. The voltage from CCD image signals forms a row of square waves whose amplitude is proportional to the photon number detected at each pixel. Commercially-available products with pixel sizes as low as 5 μm and pixel numbers at a maximum of 4,096 × 4,096 can be used. As shown in Sect. 4.4.1 (4), dark noise is reduced for scientific measurements by placing the CCD on a Peltier device and cooling the element to approximately −30–100 °C. The Peltier device is a semiconducting element in which passing an electric current in the dissimilar metal joint induces the Peltier effect, generating heat on one side and cooling on the other. Furthermore, there are CCD cameras that use liquid nitrogen for cooling. Condensation and frost can form when temperature differences between the exterior and the element are large, so the camera head is placed in a vacuum container and sealed to prevent these effects.

The fundamental imaging principles of a CCD camera are based on the *internal photoelectric effect*; where electrons in a valence band jump over a forbidden band and are excited into the conduction band when light is irradiated on a semiconductor, resulting in the increase of conduction electrons within the material interior. The basic operation of the CCD camera is to accumulate the generated signal charge within the semiconductor, transferring and reading them out to necessary locations as required.

CCD cameras mainly use photodiode or photo-*metal oxide semiconductor* (*MOS*) elements. Figure 4.73 shows a schematic of a *MOS capacitor*. An insulating layer made of silica is distributed on top of a p-type semiconducting substrate and either a thin metallic electrode (e.g. polysilicon film) or transparent electrode made of an indium tin oxide (ITO) is placed on top of this. The initials of these three layers comprise the abbreviation "MOS." Applying a positive voltage to a given electrode

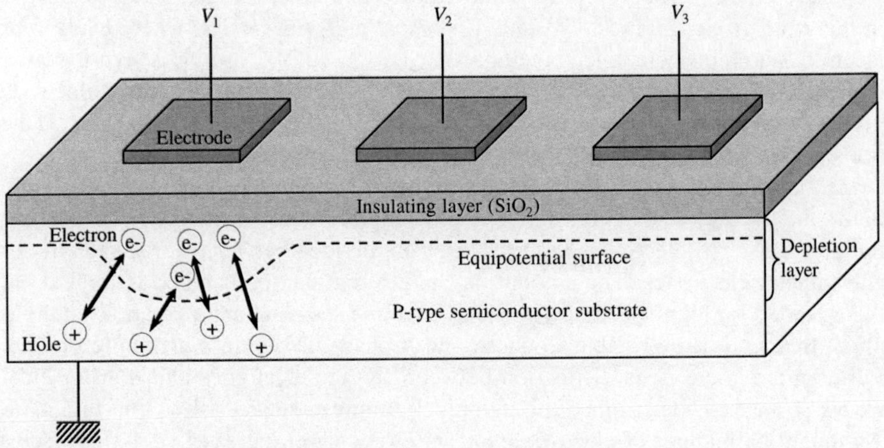

Fig. 4.73 Schematic of the basic structure of a CCD with a MOS capacitor

results in the positive hole within the p-type semiconductor moving away from the electrode, creating a *depletion layer* immediately below the electrode. The p-type semiconductor near the electrode takes a high potential and a *potential well*, which is surrounded by ground potential areas, is formed. Meanwhile, an n-type semiconductor is implanted within a p-type semiconductor at the photodiode, thereby generating a diode. Applying a reverse bias to this, results in the separation of the photoelectron from the positive hole and its accumulation in the n-type semiconducting layer. Figure 4.73 shows that the silicon crystals at the interface between the silicon and insulating layers are truncated and that *dangling bonds* of the silicon atoms are present. Photoelectrons are captured by having high-density localized levels existing in the forbidden bands, where noise is also more easily generated. An embedded channel structure, wherein the potential well is created slightly away from the interface by placing an n-type silicon layer between the silicon and insulating layers, is used.

The basic principles of charge transfer are shown in Fig. 4.74. Applying a positive 10 V electric potential on the central electrode after applying a positive 10 V electric potential on the left electrode results in the electrons distributed across the vicinities of both electrodes. Subsequently reducing the electric potential of the left electrode to 2 V results in the accumulation of electrons near the central electrode. Furthermore, sequential repetitions of this operation on the right electrode results in the charge being transferred to the right side of the figure along the semiconducting surface. In this manner, the charge can be transferred with the electrode as a single unit. There are three mechanisms by which electrons are transferred: *self-induced drift*, in which electron drift occurs due to electric potential gradients generated from the localization of charges; *thermal diffusion*; and *fringing-field drift*, due to electric potential gradients generated by applied voltages.

However, as in Figs. 4.75 and 4.76, it is often the case in actual practice that 2 to 4 electrodes are wired for a single pixel and charge transfers are conducted through a 4-phase clock pulse, where the same voltage is applied in connected electrodes in the same group. Figures 4.75 and 4.76 show a *4-phase drive*, where every four electrodes are connected. This method is widely used. At time t_1, a signal charge is present at electrode groups C and D. At time t_2, the charge spreads to electrode group B and the amount at electrode group D conversely decreases. At time t_3, the transfer of a single electrode's worth is completed compared to time t_1.

A 2-phase drive has the doping of impurities on the semiconductor immediately below the electrode and its concentration gradients are used to generate electric potential gradients. The electrodes in these cases are divided between storage electrodes and transfer electrodes. The amount of charges that can be handled is limited but this is suited for high-speed operations as a simple structure and operations at high-clock frequency are possible; 3-phase and 4-phase drives do not require doping. Furthermore, there is no distinction between storage electrodes and transfer electrodes. Transfer is determined through clock timing sequences shown in Fig. 4.75. Although the number of electrodes and wires per pixel increases, a 4-phase drive has the advantage of an increased number of charges that can be handled as this

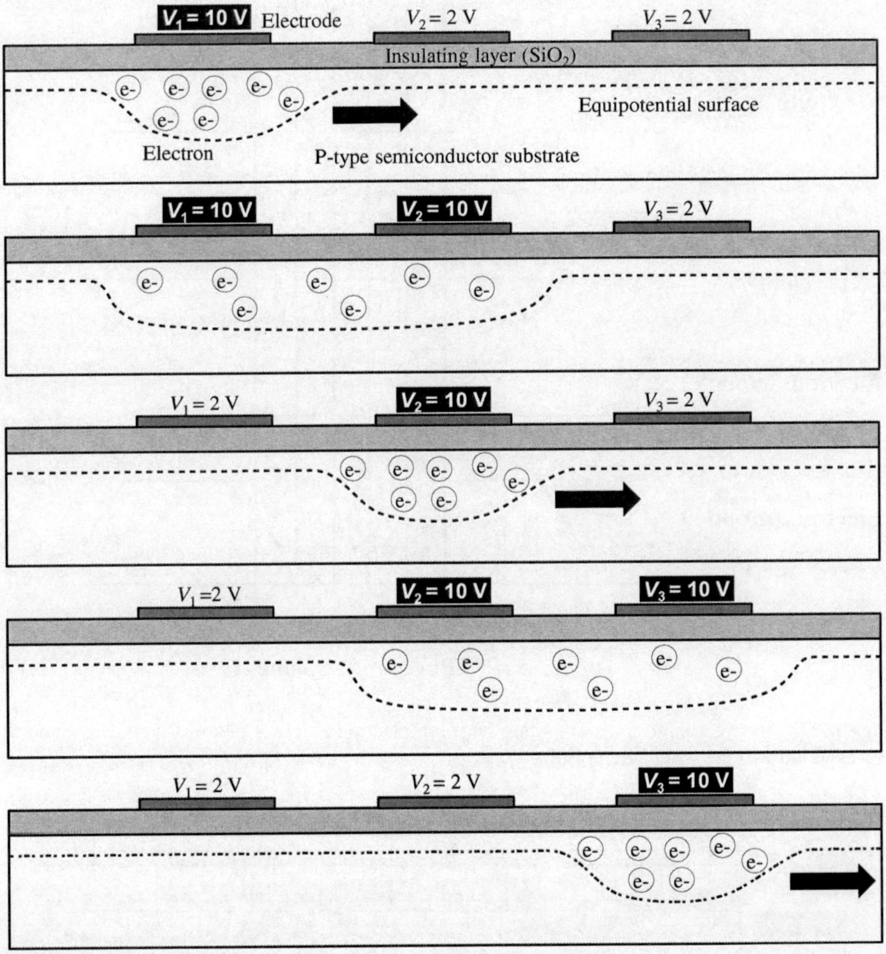

Fig. 4.74 Schematic showing charge transfer between electrodes in a CCD made from a MOS capacitor

corresponds to about half of the overall surface area. Furthermore, transfer in both directions becomes possible.

Figure 4.77 shows four representative electron transfer systems when looking at the overall CCD camera. The *interline transfer* (*IT*) system in Fig. 4.77a has a relatively complex structure in which the light-receiving section and transfer section are independent of each other. The light-receiving element and the light-shielded vertical transfer CCD comprises a single pixel, whose fill factor is relatively small. The signal charge of each pixel following exposure is promptly transferred to the adjacent vertical transfer CCD through the transfer gate. This transfer is instantaneously conducted for all pixels. Afterward, this is transferred row by row to the horizontal transfer CCD and output after converting from a charge to a voltage in the

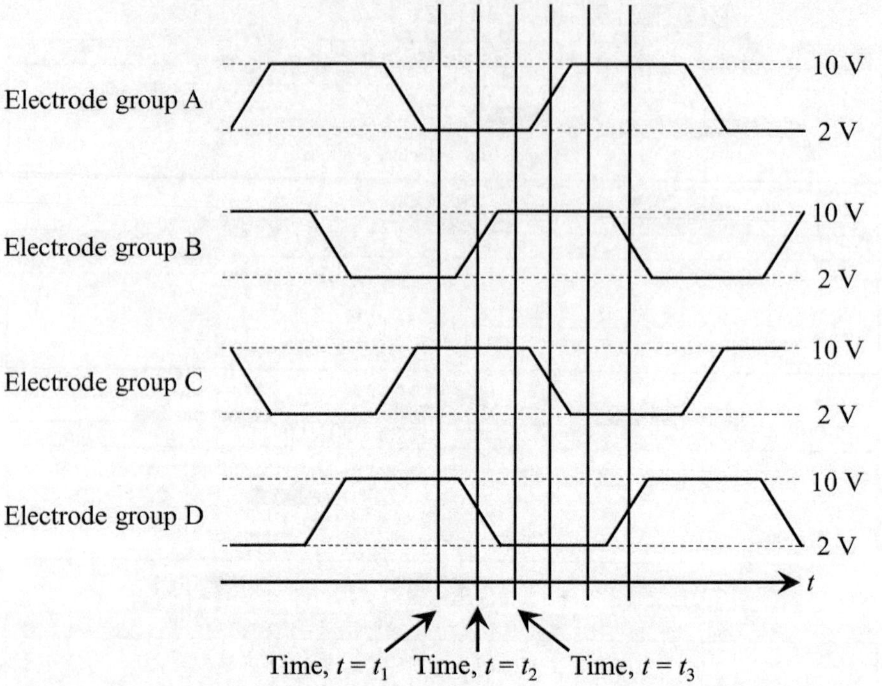

Fig. 4.75 Chart of charge transfer timing between electrodes in a 4-phase drive CCD camera, corresponding to the schematic in Fig. 4.76

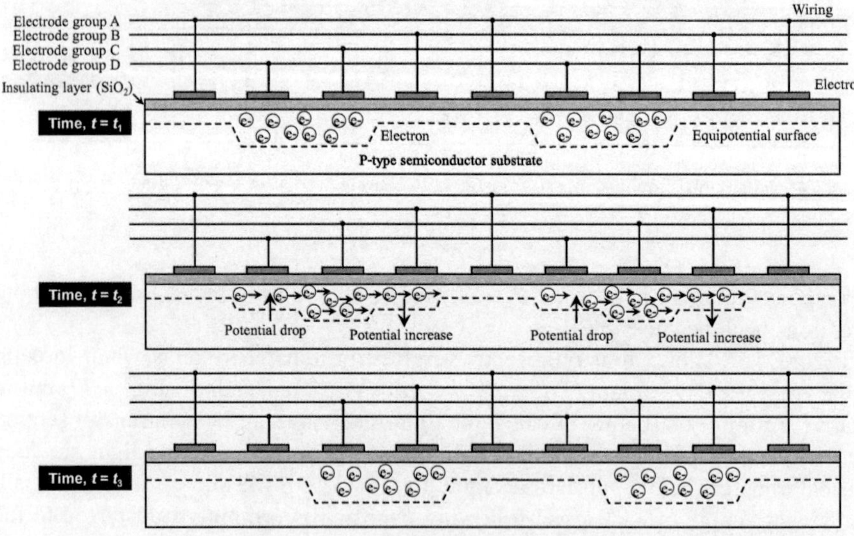

Fig. 4.76 Schematic showing the charge transfer between electrodes in a 4-phase drive CCD camera; time corresponds to the timing chart in Fig. 4.75

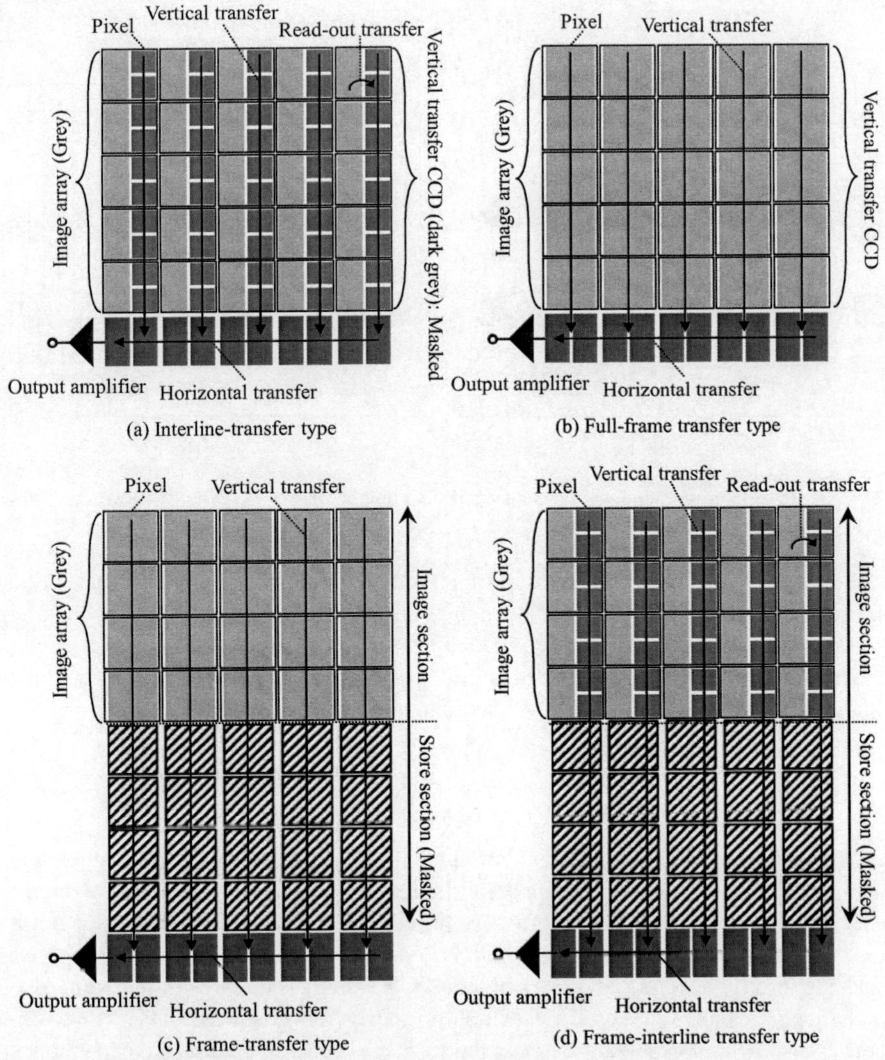

Fig. 4.77 Schematic showing the representative charge transfers in a CCD camera

output circuit. This transfer is conducted during the next exposure, so a shutter is not required. The size can be reduced as an accumulation section is not required, such as in a frame transfer method. IT-CCD cameras with minimal dead time specialize in continuous exposure and are also used in video and digital cameras. They have the disadvantage of generating *smear* due to charge leaking into the vertical transfer CCD. Smear refers to white thread-like artifacts in the vertical direction, which are generated due to signal charges caused by strong incident light entering adjacent pixels or the CCD transfer region and blurring the image. Furthermore, a micro-lens

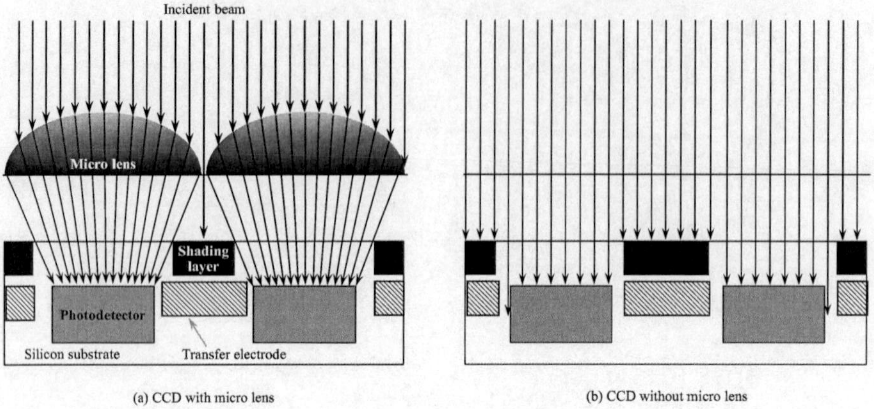

Fig. 4.78 Schematic showing the focusing effects of a micro lens

as shown in Fig. 4.78 is installed as the shielded region in an IT-CCD camera is wide and sensitivity is improved by focusing the light incident to the shielded section and guiding it to the light-receiving section.

The basic principles of the output circuit are to convert the charge Q into a voltage change ΔV_{out} between both ends of the condenser with a capacity C_{C}.

$$\Delta V_{out} = \frac{Q}{C_c} \tag{4.44}$$

Conversions using a condenser with a small capacity are beneficial as they can obtain a high voltage. Generally, an amplifier referred to as a *floating diffusion amplifier* (*FDA*) is used in CCD cameras. As previously mentioned, the read-out noise becomes predominant when the element is cooled to an extent where the dark noise can be ignored. The read-out noise from the CCD camera is mainly *thermal noise* due to channel resistance of MOSFET, which primarily comprises the FDA. Moreover, there is also *reset noise* and *1/*f *noise*; the former is generated when the capacitance of the FDA is reset and is due to fluctuations in reset electric potential due to the mixing-in of thermal noise induced by channel resistance. As can be observed in the following equation, the reset noise is proportional to the absolute temperature and condenser capacity [62] and is also referred to as *kTC noise*:

$$Q_{kTC} = \sqrt{kTC_C} \tag{4.45}$$

Here, Q_{kTC} is the amount of electric charge of noise due to the reset noise and k is the Boltzmann constant; in other words, a lower condenser capacity results in a further suppressed amount of electric charge of noise and conversely increased noise voltage. The electric charge is completely transferred in the case of a CCD camera and the kTC noise is generally not generated in the scanning region. CCD cameras can endogenously be considered to have low noise compared to CMOS cameras. Furthermore, the reset noise is removed by an external circuit referred to as a *correlated double sampling (CDS) circuit*, which takes advantage of the fact that the reset noise is present in the signals of both the reset and data sections, utilizing the difference between the sections. Meanwhile, the $1/f$ noise occurs due to the silicon interface state. As decreases in noise amount are inversely proportional to the frequency f, this noise is particularly problematic in the low-frequency region. This can also be reduced through the CDS circuit.

Incidentally, there is an *electron-multiplying CCD* (*EM-CCD*) camera, which has high sensitivity and an image multiplication functionality due to specialized output registers formed on its chip. The EM-CCD camera was separately developed in 2002 by Texas Instruments and E2V. EM-CCD cameras are not frequently used in X-ray tomography but are suitable for weak light and high-speed imaging applications. In an EM-CCD camera, the electrons in the bottom-most row of the detector are transferred to the multiplication register for each pixel, where the electron can be multiplied by a factor of up to several thousand while still suppressing the read-out noise [63]. This multiplication register is made of several hundred rows of registers. Electron multiplication is conducted through impact ionization caused by a high electric field created by applying a voltage higher than that of a standard horizontal transfer electrode [63]. The read-out noise can be reduced to less than one through electron multiplication [63], but the dark current is amplified alongside the signal. The *clock induced charge (CIC) noise*, generated when the charge drift occurs from the detection element due to electron multiplication, also becomes a new source of noise [63].

The *full-frame transfer (FF)* type in Fig. 4.77b uses virtually all of its surfaces as light-receiving sections. The signal charge is accumulated in the potential well of the light-receiving section during exposure. This is then transferred downward row by row with a vertical transfer CCD and outputted pixel by pixel with a horizontal transfer CCD. Vertical transfer of the next row is conducted once this is complete. The structure is simple, a large number of pixels is possible, and the pixel size can be increased. There is no requirement to set up a transfer region separately from the light-receiving section and the charge-accumulating region can be made wider than in the interline transfer method. For this reason, this technique has the advantages of both high sensitivity and dynamic range and is widely used for scientific measurements. The next exposure cannot be conducted until the output of all pixels is completed, so the read-out time becomes dead time. For this reason, this is not suitable for consecutive exposures like those of a video camera and is instead primarily used for measurements with a low frame rate. Furthermore, a mechanical shutter must be installed to prevent incident light during readout. This results in limits to the control of exposure time, however, which can also result in malfunction.

The *frame transfer* (*FT*) type in Fig. 4.77c has a completely shielded accumulation section with the same pixel number as the light-receiving section. After exposure, the signal charges of all the pixels are simultaneously transferred to the accumulation section and the rows are read out one by one similar to the full-frame transfer type. The

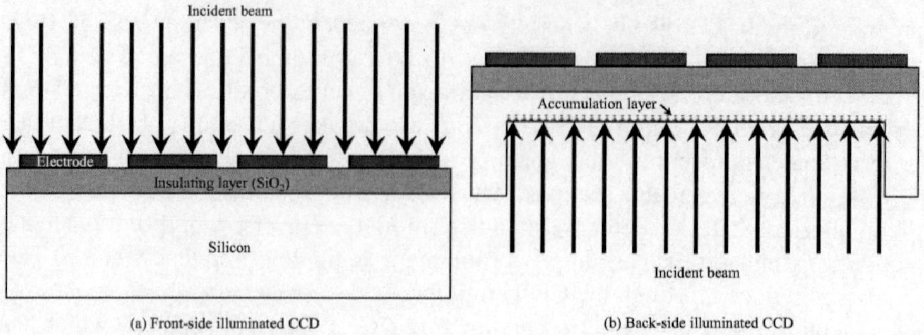

Fig. 4.79 Schematic of front-side and back-side illuminated CCD structures

transfer from the light-receiving section to the accumulation section is conducted at a high speed and the charge production due to light received during transfer is limited. Furthermore, this also has the advantage of allowing for the next exposure during readout from the accumulation section. The total dead time is low and this can also be used for video imaging. Meanwhile, disadvantages include a larger chip size due to the accumulation section and the possibility of smears. The full-frame transfer and frame transfer methods both have a structure that allows for the light transmission of transfer electrodes. There are countermeasures for improving sensitivity that can be implemented, such as using a transparent electrode or a *back-side illuminated* structure.

Figure 4.79 shows a comparison between the front-side and back-side illuminated structures. The incident light in the front-side illuminated CCD camera in Fig. 4.79a is absorbed by the electrode and reflected by the protective screen and oxide screen (e.g. boron phosphor silicate glass (BPSG) on the front-side), greatly reducing its quantum efficiency. Setting up a back-side illuminated structure such as in Fig. 4.79b can greatly increase the quantum efficiency from approximately 40%–90%, as shown in Fig. 4.80 [64]. Furthermore, the back-side illuminated camera can also measure ultraviolet rays (400 nm wavelength and below) and some near-infrared rays (750–1400 nm wavelength), which do not reach the light-receiving element of the front-side illuminated camera. However, another potential well, shown in Fig. 4.81, exists on the back-side of the CCD camera, where the charge cannot be transferred to the front-side electrode [65]. Ion implantation is conducted on the back-side with this in mind and an internal potential, which allows for charge to move from the back-side to the front-side, is generated [65]. The front-side is a p-type semiconducting region with high impurity concentrations referred to as an *accumulation layer*. The substrate thickness is reduced to around 20 μm in back-side illuminated structures to ensure acceptable resolution [65].

The *frame interline transfer* (*FIT*) type in Fig. 4.77d has pixels with the same structure as the interline transfer type and accumulation section as the frame transfer type. Signal charge can be instantaneously transferred to the completely shielded accumulation section through high-speed frame transfer. Furthermore, smear discharge

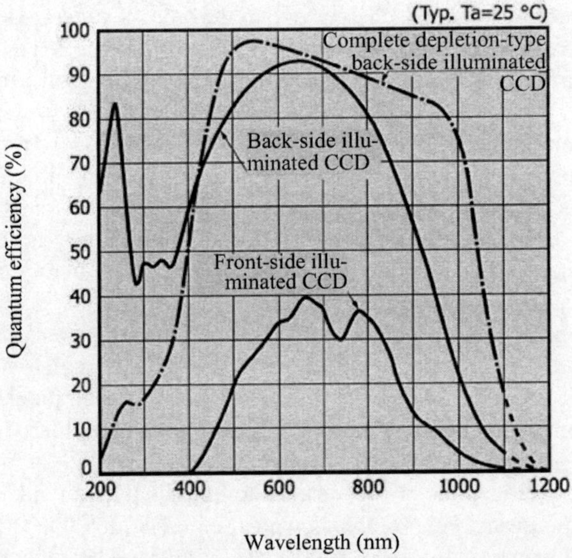

Fig. 4.80 Comparison of the spectral sensitivity characteristics of a backside and front-side illuminated CCD (courtesy of Hamamatsu Photonics)

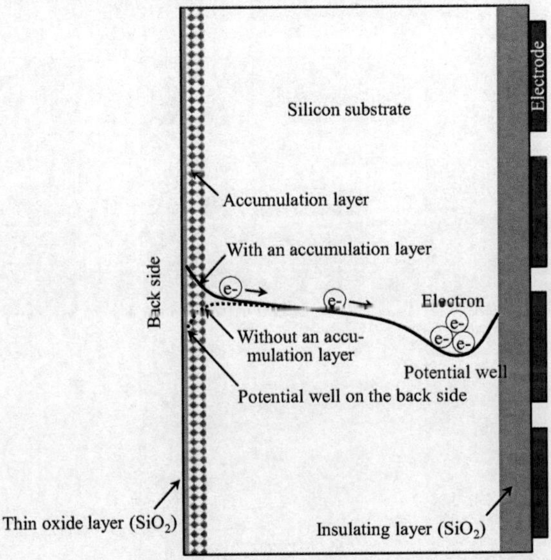

Fig. 4.81 Back-side illuminating CCD structure, internal electric potential, and charge movement

transfer is conducted to remove charges due to the smears generated during vertical transfer before readout transfer. Due to these aspects, this type has the characteristic of generating virtually no smear; its chip size is large and its power consumption is high.

Here, we examine the binning described in Sect. 4.4.1 (4) from the perspective of charge transfer. Figure 4.82 shows a schematic of the basic process when 2 × 2 binning is conducted on a full-frame transfer-type CCD camera. Transfer is conducted in the vertical direction and two columns' worth of signal charge is accumulated in the horizontal transfer CCD. When horizontal transfer is subsequently conducted, two rows' worth of signal charge is accumulated in the output node, enabling the signal charge in 2 × 2 pixels to be processed as a sum. Furthermore, the added number can be increased up to 2 × 2, 4 × 4, 8 × 8, etc. The readout is a single session compared to the numerical additions conducted after the readout of each pixel, which has the advantage of controlling the read-out noise. This can be particularly effective when the signal level is low.

Finally, the specifications of the various commercially-available CCD cameras used for scientific measurements are shown as reference in Tables 4.7, 4.8 through 4.9. Here, CCD cameras for direct X-ray detection, EM-CCD cameras, and other cameras in which the element is cooled, are displayed. Furthermore, Fig. 4.83 shows photographs of the external appearance of representative CCD cameras for scientific

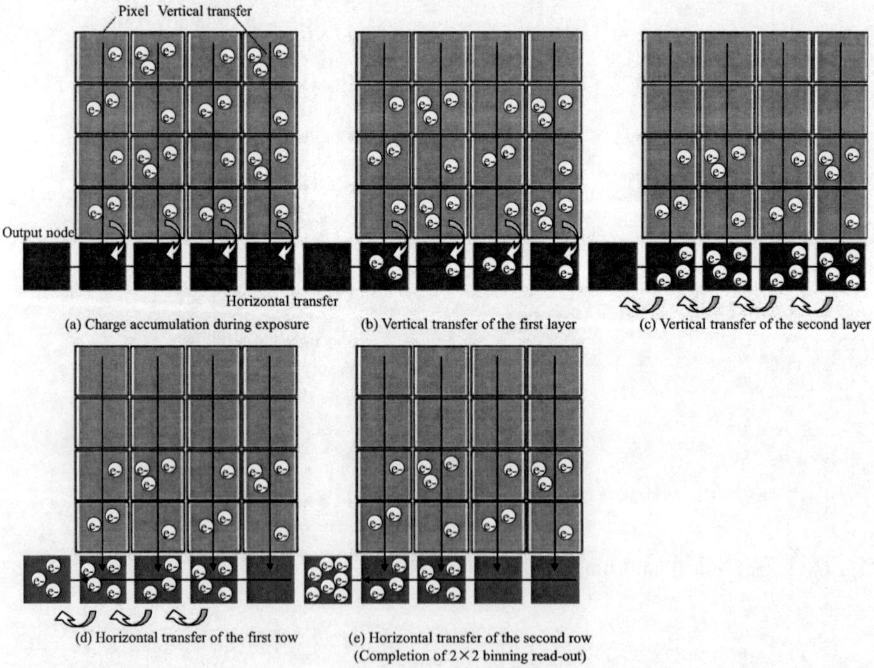

Fig. 4.82 Basic process when conducting 2×2 binning with a full-frame transfer-type CCD camera

Table 4.7 Representative cooled CCD cameras that are commercially available and their specifications

Manufacturer	Roper Technologies			PCO.AG		
Type	PyLoN 2 KB Ex	Blaze 400-LD	Sophia 2048B Ex	pco.dimax HS1	pco.dimax HS4	Pco.dimax CS1
Number of pixels	2048 × 512	1340 × 400	2048 × 2048	1000 × 1000	2000 × 2000	1296 × 1024
Pixel size (μm)	13.5 × 13.5	20 × 20	15 × 15	11 × 11	11 × 11	11 × 11
Back-side/Front-side illumination	Back side	Back side	Back side	Front side	Front side	Front side
Maximum quantum efficiency (%)	97	97	97	50	50	50
Read-out rate (Frames/sec)	3.2	35	3.2	7039	2277	3086
Read-out noise (Electrons-rms)	3.5 @ 50 kHz	2.5 @ 100 kHz	4.5 @ 100 kHz	23	23	22
Cooling system	Liquid nitrogen	Peltier elem. + Air/Water	Peltier elem. + Air/Water	Nil	Nil	Nil.
Minimum cooling temperature (°C)	120	−95/−100	−90/−90	Room temperature	Room temperature	Room temperature
Dark current (Electrons/pixel/sec)	0.1 @ −120 °C	0.0005 @ −100 °C	0.00025 @ −90 °C	530 @ 20 °C	530 @ 20 °C	530 @ 20 °C

(continued)

Table 4.7 (continued)

Manufacturer	Roper Technologies			PCO.AG		
Type	PyLoN 2 KB Ex	Blaze 400-LD	Sophia 2048B Ex	pco.dimax HS1	pco.dimax HS4	Pco.dimax CS1
Full well capacity (Electrons)	800,000	180,000	150,000	36,000	36,000	36,000
Dynamic range (bit)	16	16	16	12	12	12
Hamamatsu Photonics	Thorlabs	BITRAN			ANDOR	
C8000-30	8051 M-USB-TE	BU-51LN	BU-62 M	BQ-83E	DF936N-FB-T2	ZYLA-5.5X-FO
640 × 480	3296 × 2472	1360 × 1024	2048 × 2048	1024 × 1024	2048 × 2048	2560 × 2160
14 × 14	5.5 × 5.5	7.4 × 7.4	7.4 × 7.4	24 × 24	13.5 × 13.5	6.5 × 6.5
Back side	Front side	Front side	Front side	Front side	Front side + FOP	Front side + FOP
90-	51% at 460 nm	60	55	75	97	60
31.4	4.5	17	16	3	0.953	100
150 @ 31.4 Hz	<10 e- at 20 MHz	–	12	13	35 @ 5 MHz	0.9 @ 200 MHz
Peltier elem. + Air cooling	Peltier element	Peltier elem. + Water cooling	Peltier elem. + Water cooling	Peltier elem. + Water cooling	Peltier elem. + Air/Water	Peltier elem. + Air cooling
+ 5	−10 @ Room temperature	−25	−35	−25	−35	0
–	0.1	–	7 @ 40 °C	15.3 @ 25 °C	0.09	0.14
30,000	20,000	16,000	44,000	200,000	100,000	30,000
12	14	16	16/12	16	16	16

Table 4.8 Representative commercially available CCD cameras for direct X-ray detection and their specifications

Manufacturer	Roper Technologies			ANDOR		
Type	PI-MTE In-Vacuum	SOPHIA-XO:2048B	PIXIS-XO:2048B	DO934P	DO936N	iKon XL "SO" 230
Number of pixels	2048 × 2048	2048 × 2048	2048 × 2048	1024 × 1024	2048 × 2048	4096 (H) × 4112 (V)
Pixel size (μm)	13.5 × 13.5	15 × 15	13.5 × 13.5	13 × 13	13.5 × 13.5	15 × 15
Energy range (keV)	1.2–30000	1.2–30000	1.2–30000	–	–	–
Maximum quantum efficiency (%)	96	96	96	–	–	95
Read-out rate (Frames/s)	0.42	3.2	0.44	4.4	0.953	>0.5
Read-out noise (Electrons-rms)	3 @ 50 kHz	4.5 @ 100 kHz	3.5 @ 100 kHz	18 @ 5 MHz	31.5 @ 5 MHz	23 @ 4 MHz
Cooling system	Peltier elem. + Water	Peltier elem. + Air/Water	Peltier elem. + Air/Water	Peltier elem. + Air/Water	Peltier elem. + Air/Water	Peltier elem. + Air/Water

(continued)

Table 4.8 (continued)

Manufacturer	Roper Technologies			ANDOR		
Minimum cooling temperature (°C)	−55	−90/−90	−70/−70	−100	−100	−75
Dark current (Electrons/pixel/s)	0.02 @ −50 °C	0.00025 @ −90 °C	0.002 @ −60 °C	0.0001 @ −100 °C	0.0001 @ −100 °C	0.0001 @–75 °C
Full well capacity (Electrons)	100,000	150,000	100,000	–	–	150,000
Dynamic range (bit)	16	16	16	16	16	16/18

Hamamatsu Photonics	BITRAN		XiMEA		Photonic Science			Raptor Photonics
C8000-30D	BK-501 X	BK-502 X	MH110XC-KK	MH160XC-KK	PSL FDS 2, 83_M	PSL FDS6, 02_M	PSL X-ray VHR 80	Eagle XO X-Ray
640 × 480	1024 × 1024	1024 × 1024	4008(H) × 2672(V)	4872(H) × 3248(V)	1940(H) × 1460(V)	2750(H) × 2200(V)	4008(H) × 2672(V)	2048 × 2048
14 × 14	13 × 13	13 × 13	9 × 9	7.4 × 7.4	4.54 × 4.54	4.54 × 4.54	16 × 16	13.5 × 13.5
0.02–10	0.1–1	1–10	5–100	5–100	5–100	5–100	5–100	0.0012–20
–	90	90	–	–	–	–	–	90

(continued)

Table 4.8 (continued)

Hamamatsu Photonics	BITRAN		XiMEA		Photonic Science			Raptor Photonics
31.4	0.6	0.6	2.1	1.4	3.7	1.7	1.8	–
100 @ 31.4 Hz	12 @ 1 MHz	12 @ 1 MHz	10	8	5–6 @ 12.5 MHz	5–6 @ 12.5 MHz	14–15 @ 10 MHz	2.3 @ 75 kHz
Peltier elem. + Air	Peltier elem. + Air	Peltier elem. + Water	Peltier elem. + Air	Peltier elem. + Air	Peltier elem. + Air	Peltier elem. + Air	Peltier elem. + Air	Peltier elem. + Air, water or liquid nitrogen
+ 5	−35	−35	+12	+12	−20	−20	−10	−90
–	500 @ 20 °C	−500 @ 20 °C	40 @ 12 °C	18 @ 12 °C	0.05 @ −20 °C	0.05 @ −20 °C	0.5 @ −10 °C	0.0004 @ −75 °C
30,000	100,000	100,000	60,000	30,000	18,000	18,000	45,000	100,000
12	16	16	14	14	14	14	16	16

Table 4.9 Representative commercially available EM-CCD cameras and their specifications

Manufacturer	Hamamatsu Photonics		Roper Technologies		Raptor Photonics
Type	C9100-23B	C9100-24B	Pro-EM-HS: 512BX3	Pro-EM-HS: 1024BX3	Falcon III
Number of pixels	512 × 512	1024 × 1024	512 × 512	1024 × 1024	1024 × 1024
Pixel size (μm)	16 × 16	13 × 13	16 × 16	13 × 13	10 × 10
Maximum EM gain	1,200	1,200	1000	1000	5000
Maximum quantum efficiency (%)	>90	>90	95	95	95
Read-out rate (Frames/s)	1076	314	61	25	34
Read-out noise (Electrons-rms)	<1 (EM gain 1200)	<1 (EM gain 1200)	<1 (EM Gain 1000)	<1 (EM Gain 1000)	0.01
Cooling system	Peltier elem. + Water	Peltier elem. + Water	Peltier elem. + Air/Water	Peltier elem. + Air/Water	Peltier elem. + Air, water or liquid nitrogen
Minimum cooling temperature (°C)	−100	−80	−80/−90	−65/−65	−100
Dark current (Electrons/pixel/s)	0.0005 @ −80 °C	0.0005 @ −80 °C	0.001 @ −70 °C	0.002 @ −55 °C	0.0002
Full well capacity (Electrons)	370,000	400,000	200,000	80,000	35000
Dynamic range (bit)	16	16	16	16	16
ANDOR		BITRAN		Nüvü cameras	
iXon-Ultra 888	iXon-Ultra 897	BU-66EM-VIS	BQ-87EM	CCD55-30	
1024 × 1024	512 × 512	1920 × 1080	512 × 512	1024 × 1024	
13 × 13	16 × 16	5.5 × 5.5	16 × 16	13 × 13	
1000	1000	20	100	5000	
>95	>95	50	90	90	
26	56	55	55	17.5	
<1	<1	1 (EM gain 20)	1 (EM gain 100)	<0.1 @ 20 MHz	
Peltier elem. + Air/Water	Peltier elem. + Air/Water	Peltier elem. + Water cooling	Peltier elem. + Water cooling	Peltier elem. + Water cooling	

(continued)

Table 4.9 (continued)

ANDOR		BITRAN		Nüvü cameras
iXon-Ultra 888	iXon-Ultra 897	BU-66EM-VIS	BQ-87EM	CCD55-30
−95	−100	−35	−50	−85
0.00011 @ −95 °C	0.00015@ −100 °C	6 @ 0 °C	400 @ 20 °C	0.0004@ −85 °C
80,000	180,000	20,000	130,000	800,000
16	16	16	16	16

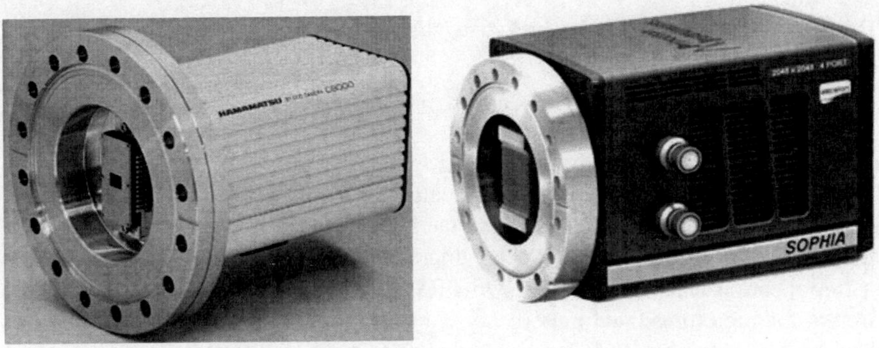

(a) (b)

Fig. 4.83 Example of commercially available CCD cameras for scientific measurements; **a** Hamamatsu Photonics C8000-30D model (cooling CCD camera courtesy of Hamamatsu Photonics) and **b** Roper Technologies PIXIS-XB model (CCD camera for direct X-ray detection courtesy of Kazuyuki Hara of Nippon Roper)

measurements that are currently commercially available. The pixel size in the table is determined by the spatial resolution required by the X-ray CT scanner, sample size, magnification due to the cone beam, and magnification due to the FOP and optical lens (discussed later). Similarly, the number of pixels determines the spatial resolution of the X-ray CT scanner through the sampling theorem discussed below.

(4) CMOS Camera

When comparing CMOS and CCD cameras, there are no differences in the readout and amplification of photoelectric conversion or signal charge. Their difference is in the readout method of the signal charge and the microstructure of the elements, which make that possible. CCD cameras are manufactured through a special process but CMOS cameras are made through a standard CMOS LSI manufacturing process. Figure 4.84 shows the relationship between the Intel CPU release year and the minimum gate length [66]. The minimum gate length has undergone a miniaturization process, wherein its length decreases by a factor of 0.7 and its surface area

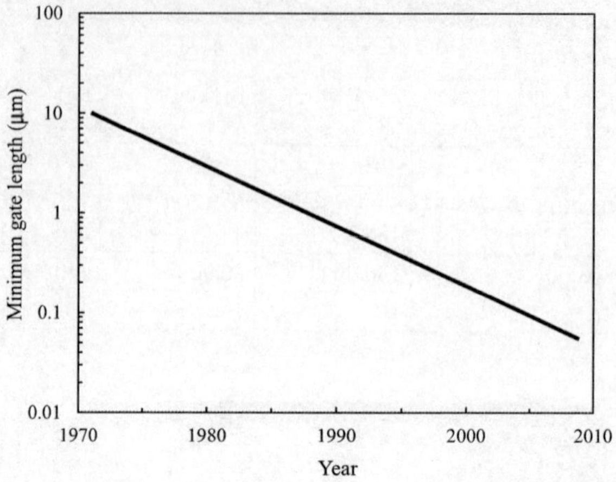

Fig. 4.84 History of miniaturization of the CMOS process [64]

halves every 2 to 3 years [66]. The CMOS camera was put to practical use in the 1990s when semiconducting micro-fabrication techniques were developed and the minimum gate length reached approximately 0.35 μm [67]. High-quality images were obtained due to the development of micro-fabrication techniques, which enabled an amplification a *field-effect transistor* (FET) to be installed in each pixel and where circuits for inhibiting fixed pattern noise were installed. Furthermore, their costs have been reduced by applying existing semiconductor manufacturing processes. For these reasons, the spread of CMOS cameras and its substitution of CCD cameras has rapidly increased in recent years. In this context, Sony announced a complete production suspension of CCD sensors in March 2015. It is anticipated that future transitions from CCD cameras to CMOS cameras will continue.

Signal charges transferred from each pixel to the output section of CCD cameras undergo impedance transformation using the *source-follower amplifier* shown in Fig. 4.85. The source-follower circuit is a basic amplifier circuit, which uses an FET. The gate terminal (G) is the input, the source terminal (S) is the output, and the drain (D) is common to both. As shown in Fig. 4.86, the source follower is referred to as

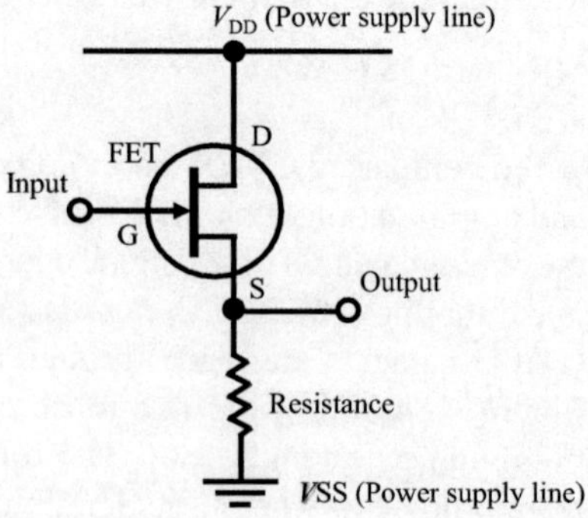

Fig. 4.85 Basic circuit of a source follower that uses an NMOS

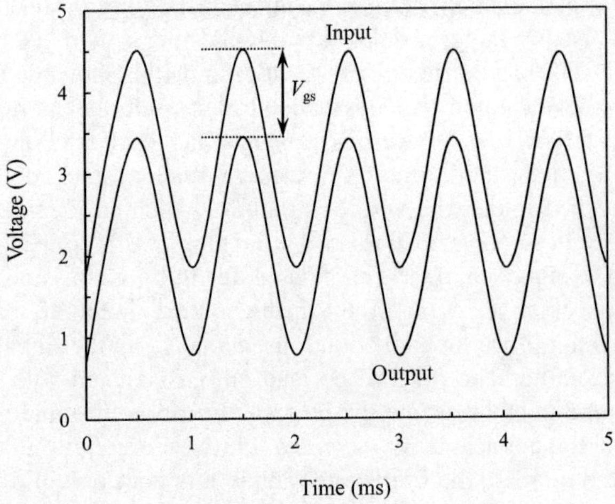

Fig. 4.86 Schematic showing the input/output voltage waveforms of the source follower circuit

such because the output signal moves as if to follow the input signal. Under ideal conditions, the input voltage V_{in} and output voltage V_{out} are equivalent, as in the following equation.

$$V_{out} = V_{in} \tag{4.46}$$

Furthermore, the input impedance is high and the output impedance decreases to a range between $1/g_m$ and several kΩ (g_m is the transconductance (basic gain of FET)), hence, the source-follower amplifier can be interpreted as a buffer amplifier. However, the V_{in}/V_{out} ratio does not become 1 in practice and is instead expressed by the following equation; here, V_{th} is the threshold voltage of the amplifying FET and G_V is the voltage gain of the source follower amplifier [67].

$$V_{out} - (V_{in} - V_{th})G_V \tag{4.47}$$

Here, G_V generally takes a value of 0.7–0.9 [67]. Furthermore, the amplification rate of the charge is between 100 to 10,000 times [67]. The signal charge can be amplified within the pixel, so the influence of noise generated in subsequent transfer stages becomes extremely small in comparison, as shown in the following equation:

$$N_t = \sqrt{G_V^2 N_{pixel}^2 + N_{readout}^2} \tag{4.48}$$

Here, N_{pixel} is the noise generated within the pixel and $N_{readout}$ is the noise generated in readout circuits and various signal-processing circuits outside of the pixel. Meanwhile, the V_{th} value varies according to each transistor and fluctuates between the range of 10–100 mV [67]. This implies that the output for each pixel varies widely for incident light of the same intensity. For this reason, the occurrence of fixed pattern noise is a larger problem in CMOS cameras than in CCD cameras.

The basic structure of a CMOS camera is shown in Fig. 4.87. An external appearance of a commercially-available element is shown in Fig. 4.88. The CMOS camera is an *active pixel sensor* (*APS*) in which each pixel has a source-follower amplifier, in which, unlike the CCD camera, readout is conducted after amplification. This is the same as the DRAM structure, which has dual-address signal lines composed of intersecting bit lines and word lines, with an FET located at each intersection. The signal charges accumulated due to light emission on each pixel are selected in the pixel selection circuit in the column direction, after which the pixel selection circuit in the row direction is used to read out the signal while conducting sequential scanning. The pixel to be read out is indicated with the (x, y) coordinates of the column and row and the post-amplification information can be read out by turning on the applicable MOS switch. Furthermore, limiting the readout to specified areas can increase the frame rate. This is a major point of difference from CCD cameras, which can only read out as indicated by the arrangement. Furthermore, the sequential readout enables this to be driven using a single low-voltage power source, thereby

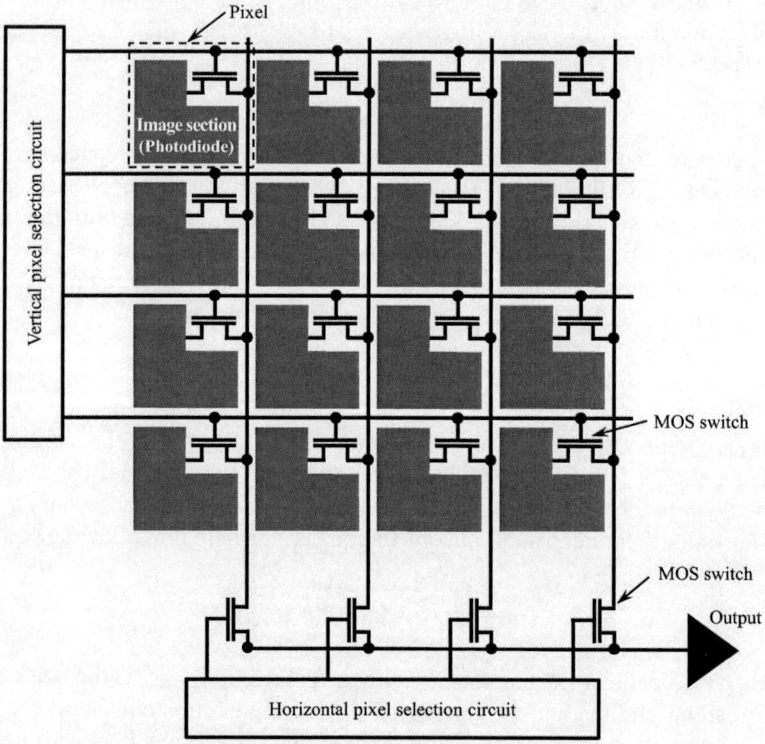

Fig. 4.87 Schematic showing the microstructure of a CMOS camera; each pixel is depicted as a passive pixel sensor without an amplifier for simplicity

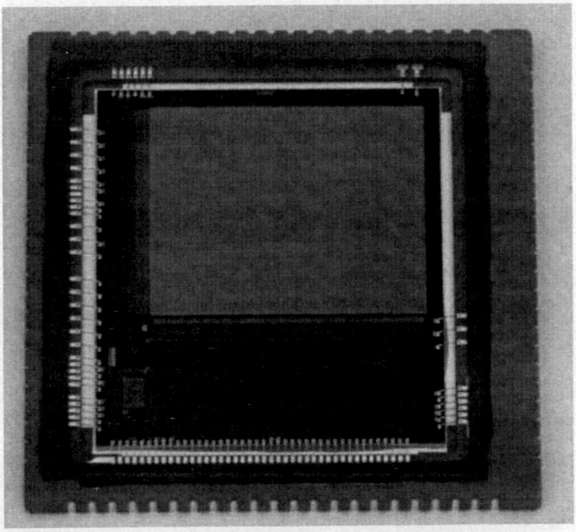

Fig. 4.88 Example of a commercially available CMOS image sensor, Hamamatsu Photonics S13101 model (courtesy of Hamamatsu Photonics)

requiring less power consumption. Incidentally, three power sources are required for the operation of the vertical and horizontal CCD in the CCD camera.

Next, a schematic of the flow process from when a single pixel receives light and obtains a charge to when a pixel signal is outputted is shown in Fig. 4.89 [62]. The photoelectric conversion of the incident light occurs at the photodiode, after which it is input in the source-follower circuit. Other than the photodiode, the pixel

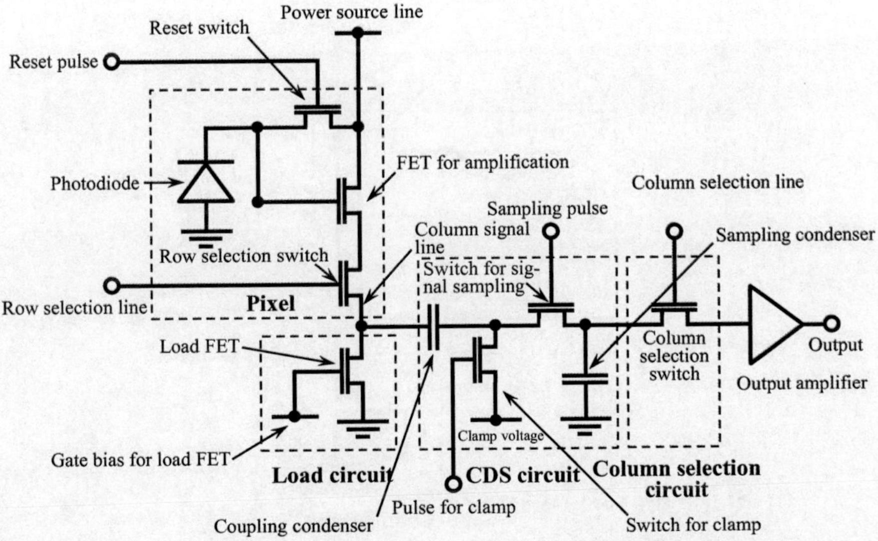

Fig. 4.89 Schematic of a basic circuit of a CMOS camera at the pixel level [65]

comprises the reset FET, which uses a reset pulse to reset the photodiode potential, and a line selection FET, which induces a conduction state between the load FET and amplification FET. After coming out of the source follower circuit, the column CDS type in Fig. 4.89 takes the difference between the signal in the reset section and the signal in the data section at the CDS circuit connected to the column signal line and the fixed pattern noise is removed. A clamp pulse is applied to the clamp FET when the signal voltage is input at the CDS circuit, and the clamp FET is turned off after the voltage at that terminal is set as the clamp voltage [67]. Regarding the second signal charge, the existence of the coupling condenser results in a potential equivalent to the change in both signals appearing in between the terminals. The difference of both signals can be obtained by retaining this potential at the sampling condenser by applying a sampling pulse in the sampling FET [67]. The signal charge retained in the sampling condenser is guided into the output amplifier due to the column selection pulse of the column selection circuit, which is then amplified to become an output image [67]. However, fixed pattern noise cannot be completely removed; the variability of the CDS circuit itself can act as the source of the fixed pattern noise, and load impedance varies depending on operation timing.

A simple structure in which each pixel is composed of three FETs is shown in Fig. 4.89; floating diffusion is set up in the photodiode in actual practice as shown in Fig. 4.90 and a fourth FET for reading out the floating diffusion between both are also frequently set up. In these cases, the reset floating diffusion potential is first outputted, after which the signal charge is transferred there, at which point the potential is outputted a second time. As both include the same reset noise, reset noise as well as fixed pattern noise can be removed from CDS circuits.

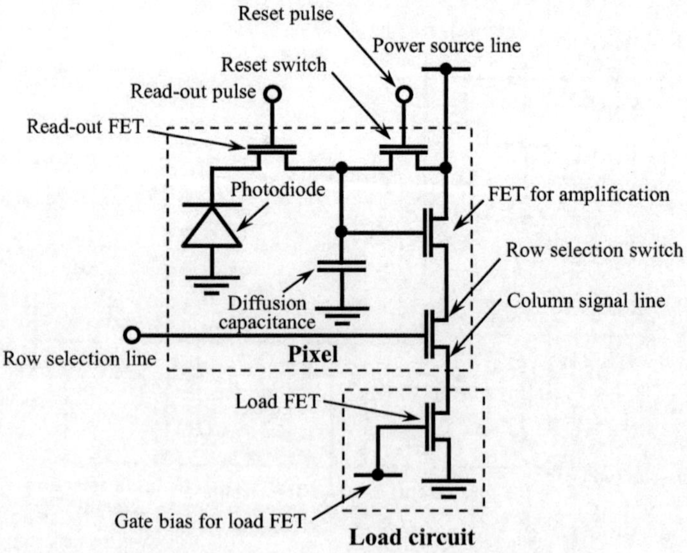

Fig. 4.90 Schematic showing the basic circuit of a CMOS camera with pixels that have 4 FETs

Incidentally, a typical CMOS sensor is a sequential exposure method by each pixel line, so there is a time lag in signal readout within the image screen. For this reason, dynamic distortion occurs when the subject movement is rapid relative to the exposure time. For example, a maximum temporal lag of 0.01 s occurs between pixels for the readout of the same image with a frame rate of 100 fps. Here, a *global shutter* functionality, which exposes all pixels simultaneously, can be realized if analog memory, which accumulates signal charge within the pixel for a duration equivalent to this lag, is prepared. Another FET must be added within the pixel to prepare the global shutter functionality, and either the fill factor of the pixel must be decreased or the size of the individual pixel must be increased. In contrast, those where resetting, accumulation, and readout are conducted for each line, as well as where the photoelectric transformation time is shifted for each line, are referred to as a *rolling-shutter* type. *Scientific CMOS* (*sCMOS*) cameras often use global shutters in addition to rolling shutters.

CMOS cameras are made using a standard CMOS LSI manufacturing process, so an AD converter can be incorporated in its interior. An AD converter can be incorporated in either the pixel, column, or final readout section. Noise is superimposed when an AD converter is inserted in the final readout section. Meanwhile, incorporating the AD converter in each pixel reduces the surface area percentage of the light-receiving area. Furthermore, variations in each pixel induce unevenness in sensitivity. Thus, setting up the AD converter in the column is currently a mainstream procedure. This has the advantage of not superimposing noise after *quantization* (approximating continuous analog data as digital data) with an AD converter. When considering the AD converter with an N_{AD}-bit resolution (*quantization bit rate*), signals below $\frac{V_{FS}}{2^{N_{AD}}}$ become information below one bit when the full voltage scale is set as V_{FS}, resulting in a *quantization error*. The noise generated from a *quantization error* can be considered random noise.

CMOS camera noise is compared with CCD camera noise and summarized in Table 4.10. The amount of dark current generation in photodiodes is also extremely high in a CMOS camera [62]. This is because a pn junction between the source/drain and substrate is used as a photodiode in the standard CMOS LSI manufacturing process; moreover, electron-hole pairs are thermally excited and generated mainly near the interface between the insulating SiO_2 screen and photodiode [62]. An embedded structure is used similar to that in CCD cameras to avoid this phenomenon. This process is thought to reduce the dark current to approximately 1/10 [62]. Of the four to five MOS FETs set up for every pixel, the noise generated by the threshold voltage V_{th} of the amplification FET has already been discussed. Much of the random noise in the pixel circuit is also due to this amplification FET [62]. Noise referred to as *random telegraph signal (RTS) noise* is prominent with decreases in the gate surface area accompanying the miniaturization of the pixel size. This occurs as a result of the entering and exiting of carriers in the silicon interface trap and appears as MOS FET drain potential fluctuations [62]. The reset noise in the CMOS camera mainly appears as the reset noise of a pixel.

Finally, we touch upon the sCMOS camera. The performance of sCMOS cameras has increased in recent years, as they have shown a tendency to replace CCD cameras

Table 4.10 Comparisons of the primary noise types present in CCD cameras and CMOS cameras

Type		CCD camera	CMOS camera
Random noise	kTC noise	Amplifier reset noise	Pixel reset noise
	Amplifier noise	1/*f* noise	1/*f* noise
		Thermal noise	Thermal noise
		–	RTS noise
		–	Output amplifier noise
	Dark noise	VCCD	–
		Photosensor	Photosensor
	Photon noise	Incidental to X-ray imaging irrespective of camera type	
Fixed-pattern noise	Dark noise	Photosensor	Photosensor
		VCCD	–
	Uneven sensitivity	Photosensor and others	Photosensor and others
		–	Variation in threshold voltage for Amplifier FET, V_{th}

for various scientific measurement applications. This is because characteristics like a high quantum efficiency, low noise, and high dynamic range have all been realized through various innovations, including an embedded-type photodiode, a micro lens, a column-parallel AD converter, (vacuum) cooling, the combined use of amplifiers with two standard high and low levels, global shutter functionalities, and back-side illumination. Specifications of various sCMOS cameras that are commercially available are summarized in Table 4.11. These should be compared with the CCD cameras and EM-CCD cameras shown in Tables 4.7, 4.8 and 4.9. Furthermore, Fig. 4.91 shows a photograph of the external appearance of a commercially-available sCMOS camera.

(5) Flat Panel Detector

The light-receiving surface size is 8–64 mm on a single size for various CCD cameras and sCMOS cameras in Tables 4.7, 4.8 and 4.9, 4.10 and 4.11. Most sCMOS cameras, in particular, have a size less than 20 mm. In contrast, flat-panel detectors are thin, have high performance and a large surface area. Light surface sections can measure from around 50 mm to up to 430 mm on one side. CCD cameras and sCMOS cameras are used in X-ray tomography at synchrotron radiation facilities, particularly in wiggler or undulator beamlines. In contrast, flat-panel detectors can be considered important detectors when imaging comparatively large samples or when placing the sample near the X-ray source to obtain high magnification. Many commercially-available X-ray CT scanner models have a flat panel detector installed as well, as in Table 6.1 shown later. Furthermore, high-performance flat panel detectors with a large effective light-receiving area size have been used in recent years and have begun

Table 4.11 Representative sCMOS cameras that are commercially available and their specifications

Manufacturer	ANDOR			PCO AG		
Type	Neo5.5	Zyla5.5	Zyla4.2 Plus	Pco.edge5.5	Pco.edge4.2	Pco.edge3.1
Number of pixels	2560 (H) × 2160 (V)	2560 (H) × 2160 (V)	2048 × 2048	2560 (H) × 2160 (V)	2048 × 2048	2048 (H) × 1536 (V)
Pixel size (μm)	6.5 × 6.5	6.5 × 6.5	6.5 × 6.5	6.5 × 6.5	6.5 × 6.5	6.5 × 6.5
Back-side/Front-side illumination	Front side	Front side	Front side	Front side	Front side	Front side
Maximum quantum efficiency (%)	60	60	82	60	82	60
Read-out rate (Frames/s)	100	100	100	100	100	50
Read-out noise (Electrons-rms)	1.3 @ 560 MHz	1.2 @ 560 MHz	1.1 @ 540 MHz	1.5 @ 95.3 MHz	1.4 @ 95.3 MHz	1.5 @ 105 MHz
Cooling system	Peltier elem. + Air/Water	Peltier elem. + Air/Water	Peltier elem. + Air/Water	Peltier elem. + Water	Peltier elem. + Water	Peltier elem. + Water
Minimum cooling temp. (°C)	−40	−10	−10	+5	+5	+5
Dark current (Electrons/pixel/s)	0.007 @ −40 °C	0.019 @ −10 °C	0.019 @ −10 °C	0.5 @ 5 °C	0.5 @ 5 °C	0.5 @ 5 °C
Full well capacity (Electrons)	30,000	30,000	30,000	30,000	30,000	30,000
Dynamic range (bit)	12/16	12/16	12/16	16	16	16
Global shutter function	○	○	×	○	×	○

(continued)

Table 4.11 (continued)

Hamamatsu Photonics	Thorlabs	Nippon Roper	SPOT IMAGING	FLI		
C13440-20CU	Quantalux CS2100M-USB	KURO 1200B	KURO 2048B	RT	KL400	KL2020
2048 × 2048	1920 (H) × 1080 (V)	1200 × 1200	2048 × 2048	2448 (H) × 2048 (V)	2048 × 2048	2048 × 2048
6.5 × 6.5	5.04 × 5.04	11 × 11	11 × 11	3.45 × 3.45	11 × 11	6.5 × 6.5
Front side	Front side	Front side	Back side	Back side	–	–
82	61	95	95	66	94	62
100	50	41	25	36	48	376
1.4 @ 100 Hz	<1 @ 74.25 MHz	1.3 @ 200 MHz	1.3 @ 200 MHz	2.1	1.5	6
Peltier elem. + Water	- (Heat dissipation structure)	Peltier elem. + Air/Water	Peltier elem. + Air/Water	Peltier elem. + Air cooling	Peltier elem. + Water	Peltier elem. + Water
−30	–	−10/−25	−10/−25	−25	−40	−40
0.006 @ −30 °C	20 @ 20 °C	0.7 @ −25 °C	0.7 @ −25 °C	0.18 @ 20 °C	0.4 @ −20 °C	1 @ 0 °C
30,000	23,000	80,000	80,000	11,000	89,000	48,000
16	16	16	16	12	16	15
×	×	×	×	O	×	O

(a) (b)

Fig. 4.91 Examples of commercially available sCMOS cameras. From Table 4.11, **a** Hamamatsu Photonics ORCA-Flash 4.0 V3 model (C13440-20CU) (courtesy of Hamamatsu Photonics) and **b** ANDOR Neo (right) and Zyla (left) (courtesy of Satoshi Hidaka of Andor Technologies, Ltd.)

to replace image intensifiers. Flat-panel detectors have the advantage of minimal image distortion when compared to image intensifiers, which are the same form of detectors with a large surface area. This is currently an important point, considering the increasing importance of the dimensional measurement applications.

The flat panel detector uses a *thin film transistor* (*TFT*) array, which serves as the backbone of the liquid crystal display. TFT is a type of FET. A semiconducting film is created on a glass substrate, into which transistors used for switching elements, gate insulation films, electrodes, or protective insulation films are incorporated. This manufacturing technique has been used for liquid crystal displays and flat-panel TVs and has been well developed. In a liquid crystal display, the liquid crystal layer is placed between the glass substrates, which comprise the TFT-formed glass substrate and the counter electrode, and the TFT functions as a switch that applies voltage on the liquid crystal in each pixel. Meanwhile, the TFT array in the flat panel detector functions as a two-dimensional readout circuit.

Incidentally, the flat panel detector can be broadly classified between a *direct conversion type* and an *indirect conversion type*. Figure 4.92 shows the mechanisms of both types. The read-out mechanisms are explained in the CMOS section (Sect. 4.4.2 (4)). The two types of photoelectric conversion mechanisms and the materials primarily used for this are detailed below.

The direct conversion type shown in Fig. 4.92a directly converts X-rays into signal charges and images them using a *photoconductive film* comprised of a 0.5–1-mm thick *amorphous selenium* (*a–Se*) semiconductor. The TFT, electrode, and signal charge accumulation condenser are arranged in a matrix form in what is referred to as a TFT array (the halftone dot mesh section in Fig. 4.92) and both the X-ray conversion screen and bias electrode are laminated to cover almost the entirety of the TFT array. The charges of the electron-hole pairs excited in correspondence with the intensity of the incident X-rays are moved to the electrodes of each pixel due to the bias voltage applied in the photoelectric conversion film, after which they are accumulated in the signal charge accumulation condenser within the TFT array. Charge transfer is

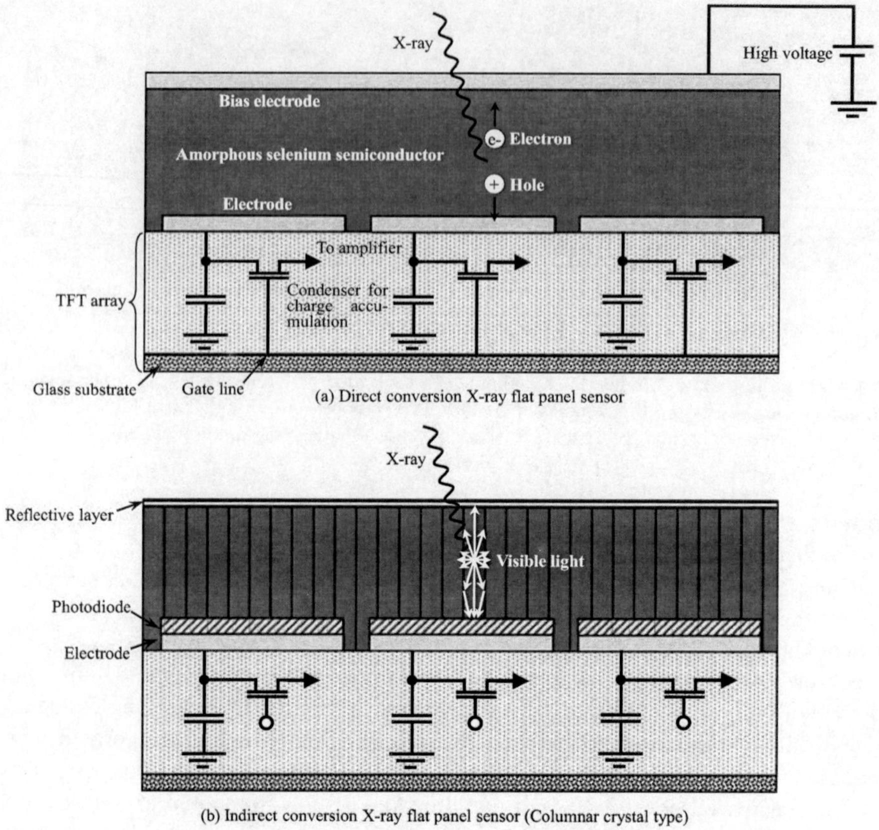

Fig. 4.92 Schematic showing the structures of two representative flat panel detectors

delayed in the direct conversion type due to the presence of energy barriers between each layer and residual images are more likely to appear. However, the structure in which a high voltage is applied, and the charge drawn into the nearest electrode is advantageous in terms of spatial resolution. A two-dimensional X-ray image can be obtained by turning the TFT in each pixel on or off and sequentially reading out the charge voltage of the condenser. This readout is similar to the CMOS camera. The amplification circuit and AD converter are located on the same substrate in the data bus line terminal and the read-out charge information is outputted as digital image information.

The materials used in a photoelectric conversion film are discussed here. *Lead iodide* (PbI_2), *lead oxide* (PbO), and thallium bromide (TlBr) are used as photoelectric conversion materials in addition to amorphous selenium; materials like *cadmium zinc telluride* ($CdZnTe$), *cadmium telluride* ($CdTe$), *cadmium selenide* ($CdSe$), and *mercury iodide* (HgI_2) are also considered promising materials in this regard [68]. Table 4.12 shows the various photoconductors that can be used for large-surface

Table 4.12 Major photoconductors that can be used in flat panel detectors and their specifications [69]

Photoconductor	Decay distance, δ(μm)		E_g (eV)	$W_\pm$ (eV)	$\mu\tau(10^{-5}$ cm^2/V)	
	20 keV	60 keV			Electron	Hole
Amorphous selenium (a-Se)[a]	49	998	2.2	45 (10 V/μm), 20 (30 V/μm)	0.03–1	0.1–6
Polycrystalline silver iodide (HgI_2)	32	252	2.1	5	1–100	0.1–1
Polycrystalline cadmium zinc telluride (Cd0.95Zn0.05Te)[a]	80	250	1.7	5	– 0	– 0.3
Polycrystalline lead iodide (PbI_2)	28	259	2.3	5	0.007	– 0.2
Polycrystalline lead oxide (PbO)[a]	12	218	1.9	8–20	0.05	Small
Polycrystalline Thallium bromide (TlBr)[a]	18	317	2.7	6.5	Small	0.15–0.3

[a]Vacuum deposition; the others: PVD

area applications (e.g. flat panel detectors) and their characteristics [69]. The δ in the table is the *decay distance*, which is the inverse of the linear absorption coefficient. The decay distance corresponds to the depth from the incident surface at which the incident X-ray is 63% attenuated. The quantum efficiency η_q for a photoelectric conversion film of thickness L_p is expressed as follows [69]:

$$\eta_q = 1 - e^{\frac{-L_p}{\delta}} \tag{4.49}$$

Using amorphous selenium with a 1000-μm thickness, for example, Table 4.12 shows that the quantum efficiency at 20 keV becomes 1. Figure 4.93 shows the X-ray energy dependency of the linear absorption coefficients of the various photoconductors shown in Table 4.12 [71]. Considering only quantum efficiency, other photoconductors are superior at X-ray energies below 12 keV. Furthermore, the *stopping power* of lead oxide or thallium bromide is also superior above the K absorption edge of selenium (12.65 keV). Equation (4.49) is as shown below when the X-ray source is not monochromatic [4]:

$$\eta_q = \frac{\int_0^{E_{max}} \Phi(E)\left(1 - e^{\frac{-L_p}{\delta}}\right) dE}{\int_0^{E_{max}} \Phi(E) dE} \tag{4.50}$$

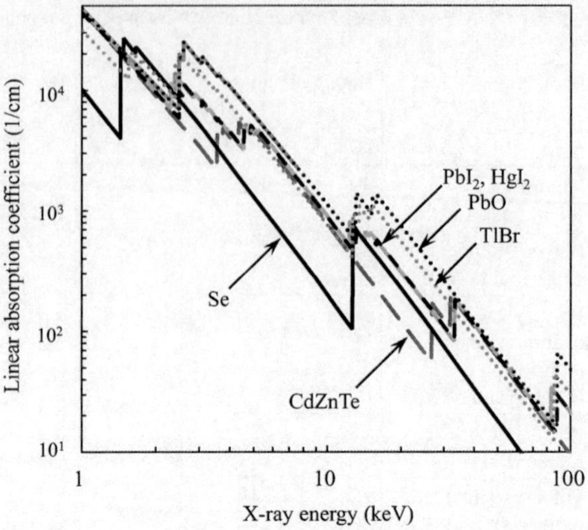

Fig. 4.93 Linear absorption coefficient of various photoconductors, which can be used in applications on large surface areas such as flat panel detectors

Here, Φ (E) is the X-ray spectrum; in other words, the number of photons entering a detector is not important for a typical X-ray detector but rather the integrated value of X-ray energies entering the detector.

X-ray attenuation is not the only important aspect of photoelectric conversion films. The E_g in Table 4.12 is the band gap. The following relational expression can be established for *electron-hole pair creation energy* $E_{\pm}$ in a polycrystalline material [69]:

$$E_{\pm} = \beta_{lc} E_g \approx 3E_g \tag{4.51}$$

Here, β_{lc} is a constant, typically with a value of 2–3 [72]. The electron–hole pair number $n_{\pm}$ of photoelectric conversion films is equivalent to the incident X-ray energy divided by the electron–hole pair production energy, as shown in Eq. (4.52). For this reason, a narrower band gap is advantageous for quantum efficiency [69].

$$n_{\pm} = \frac{E}{E_{\pm}} \tag{4.52}$$

According to Table 4.12, for example, 1000 electron-hole pairs would be produced for a single X-ray photon of 45 keV in an electric field with strength 10 V/μm.

Meanwhile, as can be observed from Table 4.12, Eq. (4.51) does not hold for amorphous selenium. In this case, the electron-hole pair creation energy is inversely proportional to the electric field F, as shown below [69]:

$$E_{\pm} = E_0 + \frac{B}{F} \tag{4.53}$$

Here, B is a constant that shows a weak dependency on X-ray energy; $B \approx 4.4 \times 10^{8}$ eV·Vm $^{-1}$ and $E_0 \approx 6$ eV in a 20–40 keV range [69]. As such, the electron-hole pair creation energy of amorphous selenium in typically used electric fields (10–20 V/μm) is extremely high, in the order of several dozens of eV. However, high quantum efficiency is obtained owing to the amount of charge produced but also requires that the generated charge does not recombine or become trapped and is efficiently collected at the electrodes. Furthermore, raising the voltage brings this closer to a standard semiconductor. However, an electron avalanche occurs prior to completely returning to normal. In Table 4.12, μ is the drift mobility and τ is the average lifespan of the charge carrier. The term $\mu\tau F$, which multiplies the electric field intensity, is the average mobility distance until the electron carrier is trapped or recombined. The $\mu\tau F$ needs to have a value higher than the photoelectron conversion film thickness for the electrodes to capture most of the generated charge.

Amorphous selenium has the advantages of low-cost, as it can be coated across wide surface areas in a low-temperature process of 60–70 °C and its ability to obtain a homogeneous response [68]. Another major advantage is its low dark current at room temperature and its resistance to radiation damage in experiments. Points of caution are that films need to be somewhat thick against high-energy X-rays given that the atomic number is relatively small ($Z = 34$) and that a high voltage (typically around 10 kV) must be applied [68]. Furthermore, high-purity amorphous selenium is straightforward to crystallize, so a 0.2–0.5% arsenic addition is included [68].

The indirect conversion type in Fig. 4.92b uses a scintillator to convert the X-ray into visible light, and then into a signal charge that corresponds to the intensity of that visible light; that this is stored with the condenser up to readout is the same as with the direct conversion type. With the indirect conversion type, the isotropic generation of visible light with the scintillator results in crosstalk and the potential for reduced spatial resolution. Thicker scintillators, in particular, result in more X-rays being absorbed and worsened spatial resolution. Furthermore, the direct conversion type is advantageous in terms of quantum detection efficiency, as shown in Fig. 4.94 [70].

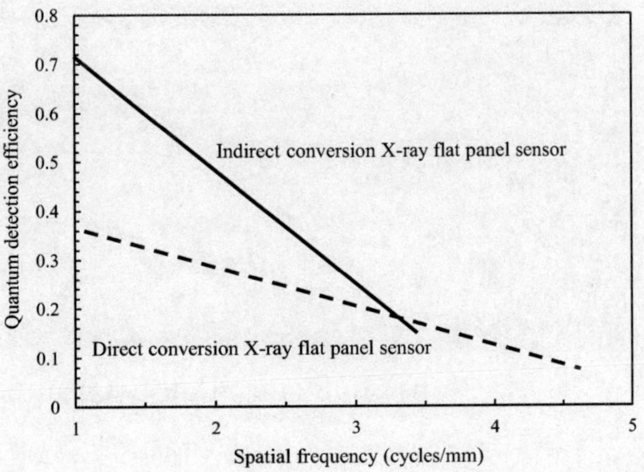

Fig. 4.94 Comparisons of quantum detection efficiency of direct conversion-type and indirect conversion-type flat panel detectors (schematic) [69]

Cesium iodide (*CsI* (*Tl*)) or *gadolinium oxysulfide* (*GOS*) (Gd_2O_2S (*Tb*)) are used for the scintillator. The powdered form of GOS is referred to as *P43*. Figure 4.95 shows cross-section photographs of these two types of scintillator materials [64]. CsI (Tl) has a primarily columnar crystal structure as seen in Fig. 4.95a) and, depending on the conditions, approximately 83% is reflected within the columnar crystal [68].

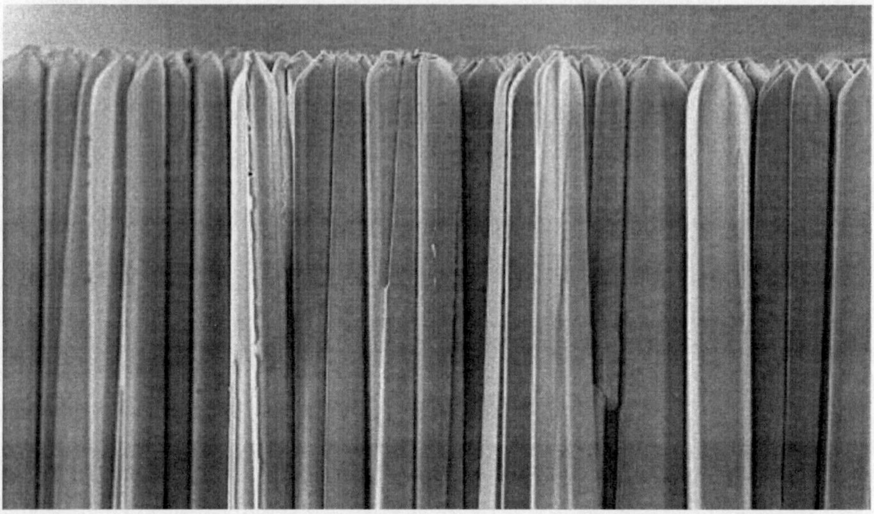

(a) CsI(Tl) scintillator (Columnar crystal)

(b) GOS scintillator (Granular crystal)

Fig. 4.95 Cross-section photograph of a scintillator (courtesy of Hamamatsu Photonics) [63]

For this reason, light is guided into the crystal wall and a certain fraction is guided into the photodiode direction. These are referred to as *structured scintillators*. CsI (Tl) scintillator can be deposited on substrates such as glass, which is then flipped and adhered to the photodiode. It can also be deposited directly on the photodiode or deposited on the fiber optic plate (discussed later) [64]. Figure 4.96 shows schematics of their cross-sections and Fig. 4.97 shows examples of commercially available products. Direct deposition suppresses fluorescent scattering and contributes to increased spatial resolution [64]. Furthermore, the columnar CsI (Tl) crystals increase in crystal diameter as the film thickness increases, which can reduce the spatial resolution by coalescence with adjacent crystals. Other points of caution with CsI (Tl) include hygroscopicity, toxicity (particularly the trace thallium iodide (TlI) amounts added as an activator in CsI (Tl)) and its relatively low mechanical strength [68]. Details on scintillation phenomena are discussed later in Sect. 4.4.3.

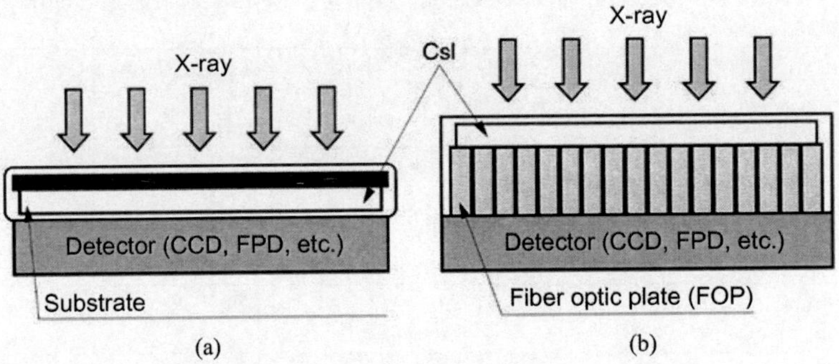

Fig. 4.96 Implementation example when using cesium iodide in a scintillator; **a** is when cesium iodide is deposited on a substrate, after which the substrate is flipped and joined with the detector; **b** is when cesium iodide is deposited on a fiber optic plate which joins the detector and scintillator (courtesy of Hamamatsu Photonics)

Fig. 4.97 Example of commercially-available cesium iodide scintillator; **a** is when this is deposited on an aluminum or carbon substrate and **b** is when this is deposited on a fiber optic plate (courtesy of Hamamatsu Photonics)

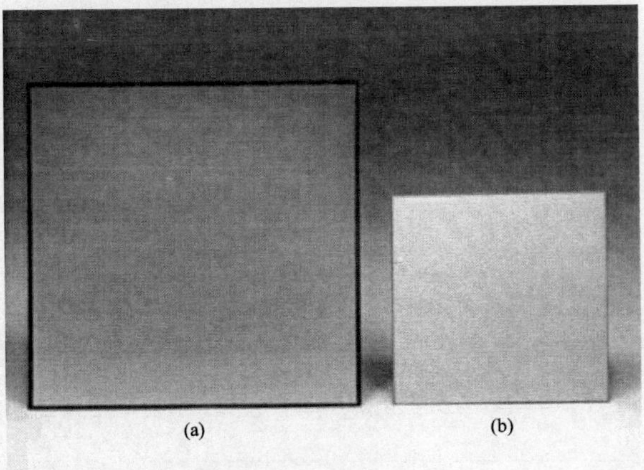

Finally, the specifications of various flat-panel detectors that are commercially available are displayed for reference in Table 4.13. These should be compared with the CCD cameras and EM-CCD cameras in Tables 4.7, 4.8 and 4.9 and the sCMOS

Table 4.13 Representative flat panel detectors that are commercially available and their specifications

Manufacturer	Hamamatsu photonics		Rayence	
Type	C7942CA-22	0505A	1215A	2020A
Number of pixels	2240 × 2344	1176 × 1104	2352 × 2944	1120 × 1120
Pixel size (μm)	50 × 50	49.5	49.5	180
Effective photodetection surface (mm)	112 × 117.2	58 × 54	116.4 × 145.7	201.6 × 201.6
Indirect conversion/direct conversion	Indirect conversion	Indirect conversion	Indirect conversion	Indirect conversion
Photoelectric conversion film/scintillator	Scintillator (CsI)	GOS/CsI	GOS/CsI	GOS/CsI
Read-out rate (Frames/s)	2	30(1 × 1) 90(2 × 2)	8(1 × 1) 32(2 × 2)	30
Read-out noise (Electrons-rms)	1100	<4	<3.4	8
Dark current (Electrons/pixel/sec)	–	403	74	–
Full well capacity (Electrons)	2,200,000	16.384	16.384	32.768
Dynamic range (bit)	–	14	14	16
ANSeeN	Pony industry	Acrorad	Xcounter (Acrorad)	
ANS-FPD4 × 2S01S	SID-A50	FPD4x2	HYDRA FX35	
262,144 pixel	237,568 pixel	237,552 pixel	3584 × 60	
100	100	100 × 100	100 × 100	
51.2 × 51.2	51.2 × 46.4	51.5 × 46.5	360 × 6	
Direct conversion	Direct conversion	Direct conversion	Direct conversion	
CdTe	CdTe	–	–	
200	5–50 (1:1)	50	200 (Frame mode) 10,000 (TDS mode)	
45,000	–	–	–	
250 fA	–	–	–	
20,000,000	–	–	–	
14	12	12	12 (Frame mode) 18 (TDS mode)	

(a) (b)

Fig. 4.98 Example of commercially available flat panel detectors; **a** Hamamatsu Photonics product (courtesy of Hamamatsu Photonics) and **b** Pony Industry SID-A50 model (courtesy of Toshiro Yamazaki of Pony Industry)

cameras in Table 4.11. Furthermore, Fig. 4.98 shows a photograph of the external appearance of a commercially available flat-panel detector.

4.4.3 Scintillator

(1) Emission Mechanism

Scintillators can be organic, plastic, or inorganic. Organic materials include naphthalene ($C_{10}H_8$), anthracene ($C_{14}H_{10}$), and trans-stilbene ($C_{14}H_{12}$). Each of their effective atomic numbers, as well as their densities, are small, measuring at around 1 g/cm^3. Their characteristics include a decay time of several ns, which is shorter than that of inorganic materials (discussed later), and low cost. Furthermore, their light emission is relatively weak compared to scintillators made of inorganic materials and their stopping power is considerably inferior, as shown in Fig. 4.99 [71]. Liquid scintillators are organic scintillators that have been dissolved in a toluene or xylene solvent. Plastic scintillators are organic scintillators that have been dissolved in plastics, such as polystyrene. These are suited for alpha- and beta-ray detection but are not used often for X-ray imaging. With this in mind, a discussion on scintillators made with inorganic materials, which is important when discussing X-ray tomography, is presented below. Generally, light emission is stronger in inorganic materials, stopping power is high as in Fig. 4.99, and linearity with regard to energy is favorable.

Light emission from inorganic materials can be explained using their discrete energy band; Fig. 4.100 shows a schematic of this. When an X-ray is irradiated on the scintillator material, the electron in the *valence band* leaves a hole in that band and is excited to the *conduction band*; a forbidden band is present between the two,

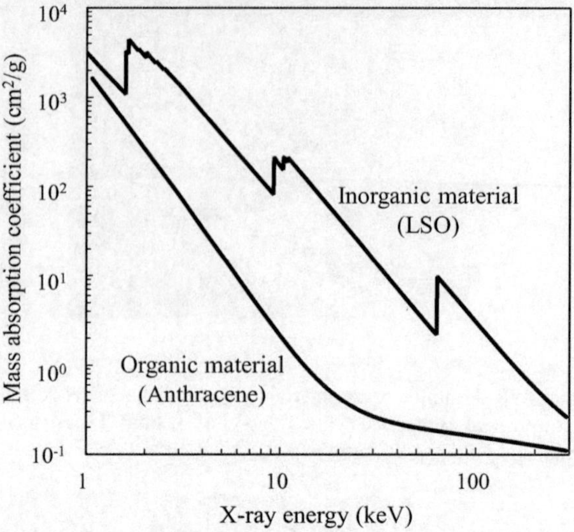

Fig. 4.99 X-ray energy dependencies of mass absorption coefficients in representative organic and inorganic material scintillators [3]; anthracene and LSO are shown as organic and inorganic materials, respectively

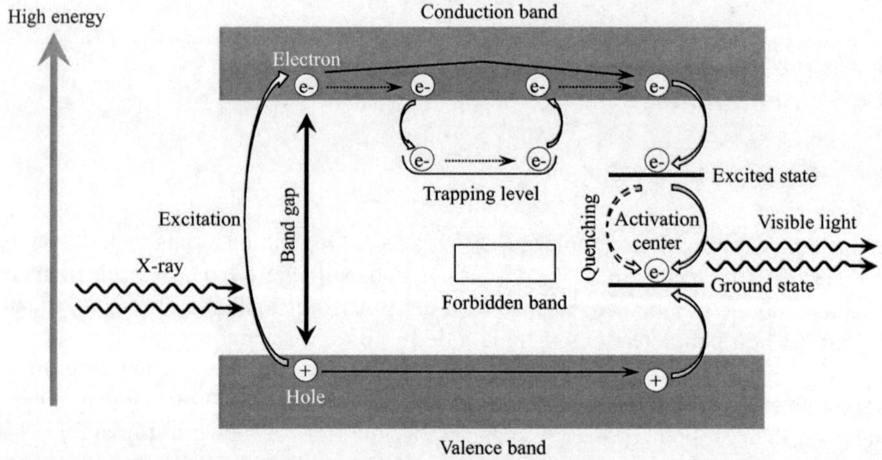

Fig. 4.100 Schematic showing the light emission mechanism in an inorganic material scintillator

where the electrons cannot exist in pure crystals. When the excited electron drops back down to the valence band and rejoins the hole, the excess energy corresponding to the band gap between the valence and conduction bands is radiated through photon discharge.

The maximum wavelength of the emission spectra is generally located in the long-wavelength side (low-energy side) relative to the maximum wavelength of the absorption spectra. This phenomenon where the incident radiation and generated radiation have different energies is referred to as *Stokes shift*. This occurs because excitation is the transition from the lowest ground state to the excited state, whereas light emission corresponds to the transition from the lowest vibrational state of the first excited state to the ground state. The generated wave can be radiated externally without being absorbed into the scintillator material through the Stokes shift. Furthermore, a larger Stokes shift results in a smaller overlapping of the incident wave and generated wave spectra; the self-absorption of the generated wave due to the scintillator also decreases.

Scintillator materials, which use the light emission from wide band gap materials with no *activator* addition, include *cadmium tungstate* ($CdWO_4$) (CWO), *barium fluoride* (BaF_2), and *bismuth germanium oxide* ($Bi_4Ge_3O_{12}$) (BGO). The light emission efficiency in these cases is generally low and the discharged energy is too large for visible light emission. The light emission mechanism in the slow-decay components of barium fluoride or cesium iodide with no activator additions is from *self-trapped excitons*, which are excitons trapped in crystal lattices that have been strained due to interactions between excitons and the lattice. The fast-decay components of barium fluoride are due to the *core-valence luminescence* (*CVL*) generated when the valence electron transitions into the core level with the lowest binding energy. This light emission has an extremely low lifespan at the ns order and is effective as a light emission mechanism of a high-speed scintillator. Furthermore, the WO_4^{2-} ions in cadmium tungstate are the center of light emission and the *charge transfer* transitions between the 2p orbitals of tungsten and its surrounding O^{2-} oxygen ions.

Scintillators that efficiently emit visible light are those that have activator doping. These types of scintillator materials have rare-earth elements like cerium (Ce), terbium (Tb), and europium (Eu), as well as thallium (Tl) and sodium (Na) ions, which act as *luminescent centers* within the forbidden band. Scintillator materials that have activator additions include CsI (Tl) and *yttrium aluminum garnet* ($Y_3Al_5O_{12}$) (*YAG*: Ce^{3+}). As shown in Fig. 4.100, the electrons in the conduction band arrive at the activation center while being trapped at the trap centers when activators are added. The electron is positioned at the excited level of the activator and holes in the valence band take trapped electrons at the ground level from the activation center, creating a vacancy. Visible light is then emitted by an electron in the excited state transitioning into this vacancy. Figure 4.101 shows the wavelength at which cesium iodide emits light [74]. As pure cesium iodide emits near-ultraviolet light, the emission can then be adjusted to the visible light range by doping with sodium to shift this to a 430 nm range or with thallium to reach a longer wavelength in the 550 nm range. Furthermore, the addition of activators has a large influence on light emission efficiency.

High-efficiency light emission accompanying the allowable 5d–4f transition occurs in trivalent rare-earth ions such as the representative activator cerium and further decays within a short period [75]. There are divalent and trivalent forms of the europium ion and its ratio varies with concentration. The decay time is longer

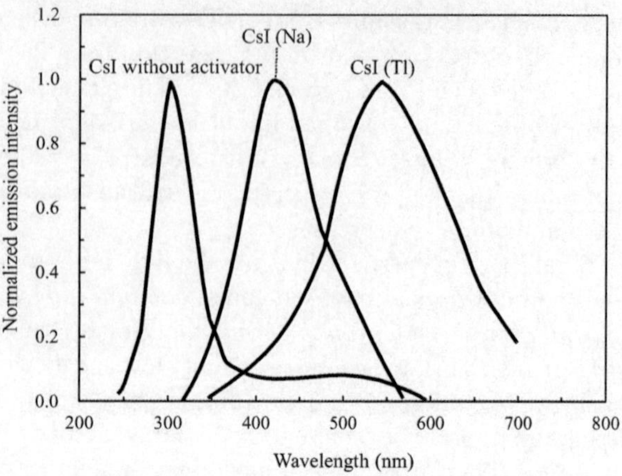

Fig. 4.101 Schematic showing the wavelength spectra of visible light irradiated by scintillators with pure cesium iodide or cesium iodide with sodium or thallium doping [73]

compared to cerium during light emission accompanying 5d–4f transitions in divalent rare-earth ions such as Eu^{2+} [75]. Meanwhile, light emission accompanying 4f–4f transitions in trivalent rare-earth ions such as Eu^{3+} is both a parity-forbidden transition and a spin-forbidden transition and its fluorescence time increases [75]. Furthermore, thallium is classified into light emission accompanying 6s6p–6s transitions [75].

Here, the following equation can be obtained from Eqs. (4.51) and (4.52) by setting the number of generated photons as N_{scin} and the number of electron-hole pairs as $n_{\pm}$ [72]:

$$N_{scin} = n_{\pm}\eta_{lc}\eta_q^{lc} = \frac{E}{E_{\pm}}\eta_{lc}\eta_q^{lc} = \frac{E}{\beta_{lc}E_g}\eta_{lc}\eta_q^{lc} \tag{4.54}$$

η_{lc} and $\eta_{\text{q}}^{\text{lc}}$ are the efficiency of light emission center excitation and quantum efficiency of the light emission center, respectively. $\eta_{\text{q}}^{\text{lc}}$ is dependent on the positional relationship of the energy band of the scintillator material and the light emission center energy level and η_{lc} is dependent on the trapping center [72]. The intensity of the visible light emitted from the scintillator needs to consider the percentage of X-rays, which transmit through the scintillator, and the absorption of the generated visible light by the scintillator itself.

(2) Characteristics Assessment

Representative inorganic scintillators and their various characteristics are summarized in Table 4.14 [76, 77].

(a) Stopping Power

The density and *effective atomic number* in the table are related to the detection efficiency and stopping power. The effective atomic number Z_{eff} is an effective parameter that expresses the attenuation behavior of the X-rays illuminated on compounds. This

Table 4.14 Representative materials for scintillators and their characteristics

Substance	Density (g/cm^3)	Effective atomic number	Refractive index	Emission peak wavelength (nm)	Photon yield (Photons/keV)	Decay time (ns)	Decay distance (μm)	
							20 keV	60 keV
BaF_2[a]	4.89	53	1.56	310	9.5	630	40	125
$Bi_4Ge_3O_{12}$ (BGO)[a]	7.13	75	2.15	480	8.2	300	9	150
$CdWO_4$[a]	7.90	64	2.25	470	13	12	14	137
NaI (Tl)	3.67	50	1.85	415	38	230	56	173
CsI (Tl)	4.51	54	1.80	540	65	1100	37	115
CsI (Na)	4.51	54	1.84	420	39	650	37	115
LiI (Eu)	4.08	52	1.96	470	11	1400	46	139
CaF_2 (Eu)	3.19	13	1.47	435	24	900	193	3024
$LaBr_3$ (Ce)	5.29	46	2.05	380	63	26	19	157
$YAlO_3$ (Ce) (YAP)	5.37	35	1.95	370	18	27	22	382
$Y_3Al_5O_{12}$ (Ce) (YAG, P46[b])	4.56	32	1.82	550	17	70	30	530
Lu_2SiO_5 (Ce) (LSO)	7.40	66	1.82	420	25	47	13	218
Gd_2SiO_5 (Ce) (GSO)	6.71	59	1.85	440	9	60	20	69
$LuAlO_3$ (Ce) (LuAP)	8.34	65	1.94	365	17	17	14	234

(continued)

Table 4.14 (continued)

Substance	Density (g/cm^3)	Effective atomic number	Refractive index	Emission peak wavelength (nm)	Photon yield (Photons/keV)	Decay time (ns)	Decay distance (μm)	
							20 keV	60 keV
$Lu_3Al_5O_{12}$ (Ce) (LuAG)	6.73	63	1.84	550	20	58	18	291
$Gd_3Al_2Ga_3O_{12}$ (Ce) (GAGG)	6.63	52	1.93	520	57	88	21	95
$Gd_3Ga_5O_{12}$ (Eu) (GGG)	7.1	53	1.96	595	44	140000	19	93
Gd_2O_2S (Tb) (P43[b] or Gadox)	7.34	60	2.2	545	60	1500	17	56
$Lu_3Al_5O_{12}$ (Pr) (LuAG)	6.71	63	1.84	310	16	26	18	292

[a]Without activator
[b]Designation for powder

is expressed as follows, where the ratio between the sum of electrons in the compound and the electron number of the ith element, which comprises the composite is set as f_i and the atomic number of the ith element is set as Z_i.

$$Z_{eff} = \sqrt[\delta]{\sum_i f_i Z_i^{\delta}} \tag{4.55}$$

The coefficient δ in the equation is equal to 2.94 when photoelectric absorption is predominant. Meanwhile, this value is 3–3.5 when the X-ray energies are in the range 100–600 keV where Compton scattering is predominant [78]. For example, water is comprised of hydrogen (atomic number 1) and oxygen (atomic number 8) and its effective atomic number is $\sqrt[2.94]{0.2 \cdot 1^{2.94} + 0.8 \cdot 8^{2.94}} \approx 7.4$. The decay distances for the two standard X-ray energies of 20 keV and 60 keV are shown in Table 4.14. This decay distance becomes an index for stopping power. Detector efficiency decreases when a scintillator with a thickness smaller than the decay distance is used. Furthermore, the spatial resolution of the X-ray CT scanner decreases when a scintillator with a thickness greater than the decay distance is used.

(b) Emission Wavelength

The *emission peak wavelength* and *photon yield* are both quantities relating to the sensitivity of the detection system. The emission peak wavelength of the scintillator generally needs to be matched with the spectral response characteristics of the photoelectric surface on the visible light detector. The efficiency of the detection system can be optimized in this manner. As previously discussed, the emission peak wavelength of the scintillator can be controlled by changing the type of activator used.

(c) Photon Yield

The photon yield L_y, which expresses the sensitivity of the scintillator, is expressed by the amount of emitted light per unit absorption energy as follows using Eq. (4.54). The units are in photon number/keV or photon number/MeV.

$$L_y = \frac{N_{scin}}{E} = \frac{\eta_{lc}\eta_q^{lc}}{\beta_{lc} E_g} \tag{4.56}$$

In other words, the photon yield is superior for materials with narrower band gaps. Band gap energies of pure substances without activator additions generally decrease in the order of fluorides, oxides, chlorides, bromides, and iodides [79]; thus, higher photon yields can be expected with materials further down this list [79]. Equation (4.51) confirms that the electron-hole pair creation energy $E_{\pm}$ in various photoelectric conversion materials satisfies the relationship $E_{\pm} = \beta_{1c} E_g \approx 3E_g$. Reports have indicated that this coefficient β_{1c} is 3 for *sodium iodide* (*NaI*), 5.6 for YAG, and 7 for *calcium tungstate* ($CaWO_4$) [80]. Slight undulations in photon yield are not frequently problematic in standard X-ray microtomography, but in contrast,

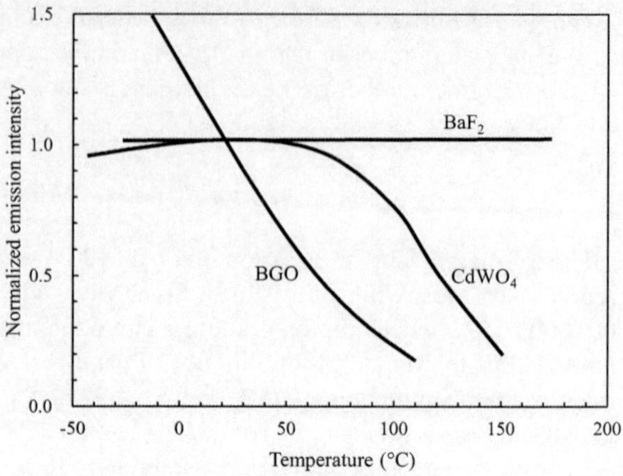

Fig. 4.102 Temperature dependencies of normalized light emission intensities of visible light emitted by scintillators comprising cadmium tungstate ($CdWO_4$), bismuth germanium oxide (BGO), and barium fluoride (each shown in Table 4.14) [80]

the linearity of photon yield to X-ray energies is important. Figure 4.102 shows the temperature dependencies of photon yield for several scintillators [81]. There are many scintillators whose light emission intensity decreases with increases in temperature, while there are some such as BGO in the figure, which vary widely, even near room temperature. Furthermore, there are those that peak within a range of approximately 80 °C from room temperature (in the case of sodium-activated cesium iodide [79]).

(d) Refractive Index

The refractive index is a necessary material characteristic when there are mismatches in scintillator material/detector junctions, which may be made of glass fibers or others. For example, as discussed later, if the refractive index n_2 of core fiber material is approximately 1.8, the refractive index of the scintillator should also be set to approximately 1.8 to prevent loss at the junction surface.

(e) Rise Time and Decay Time

The decay time τ_d expresses the time necessary for the decay of light emission intensity, which is important for applications in high-speed imaging. This is the time necessary for an electron to return to the ground state from the excited state. The decay time generally follows the law shown below and has the relationship $\tau_d \approx \lambda^2$ [80]:

$$\tau_d = \frac{cm\lambda^2}{8\pi e^2 f n_{scin}}\left(\frac{3}{n_{scin}^2 + 2}\right)^2 \tag{4.57}$$

Here, c is the speed of light, m is the electron mass, λ is the wavelength, n_scin is the refractive index of the scintillator material, and f is the vibrator strength of the transition; in other words, scintillators that emit ultraviolet light and purple visible light are suited for high-speed imaging. Furthermore, the light emission intensity $I(t)$ of the scintillator expresses decay following an exponential function, as in the following equation:

$$I(t) = I_0 e^{\left(\frac{-t}{\tau_d}\right)} \quad (4.58)$$

Here, I_0 is $I(t)$ at time $t = 0$. For example, the decay time of thallium-activated cesium iodide is approximately 1100 ns according to Table 4.14 but the rise time is about 2 orders of magnitude smaller at approximately 20 ns. For this reason, the decay time is a larger issue in actual practice. Increases in activator concentration also decrease the rise time [79]. As shown in Table 4.14, scintillators with thallium doping have a high photon yield but a lengthy decay time. Meanwhile, those with cerium doping have a comparatively shorter decay time. Actual decay behavior data with *gadolinium aluminum gallium garnet* (*GAGG (Ce)*) ($Gd_3Al_2Ga_3O_{12}$) as an example are shown in Fig. 4.103.

Generally, a single scintillator material also has multiple metastable states, and the reported decay times shown in the data in Table 4.14 are the result of adding these together. In actual practice, the scintillation of several inorganic scintillators are the sum of the fast-decay component and slow-decay component. This type of decay behavior is schematically shown in Fig. 4.104. When considering the decay behavior in such cases, Eq. (4.58) should be summed as a percentage of the intensities of each material, as shown in the following equation:

$$I(t) = I_0^f e^{\left(\frac{-t}{\tau_d^f}\right)} + I_0^s e^{\left(\frac{-t}{\tau_d^s}\right)} \quad (4.59)$$

Here, the superscripts f and s each express the fast and slow components, respectively. For example, barium fluoride has fast components of 0.8 ns in addition to the

Fig. 4.103 GAGG (Ce) decay properties (courtesy of Hiroki Sato of Furukawa Scintitech)

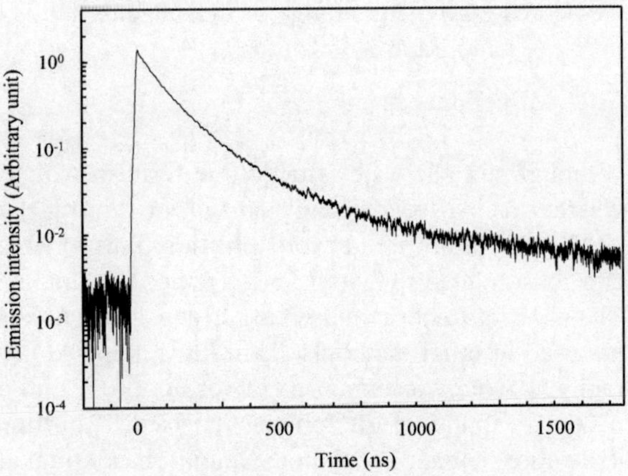

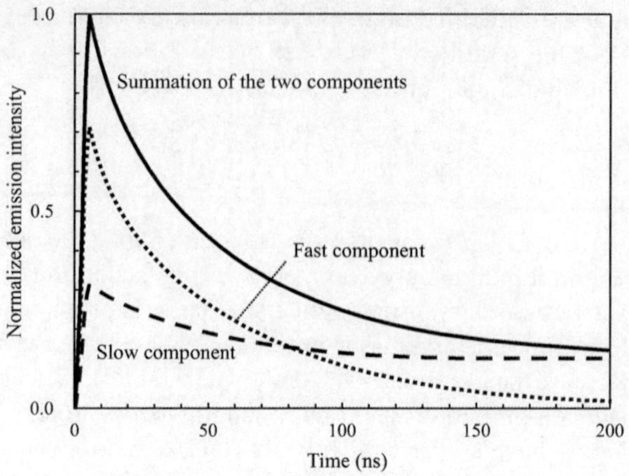

Fig. 4.104 Schematic of light emission behavior in scintillator materials comprising components with fast and slow light emission decay

slow components of 630 ns shown in Table 4.14; thallium-activated cesium iodide has a fast component of 3.34 ns as well [79].

(f) Afterglow

The phenomenon referred to as *afterglow* is phosphorescence that occurs due to the thermal release of electrons from a trap and can last as long as several ms. The origin of the trap is from crystal defects and impurities and its extent varies widely even in the same material according to its manufacturing process or heat treatment. This can also depend on impurity element concentration. The extent of afterglow can also vary widely depending on the material; for example, thallium-activated sodium iodide reaches a maximum of 5% after 6 ms of elapsed time, whereas this can be constrained to 0.005% for BGO (after 3 ms elapsed time) [79].

As decay and afterglow both result in time lags during imaging, this can be a problem when trying to achieve a high frame rate. Furthermore, afterglow results in the decreased dynamic range of the detector and is thus an important characteristic in X-ray tomography.

(g) Miscellaneous

Although not shown in a table, the form in which a scintillator can be used (e.g. whether it is a powder, bulk crystal, or thin film) is also important. For example, GOS scintillators are network-like thin films in which a binder is added to a powder and consolidated so that it can be formed to a thickness of approximately 5–100 μm. The packing fraction in this case is considerably lower than 1. Crystals or thin films are used in other materials. Thin films are used as a general rule for microtomography to avoid worsening spatial resolution. A film thickness that corresponds to the necessary spatial resolution must be used. Thin-film fabrication includes pulse laser deposition, spattering methods, liquid-phase epitaxial methods, sol-gel methods, and cutting/polishing methods. Furthermore, columnar crystals can be formed to increase the spatial resolution, as in the cesium iodide introduced in Sect. 4.4.2 (5).

Localized debonding can occur due to long-term X-ray irradiation when a thin film is created on a substrate, which can result in visible irregularities in the image. This ease by which scintillator damage can occur is also a problem in actual practice. Moreover, materials like sodium iodide, *lithium iodide* (*LiI*), cesium iodide, and lanthanum bromide ($LaBr_3$) have hygroscopic and deliquescent characteristics. For this reason, these are made commercially available by being placed in a case and cut off from the atmosphere, as shown in Fig. 4.105.

(3) Characteristics of Various Scintillator Materials

(a) Thallium-activated Sodium Iodide

The fact that thallium-activated sodium iodide is a scintillator material with a high photon yield was discovered in 1948 by the U.S. physicist Hofstadter, who later won the Nobel Prize in Physics for his research on high-energy electron scattering with linear accelerators. Sodium iodide is an ion crystal with a sodium chloride-type crystal structure, whose {100} plane is the *cleavage* plane, exhibiting brittle fracture. Today, many scintillator materials, which are superior in terms of photon yield, have been discovered and applied to actual practice, as shown in Table 4.14. However, large-scale crystals can be fabricated inexpensively using the Bridgman-Stockbarger method and currently sodium iodide remains widely used for gamma cameras. In addition to the components with a 230-ns decay time, as shown in Table 4.14, slow components with a 150-ms decay time are included at a percentage comprising 9% of the total photon yield [76]. Sodium chloride without an activator has a short decay time at approximately 100 ns but its photon yield is low. However, cooling this to liquid nitrogen temperatures generates photon yields that are at the same level as thallium-activated materials at room temperature [76].

(b) Thallium- or Sodium-Activated Cesium Iodide

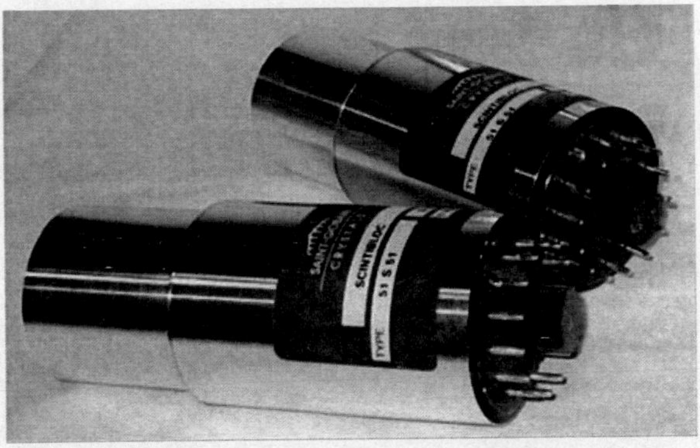

Fig. 4.105 Example of a commercially available inorganic scintillator; Saint-Gobain lanthanum bromide scintillator (courtesy of Seiko EG & G)

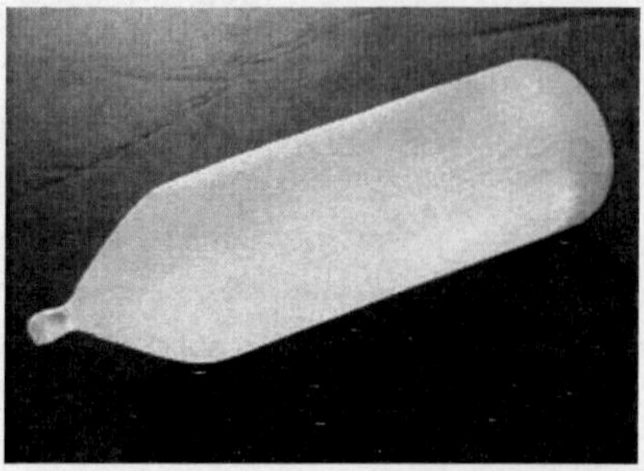

(a) GAGG (Ce) ingot grown by the Czochralski method

(b) LuAG (Pr) ingots grown by the Czochralski method

(C) GAGG (Ce) scintillator cut down from an ingot

Fig. 4.106 Examples of commercially available inorganic scintillators. Furukawa Scintitech GAGG (Ce) and LuAG (Pr) (courtesy of Hiroki Sato of Furukawa Scintitech)

The ion radius of cesium is larger than sodium and cesium iodide has a cubic crystal structure in which the fluoride ions are at the corner of the cube and the cesium ion is in the center. This structure does not have any cleavage planes and exhibits some degree of flexibility. It is also easier to handle due to its weaker deliquescence compared to sodium iodide. Sodium iodide and cesium iodide both have a reduced photon yield at X-ray energies lower than 10 keV. The photon yield of cesium iodide in particular rapidly decreases [76]. This has the characteristic of having a lengthy decay time even after undergoing thallium or sodium doping. Meanwhile, the photon yield without any doping drops to 1/10 or lower but the decay time shortens by several dozen factors to approximately 10 ns.

Cesium iodide uses an aluminum or amorphous carbon substrate and depositing on these can form a columnar crystal, introduced in Sect. 4.4.2 (5). For this reason, they are often used in flat-panel detectors. Please refer to Sect. 4.4.2 (5) for other characteristics.

(c) Europium-Activated Lithium Iodide

The two isotopes ^{6}Li and ^{7}Li are present in europium-activated lithium iodide. Although the abundance ratio of ^{6}Li is 7.5%, reactions occur between ^{6}Li and thermal neutrons, therefore, europium-activated lithium iodide is important for the detection of thermal neutrons.

(d) Europium-Activated Calcium Fluoride and Barium Fluoride

Europium-activated *calcium fluoride* (CaF_2) is produced naturally as fluorite. It has the characteristics of phosphorescence and light emission with the application of heat. It has a fluorite-type cubic crystal structure in which fluoride ions fit inside the tetrahedral gaps of the face-centered cubic lattice made of calcium ions. Calcium fluoride is used with europium doping as a scintillator material. Calcium fluoride has low stopping power and a significantly long decay time due to its small effective atomic number. Furthermore, its cleavage plane is the {111} plane and it is easily susceptible to brittle fracture. For these reasons, its range of applications is relatively narrow [76]. Moreover, as it has no deliquescence and low vapor pressure, it can be used in special environments such as vacuum.

There were several publications on barium fluoride in the 1980s. It has a fluorite-type structure similar to calcium fluoride and its cleavage planes are the same. It is fabricated through the Bridgman-Stockbarger method. This material is used as a scintillator without an activator, and as an optical material for high-output infrared lasers; it is also known to be deliquescent. For a scintillator, in addition to the slow-decay component shown in Table 4.14, the extreme fast-decay (0.6 ns) component exists at 20% intensity in the ultraviolet wavelength range [76, 82]. The former is due to light emission from self-trapped excitons and the latter is due to Auger-free luminescence. The presence of the slow-decay component hinders the application of this fast-decay component and large effective atomic number. Heating barium fluoride to 200 °C reduces the peak intensity of the slow-decay component to below 1/10 its original value, whereas the fast-decay component has no temperature dependency and is virtually unchanged. Reports have also indicated applications of these characteristics.

(e) Cerium-Activated Lanthanum Bromide

Of the crystals that use lanthanoids, the only elements that are transparent and emit light under bulk conditions are lanthanum, gadolinium, and lutetium. Lanthanum bromide was reported in 2002 by van Loef of the Netherlands [83]. Alongside *gadolinium gallium garnet* ($Gd_3Ga_5O_{12}$) (*GGG*) and *lutetium oxyorthosilicate* (Lu_2SiO_5) (*LSO*), these new scintillator materials have been developed from the 1990s to the 2000s [84]. Lanthanum bromide has a hexagonal crystal structure referred to as a UCl_3 type. This has a high photon yield, short decay time, and some degree of stopping power. Therefore, it is considered a superior scintillator material with regard to its high-speed response. It has further advantages of not being dependent on temperature and having a virtually fixed photon yield.

Cerium-activated *lanthanum chloride* ($LaCl_3$ (Ce)), which is also a halide-group crystal, has the same type of crystal structure as lanthanum bromide and is a scintillator material with a similar short light emission lifespan, high density, and high photon yield.

Although this material has no cleavage planes, it has a deliquescence that is even stronger than sodium iodide [76], therefore, it needs to be used while sealed in a case made of materials such as aluminum. Furthermore, thermal stresses are generated during heating and cooling due to the anisotropy of the hexagonal crystal, which can result in its fracture. Care must be taken with compounds that include lanthanum as 789 keV and 1436 keV γ-rays are emitted from naturally-occurring ^{138}La, which has an abundance of 0.09% [84]. Furthermore, α-rays emitted from ^{227}Ac, which cannot be removed even with lanthanum refining, form a continuous spectrum between 1850–3000 keV [84]. Care must also be taken with regard to the usage of lanthanum given that their Kα-rays are 33.436 keV.

(f) BGO

All scintillator materials discussed from here onwards are oxides. The hygroscopic properties, strength, and toughness of these materials are mostly stable compared to the iodides discussed above. BGO, which was discovered in 1973 by Weber et al., does not use any activators and functions as a scintillator through the energy transitions of the bismuth ions (Bi^{3+}) itself. Its effective atomic number is the highest among those shown in Table 4.14 and it has superior stopping power. Furthermore, it has minimal afterglow. However, its photon yield is the lowest among those shown in Table 4.14 and its refractive index is considerably high. In addition, it is difficult to obtain a flat surface with BGO; [76] it has the added disadvantage of a lengthy decay time. As seen in Fig. 4.101, its light emission intensity has a strong temperature dependency and its photon yield further decreases under high temperature. Meanwhile, cooling it to liquid nitrogen temperatures enables it to function effectively as a scintillator [76]. Therefore, its usage is limited to specialized uses with high-energy X-rays or γ-rays. It is fabricated through the Czochralski method, after which it is cut and polished.

(g) Cerium-Activated Perovskite-Type Aluminum Composite Oxides ($ReAlO_3$)

Yttrium aluminum perovskite ($YAlO_3$) (*YAP*) and *lutetium aluminum perovskite* ($LuAlO_3$) (*LuAP*) belong to these groups. These have a perovskite structure similar to barium titanate ($BaTiO_3$). One type of metallic atom is present in each vertex of the unit lattice of the orthorhombic crystal, another type of metallic atom is located on the body center of the unit lattice, and the oxygen is located in the face-center of the cubic crystal.

YAG (Nd), which has neodymium doping, is the most often used with industrial-use lasers, but YAP (Nd) also has similar applications. As a scintillator material, YAP (Ce) has the advantages of a shorter decay time and shorter emission peak wavelength compared with those of YAG (Ce). LuAP (Ce), which exchanges yttrium with lutetium, has an effective atomic number that increases by a factor of about two, giving it superior stopping power. This material was first reported by Moses et al. in 1995. This has an even shorter decay time when compared to YAP (Ce), as well as a feature where the fast-decay component with a decay time of 17 ns, shown in Table 4.14, comprises 80% of the total light emission intensity [76]. However, care must be taken in that lutetium exists in the form of naturally-occurring ^{175}Lu, which is stable and has an abundance of 97.41%, but also radioactive ^{176}Lu, which has an abundance of 2.59% and a half-life of 3.8×10^{10} years. There have also been reports on $Lu_xY_{1-x}AlO_3$ that replace a part of lutetium with yttrium to stabilize the crystal growth process. However, these have also indicated a problem wherein the generated light is absorbed within the scintillator material interior due to this replacement [76].

(h) Garnet-Type Aluminum Complex Oxide ($Re_3(Al, Ga)_5O_{12}$ ∖ $Re_3Al_2Ga_3O_{12}$)

YAG and *lutetium aluminum garnet* ($Lu_3Al_5O_{12}$) (*GGG*), LuAG, and GAGG belong to this group. YAG (Ce) was reported in 1967 by Blasse et al. and is a scintillator material that has seen widespread use. A garnet structure has a cubic unit lattice with eight unit-lattices of $A_3B'_2B''_3O_{12}$, where ion A is a dodecahedral coordination, ion B' an octahedral coordination, and ion B″ a tetrahedral coordination. In the case of YAG, the A site is the yttrium ion and the B′ and B″ sites are aluminum ions. Lutetium is used in place of yttrium for LuAG. Care must be taken for radioisotopes for this reason, as mentioned above. Furthermore, GGG has gadolinium and gallium placed instead of the yttrium and aluminum, respectively, in YAG. Furthermore, GAGG has aluminum present in the B' ion of GGG.

As shown in Table 4.14, garnet-type aluminum composite oxides have an extremely long light emission peak wavelength and a relatively small refractive index. As observed in the spectral response characteristics of the CCD camera shown in Figs. 4.61 and 4.80, there are many detectors with a quantum efficiency peak of approximately 600 nm or higher. The garnet-type aluminum composite oxide is considered advantageous in such circumstances. Figure 4.107 shows the light emission spectra of GAGG (Ce) as examples. YAG is also used in the form of powder with a diameter up to 1 μm, referred to as *P46* in these cases. Compared with P43 (discussed later), this has the characteristics of short decay time and long light emission peak wavelength. Furthermore, those with praseodymium (Pr) doping instead of cerium have an even shorter light emission wavelength. As shown in Eq. (4.57), Pr-doping considerably reduces the decay time, making it a scintillator material even more suited for high-speed applications; LuAG (Pr) is shown as an example of this in Table 4.14. LuAG has a superior stopping power than YAG (Ce) and the superiority of both GGG (Eu) and GAGG (Ce) increases with higher energy. Meanwhile, YAG (Ce), LuAG (Ce), and LuAG (Pr) all have average photon yield levels, as can be observed in Table 4.14. GGG (Eu) and GAGG (Ce) levels are several times higher and are considerably more favorable.

Compared to YAG (Ce), GGG (Eu) has the characteristics of a shorter afterglow and a considerably higher dynamic range, with values between 16–17 bits in the detector as opposed to the 10 bits of YAG (Ce) [85]. However, the disadvantage of an extremely long decay time exists. Generally, materials doped with cerium have a shorter decay time than those doped with europium and are suited for microtomography cases where the number of projections is large and short exposure time is required.

As shown in Fig. 4.106, a single crystal is obtained using the Czochralski method and then fabricated after cutting and polishing when a bulk scintillator is necessary. Meanwhile, reports have indicated examples with LuAG (Ce) where a 2.9 μm-thick

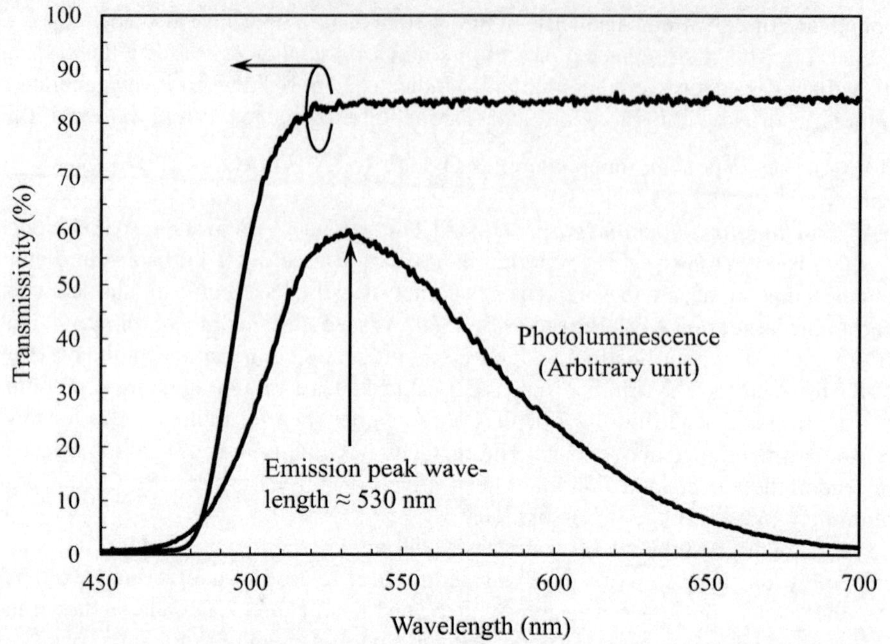

Fig. 4.107 Transmissivity and light emission spectrum of GAGG (Ce) (5-mm thickness) (courtesy of Hiroki Sato of Furukawa Scintitech)

LuAG (Ce) film with 0.07% Ce-doping was attached on top of a 150-μm-thick YAG substrate [86]. Approximately 3% of europium is needed for GGG (Eu) and reports have indicated examples where a 2-μm thick film is deposited on a 170-μm GGG substrate with no activator [86]. A similar procedure is also used for GAGG (Ce), where single-crystal films are created for both using a liquid-phase epitaxial method [86]. A 1–25-μm thick thin-film scintillator material often comprises thin films according to the necessary spatial resolution or the X-ray energy used on the transparent substrate. Having a lattice constant mismatch between the substrate and scintillator material less than 1% is important in forming a high-quality scintillator layer.

(i) Rare-Earth Silicates (Re_2SiO_5)

These correspond to LSO, *gadolinium oxyorthosilicate* (Gd_2SiO_5) (*GSO*), and *yttrium oxyorthosilicate* (Y_2SiO_5) (*YSO*); $Re^1_xRe^2_{(2-x)}SiO_5$-type LGSO and LYSO also belong to this group. GSO (Ce) crystals have a light-yellow color when Ce concentrations are high. Takagi and Fukazawa from Hitachi Chemical, Co., Ltd. showed in 1983 that GSO (Ce) could be used as a scintillator by trace additions of zirconium [87]. Afterward, Melcher from Schlumberger Ltd. reported in 1992 on an LSO (Ce) scintillator that replaced gadolinium with lutetium [88]. Compared

Fig. 4.108 Example of a commercially available fiber optic plate and fiber optic taper (photograph provided by Edmund Optics Japan)

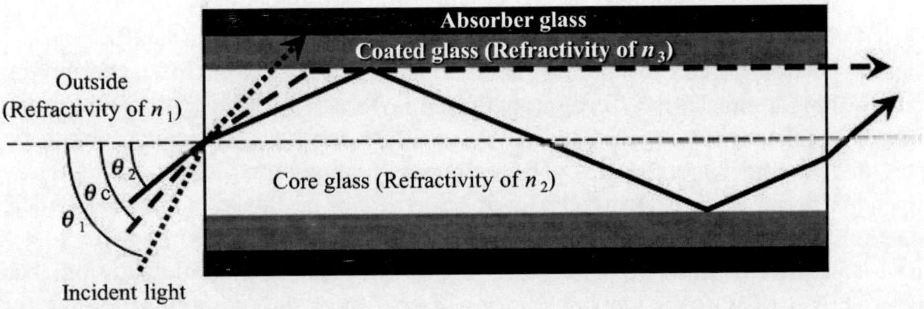

Fig. 4.109 Schematic of the light propagation behavior when visible light is incident to the fibers in the fiber optic plate

to garnet-type aluminum composite oxides, both materials had light emission peak wavelengths that were shorter by 100 nm and had superior stopping power.

The Si included in Re_2SiO_5-type silicates is tetracoordinated due to its sp3 hybrid orbital and its monomer is a regular SiO_4 tetrahedron. For this reason, rare-earth silicates have a structure where the regular SiO_4 tetrahedron encloses the lutetium or gadolinium atoms. The chemical formula for GSO and LSO are both Re_2SiO_5 but its crystal structure and characteristics are different. GSO belongs to the monocline phase of the space group P2 1/C and its cleavage plane is the (100) plane. The thermal expansion coefficient in the [010] axis is several times larger than in other directions, so brittle fracture due to cleavage can occur during fabrication. Meanwhile, LSO is a monocline of the space group C2/c and has no cleavage planes.

The photon yield of GSO (Ce) is maximal when cerium is 0.5 at.%; the crystal becomes colored when cerium is increased to 1.5 at.%. The decay time decreases when cerium concentration increases. This material has the characteristics of a long rise time at 10–20 ns and a relatively short decay time at 60 ns. Either the Czochralski method or Bridgman-Stockbarger method is used for fabrication.

LSO (Ce) has a shorter decay time than GSO and favorable photon yield. However, the ^{176}Lu included in lutetium generates 240 cps of background noise per cm^3 of LSO [88]. Background due to other radioisotopes caused by Lu_2O_3 used in this context also exists [76]. Furthermore, LSO has been reported to have afterglow lasting for several seconds. The Czochralski method is used for its fabrication. Other than LSO, LYSO has been developed, applying the simple fabrication and low-cost advantages of YSO and partially replacing lutetium with yttrium [76]. The concentration of yttrium in this case can be freely adjusted to 5–70%. The light emission peak wavelength and decay time of LYSO do not significantly vary with LSO; however, significant decreases in stopping power are unavoidable due to the large atomic number differences between lutetium and yttrium. Meanwhile, reports have indicated that the additions of combinations of rare-earth elements have improved the distribution of cerium and increased photon yield [89].

(j) Terbium-Activated GOS

The GOS crystal structure has trigonal symmetry, comprised of three sulfur ions and four oxygen ions joined around the gadolinium ion and the two enclosed gadolinium ions forming a unit lattice. Terbium-activated GOS is a scintillator that applies the light emission that accompanies the 4f–4f transitions due to the trivalent rare-earth ion Tb^{3+} and has the characteristics of an extremely high photon yield and long decay time. Furthermore, it is relatively low-cost as it can be fabricated with chemical reactions between 88% gadolinium oxide and 12% sulfur. Protium, which creates the trivalent rare-earth ion Pr^{3+} or protium with cerium or fluoride additions can be used as activators, as well. Additions of cerium or fluoride aim to reduce the afterglow in exchange for photon yield. Even after considering their limited packing fraction, due to being used in powder form, and their semi-transparency, they are still advantageous in terms of light emission efficiency. A green light emission peak is observed at a 545 nm wavelength when terbium is used as an activator. There are issues with particle diameter/grain diameter due to its use as thick films or as a powder/polycrystal; moreover, there are often limitations due to spatial resolution.

(k) Cadmium tungstate

Cadmium tungstate is an MWO_4 compound referred to as wolframite, which includes a divalent transition metal M and hexavalent tungsten and is monoclinic. The cadmium and tungsten atoms are placed in the center of an extremely distorted octahedron comprised of oxygen atoms. The W-O bonds include 4 short bonds and 2 long bonds. These long and weak bonds, particularly in single-crystal materials, are susceptible to brittle fracture due to cleavage in the (010) plane. Cadmium tungstate is primarily used as thin films with a 50–60-μm thickness. The sol-gel process is applied when thinner scintillators are used.

As mentioned above, cadmium tungstate emits light due to charge transfer. Cadmium tungstate as a scintillator material has a short decay time, no problems with afterglow, and relatively favorable stopping power.

4.4.4 Coupling Between the Camera and Scintillator

Multiple optical lens combinations or glass fibers are used when combining a camera with a scintillator to efficiently guide visible light or to magnify/demagnify an image. An objective lens for microscopes is used for the former. Optical elements that use glass fiber bundles (Fig. 4.110) will primarily be discussed below.

Of the optical devices that use glass fibers, the plate on the left in Fig. 4.108 is referred to as a fiber-optic plate (FOP) and the tapered device on the right in Fig. 4.108 is referred to as a fiber-optic taper (FOT). The material in fiber-optic plates comprises a high-refractive-index glass fiber core inserted in a low-refractive-index coated glass tube, with the gaps between the glass fibers in the core filled in with absorbent glass fibers. This repeatedly undergoes a process of heating the bundle to 500–700 °C and drawing it. The thin glass fiber clumps obtained are cut, processed, and polished, from which a fiber optic plate is formed. In the case of the fiber optic taper, both ends of the cylindrical billet are retained and the central part is stretched while heated. This stretching is stopped when the central neck area obtains the desired magnification, from which two fiber optic tapers can be obtained by cutting at the center. Images are transferred at a magnification factor of 1 between the incidence and emission surfaces of the fiber optic plate. The fiber diameters within the final product are between 6–25 μm and are selected depending on the necessary spatial resolution. Meanwhile, the

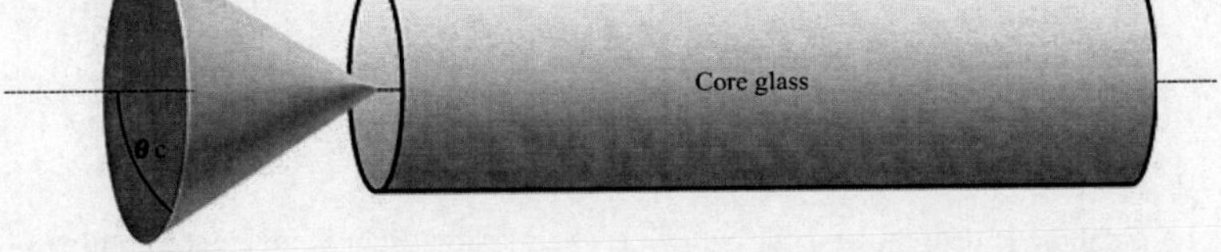

Fig. 4.110 Schematic of the conical region under which the incident light must fall for total reflection to occur within the inner fiber surface when visible light is incident to the fiber of the fiber optic plate

fiber optic taper can magnify or demagnify an image by a factor corresponding to the ratio of the larger and smaller diameters of the fiber optic taper. Generally, the diameter of the end surface is made to match the sensor sizes of a CCD camera or CMOS camera.

Figure 4.109 shows a schematic of the visible light incident to a single fiber. Light incident to the fiber axis at a large angle θ_1 (dotted line arrow in the figure) enters the coated glass from the core glass and then the absorbent glass, whereupon it is absorbed. For this reason, it does not influence adjacent fibers by producing crosstalk. Light incident at an angle θ_c (dashed line arrow in the figure) is propagated along the interface between the core glass and coated glass after entering the former. The term θ_c is the critical angle and light incident at an angle θ_2 smaller than θ_c (solid line arrow in the figure) is propagated within the fiber while repeatedly experiencing total reflection within the inner core glass surface and is guided to the other end surface. In other words, light received on the inner side of the cone in Fig. 4.110 will be efficiently transferred due to the fiber. This critical angle is expressed as follows, with n_1–n_3 as the refractive indices of the external environment, fiber core, and coating material, respectively:

$$NA = \sqrt{n_2^2 - n_3^2} = n_1 sin\theta_c \tag{4.60}$$

Here, *NA* is the *numerical aperture*. Its dependency on the refractive index difference between the core material and coating material can be seen. The numerical aperture of commercially-available fiber optic plates is between 0.35 to 1.0.

Meanwhile, fiber optic tapers are as shown in Fig. 4.111. In the case of Fig. 4.111, the image is magnified by being incident to the small-diameter side and coming out of the large-diameter side. The following relational expression is established when the incidence and emission angles are respectively set as θ_{in} and θ_{out}, and the diameters on the incidence and emission sides are respectively set as d_{in} and d_{out}:

$$d_{in} sin\theta_{in} = d_{out} sin\theta_{ou} \tag{4.61}$$

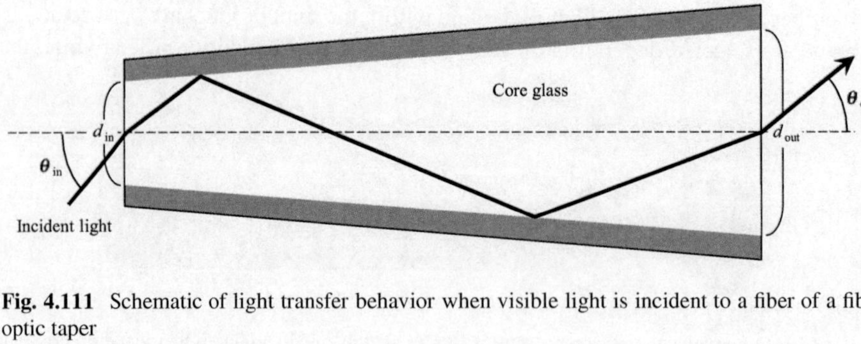

Fig. 4.111 Schematic of light transfer behavior when visible light is incident to a fiber of a fiber optic taper

In other words, the beam narrows in the emission side. Conversely, incidence from the large-diameter side results in the beam being magnified when emitted relative to the incident side.

Images must be minimized with the fiber optic taper when the scintillator size is larger than the sensor size in CCD or sCMOS detectors, such as in industrial X-ray CT scanners. Meanwhile, images produced by X-ray microtomography using synchrotron radiation may conversely need to be magnified. We investigate the former case and analyze efficiency under two scenarios where either an optical lens or fiber-optic taper is used. First, the coupling efficiency η_{OL} when the optical lens is used is as shown below when a powder scintillator is used (assuming a Lambert light source) [91]:

$$\eta_{OL} = \frac{T_L}{1 + 4f^2(1+m)^2} \tag{4.62}$$

Here, T_{L} is the lens transmissivity, f is the f-value determined by dividing the focal distance by the effective diameter, and m is the minification ratio. Generally, T_{L} is 0.7–0.8 at a wavelength of 545 nm [91]. The f-value uses a lens of approximately 1.2. Furthermore, m is approximately 1 to 10. Meanwhile, the transmission efficiency η_{FOT} of the fiber-optic taper is expressed as follows for the same Lambert light source [91]:

$$\eta_{FOT} = \left(\frac{1}{m}\right)^2 \left(\frac{\left(n_2^2 - n_3^2\right)^{1/2}}{n_1}\right)^2 T_F(1 - L_R)F_c \tag{4.63}$$

Here, T_{F} is the transmissivity of the fiber core, L_{R} is the loss at the surface due to Fresnel diffraction, and F_{c} is the fill factor of the fiber core. T_{F} is expressed as follows, with the transmissivity of the fiber core material as μ_{F} and fiber length as l_{F}:

$$T_F = e^{-\mu_F l_F} \tag{4.64}$$

Generally, T_{F} is roughly 0.8, L_{R} is an extremely small value, and F_{c} is approximately 0.85 [91]. Here, we assume $L_{\mathrm{R}} = 0$, $n_1 = 1.0$ (in air), $n_2 = 1.8$, and $n_3 = 1.5$. Figure 4.112 shows a comparison between η_{OL} and η_{FOT}. The numerical aperture, in this case, is approximately equal to 1. When the minification ratio is 2 (i.e. magnification is 0.5), the transmission efficiencies of the fiber-optic taper and optic lens are 16.8% and 1.4%, respectively. The fiber-optic taper is approximately 12 times more efficient than the optic lens. Furthermore, the practical-use range of the fiber-optic taper is below a minification ratio of about 3 when considering the transmission efficiency. In this manner, the photon number per pixel can be increased and the detection system sensitivity can be improved by using the fiber-optic taper. Conversely, the coupling efficiency in the optic lens approximates 0.2 when images must be magnified through coupling during X-ray microtomography using synchrotron radiation; this becomes a level for practical use.

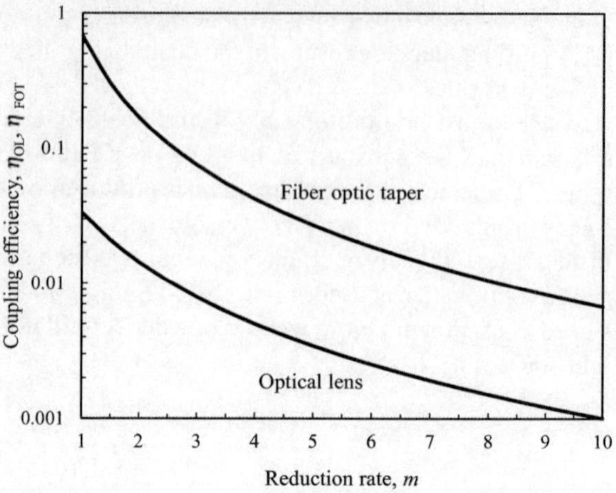

Fig. 4.112 Comparison of coupling efficiencies when using either an optical lens or fiber optic taper

However, considerable accuracy is required for outfitting a fiber optic taper or plate on a detector [85]. CCD cameras are frequently used in a cooled state and need to be outfitted so that they can withstand thermal cycles between room temperatures and operational temperatures. Furthermore, spot and line blemishes caused by the fabrication process (low-sensitivity areas), as well as image distortion, are disadvantages that should be considered.

Glass deterioration due to the X-rays, which transmit through the scintillator, becomes an issue when using an optic lens. For example, Akutagawa et al. reported that the glass surface layer was discolored into a brown color when they were irradiated with soft X-rays that had an energy of 1.24 keV for several hours [90]. Furthermore, reports have indicated that significant irradiation damage appeared in lenses when X-ray energy was 65 keV and exposure 70 s [85]. Glass discoloration is considered removable to some extent using ultraviolet irradiation [73]. As shown in Fig. 4.113, an effective countermeasure against optic glass damage/deterioration is to divide visible light and X-ray paths using visible light mirrors, ensuring that X-rays do not pass through an optic lens. This takes advantage of the fact that X-rays are simply transmitted through the optic lens without being reflected. This countermeasure is widely used in synchrotron radiation facilities. As shown in Fig. 4.114, reports have indicated that ESRF has implemented countermeasures using combinations of convex and concave mirrors [85]. This takes advantage of the fact that X-rays propagate in a virtually straight line, whereas the visible light generated from a scintillator tends to expand outwards. The countermeasures shown in Figs. 4.113 and 4.114 are effective for preventing the damage/deterioration of visible light cameras through the irradiation of X-rays, which have been transmitted through the scintillator.

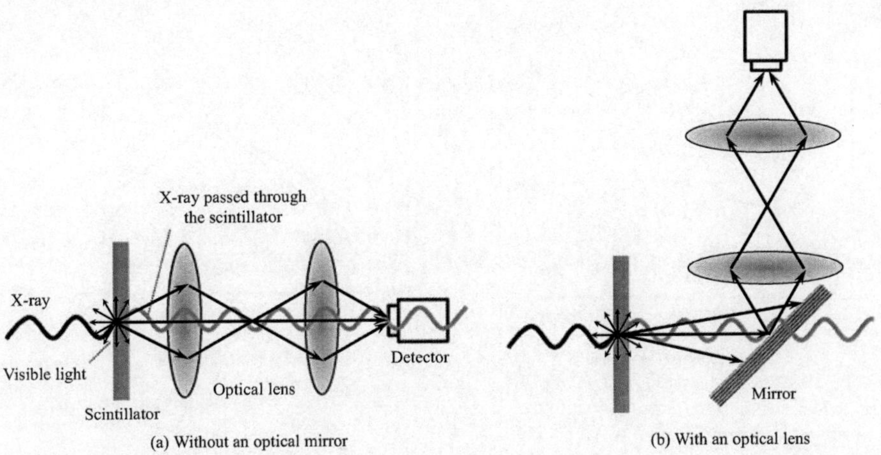

Fig. 4.113 Schematic showing the coupling between the visible light detector, which uses an optical lens, and the scintillator

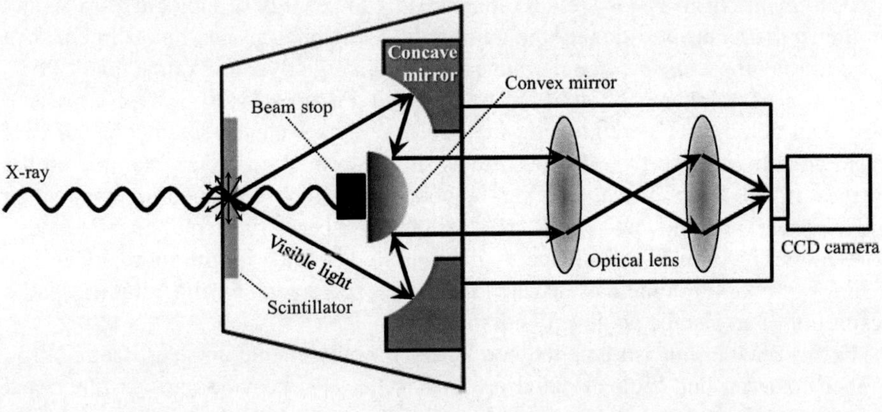

Fig. 4.114 Schematic of a detection system that uses a reflective optical system developed at ESRF [85]

4.4.5 Photon Counting Measurements

The difference between standard X-ray CT scanners and *photon counting*-type (also referred to as "counting-type scanners") X-ray CT scanners is whether they use a *photon-counting detector* (*PCD*). The detectors discussed until now in Sect. 4.4 are so-called *energy integrating detectors* (also referred to as just "integrating detectors"). Energy integrating detectors provisionally convert incident X-rays into visible light, which is then converted to photoelectric charge. As shown in Fig. 4.115a, the amount of photoelectric charge is measured by integrating across only the exposure time and setting that integrated value as the X-ray intensity. Noise influences

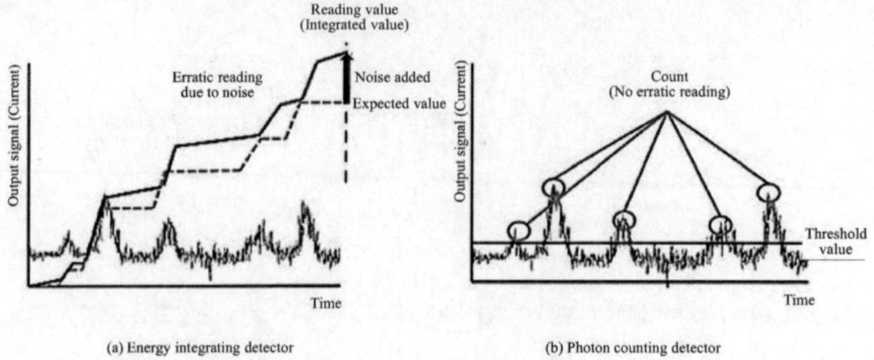

Fig. 4.115 Detection output signal in the energy integrating detector and photon counting detector (courtesy of Akifumi Koike of ANSeeN)

may be strong in this case, particularly when noise is high. Meanwhile, PCDs count each of the photoelectrons generated by incident X-rays, as shown in Fig. 4.115b. Measurements from PCDs are not influenced by the energy of incident X-rays, and an image that is proportional to the number of X-ray photons can be obtained. As a general rule, there is no dynamic range limit and a high dynamic range that follows the measured number of photons can be obtained. Figure 4.116 shows a comparison of images obtained at low light intensity. A cleaner image can be obtained with PCDs even when there is sufficient exposure, as shown in the figure (see example on the furthest right). Favorable images can be obtained with photon-counting even when the exposure drops down to 1/800, whereas the noise becomes so high that the image is no longer recognizable with energy integrating detectors. Furthermore, PCDs can achieve a favorable image even when the pixel size is smaller, thus, higher spatial resolutions can also be achieved with this.

In this manner, measurements with PCDs not only enable noise/artifact control, higher contrast, and higher spatial resolution, but also provide various functional additions to X-ray tomography, including the measurement of X-ray photon energies

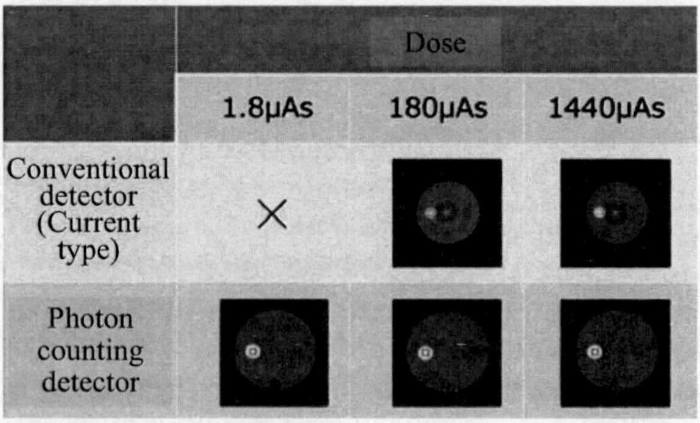

Fig. 4.116 Relationship between the 3D image (cross-section) and dose between an energy integrating detector (conventional) and photon counting detector. The dose is expressed in units of tube current (A) multiplied by exposure time (s) (courtesy of Akifumi Koike of ANSeeN)

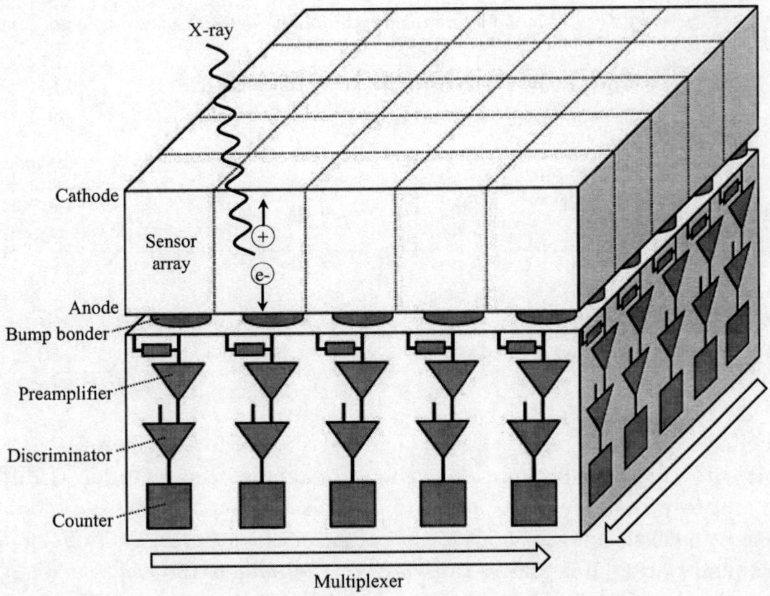

Fig. 4.117 Schematic of the photon counting detector structure

or the mapping of elemental concentrations. In the future, this is suited to become an important device in both medical-use and industrial X-ray CT scanners [92].

(1) Structure

Figure 4.117 shows a schematic of the PCD structure. Semiconductors, which can directly convert X-rays into photoelectric charge (e.g. *cadmium telluride* (CdTe)), are generally used as sensors and a signal processing circuit for photon counting is added. Electron-hole pairs are excited through X-ray incidence. A bias voltage of approximately several hundred V is applied to the sensor and the photoelectric charge is captured by an electrode with a positive voltage. The application of high voltages prevents the rejoining of electrons and holes in the sensor interior, which results in higher measurement accuracy. The semiconductor does not need to be physically divided and the pixel size of the detector can be set up using a smaller electrode. The obtained signal enters an integrated circuit referred to as an *application-specific integrated circuit* (*ASIC*) and amplified with a preamplifier. The number of photons is counted after passing through the pulse formation circuit and divided between pulse and noise using a threshold. This *discriminator* is also used in X-ray photon energy measurements.

(2) Semiconductor

Cadmium zinc telluride and silicon are used in addition to cadmium telluride in direct conversion-type semiconductor detectors used in PCDs. Mercury iodide and gallium

Table 4.15 Representative semiconducting materials used in photon counting detectors and their characteristics [92]

Material	Density (g/cm^3)	Decay distance, δ (μm)			E_g (eV)	$\mu\tau$ (cm^2/V)		Electron-hole pair creation energy (eV/pair)
		Effective atomic number	20 keV	60 keV		Electron	Hole	
Cadmium telluride (CdTe)	5.85	50	36	107	1.44	3.3×10^{-3}	2×10^{-4}	4.43
Cadmium zinc telluride (Cd0.9Zn0.1Te)	5.78	49	35	110	1.572	3–5 $\times 10^{-3}$	5×10^{-5}	4.64
Silicon (Si)	2.33	14	438	5593	1.12	>1	1	3.62

arsenide have also been reported. The characteristics of these materials are shown in Table 4.15 [93] (the $\mu\tau$ (cm^2/V) value of cadmium zinc telluride is different from that shown in Tables 4.12 and 4.15 but this is thought to be due to material differences and has, therefore, been left as per the original source. Cadmium telluride and cadmium zinc telluride have a stopping power similar to LSO and LuAG, as well as the cesium iodide in Table 4.14. Cadmium telluride has what is referred to as a sphalerite structure. A sphalerite structure is a cubic system where the telluride forms the face-centered cubic lattice and cadmium is placed in tetracoordinated positions of a regular tetrahedron type. Furthermore, the single crystal cleaves along the {110} plane. In contrast, cadmium zinc telluride has a structure where approximately 10% of the cadmium in cadmium telluride is replaced with zinc.

As shown in Eq. (4.51), the electron-hole pair creation energies of these materials are approximately 3 times that of the band gap energy. For example, cadmium telluride has an electron-hole pair creation energy of 4.43 eV/pair, and an average of 4500 electron-hole pairs are created for a single X-ray photon incident at 20 keV of energy. Furthermore, the product of electron lifespan and mobility $\mu_e\tau_e$ is approximately 10 $^{-3}$ cm^2/V, which is several orders of magnitude smaller than the value in silicon. $\mu_h\tau_h$ for the hole becomes 1–2 orders of magnitude smaller compared to $\mu_e\tau_e$. Furthermore, $\mu_e\tau_e$ and $\mu_h\tau_h$ multiplied by the electric field strength F correspond to the average transfer distance until the holes and charge are trapped or rejoined. The thickness of the sensor section must be smaller than $\mu_h\tau_h F$ to capture most of the produced charge at the electrode. The charge collection efficiency η_{cc} at a distance from the cathode z is expressed as follows, with the thickness of the sensor section as L_s [93].

$$\eta_{cc} = \frac{\mu_e\tau_e F}{L_s} \tag{4.65}$$

When taking the $\mu_e\tau_e$ and $\mu_h\tau_h$ values of cadmium telluride from Table 4.15 and setting L and F as 100 μm and 1 V/μm, respectively, $\eta_{cc} = 99.9\%$ (contribution of the first term in Eq. (4.65) is approximately 90%) at $z = 10$ μm, $\eta_{cc} = 99.3\%$ at $z =$

50 μm (middle of the thickness direction), and $\eta_{cc} = 98.0\%$ (contribution of the first term in Eq. (4.60) is approximately 10%) at $z = 90$ μm. The differences in charge collection efficiency between the two sides of this thin film show that a higher value is observed on the side where the X-ray is incident on the cathode side and where the charge is collected from the anode side [93].

A phenomenon known as *charge sharing*, where a single photon supplies charge to multiple pixels, occurs in PCDs. Electron sharing induces errors in the X-ray photon count and makes energy information inaccurate. It is important to consider the ratio L_s/w_s between pixel size w_s and L_s and ensure that L_s/w_s is not too large (i.e. the pixel does not become too small) to prevent charge sharing. For example, reports have indicated energy resolution worsened at $L_s/w_s = 10$ and that energy could not be resolved at $L_s/w_s = 18$ when X-rays with 60 keV energy were detected with cadmium telluride [94].

(3) Measurements of X-Ray Energy Spectra

The energy of incident X-rays and photoelectric charge number are generally proportional to one another. The energy of X-ray photons can thus be distinguished when measuring the amount of charge within a range where this proportional relationship is present. As shown in the schematic in Fig. 4.118, a histogram of the X-ray photon

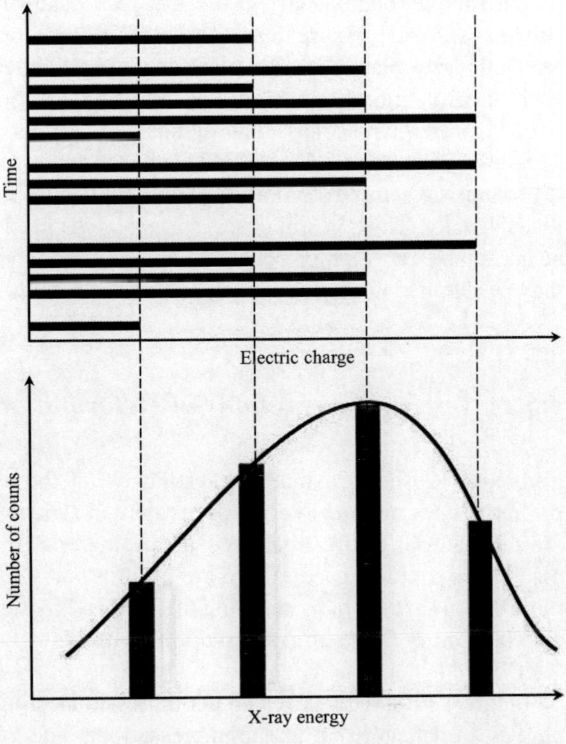

Fig. 4.118 Schematic of the basic principles of X-ray energy spectra measurement using the photon counting detector

energy can be obtained when measurements are continued for a given period. This becomes the energy spectrum of the incident X-ray. Furthermore, projection data for each energy range can be used to highlight the contrast of specified elemental components. Further advancing this, energy discrimination allows for the implementation of absorption edge subtraction imaging methods (discussed later) without using monochromatic X-rays; moreover, 3D mapping of alloy element concentration distributions in a material can be conducted in a simple manner. Furthermore, artifacts (beam hardening) can be controlled by applying X-ray energy discrimination.

4.5 In Situ Observation Devices

X-ray CT scanners have a large advantage in that they can accurately measure the internal structure or outer appearance of actual materials. Furthermore, it is not an exaggeration to state that directly and accurately observing changes in the internal structure of actual materials or parts due to external disturbances can only be completed with X-ray CT scanners. The disturbance to be studied must be recreated on the sample rotation stage of the X-ray CT scanner to realize these observations. This section introduces various devices that make this possible. Many of the devices introduced here are for specialized setups.

The sample rotation stage of the X-ray CT scanner is placed in room temperature and atmospheric pressure in most cases. Furthermore, the X-ray source and detector are frequently placed over several dozen millimeters away from the sample. As such, there is enough space for setting up in situ observational devices compared with electron microscopes. As previously mentioned, the positional accuracy of the sample rotation stage is determined by the spatial resolution necessary from the X-ray CT scanner. A sample rotation stage with higher accuracy generally results in a lower load capacity. Thus, in situ observation devices need to be set up and placed within an acceptable range of load capacity in the sample rotation stage. Furthermore, these may be supplied in an optional or general-use form for industrial X-ray CT scanners.

4.5.1 In Situ Observations of Deformation/Fracture Behavior

Figure 4.119 shows a material testing rig at the author's laboratory, which has primarily been used for X-ray tomography in synchrotron radiation facilities. These are all material testing rigs used in the material science/mechanical engineering disciplines. In most cases, these are used in *time-lapse* in situ *observations*, where observations are made by stopping the load and fixing the deformation, but these can also be used in *full in situ observations* with high-speed tomography, introduced in Sect. 5.3.

Figure 4.119a shows a tension/compression testing machine used since 2003 that applies an actuator that uses compressed air. The load frame which supports load

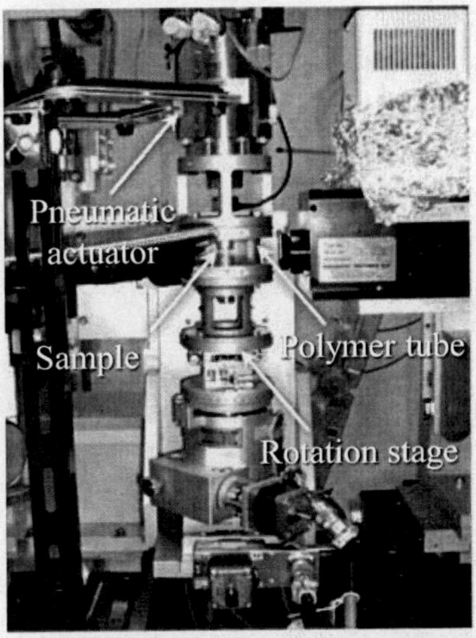

(a) Pneumatic tension/compres-sion testing machine

(b) Miniature testing machine for the imaging CT

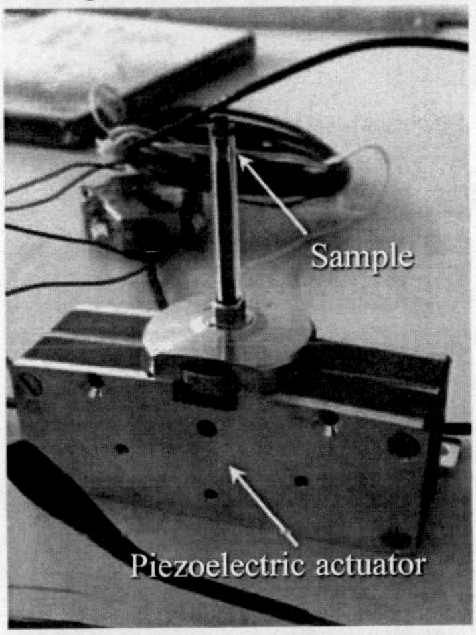

(c) Piezoelectric fatigue testing machine

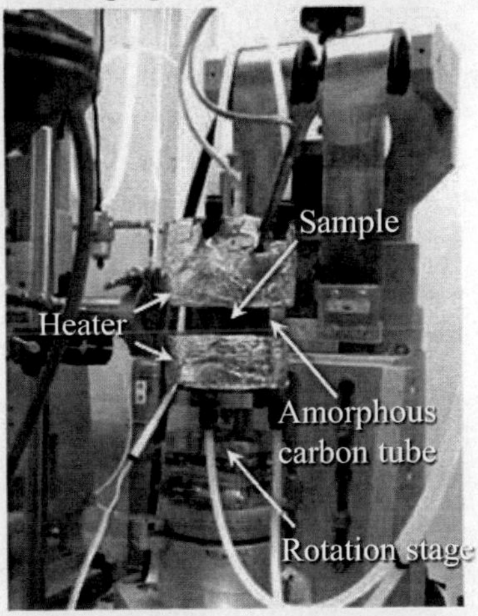

(d) High temperature tension/creep testing machine

Fig. 4.119 Various material testing machines that can be set up on a rotation

does not take the form of a pillar as in typical material testing machines but rather that of a cylinder. The material used is a polymer (polycarbonate) and its thickness is 5 mm. This structure and material satisfy the two conditions of being able to image under consistently equal conditions even when the sample is rotated and with minimal X-ray absorption. Tension, compression, and fatigue tests are conducted with this testing machine under loads of up to 2 kN.

We now consider the implementation of tensile tests of high-strength steel with generated loads of 500 N using this testing machine. For a 0.7 mm by 0.7 mm square area in a gauge section, loads of 1020 MPa and 0.3 MPa are applied to the miniature sample for synchrotron radiation-based X-ray tomography and polymer load frame, respectively. The stress applied to the load frame appears extremely small at first glance. However, the modulus of elasticity of the polycarbonate is 2.25 GPa and that of the steel is larger by a factor of approximately 100. For this reason, the elastic elongation of the steel sample in the 1-mm distance gauge length is approximately 4.9 μm, whereas the elastic deformation of the approximately 40-mm high polycarbonate part is roughly 5.7 μm. This is far removed from typical material tests, where the sample device stiffness is extremely high and does not deform. In contrast with this deformation of the testing machine, the material testing rig in Fig. 4.119a is controlled to a total weight of 5 kg and both the eccentricity and surface runout amount of the sample rotation stage are minimized to maintain a spatial resolution of approximately 1 μm. This testing device is available at BL20XU in SPring-8; interested parties should contact the author. Furthermore, the load capacity of the sample rotation stage becomes even more strict when high spatial resolutions are obtained with X-ray microscopy. Figure 4.119b is a manual testing machine that was miniaturized by removing as much weight as possible from the machine.

A fatigue testing machine with a piezoelectric actuator is used when strokes are not necessary, an example is shown in Fig. 4.119c. Furthermore, a testing machine with a heater attached is used for observations at high temperatures, as shown in Fig. 4.119d. These have been used to successfully conduct in situ observations of creep deformation behavior in aluminum alloys at a temperature of 500 °C [95]. An amorphous carbon cylinder whose X-ray absorption is as small as a polymer is used as a load frame in these cases.

4.5.2 *In Vivo Observations of Living Bodies*

Drug development requires animal experimentation with small animals whose specified genes can be easily altered; moreover, it is important to be able to image the insides of their bodies while still alive at a high resolution. Figure 4.120 shows a schematic where synchrotron radiation was used as an X-ray source to conduct dynamic 3D observations of the coronary arteries and bronchioles in living rats and mice [96]. To conduct in vivo *observations* (referring to the observations of living bodies) in small animals, their airway pressure and electrocardiograms must be constantly monitored and transmission images are continuously taken while coordinating the opening/closing of the X-ray shutter and exposure start timing with airway pressure and electrocardiogram signals. This can significantly reduce motion artifacts due to small movements of the living body. In these cases, high-speed X-ray shutters have been introduced to minimize the radiation dose in animals as much as possible.

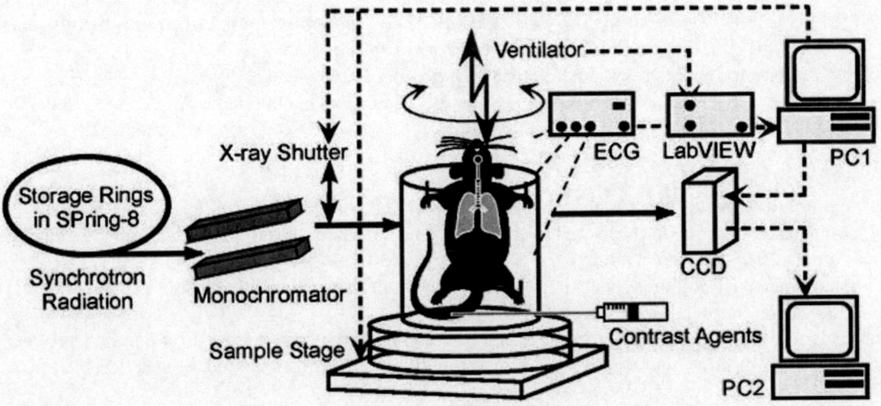

Fig. 4.120 Overview of high-resolution in-vivo CT system (courtesy of Professor Toshihiro Sera of Kyushu University)

References

1. K. Sanderson, Nature digest, **5** (2008). https://doi.org/10.1038/ndigest.2008.081211
2. C.G. Camara, J.V. Escobar, J.R. Hird, S.J. Putterman, Nature **455**, 1089–1092 (2008)
3. C.A. Clark, Los alamos/tribogenics create highly portable imaging system, Los Alamos Daily Post (2013). June 26. http://www.ladailypost.com/content/los-alamostribogenics-create-highly-portable-imaging-system
4. J.M. Boone, J. Beutel, H.L. Kundel, R.L. VanMetter (eds.). *Handbook of Medical Imaging*, Bellingham, Washington, SPIE Press, 1, 78 (2000)
5. H.A. Kramers, Phil. Mag. **46**, 836–871 (1923)
6. T.M. Buzug, *Computed Tomography: From Photon Statistics to Modern Cone-Beam CT* (Springer, Berlin, Germany, 2008)
7. Chemical Society of Japan, Ed.: Structure of Materials III: Diffraction, Fifth Series of Experimental Chemistry Vol. 11, Maruzen, (2006), ISBN 9784621073100
8. International Tables for X-Ray Crystallography, Vol.3, Physical and Chemical Tables, editors C.H. Macgillavry and G.D. Rieck, D. Reidel Publishing Company, (1985). (published online http://xdb.lbl.gov/xdb.pdfhttp://xdb.lbl.gov/xdb. Accessed August 2017
9. Japan Institute of Metals and Materials, Ed.: Metals Databook, Revised 4th Edition, Maruzen, Co., Ltd., (2004), 228
10. O.W. Richardson, Philosophical of the Cambridge Philosophical Society **11**, 286–295 (1901)
11. M.N. Avadhanulu, P.G. Kshirsagar, *A Textbook of Engineering Physics* (S Chand & Co., Ltd., New Delhi, 1992), p. 348
12. M. Hagino, Surf. Sci. **8**, 472–479 (1987)
13. T.M. Buzug, *Computed tomography: from photon statistics to modern cone-beam CT* (Springer, Berlin, Germany, 2008), pp. 18–19
14. L.A.G. Perini, P. Bleuet, J. Filevich, W. Parker, B. Buijsse, L.F.Tz. Kwakman, *Review of Scientific Instruments*, 88 (2017), 063706-1-10
15. G.G. Poludniowski, P.M. Evan, Med. Phys. **34**, 2164–2174 (2016)
16. T. Fujita, H. Anno, Toshiba Rev. **66**, 24–28 (2011)
17. E. Lassner, W.-D. Schubert, *Tungsten: Properties, Chemistry, Technology of the Element, Alloys, and Chemical Compounds* (Kluwer Academic/Plenum Publishers, New York, 1999), pp. 256–258
18. Y. Mutoh, K. Ichikawa, K. Nagata, J. Soc. Mater. Sci. Japan **40**, 882–888 (1991)

19. R.L. Sproull, W.A. Phillips, *Modern Physics: The Quantum Physics of Atoms, Solids, and Nuclei*, 3rd edn. (Dover Publications, New York, 1980), p. 455
20. S. Yamamoto, H. Aoyama, Toshiba Rev. **69**, 57–59 (2014)
21. H. Sugie, M. Tanemura, V. Filip, K. Iwata, K. Takahashi, F. Okuyama, Appl. Phys. Lett. **78**, 2578–2580 (2001)
22. W. Sugimoto, S. Sugita, Y. Sakai, H. Goto, Y. Watanabe, Y. Ohga, S. Kita, T. Ohara, J. Appl. Phys. **108**, 044507 (2010)
23. Comcraft website: http://www.comcraft.co.jp/products/05.html. Accessed August 2017
24. Matsusada Precision website: https://www.matsusada.co.jp/product/xm/xunit/. Accessed August 2017
25. Hamamatsu Photonics website: http://www.hamamatsu.com/jp/ja/product/category/1001/3028/index.html. Accessed August 2017
26. Japanese Inspection Instruments Manufacturers' Association: Association-handled product information. http://www.jima.jp/content/assen.html#assen_d. Accessed August 2017
27. M. Ito, Radioisotopes **52**, 699–703 (2003)
28. M.M. Nasseri, Nuclear Eng. Technol. **48**, 795–798 (2016)
29. R.W. Hamm, M.E. Hamm, *Industrial Accelerators and their Applications* (World Scientific, Singapore, 2012), pp. 307–369
30. H.E. Martz, C.M. Logan, D.J. Schneberk, P.J. Shull, *X-Ray Imaging: Fundamentals* (CRC Press, Boca Raton, Industrial Techniques and Applications, 2017), pp. 176–177
31. G. Claus, *Introduction to radiation protection: practical knowledge for handling radioactive sources* (Springer, Berlin, 2010), pp. 323–325
32. J. Davis, P. Wells, Industrial Metrology **2**, 195–218 (1992)
33. R. Hoffmann, T. Kögl, Flow Meas. Instrum. **53**, 147–153 (2017)
34. M.S. Rapaport, A. Gayer, NDT&E International **24**, 141–144 (1991)
35. D.C. Copley, J.W. Eberhard, G.A. Mohr, JPM, 15–26 (1994)
36. C. Rizescu, C. Besliu, A. Jipa, Nuclear Instrum. Methods Phys. Res. A **465**, 584–599 (2001)
37. SPring-8 website: http://www.spring8.or.jp/ja/about_us/whats_sp8/facilities/accelerators/ http://www.spring8.or.jp/ja/about_us/whats_sp8/facilities/accelerators/, etc. Accessed September 2017
38. K. Kora, editor-in-chief: Synchrotron Radiation in Practical Use, Council for the Promotion of Photon Science Technologies, Nikkan Kogyo Shimbun (1997)
39. T. Kaneyasu, Y. Takabayashi, Y. Iwasaki, S. Koda: Proceedings of The 12th International Conference on Accelerator and Large Experimental Control System, Kobe, pp. 307–309 (2009)
40. Y. Takashima, M. Hosaka, N. Yamamoto, K. Takami, T. Takano, A. Mano, H. Morimoto, M. Katoh, Y. Hori, S. Sasaki, S. Koda, Y. Takeda: Proceedings of the 10th Annual Meeting of Particle Accelerator Society of Japan, Nagoya, pp. 385–387 (2013)
41. H. Yonehara, Journal of the Japanese society for synchrotron. Radiat. Res. **16**, 178–185 (2003)
42. T. Asaka, H. Dewa, H. Hanaki, T. Kobayashi, A. Mizuno, S. Suzuki, T. Taniuchi, H. Tomizawa, K. Yanagida, Proceedings of EPAC 2002, Paris, France, pp. 2685–2687 (2002)
43. H. Yonehara: SPring-8 user information, highlights, 2, 1–14 (1997)
44. K. Wille, Rep. Prog. Phys. **54**, 1005–1068 (1991)
45. Council for the Promotion of Photon Science Technologies (K. Kora, Ed.): Synchrotron Radiation in Actual Use, Nikkan Kogyo Shimbun (1997)
46. H. Wiedemann, *Synchrotron Radiation* (Springer, Berlin, 2003)
47. S. Kikuta, *X-Ray Scattering and Radiation Science* (Tokyo University Press, Fundamentals, 2011)
48. H. Tanaka, H. Ohkuma: SPring-8 user information, 298–304 (2003)
49. T. Tomimasu, *Synchrotron Radiation Technologies.* Kogyo Chosakai Publishing pp. 131–133 (1990)
50. Y. Sakurai, H. Kitamura, *SPring-8 User Information*,pp. 6–8 (1997)
51. K. Uesugi, *(Riken SPring-8 Center): Personal Communication* (2017)
52. H. Ohashi, K. Hirano (Eds.), Introduction to radiation beamline optics—for first-time radiation users. Japanese Society for Synchrotron Radiation Research (2008)

53. Y. Kashihara, JAERI-M Report, 91-008, Japan Atomic Energy Research Institute (1991)
54. E.L. Nickoloff, H.L. Berman, Radiographics **13**, 1337–1348 (1993)
55. C. Engblom, Overview of some feedback- & control systems in synchrotron SOLEIL. In The 11th International Workshop on Personal Computers and Particle Accelerator Controls (PCaPAC), Brazil (2016)
56. A. King, N. Guignot, P. Zerbino, E. Boulard, K. Desjardins, M. Bordessoule, N. Leclerq, S. Le, G. Renaud, M. Cerato, M. Bornert, N. Lenoir, S. Delzon, J.-P. Perrillat, Y. Legodec, J.-P. Itié, Rev. Sci. Instrum. **87**, 093704 (2016)
57. H. Toda, M. Kobayashi, Y. Suzuki, A. Takeuchi, K. Uesugi, Microscopy **44**(2009), 199–205 (2009)
58. S.M. Gruner, J.R. Milch, G.T. Reynolds, IEEE Trans. Nucl. Sci. **25**, 562–565 (1978)
59. C. Ponchut, Journal Synchrotron Radiation **13**, 195–203 (2006)
60. K. Ito, Y. Amemiya, N. Sakebe, Development of Array-CCDX line detector and protein crystal structure analysis applications. In Proceedings of the Joint Symposium on the 13th Annual Meeting of Japan Synchrotron Radiation Society and Synchrotron Radiation Science, 9-P-60 (2001)
61. M. Tomizawa, J. Japanese Soc. Non-destructive Inspection **63**, 221–231 (2014)
62. M. Abe (Contr.), A. Ota (Ed.): Latest Trends in CMOS Image Sensors—From High Performance, High Functionality, to Application Developments,CMC Publishing, 10 (2014)
63. D.J. Denvir, C.G. Coates, Proc. SPIE., *Biomedical Nanotechnology Architectures and Applications*, 4626, 502–512 (2002)
64. Hamamatsu Photonics: Photosemiconductor Element Handbook, https://www.hamamatsu.com/resources/pdf/ssd/05_handbook.pdf. Accessed January 2018
65. Y. Kimura, Microscopy **45**, 257–263 (2010)
66. H. Shibata, *Design Wave Magazine*, 128–129 (2009)
67. T. Kuroda, *Journal of the Institue of Image Information and Television Engineers*, 68, 216–222 (2914)
68. S. Risticj, Proceedings of Goran Conference on medical physics and biomedical engineering. Macedonia, 65–71 (2013)
69. S. Kasap, J.B. Frey, G. Belev, O. Tousignant, H. Mani, J. Greenspan, L. Laperriere, O. Bubon, A. Reznik, G. DeCrescenzo, K.S. Karim, J.A. Rowlands, Sensors **11**, 5112–5157 (2011)
70. T. Yamasaki, J. Soc. Photographic Sci. Technol. Japan **70**(separate issue), 24–25 (2007)
71. Physical Measurement Laboratory (PML), National Institute of Standards and Technology (NIST): X-Ray Form Factor, Attenuation and Scattering Tables, NIST, U.S. Commerce Department, USA, (2011). https://www.nist.gov/pml/x-ray-form-factor-attenuation-and-scattering-tables. Accessed January 2018
72. G. Bizarri, J. Cryst. Growth **312**, 1213–1215 (2010)
73. Y. Suzuki (part-time lecturer at Tokyo University, formerly of Japan Synchrotron Radiation Research Institute), personal communication
74. T. Jing, A. Goodman, J. Drewery, G. Cho, W.S. Hong, H. Lee, S.N. Kaplan, V. Perez-Mendez, D. Wildermuth, Nucl. Instrum. Methods Phys. Res., Sect. A **368**, 757–764 (1996)
75. A. Yoshikawa, FBNews No. 463, 1–6 (2015)
76. G.F. Knoll, *Radiation Detection and Measurement*, 4th edn. (Wiley India, Noida, Uttar Pradesh, 2011)
77. K. Saegusa, T. Irifune, M. Fukushi, H. Saitoh, G. Nakaya, *Basic Radiation Metrology*, Iryo Kagaku Co. (2001)
78. S.R. Cherry, J.A. Sorensen, M.E. Phelps, *Physics in Nuclear Medicine*, 4th edn. (Elsevier, Philadelphia, 2012)
79. P.A. Rodny, *Physical Processes in Inorganic Scintillators* (CRC Press, Boca Raton, FL, 1997)
80. G. Blasse, B.C. Grabmaier, *Luminescent materials* (Springer-Verlag, Berlin, 1994)
81. C.L. Melcher, J.S. Schweitzer, A. Liberman, J. Simonetti, IEEE Trans. Nucl. Sci. **32**, 529–532 (1985)
82. M. Laval, M. Moszynski, R. Allemand, E. Carmoreche, P. Guinet, R. Odru, J. Vacher, Nucl. Instrum. Meth. Phys. **206**, 169 (1983)

83. E.V.D. van Loef, P. Dorenbos, C.W.E. van Eijk, K.W. Kraemer, H.U. Guedel, Appl. Phys. Lett. **79**, 1573 (2001)
84. T. Suzuki (National Institute of Radiology Sciences): *Recent Trends in Scintillators*, 6th Annual JRSM Symposium, 120–122 (2011)
85. H. Graafsma, T. Martin, *Advanced Tomographic Methods in Materials Research and Engineering*, ed. John Banhart, Oxford University Press, (2008), Section 10
86. M.S. Alekhin, J. Renger, M. Kasperczyk, P.-A. Douissard, T. Martin, Y. Zorenko, D.A. Vasilév, M. Stiefel, L. Novotny, M. Stampanoni, Opt. Express **25**, 1251–1261 (2017)
87. K. Takagi, T. Fukazawa, Appl. Phys. Lett. **42**, 43–45 (1983)
88. C.L. Melcher, J.S. Schweitzer, Nuclear instruments and methods in physics research section a: accelerators. Spectrometers Detectors Assoc. Equip. **314**, 212–214 (1992)
89. L. Qin, G. Ren, S. Lu, D. Ding, H. Li, IEEE Trans. Nucl. Sci. **55**, 1216–1220 (2008)
90. J. Akutagawa, D. Yamamoto, W. Pong, J. Electron Spectrosc. Relat. Phenom. **82**, 75–77 (1996)
91. S. Hejazi, D.P. Trauernicht, Med. Phys. **24**, 287–297 (1997)
92. K. Kan, Y. Imura, H. Morii, K. Kobayashi, T. Minemura, T. Aoki, World J. Nuclear Sci. Technol. **3**(2013), 106–108. https://doi.org/10.4236/wjnst.2013.33018. published online
93. P. Russo, A.D. Guerra, Chapter 2 "Solid-State Detector for Small-Animal Imaging", Molecular Imaging of Small Animals Instrumentation and Applications, Edited by H. Zaidi, Springer, New York (2014)
94. K. Spartiotis, A. Leppänen, T. Pantsar, J. Pyyhtiä, P. Laukka, K. Muukkonen, O. Maännistö, J. Kinnari, T. Schulman, Nucl. Instrum. Methods Phys. Res., Sect. A **550**, 267–277 (2005)
95. H. Toda, Z.A.B. Shamsudin, K. Shimizu, K. Uesugi, A. Takeuchi, Y. Suzuki, M. Nakazawa, Y. Aoki, M. Kobayashi, Acta Mater. **61**, 2403–2413 (2013)
96. T. Sera, H. Yokota, K. Fujisaki, K. Fukasaku, H. Tachibana, K. Uesugi, N. Yagi, R. Himeno, Phys. Med. Biol. **53**, 4285–4301 (2008)

Chapter 5
Applied Imaging Methods

Absorption and phase contrasts were previously discussed in Chap. 2 as part of the discussion on the fundamentals of X-ray imaging. In this chapter, we will expand on the methods used in actual practice and discuss the various applied imaging techniques, including X-ray tomography, which uses phase contrast or *imaging optical systems*; fast X-ray tomography; and three-dimensional (3D) mapping of chemical composition or crystal orientation of various compounds. These imaging techniques are primarily used in X-ray tomography at synchrotron radiation facilities; however, they will become increasingly important for industrial X-ray computerized tomography (CT) scanners in the future.

5.1 Imaging-Type X-ray Tomography

Achieving a high spatial resolution with X-ray tomography requires considering all the factors associated with spatial resolution, such as the radiation source, sample, positioning stage, and detector, which constitute the X-ray CT scanner. As shown later in Eqs. (5.29) and (7.30), the single factor with the lowest accuracy acts as the rate-determining step for spatial resolution in the imaging system. Moving sequentially from the upstream X-ray source, these factors include: ① effective focal point size of the X-ray source (only for industrial X-ray CT scanners), ② Nyquist frequency determined by the pixel size of the detector and the rotation step of the sample rotation stage, ③ the positional accuracy of the positioning stage, e.g., the sample rotation stage, ④ drift of the sample, etc., ⑤ Fresnel diffraction of the X-ray, ⑥ scintillator thickness or the stopping power of the scintillator material, and ⑦ the spatial resolution of the detector system as stipulated by factors, such as the diffraction limit of the visible light.

H. Toda, *X-Ray CT*,
https://doi.org/10.1007/978-981-16-0590-1_5

In the case of projection X-ray tomography, the physical limit of the spatial resolution is marginally less than 1 μm. However, in this section, we consider techniques that achieve high spatial resolutions exceeding this limit. Let us see how these techniques resolve the factors that affect spatial resolution and achieve their high performances. First, factor ① is not a problem in large-scale synchrotron radiation facilities that use parallel beams. Furthermore, as shown in Sect. 4.1.2 (5), industrial X-ray CT scanners can achieve a focal point size of 250 nm with microfocus X-ray tubes that are currently in use; moreover, transmission images with a maximum spatial resolution of 60 nm can be obtained using the electron beams discussed in Sect. 4.1.2 (1). Next, factor ② can be resolved by selecting detectors with a large number of pixels and optimal magnification projection and imaging conditions for the imaging optical systems used following the conversion to visible light. For example, the CCD cameras listed in Tables 4.7, 4.8 and 4.9 and sCMOS cameras listed in Table 4.11 have minimum pixel size values of 5.5 μm and 6.5 μm respectively. Furthermore, as discussed in Sect. 4.4.1 (3), it is standard practice to use a scintillator to convert X-rays initially into visible light before detection instead of directly beaming X-rays into CCD cameras or sCMOS cameras to obtain a wide dynamic range. It is possible to achieve effective pixel sizes of less than 0.5 μm (corresponding to a maximum effective spatial resolution of 1 μm) after converting X-rays into visible light using fiber optic tapers or optical lenses to create an image on the detector. For factor ③, a maximum eccentricity of 70 nm can be achieved in the sample rotation stage through equipment selection, as discussed in Sect. 4.3.1. Finally, for factor ④, temperature control and aging can be used as resolution strategies. Through these techniques, factors ①–④ are not necessarily insurmountable barriers in achieving a spatial resolution surpassing 1 μm.

Considering factor ⑤, we calculate the radius of the first Fresnel zone using Eq. (2.32). $r_{\mathrm{FZ}} \approx 1$ μm can be attained under conventional experimental conditions where a monochromatic X-ray with an energy of 20 keV is used and the sample–detector distance is 16 mm; the physical limit of spatial resolution is reached in projection-type X-ray tomography. Interference fringes due to Fresnel diffraction are observable in the images in these cases. In addition, for factor ⑥, considerably surpassing a spatial resolution of 1 μm in projection-type tomography is not practical because of the limits of stopping power. Furthermore, the *Rayleigh diffraction limit* δ_{DL} associated with factor ⑦ is expressed using the following equation:

$$\Delta_{DL} = 0.61\frac{\lambda}{NA} \tag{5.1}$$

Here, *NA* is the numerical aperture and λ is the wavelength of the light. The use of the peak light emission wavelength (520–550 nm) of the LuAG (Ce) and GAGG (Ce) scintillators and the numerical aperture (0.4) of the optical lens recently used by the author at SPring-8 yields $\delta_{\mathrm{DL}} \approx 0.8$ μm; this also serves as a barrier to surpassing a spatial resolution of 1 μm. In such cases, to resolve factors ⑤–⑦, magnification

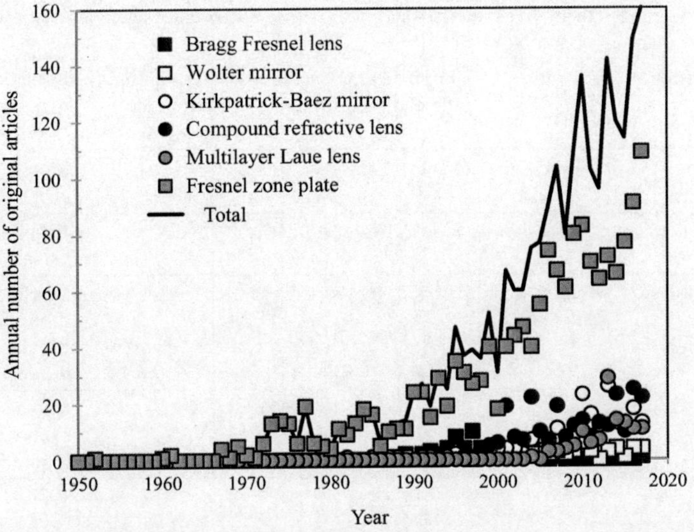

Fig. 5.1 Annual changes in the number of original publications relating to the basics and applications of X-ray imaging using various X-ray focusing optics

projections must be conducted via the X-ray imaging optical system before the X-ray is converted to visible light. To accomplish this, two methods of positioning *X-ray focusing optics*—either between the X-ray source and the sample or between the sample and the detector—can be considered. Reports have indicated that the latter is more effective for obtaining high spatial resolutions [1]. Various X-ray focusing optics have been proposed to realize this, and they have been used in practice.

Figure 5.1 shows the annual changes in the number of original articles with various X-ray focusing optics terms included in the keywords or abstract based on a search of the reference database (SCOPUS). Studies on focusing optics at synchrotron radiation facilities and applied imaging experiments, which use various focusing optics, are both included in the articles counted in Fig. 5.1. With regard to X-ray focusing optics, 2,327 articles have been published on the subject from 1952 to 2017. Articles concerning Fresnel zone plates account for approximately 70% of this total. Additional studies have been conducted on *compound refractive lenses*, *Kirkpatrick–Baez mirrors*, and *multilayer Laue lenses* from 2000, 2004, and 2009 onward, respectively. Table 5.1 presents a simplified summary of the characteristics of the various X-ray focusing optics. Section 5.1 discusses each focusing optic in the order of its importance, as shown in Fig. 5.1.

Table 5.1 Summary of the characteristics of primary X-ray focusing elements used in imaging X-ray micro-/nano-tomography

X-ray focusing device	Principle	Chromatic compactness efficiency aberration			Deviation from an optical axis	Remarks
Compound refractive lens	Refraction	Good	Yes	– 50%	No	Simple and low cost
Fresnel zone plate	Diffraction	Good	Yes	– 10%	No	Excellent versatility
Kirkpatrick-Baez mirror	Total reflection	Bad	No	– 75%	Yes	Considerable time and efforts for alignment needed
Multilayer Laue lens	Bragg diffraction	Good	Yes	– 80%	No	Coma aberration cannot be ignored

5.1.1 *Imaging Optical Systems Using a Fresnel Zone Plate*

(1) Basic Principles

Figure 5.2 shows a microscope photograph of a Fresnel zone plate, whereas Fig. 5.3 shows its schematic. The Fresnel zone plate is a disk-shaped diffraction grating featuring unevenly spaced diffraction intervals that gradually become narrower moving from the center to the outer edge. The gratings consist of concentric circles; in practice, they are made by alternating bands of high and low X-ray absorption.

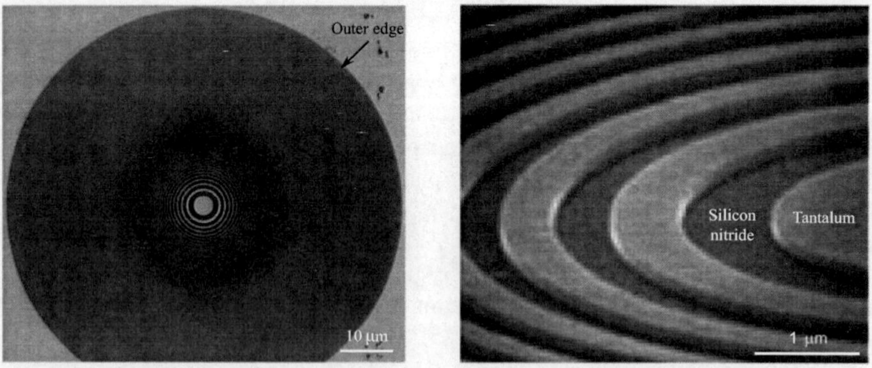

(a) Optical microscope image

(b) Scanning electron microscope image of the central part

Fig. 5.2 **a** Optical microscope image of the Fresnel zone plate and its **b** scanning electron microscope image. **a** Only the central gratings can be seen due to spatial resolution of the optical microscope image. **b** A 3D structure that corresponds to the schematic shown in Fig. 5.3 can be observed

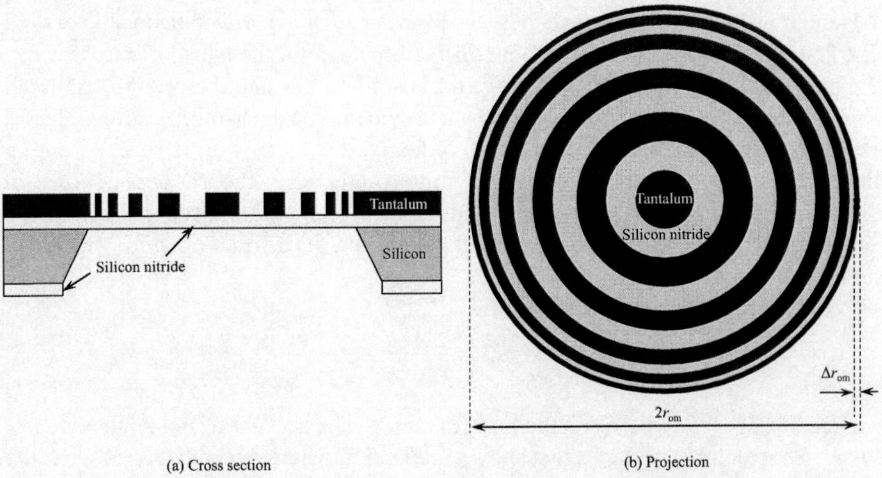

Fig. 5.3 **a** Cross-section and **b** projection image schematic of the Fresnel zone plate

Figure 5.4 illustrates why the grating intervals become gradually narrower. First, Fig. 5.4a shows a simple case where a parallel monochromatic X-ray beam is directed on an equally spaced grating. As expected, the X-ray passes through the adjacent grating; the X-ray waves strengthen each other through interference when the path difference of X-rays diffracting in the same direction becomes integer multiples of the wavelength.

$$d\sin\theta_m = m\lambda \tag{5.2}$$

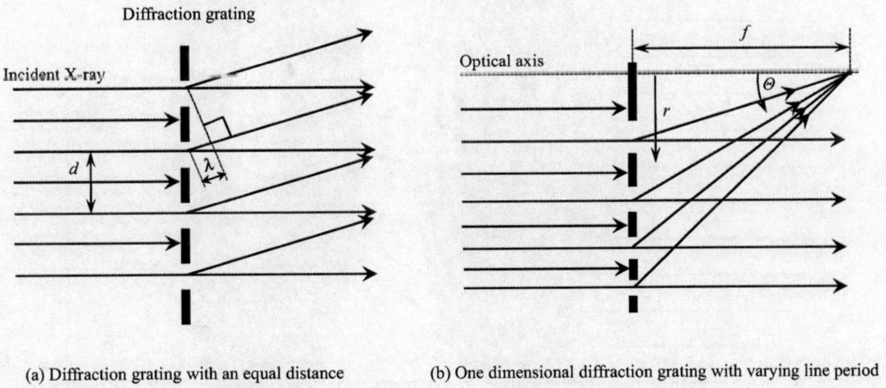

Fig. 5.4 Differences in X-ray diffraction behavior in a diffraction grating with equal intervals and when a parallel beam is illuminated into diffraction grating with steadily changing intervals

Here, d is the grating interval and m is the *order of diffraction* with a value of 0, ± 1, ± 2, ± 3,... Furthermore, θ_m is the diffraction angle of the mth-order light wave. A zero-order light wave refers to a beam transmitted through the grating unaltered without being diffracted. From Eq. (5.2), it is evident that a narrower grating interval results in a larger diffraction angle. Accordingly, the X-ray can be focused onto a single point by skillfully adjusting the grating interval in the 2D cases shown in Fig. 5.4b. Specifically, a first-order focused wave can be achieved by setting the grating interval d_r at the radius from the light axis on the diffraction grating r and an angle position of Θ.

$$d_r = \frac{\lambda}{sin\Theta} \tag{5.3}$$

Here, $\tan\Theta = \frac{r}{f}$,where f is the *focal distance*. The actual Fresnel zone plate has a disk-like structure similar to that shown in Fig. 5.5, where the light axis is set as the center and a one-dimensional (1D) diffraction grating as shown in Fig. 5.3b is rotated around the optical axis. When only a first-order wave is used, an aperture plate with a round hole referred to as an *order-sorting aperture* (*OSA*) is placed between the Fresnel zone plate and the focal spot, as shown in Fig. 5.5. In this case, an OSA with

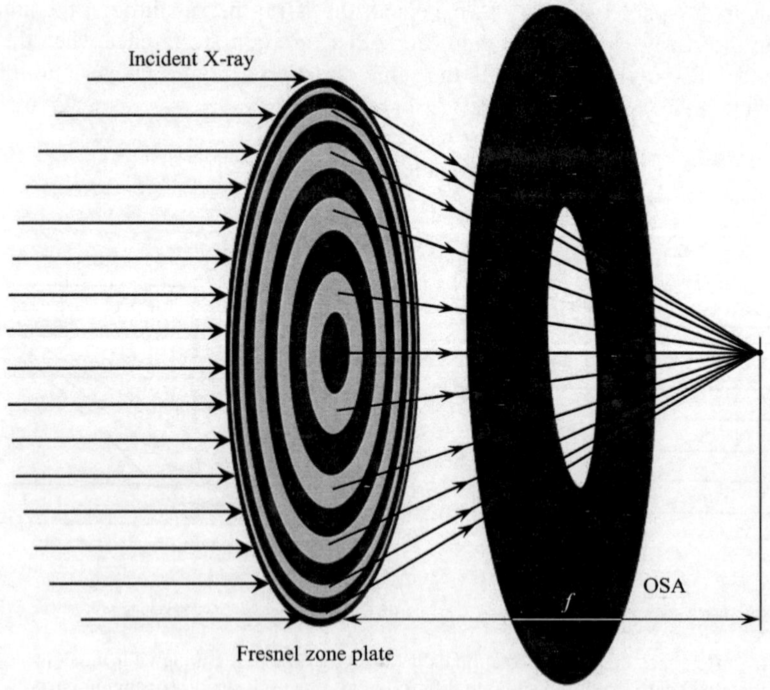

Fig. 5.5 Schematic of a parallel beam being focused using a Fresnel zone plate

a small aperture should be placed near the focal spot. Furthermore, a *center beam stop* is either placed upstream of the Fresnel zone plate or attached to the Fresnel zone plate to remove the zero-order wave.

(2) Fabrication Method

In most cases, the Fresnel zone plate is fabricated using *electron-beam lithography* [2]. First, as shown in Fig. 5.3a, a tantalum or gold metallic film is affixed on top of a supporting film of either silicon nitride or silicon carbide formed on top of silicon; this structure is then coated with a silicon oxide film. A photosensitive material referred to as an electron-beam resist is then applied on top of this structure. Next, an Fresnel zone plate pattern is drawn on the resist using electron-beam lithography. A silicon oxide pattern is fabricated with reactive-ion etching methods on the obtained resist pattern. With this silicon oxide pattern as a mask, the pattern is transcribed on to the metal portion using ion etching. Subsequently, a window is opened on the rear-side supporting film, and the Fresnel zone plate is completed by removing the silicon parts on the outer edges of the zone patterns, which served as the frame. The aspect ratio (ratio of pattern height (depth) and line width) obtained through processing using this method is 10–15 for a line width of 100 nm [3].

As discussed later, in practice, the metal portions of Fresnel zone plates do not completely stop X-rays but instead are designed to increase the diffraction efficiency by allowing a certain degree of X-ray transmission [1]. The metal portion thickness, t_{FZP}, in this case is expressed as follows [1]:

$$t_{FZP}\delta = \frac{\lambda}{2}. \tag{5.4}$$

Here, δ is the real part of the complex refractive index expressed in Eq. (2.19). Processing with a large aspect ratio is required when t_{FZP} is large; hence, materials with a minimal δ (e.g., high-density metals like tantalum or gold) are used. For example, the necessary thicknesses of tantalum for X-ray energies of 8, 20, and 30 keV are 1.8, 4.4, and 6.7 μm, respectively. In other words, maximum aspect ratios of several dozen are required for groove processing in the outermost grid with widths of several dozen to several hundred nanometers, making processing in this region extremely difficult. Therefore, Fresnel zone plates have so far been used exclusively for low X-ray energies below 10 keV. However, trials have been conducted at SPring-8 with imaging-type X-ray micro/nanotomography using high-energy X-rays of 20 and 30 keV in 2017 and 2018, respectively, and research results continue to be obtained for aluminum, iron, and titanium alloys [4]. The use of *apodized Fresnel zone plates* has been crucial in achieving these results.

(3) Shape Details and Performance

Spherical aberration can be ignored when $f \gg n\lambda/2$; the radius r_{n} of the nth zone from the center can be approximated as follows:

$$r_n^2 = n\lambda f \tag{5.5}$$

Quantitatively, the spatial resolution of the diffraction limit is reached when $f \gg 2n^2\lambda$. The term "spherical aberration" refers to how the focal spot will focus on different positions on the optical axis depending on the height of the incident X-ray (distance from the optical axis), blurring the image. The focal distance is expressed as follows, with the grid interval on the outermost edge (*outermost zone width*) Δr_{om}, number of zones N_{FZP}, and radius of the outermost zone r_{om}:

$$f = \frac{4N_{FZP}\Delta r_{om}^2}{\lambda} = \frac{2r_{om}\Delta r_{om}}{\lambda} \tag{5.6}$$

Here, we can observe that the focal distance is dependent on the wavelength. This is referred to as *chromatic aberration*. In actual experiments, the wavelength of the X-ray to be used is first determined based on factors such as the composition and diameter of the sample. The focal distance can be subsequently calculated using the diameter, number of zones, and outermost zone width of the Fresnel zone plate if Eq. (5.6) is used. The radius r_1 of the central disk region and the Fresnel zone plate diameter $2r_{\mathrm{om}}$ are determined as follows:

$$r_1^2 = 2r_{om}\Delta r_{om} \tag{5.7}$$

$$2r_{om} = 4N_{FZP}\Delta r_{om} \tag{5.8}$$

For Fresnel zone plates, the outermost zone radius is considerably smaller than the focal distance; hence, the numerical aperture *NA* is determined as follows:

$$NA = \frac{r_{om}}{f} = \frac{\lambda}{2\Delta r_{om}} \tag{5.9}$$

The numerical aperture is defined by Eq. (4.60). The spatial resolution δ_{FZP} of the Fresnel zone plate itself for the *m*th-order diffracted light is expressed as follows, based on Eqs. (5.1) and (5.9):

$$\Delta_{FZP} = \frac{1.22\Delta r_{om}}{m} \tag{5.10}$$

In other words, a Fresnel zone plate with a small outermost zone width should be created to achieve imaging with high spatial resolution. This requires microfabrication at a level of several dozen to several hundred nanometers. In actual practice, Fresnel zone plates with an outermost zone width of approximately 25 nm for hard X-rays are currently commercially available [5, 6]. The availability of plates with an outermost zone width of approximately half the aforementioned value, i.e., 12 nm, has also been reported [7].

Equation (5.10) also indicates that spatial resolution is not dependent on X-ray energy. However, considering the chromatic aberration in Eq. (5.6), the monochromaticity of the X-ray, to an extent indicated by the following equation, is necessary to achieve the spatial resolutions shown in Eq. (5.10):

$$\frac{\Delta\lambda}{\lambda} \leq \frac{1}{N_{FZP}} \tag{5.11}$$

The number of zones N_{FZP} typically used is between 300 and 3000. For the high-resolution imaging beamline BL20XU in SPring-8, the wavelength bandwidth $\Delta\lambda/\lambda$ of the synchrotron radiation obtained from the undulator using Eq. (4.29) is less than 5.8×10^{-3}; it becomes less than 10^{-4} after spectroscopy with a monochromator. The wavelength bandwidth for this synchrotron radiation is obtained for the center of irradiation in an ideal undulator and is realistically approximately 0.01–0.02 due to ring emittance, energy width, and the spread of the take-off angle range. As such, monochromatization using a monochromator is essential while using a Fresnel zone plate.

The *depth of focus* (*DOF*) refers to the allowable range (distance) of blur generated by the misalignment of the detector in the optical axis direction from the geometrically correct imaging surface position. The *DOF* stipulates the sample size that can be imaged in X-ray tomography. It is expressed as follows:

$$DOF = \frac{\delta_{FZP}^2}{0.61\lambda} \tag{5.12}$$

This signifies that, as the spatial resolution increases, the observable sample size rapidly decreases. For example, for an X-ray energy of 20 keV, the *DOF* rapidly decreases from 161 μm to 15 μm when the spatial resolution increases from 100 nm to 30 nm. The issue of fitting the sample diameter (thickness) inside this *DOF* typically presents a limitation for high spatial resolution. Meanwhile, with electron-beam tomography, the maximum thickness of the sample film is several micrometers, even when a large-scale transmission electron microscope with an ultra-high accelerating voltage of over 1 MV is used. Although the *DOF* values for electron microscopes are considerably greater than those for X-ray microscopes at the same spatial resolution, their full *DOF* cannot be utilized owing to the extremely weak penetrating power of the electron beam. As such, there is sufficient significance in pursuing higher resolutions with imaging-type X-ray tomography, even when there are limitations in sample size due to Eq. (5.12).

When an imaging optical system using Fresnel zone plates is assembled, its diffraction efficiency is an important factor directly linked to aspects such as exposure time. The diffraction efficiency for the *m*th-order diffracted light η_{FZP} is expressed as follows when the large-absorption component (tantalum component) is used to shield X-rays completely as shown in Fig. 5.2:

$$\eta_{FZP} = \begin{cases} 0.25\ m = 0 \\ \frac{1}{m^2\pi^2}\ \text{odd} - \text{order} \\ 0\ \text{even} - \text{order} \end{cases} \quad (5.13)$$

In other words, the diffraction efficiencies of zero-order, first-order, and third-order light are approximately 25%, 10%, and 1.1%, respectively. Additionally, –1 (negative first)-order light is diffracted at an efficiency of 10% outward. Furthermore, the even-order diffractions cancel each other out at the focal spot locations. In total, 50% of the incident light contributes to the diffraction of each order; the remaining 50% is absorbed by the Fresnel zone plate.

To increase the diffraction efficiency, the X-ray should be transmitted and the beam phase should be modulated instead of just blocking the ray at the large-absorption area in Fig. 5.4 [1]. To this end, the large-absorption area should be replaced with a low-absorption material whose phase differs by π from that of the original low-absorption area. This would ideally increase the efficiency of odd-order diffracted waves in the Fresnel zone plate to $\frac{4}{m^2\pi^2}$ [1]. For example, the diffraction efficiency for a first-order wave corresponds to a value of approximately 40%. However, no material is capable of changing the phase of X-rays without a degree of absorption; therefore, in practical use, the diffraction efficiency is lower than the aforementioned value due to X-ray absorption. The optimal Fresnel zone plate thicknesses shown in Eq. (5.4) drops to roughly 90% and 80% when the real/imaginary part ratios of the complex refractive indices β/δ are equal to 0.2 and 0.5, respectively [8].

As shown in Fig. 5.3a, a theoretical diffraction efficiency of 100% can be obtained from refraction effects if the square-shaped grating forms are set as curves. This type of Fresnel zone plate is called a "*Kinoform type*" plate. A schematic of such a plate is shown in Fig. 5.6a. Fabricating curved forms is difficult; however, even the

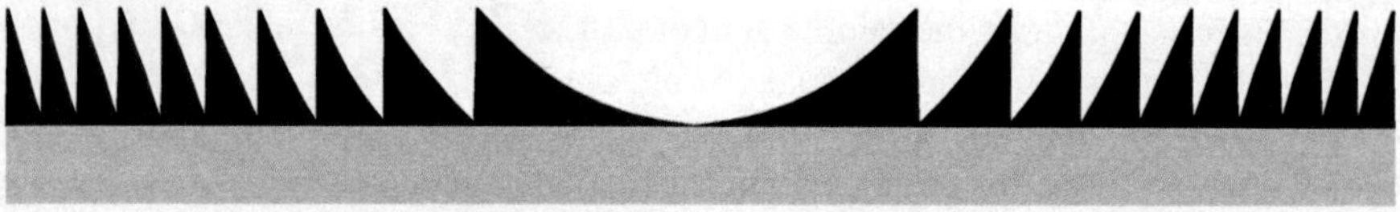

(a) Kinoform-type Fresnel zone plate

(b) Pseudo Kinoform-type Fresnel zone plate

Fig. 5.6 Schematic of a Kinoform-type Fresnel zone plate and a pseudo-Kinoform-type Fresnel zone plate using a step-like form

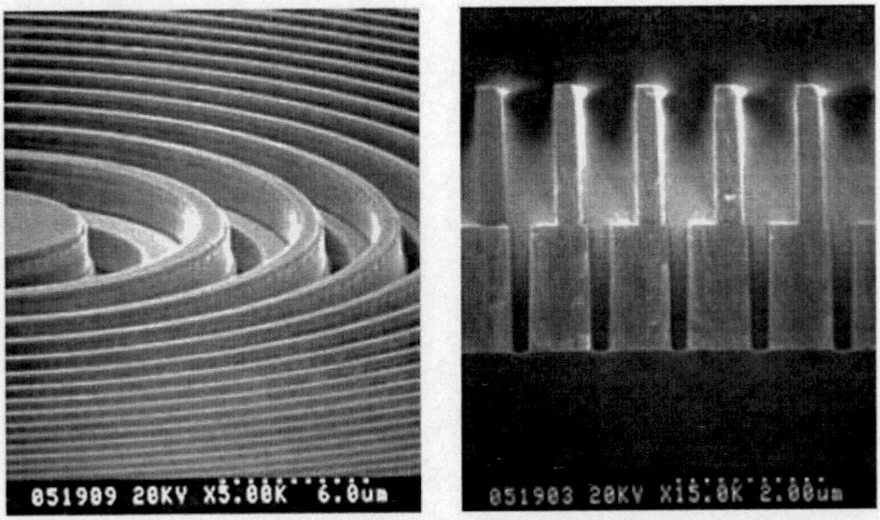

Fig. 5.7 Scanning electron microscope images of the external appearance of a Kinoform-type Fresnel zone plate (left) and its cross-section (right) (courtesy of Seiichi Itabashi from NTT Advanced Technology Co.)

step-like forms shown in Fig. 5.6b can achieve a high efficiency. Figure 5.7 shows the observations of a Kinoform-type Fresnel plate manufactured by NTT Advanced Technology Co. using a scanning electron microscope.

As discussed earlier, the aspect ratio of the outermost zone when using high-energy X-rays of approximately 20–30 keV for imaging metal materials is several dozen; consequently, fabrication becomes extremely difficult. Takeuchi et al. reported an apodized Fresnel zone plate that can be used for high-energy X-rays [9]. As shown in Fig. 5.8, an apodized Fresnel zone plate has grooves that gradually become shallower as they approach the outer circumference. This results in a beam with an intensity distribution within the plane vertical to the optical axis that approaches a Gaussian distribution [9]. The outermost zone grooves are suitably shallow in apodized Fresnel zone plates; hence, the difficulties associated with their fabrication are considerably reduced. Other advantages include the ability to achieve high diffraction efficiencies without significantly sacrificing the spatial resolution for high-energy X-rays [9]. Takeuchi et al. reported that the seemingly complex structure shown in Fig. 5.8 can

Fig. 5.8 Schematic of an apodization Fresnel zone plate

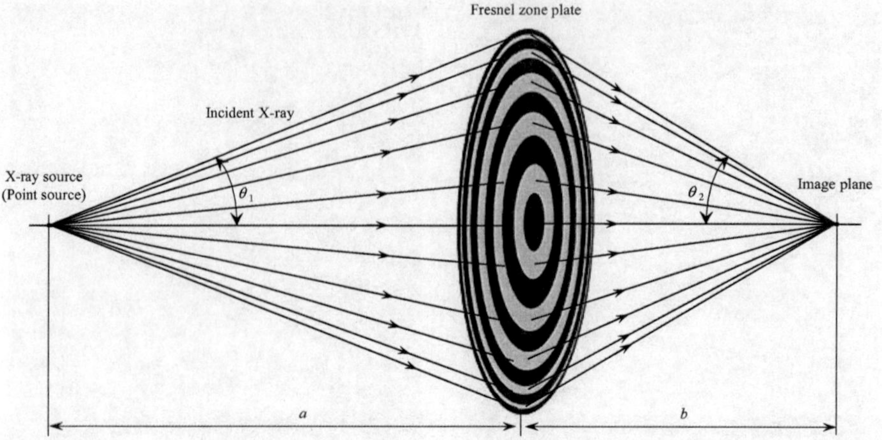

Fig. 5.9 Schematic of spherical waves emitted from a point source being focused using a Fresnel zone plate

be achieved merely with conventional fabrication processes for Fresnel zone plates and adjustments in etching conditions, indicating that this technique could be deemed suitable for practical use.

Subsequently, we consider the case where a point source like an X-ray tube is placed on the optical axis at a distance a from the Fresnel zone plate and where a spherical wave instead of a parallel beam is illuminated, as shown in Fig. 5.9. Diffraction similar to Fig. 5.5 is produced owing to the Fresnel zone plate, and the waves are focused on the optical axis at a distance b from the Fresnel zone plate. In this case, the following *lens formula* is established using the *paraxial approximation* ($r_{\text{om}} \ll a, b$):

$$\frac{1}{f} = \frac{1}{a} + \frac{1}{b} \tag{5.14}$$

Additionally, $\cos\theta = 1$ in the X-ray tube as $\theta << 1$. The sine conditions can be approximately established. Therefore, θ_1 and θ_2 shown in Fig. 5.9 can be approximated to satisfy the *Abbe sine condition*.

$$\frac{\sin\theta_1}{\sin\theta_2} = \text{const.} \tag{5.15}$$

This is equivalent to an infinitely thin lens in the case of visible light. Images of object points at a distance from the optical axis are also not distorted, signifying that a favorable imaging optical system can be achieved.

(4) Illumination System

The spatial resolution of an X-ray microscope can be improved via a suitable illumination system in imaging optical systems that use a Fresnel zone plate. From Eqs. (5.9) and (5.10), we observe that the spatial resolution when considering diffraction in a Fresnel zone plate is dependent on the numerical aperture of the Fresnel zone plate. Moreover, the spatial resolution when considering the illumination system is dependent on the *convergent angle* of the illumination system beam [1]. In other words, both lens apertures should be matched. When considering spatial resolution, under optimal conditions, the numerical aperture of the zone plate for illumination (*condenser zone plate*) is equal to 1.5 times the numerical aperture of the Fresnel zone plate [1]. The spatial resolution, in this case, is approximately expressed as 0.57 λ/NA, which is a considerable improvement over the value of 0.82 λ/NA when the parallel beam is used by itself as an incident beam [1]. Furthermore, Takeuchi et al. showed experimentally that a field of view approximately half the diameter of the Fresnel zone plate could be achieved while only slightly sacrificing the spatial resolution by setting the numerical aperture of the condenser zone plate to half of that of the Fresnel zone plate [10].

Illumination systems include *critical* and *Köhler illuminations*, as shown in Fig. 5.10 [1]. In the critical illumination shown in Fig. 5.10a, a light source image is minimized and formed on the surface to be illuminated. Although the illumination

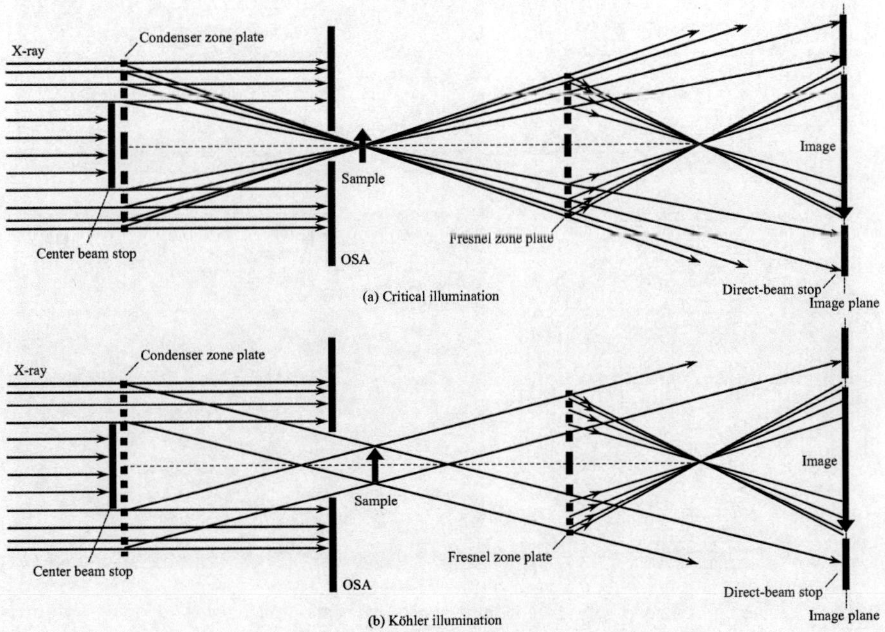

Fig. 5.10 Schematic of two illumination types of image optical systems using Fresnel zone plates [1]

efficiency is high, this technique is not often used in X-ray imaging, as the field of view becomes narrow at several micrometers [1]. Conversely, the Köhler illumination shown in Fig. 5.10b features X-rays transmitting through the condenser zone plate deflected at the same angle; there is no reduction in the light-source image produced [1]. An example of a condenser zone plate used in Köhler illumination is shown in Fig. 5.11 [11]. The X-ray intensity is inversely proportional to the distance from the optical axis as shown in Fig. 5.11b when a ring-shaped grating with equal intervals is used as shown in Fig. 5.11a. Therefore, a *diffuser* that equalizes X-ray intensity needs to be inserted. A comparatively homogeneous X-ray intensity across the entire field of view is obtained as shown in Fig. 5.11d when using a *sector condenser zone plate* comprising eight trapezoidal regions, as shown in Fig. 5.11c [11]. However, speckle noise is produced when the incident wave is coherent; therefore, either the sector condenser zone plate itself needs to be rotated around the optical axis or a diffuser needs to be placed between the sector condenser zone plate and the sample [11]. Furthermore, the transmission image acquisition and rotation need to be synchronized

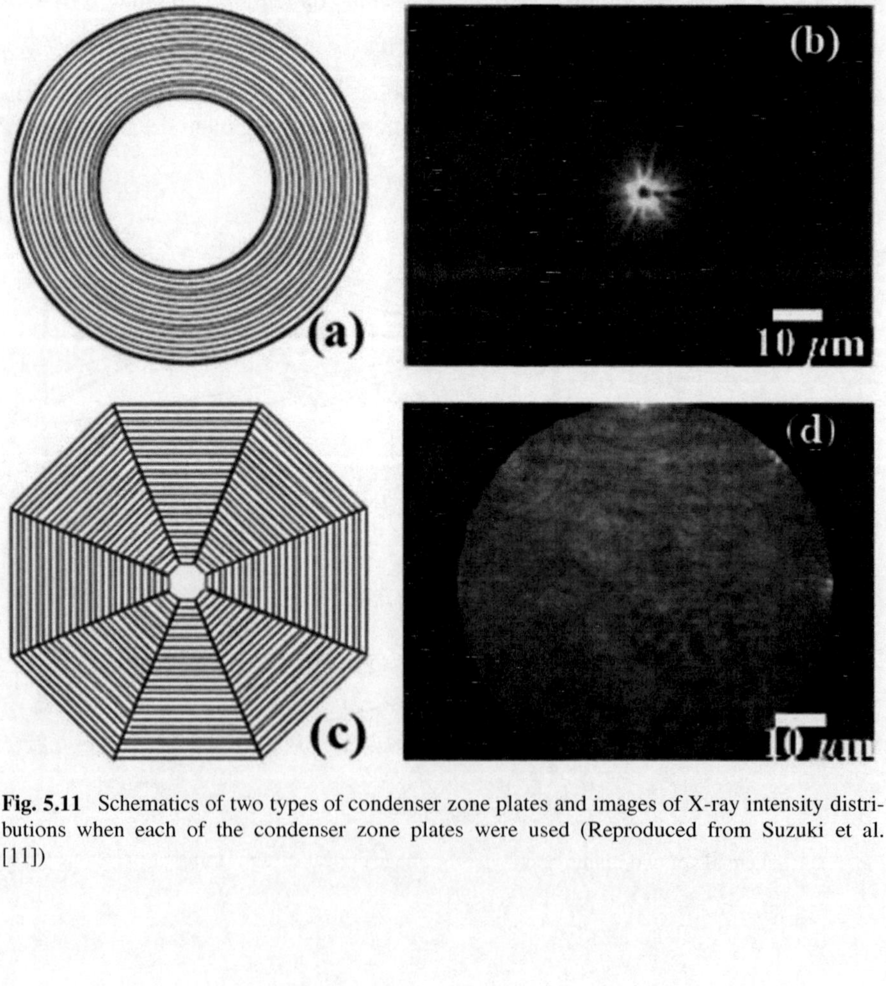

Fig. 5.11 Schematics of two types of condenser zone plates and images of X-ray intensity distributions when each of the condenser zone plates were used (Reproduced from Suzuki et al. [11])

when rotating the sector condenser zone plate. The sector condenser zone plate used in reference [11] has octagonal gratings at intervals of 400 nm with an inscribed circle of diameter 1 mm; tantalum with a thickness of 1.6 μm is used as the material [11].

Micro-/nano-tomography, which uses a Fresnel zone plate, can be used with commercially available X-ray CT scanners for industrial use [12]. Illumination systems that use a condenser zone plate are used in these cases as well.

5.1.2 *Imaging Optical Systems Using a Mirror*

The total reflection of an X-ray was observed in Figs. 2.15 and 2.16 in Chap. 2. This can be applied to form an imaging optical system. Representative examples of this include the Kirkpatrick–Baez and *Wolter mirrors*. (As a side note, Baez's daughter is the world-famous folk singer Joan Baez.)

We will first describe the basic principles of mirror-based focusing. Subsequently, each mirror will be discussed.

(1) Basic Principles of Mirror-based Focusing

A *spheroid* (more accurately, an ellipsoid of revolution) and a *paraboloid* could be termed the ideal forms of a mirror for the total reflection of an X-ray. The former focuses X-rays from a point source to a single point; the latter focuses parallel beams to a single point. This is referred to as "two-dimensional focusing." By contrast, let us consider focusing in a single direction i.e., either the curvature in the optical axis direction (radius of curvature r_m) or the curvature in the direction vertical to the optical axis (radius of curvature r_s), as shown in Fig. 5.12. A surface whose plane is elliptically curved in only one direction is referred to as the *elliptic cylinder mirror*; the act of focusing X-rays linearly using an elliptic cylinder mirror is called "one-dimensional focusing." Here, the following relationship is established when d_{in} is set as the distance from the light source to the mirror, d_{ref} is set as the distance from the mirror to the image point, and θ_{in} is set as the glancing angle of the X-ray [13]:

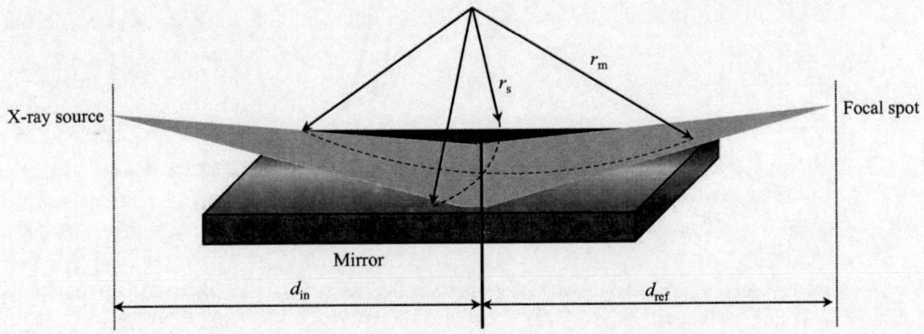

Fig. 5.12 Schematic of two-dimensional X-ray focusing with the curvatures following or vertical to the optical axis

$$r_m = \frac{2d_{in}d_{ref}}{(d_{in} + d_{ref})sin\theta_{in}} \tag{5.16}$$

$$r_s = \frac{2d_{in}d_{ref}sin\theta_{in}}{(d_{in} + d_{ref})} \tag{5.17}$$

The term r_s typically takes a value between several dozen and hundred millimeters, whereas r_m takes a much larger value of 100 m–10 km [13]. As such, the curvature in the optical axis direction can be realized utilizing the elastic bending deformation of the planar mirror. Conversely, the curvature in the direction vertical to the optical axis is formed through surface machining of the mirror. The radius of curvature is extremely large for the curvature in the optical axis direction; therefore, the errors are small even if the elliptic cylinder mirror is approximated as a spheroid.

Incidentally, the influence of surface roughness on the mirror reflectivity R can be determined by multiplying an attenuation factor due to surface roughness to the reflectivity R_0 of an ideal mirror surface [13].

$$R = R_0 e^{\left[-\left(\frac{4\pi\sigma_{rms}}{\lambda}\right)^2\right]sin^2\theta_{in}} \tag{5.18}$$

Here, σ_{rms} is the root-mean-square (rms) surface roughness, calculated as $\sigma_{\text{rms}} =$ 0.1–6 nm [14], as shown in Fig. 5.13. Furthermore, $\theta_{\text{in}} = 4$ mrad [14], as per Fig. 5.13; under these conditions, this mirror functions as a low-pass filter that eliminates values

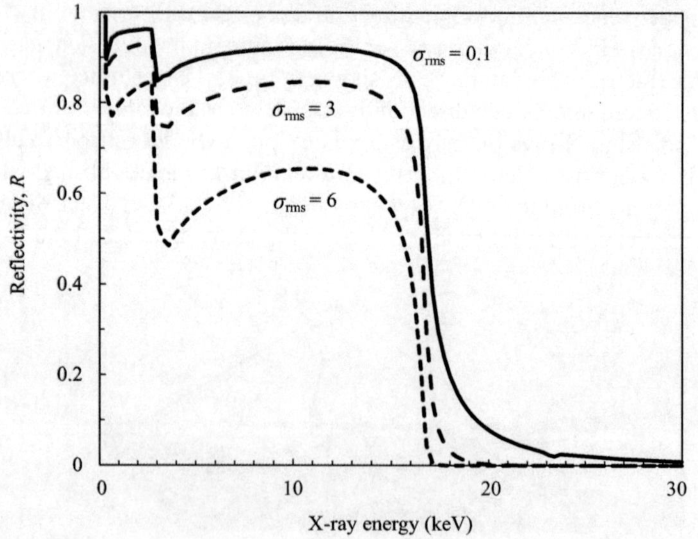

Fig. 5.13 X-ray energy dependency of X-ray mirror reflectivity. Changes when surface roughness (root-mean-square surface roughness) was varied are shown

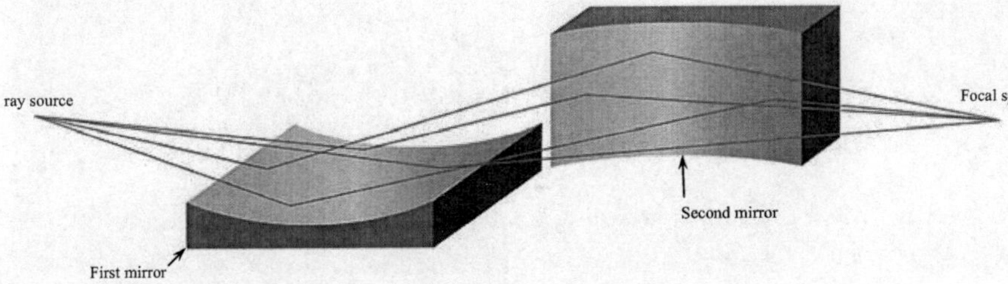

Fig. 5.14 Schematic of focusing using a Kirkpatrick-Baez mirror

above approximately 18 keV. The influence of surface roughness is extremely large, and reducing the rms surface roughness to a value below 1 nm is important from a utility standpoint.

In addition to mirrors with total reflection, multilayer film mirrors are frequently used. Multilayer films involve the lamination of several dozen to several hundreds of layers of thickness 1–10 nm made of materials with different refractive indices in an alternating manner [13]. As with crystal diffraction that occurs due to the width of spacing on the crystal lattice surface, this structure strongly reflects only X-rays whose wavelengths satisfy the diffraction conditions of the thickness of each layer. This phenomenon is used in multilayer Laue lenses (discussed later). In several cases, the glancing angle for the X-ray mirror is set at an angle at which first-order diffraction peaks are formed. Thus, multilayer mirrors can not only reflect X-rays but also conduct simultaneous spectral diffraction.

It is extremely difficult to fabricate a mirror with a spheroid or paraboloid surface accurately. As an alternative, optical systems in which two cylinder or elliptic cylinder mirrors are positioned perpendicular to each other are frequently used. This is illustrated with the Kirkpatrick–Baez mirror system in Fig. 5.14. Additionally, note that the study by Kirkpatrick and Baez illustrated only cases where a spherical surface was used. The present text will broadly apply the "Kirkpatrick–Baez" term to systems featuring two perpendicular mirrors. The radius of curvature is fixed when machining a mirror curved in the direction vertical to the optical axis through grinding; hence, the focal distance changes as the glancing angle changes. Therefore, mirrors curved in the optical axis direction are often used.

(2) Kirkpatrick–Baez Mirror Systems

A Kirkpatrick–Baez mirror system involves two perpendicular mirrors, with the first mirror focusing light in a single dimension, and the second mirror focusing the light in the direction opposite to that of the first mirror. An example of the mirrors used in this system is shown in Fig. 5.15a. In Kirkpatrick–Baez mirror systems, the spatial resolution is dictated by mirror misalignment and spherical aberration. The mirror surface is made elliptical instead of cylindrical to reduce spherical aberration. A curved mirror surface is obtained by actively controlling the smooth mirror shape with a bender similar to that shown in Fig. 5.15b or by grinding the mirror surface

(a) Mirror (b) Bender

Fig. 5.15 Photographs of a commercially-available X-ray mirror and its bender; products from the French company WinlightX (courtesy of Mr. Ibara of Toyama Co.)

into that of an elliptic cylindrical mirror beforehand. A bender is primarily used in the formation of cylindrical surfaces, but clamping both ends and asymmetrically controlling the bending moments on both sides can approximate an elliptic cylindrical surface. A Kirkpatrick–Baez mirror system is advantageous as it can completely remove astigmatism because of the use of two mirrors. Furthermore, as conventional surface machining techniques can be applied and the form of the mirror system can be controlled using a bender, this system is practical. However, it requires a rather large-scale setup, including cooling systems, particularly when a bender is used. Furthermore, mirrors cannot be easily inserted into or withdrawn from the optical axis, making alignment adjustment challenging. Another disadvantage hampering the utility of the system is the deviation of the beam from the optical axis. The long focal distance is another factor that further increases the scale of the setup. Therefore, this technique has not been used in industrial X-ray CT scanners until now.

A major characteristic of the Kirkpatrick–Baez mirror system is that it does not have chromatic aberration. This is the same as with the Wolter mirror system discussed below. As seen previously, for a Fresnel zone plate, the focal distance changes with the X-ray wavelength (shown in Eq. 5.6). Conversely, the reflection angle in the case of total reflection is not dependent on the X-ray wavelength, and the focal distance does not change. However, unlike in a Fresnel zone plate, the limit of spatial resolution varies with the wavelength. Furthermore, as shown in Fig. 2.16, the critical angle of total reflection decreases when the X-ray energy increases; therefore, this system can also be used as a low-pass filter. Hence, synchrotron radiation facilities use Kirkpatrick–Baez mirror systems not only for focusing but also for eliminating higher-order harmonics from undulators or high-order light from double-crystal monochromators.

The mirror substrate is made with fused silica or silicon carbide, on top of which a nickel, metal, or platinum coating is applied. The surface is sufficiently polished. Recently, *elastic emission machining* (*EEM*) methods, which enable fabrication at an accuracy below 1 nm, have been used [15].

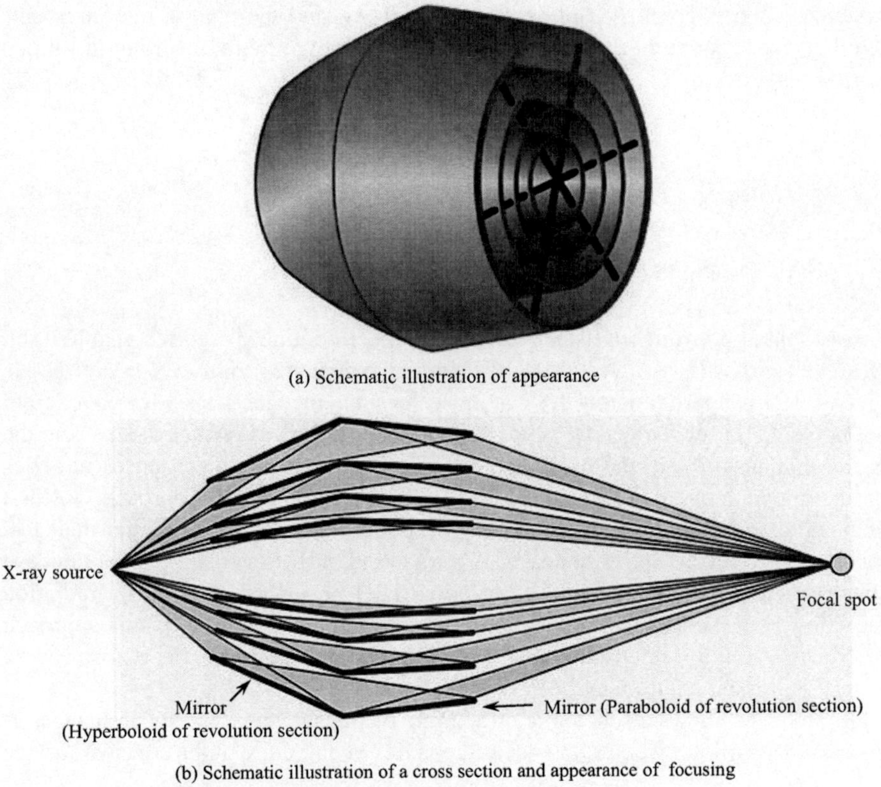

Fig. 5.16 Schematic of a Wolter mirror and focusing at its cross-section

(3) Wolter Mirror Systems

A schematic of a Wolter mirror system is shown in Fig. 5.16. The inner surface of the Wolter mirror is aspherical and is structured so that a hyperboloid surface and either a spheroid or paraboloid surface share a single focal spot. Two reflections are used to focus on the single focal spot. The Wolter mirror system is believed to satisfy the Abbe sine condition approximately [16], and superior image formation and high focusing efficiency can be expected. *Comas* (recently also referred to as "comatic aberrations" and signifying those aberrations in which light from single-point sources outside the optical axis does not converge to a single point), which are unavoidable with the use of a single mirror, can be minimized using two mirror surfaces. They can also be described as aberrations caused by the changes in magnification that occur at each point where the X-ray hits the focusing element. Furthermore, using two reflections decreases the focal distance by approximately half compared with using only one paraboloid mirror. However, the accurate fabrication of the inner Wolter mirror system, which is composed of aspherical surfaces, and the precise measurements required for accurate mirror fabrication are extremely difficult, resulting in limited practicality.

5.1.3 Complex Refractive Lens

(1) Basic Principles of X-ray Lens

Lenses that apply refractive phenomena have been routinely used for visible light. However, as discussed in Sect. 2.2.1, the amount of deviation from an X-ray refractive index n of 1 is approximately 10^{-5} at most. As such, the lenses used for visual light are not suitable for X-rays. Complex refractive lenses act as refractive lenses in the X-ray domain through the placement of a large number of single lenses in series on the optical axis, as shown in Fig. 5.17. The potential of this approach was first demonstrated by Snigirev in 1996 [17]. As the refractive index is smaller than 1 in this approach, each lens is concave. A paraboloid surface is used for the concave shape when the system is used with the parallel beams of synchrotron radiation facilities. However, the mathematically rigorous general solution for this approach is a quartic solution referred to as a Cartesian oval. Furthermore, the ellipse can be a strict particular solution in the case of a planar wave.

Figure 5.18 shows a schematic of a single biconcave lens. The focal distance is expressed as follows when r_{RL} is set as the radius of curvature of the concave surface [3].

$$f_1 = \frac{r_{RL}}{2(1-n)} = \frac{r_{RL}}{2\delta} \tag{5.19}$$

The focal distance f_{CRL} of the complex refractive lens is expressed as follows when the number of single biconcave lenses that are aligned in series is N_{CRL} as shown in Fig. 5.17.

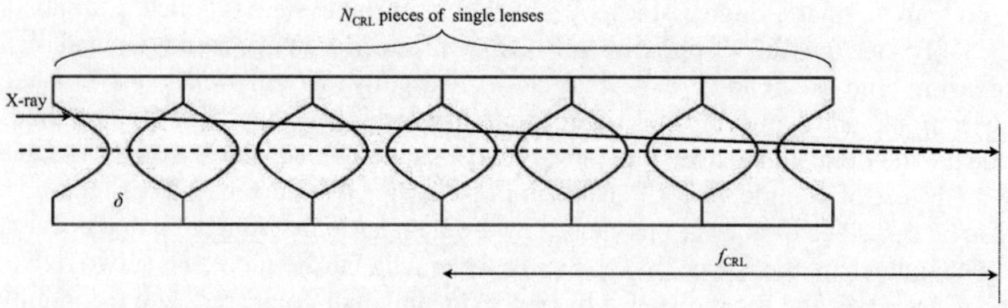

Fig. 5.17 Schematic of a complex refractive lens

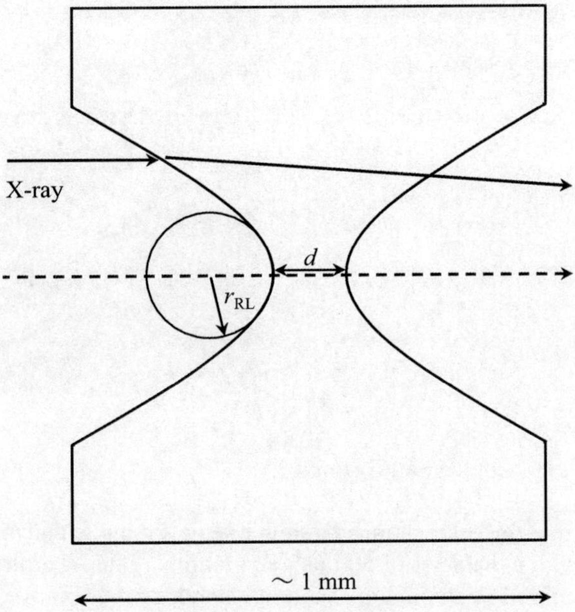

Fig. 5.18 Schematic of a single biconcave lens that comprises a complex refractive lens

$$\frac{1}{f_{CRL}} = \frac{1}{f_1} + \frac{1}{f_2} + \cdots \frac{1}{f_{N_{CRL}}} = \frac{2N_{CRL}\delta}{r_{RL}} \tag{5.20}$$

Consequently, the focal distance of the complex refractive lens becomes $\frac{r_{RL}}{2N_{CRL}\delta}$. It becomes shorter as the radius of curvature of the concave surface becomes smaller and as more lenses are aligned. For example, opening circular apertures with a diameter of 500 μm in aluminum and setting a focal distance of 1 m requires aligning 92 and 208 single lenses at the X-ray energies of 20 and 30 keV, respectively, as shown in Eq. (5.20). However, the X-ray absorption will increase if too many lenses are aligned. As shown in Fig. 5.17, the absorption effect becomes more pronounced with increasing vertical distance from the optical axis. Therefore, the effective aperture is restricted, decreasing the spatial resolution. Hence, when designing X-ray refractive lens, X-ray energies and lens materials are selected so that the δ/β is as large as possible, considering the complex refractive index [13]. As shown in Eq. (2.21) and Fig. 2.17, δ/β increases as the X-ray energy increases or as the element becomes lighter. Therefore, complex refractive lenses require a combination of materials with low X-ray absorption (e.g., aluminum, carbon, lithium, beryllium, or boron) and high-energy X-rays. Furthermore, the thickness of the lens itself must be restricted to a level of several dozen micrometers.

As with the Kirkpatrick–Baez mirror system, a single lens should be aligned while rotating it by 90° to obtain the point focus, which is important for X-ray imaging. Examples of commercially available complex refractive lenses with over 10 years of operation duration at SPring-8 are shown in Fig. 5.19. The single lens arrangement can be confirmed in Fig. 5.19.

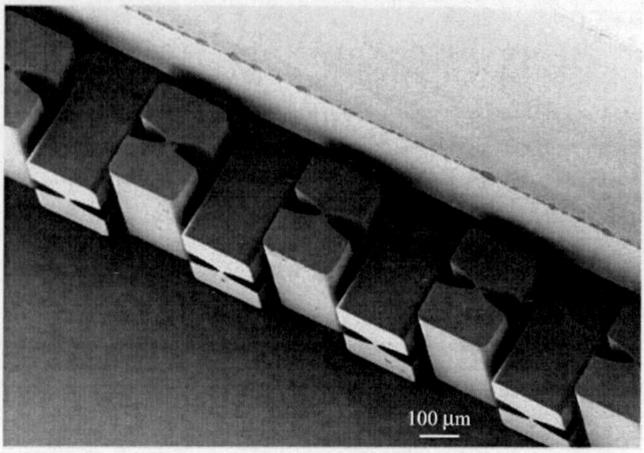

Fig. 5.19 Photograph of the external appearance of a commercially available X-ray complex refractive lens; complex refractive lens with SU-8 polymer material created by the Karlsruhe Institute of Technology (IMT) in Germany (courtesy of ASICON Co.)

(2) Application Example

Snigirev [17] examined a simple setup similar to that in Fig. 5.18, where 30 apertures with a diameter of 500 μm and minimal inter-aperture thickness d of 25 μm were opened in aluminum. The total length of the complex refractive lens was approximately 19 mm. This setup was used to conduct experiments at ESRF with an X-ray energy of 14 keV and a focal distance of 1.8 m. A favorable aperture diameter range of 250–600 μm was determined, with absorption at the thick-lens section stipulating the gain. This conveniently corresponds with the size of X-ray beam used in synchrotron radiation facilities.

Shortly afterward, Lengeler et al. reported using a paraboloid complex refractive lens and focusing the X-ray energy of 5–200 keV to achieve a spatial resolution of 300–500 nm [18]. In this case, focal distance adjustments were achieved by changing the single mirror stack number N_{CRL}. The same group also applied complex refractive lenses to X-ray tomography and achieved a sub-micron spatial resolution [19].

Additionally, Dudchik et al. prepared a complex refractive lens by inserting epoxy into a glass capillary with an inner diameter of 200 μm, creating an air bubble that was then used in experiments at APS [20]. It was also concluded that complex refractive lenses were suitable for imaging using X-ray tubes [20]. Thus, complex refractive lenses can be fabricated with common materials in the form of simple structures; therefore, various mechanisms can be devised for creating a low-cost, general-use X-ray lens.

5.1.4 Multilayer Laue Lens

The multilayer Laue lens is an X-ray focusing element based on new fabrication methods. It is a high-performance imaging element that can achieve a large aperture for X-rays over 10 keV and a focal point size below 10 nm [21]. Typically, a structure

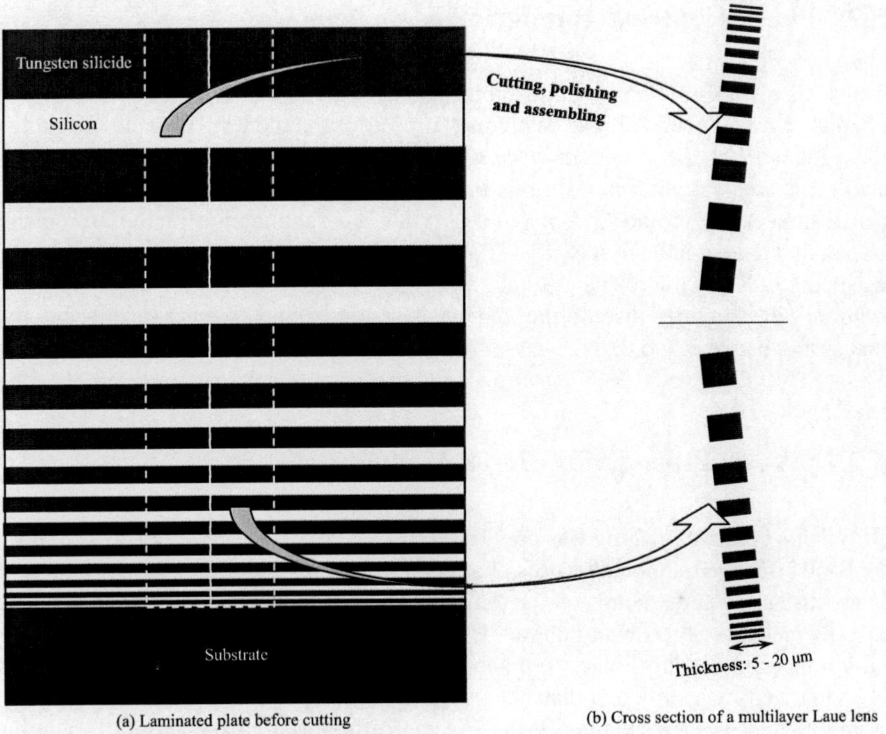

Fig. 5.20 Schematic of the multilayer Laue lens fabrication process through lamination and cutting

with an aspect ratio of approximately 10 to 15 is created on a Fresnel zone plate using electron beam exposure. Meanwhile, for multilayer Laue lenses, multilayer film lamination is first performed on a smooth surface using DC magnetron sputtering, as shown in Fig. 5.20. A high spatial resolution can be assured through sequential lamination from the outermost width. Subsequently, a multilayer Laue lens is formed by cutting the multilayer film formed on the substrate in the vertical direction, and then joining the pieces after polishing. This enables the fabrication of a structure with a maximum aspect ratio of several thousand [22].

In 2008, Kang et al. first reported the multilayer Laue lens shown in Fig. 5.20. They used tungsten disilicide (WSi_2) in the layers with higher electron density and silicon for the layers with lower electron density to create a multilayer Laue lens with an outermost zone width of 5 nm and thickness of 15 μm. This lens was used to focus one-dimensional X-rays with an energy of 19.5 keV to obtain a focal point size of 16 nm. In contrast with the Fresnel zone plate, high spatial resolution and a high diffraction efficiency of approximately 31% were obtained by satisfying the Bragg condition $2d\sin\theta = n\lambda$ for X-ray diffraction.

5.2 Phase-Contrast Tomography

The fundamentals of phase-contrast tomography were discussed in Sect. 2.2; in this chapter, we will focus on the techniques used to implement it. Broadly speaking, these include ① X-ray *propagation-based methods*; ② *Zernike phase-contrast microscopes*; ③ methods that use various *interferometers* such as crystals, prisms, and diffraction gratings; and ④ *X-ray holography*. The application of the aforementioned methods in industrial X-ray CT scanners has been limited, with their primary use being in X-ray micro-/nano-tomography at synchrotron radiation facilities. This section will primarily discuss the widely used approaches ① and ② while briefly discussing items ③ and ④.

5.2.1 X-ray Propagation-Based Methods

Refraction contrast imaging was previously discussed in Sect. 2.2.2 (2). This important and effective method also uses phase contrast. Its technique is simple; hence, it is not discussed here; details on the topic can be found in Chap. 2. This chapter will discuss methods of retrieving phase distribution information from sample transmission images in a more proactive manner. To this end, the sample and the detector are separated by a sufficient distance; an image where the sample interior interface is highlighted is then obtained using Fresnel diffraction. Next, phase information is retrieved by solving the *transport of intensity (TIE) equation*, which expresses X-ray wavefront deformation. This method could be considered a superior approach given that specialized devices such as interferometers are not required and that it is straightforward to perform.

(1) Transport of Intensity Equation

The coordinates are first defined as shown in Fig. 5.21. The distance from the sample to the detector is defined as z, and the X-ray intensity at that position is expressed as I (x, y, z). Furthermore, the sample phase distribution is set as ϕ (x, y). The TIE equation at this point is expressed as follows [23]:

$$k\frac{\partial I(x,y,z)}{\partial z} = \nabla_{\perp} \cdot [I(x,y,z)\nabla_{\perp}\phi(x,y)] \tag{5.21}$$

Here, k is the wave number and is set as $k = \lambda/2\pi$. Furthermore, $\nabla_{\perp}$ is the vector differential operator in the (x, y) plane and is expressed as follows:

$$\nabla_{\perp} = \left(\frac{\partial}{\partial x}, \frac{\partial}{\partial y}\right) \tag{5.22}$$

Equation (5.21) shows that the differential in the X-ray intensity propagation direction is related to the phase distribution. It may be difficult to understand the

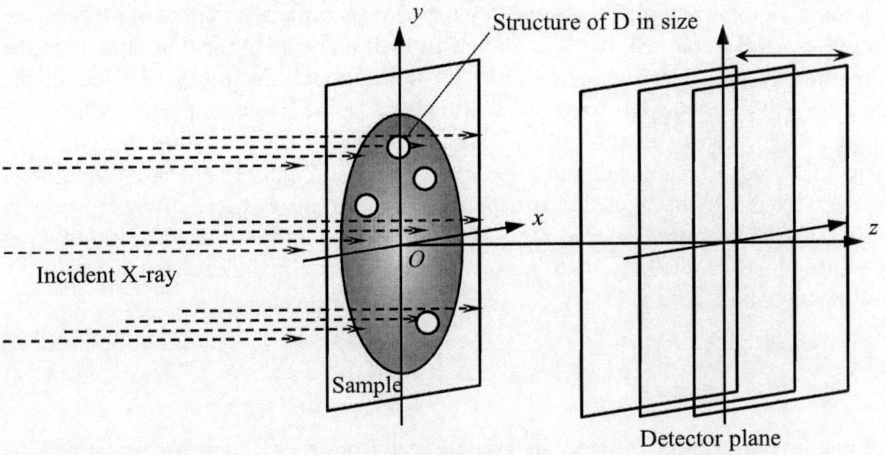

Fig. 5.21 Hypothetical case where X-rays are incident to a sample and the transmitted Xrays are observed with a two-dimensional detector. X-ray intensities are measured with a detector placed at a distance z from the (x, y) plane of the sample location

physical significance of this equation; however, the schematic in Fig. 5.22 should provide a fundamental understanding [24]. The local X-ray intensity measured at the detector position will decrease with X-ray propagation in the domain where the

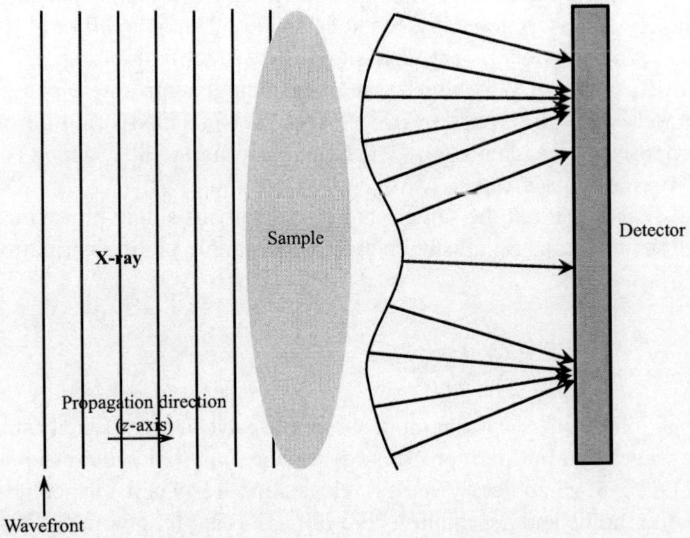

Fig. 5.22 Schematic showing that X-ray transmission through a sample can induce local changes in measured X-ray intensity by changing the X-ray wavefront when the parallel beam of an X-ray is incident to a sample

curvature of the wavefront is positive (convex), as shown in the middle of the detector in Fig. 5.22. By contrast, the X-ray intensity will increase in the domain where the curvature of the wavefront is negative, as in the upper and lower sections of the detector. However, the X-ray intensity differential in the X-ray propagation direction shown on the left-hand side of Eq. (5.21) cannot be measured directly. Thus, the detector is moved backward and forward as shown in Fig. 5.21 at $z = z_1$ and $z = z_2$ to measure the X-ray intensity distributions at different positions along the optical axis. The partial differential in the z-direction is then approximated by calculating the difference in intensity distributions between those positions, as shown in the following equation:

$$\frac{\partial I(x, y, z)}{\partial z} \cong \frac{I(x, y, z_1) - I(x, y, z_2)}{\Delta z} \tag{5.23}$$

Phase-contrast tomography can be conducted from X-ray intensity measurements if the phase distribution ϕ (x, y) can be determined from Eq. (5.21).

(2) Single-distance Phase Retrieval Method

Obtaining multiple images by moving the detector back and forth is a complex and time-consuming procedure; it requires accurate positioning with regard to data processing. A simple and robust experimental method can be developed if a single X-ray transmission image is used and if *phase retrieval* can be conducted by solving the TIE equation under a given assumption. Numerous techniques that facilitate this process have been reported [25]. Examples include the Bronnikov method, which assumes no X-ray absorption [25]; methods proposed by Burvall et al. [25] or by Suzuki et al. [26], in which X-ray absorption is considered minimal; and a method proposed by Burvall et al. [25], which assumes a homogeneous single-phase material. These methods were all developed from 1999 to 2006. This section introduces the method proposed by Paganin from CSIRO in Australia in 2002, which is currently widely used as a *single-distance phase retrieval method* [27].

First, if we assume that the sample is a homogeneous single-phase material, the following relationship is established between the sample phase distribution and the thickness t (x, y) [28]:

$$\phi(x, y) = \frac{-2\pi\delta}{\lambda} t(x, y) \tag{5.24}$$

This equation assumes a planar monochromatic wave or an extremely small curvature of the wavefront. Furthermore, this equation assumes a large Fresnel number as defined in Eq. (2.31): in other words, it is relevant in cases where the geometric optics approximation holds and a domain where refraction is predominant. This signifies that the following *near-field condition* holds when focusing on the internal structure of the sample set as the observation subject and its size is set as D, as shown in Fig. 5.21.

$$z \ll \frac{D^2}{\lambda} \tag{5.25}$$

Under these conditions, Paganin demonstrated that the sample thickness distribution can be defined with the following equation:

$$t(x, y) = \frac{-1}{\mu} ln\left(F^{-1}\left\{ \frac{F\left[\frac{I(x,y)}{I_0(x,y)}\right]}{1 + z\delta\mu^{-1}(u^2 + v^2)} \right\}\right) \tag{5.26}$$

Here, μ is the linear absorption coefficient, $I_0\,(x, y)$ is the incident X-ray intensity at the detector position, and (u, v) represents the reciprocal-space Cartesian coordinates of the real space (x, y). The aforementioned equation holds even when the sample is not homogeneous if the real-/imaginary-part ratio β/δ of the complex refractive index is fixed. The phase distribution can be then determined as follows using Eq. (5.24) [25]:

$$\phi(x, y) = \frac{1}{2} ln\left(F^{-1}\left\{ \frac{F\left[\frac{I(x,y)}{I_0(x,y)}\right]}{\frac{\beta}{\delta} + \left(\frac{\lambda z}{4\pi}\right)(u^2 + v^2)} \right\}\right) \tag{5.27}$$

This equation shows that the image is filtered in the reciprocal space, and subsequently, an inverse Fourier transform is applied to obtain the phase distribution.

Incidentally, a single-distance phase retrieval algorithm based on Paganin's method was released in February 2014 as a plug-in for ImageJ, a free image-processing software program widely used for image processing in scientific research [29]. If an X-ray transmission image obtained with a slightly longer camera length setting is available, the use of software and plug-ins is a simple and easy method of testing phase contrast-imaging effects. However, as shown in the following application example, internal structures that cannot be visualized with the absorption contrast require the selection of suitable conditions for imaging and phase retrieval to conduct visualization at a sufficiently high contrast to distinguish the internal structure from the matrix using image processing.

Paganin et al. have also applied the algorithm in Eq. (5.27) to dual-phase materials such as copper-dispersed polycarbonate to demonstrate its effectiveness [30]. This is because only the refractive index difference between the two phases influences the image contrast. Similarly, Fig. 5.23 shows an example of the application of this method to visualize a ferrous material with two phases, referred to as "DP steel," conducted at a synchrotron radiation facility [31]. Although there are virtually no differences in the chemical compositions of the ferrite and martensite phases in steel, their densities—i.e., 7.87 g/cm^3 and 7.76 g/cm^3, respectively—vary by as much as 1.4%. Paganin's method uses this slight density difference to distinguish between the two phases. Studies have been conducted on the relationship between the microstructure and fracture behavior of this material.

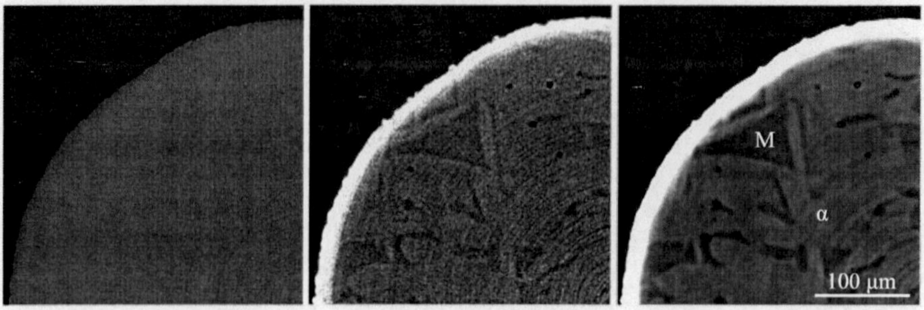

(a) Before phase retrieval ($z = 65$ mm) (b) Before phase retrieval ($z = 700$ mm) (c) After phase retrieval ($z = 700$ mm)

Fig. 5.23 3D reconstruction of images of DP steel composed of ferrite (α in figure) and martensite (M in figure) taken at SPring-8. Approximately 1/4 of a disk sample was shown as a virtual cross-section; **a** was optimized for absorption contrast tomography at a camera length of 65 mm; **b** and **c** show optimized values at a camera length of 700 mm, where the contrast and spatial resolution in the two phases were balanced for phase contrast tomography [31]

Compared with other phase-contrast tomography methods, the single-distance phase retrieval method has the disadvantages of a relatively low contrast and, as Eq. (5.26) functions as a low-pass filter, a significantly reduced spatial resolution. For example, the research shown in Fig. 5.23 has a spatial resolution of approximately 1 μm when the camera length is short, as shown in Fig. 5.23a. The measured spatial resolution decreased to approximately 2.8 μm at the interface of the sample interior when the image was acquired by moving the detector to an optimal camera length and conducting phase-retrieval processing, as shown in Fig. 5.23c. Therefore, the research in reference [31] repeatedly conducted imaging measurements with high-contrast/low-spatial-resolution phase-contrast tomography to observe the microstructure and low-contrast/high-spatial-resolution absorption-contrast tomography to observe material damage while applying loads to the material. The fracture behavior of materials was assessed by combining these methods.

Although materials can vary to a certain extent even within the same type of steel, Fig. 5.24 shows the spatial resolution, contrast, and S/N ratio when dual-phase stainless steel was imaged under the same experimental conditions at the same synchrotron radiation facility [32]. There is a trade-off between the contrast between the two phases and the spatial resolution; in other words, increasing the contrast results in decreasing spatial resolution. In this case, the spatial resolution is equal to $\sqrt{zD}$. Hence, in Refs. [31, 32], the optimal camera length was selected beforehand, considering the balance between spatial resolution and contrast. Incidentally, the two ferrite and austenite phases that compose this material have densities of 7.65 g/cm^3 and 7.71 g/cm^3 respectively. The density difference between these two phases is even smaller than that in the DP steel shown in Fig. 5.23, with a difference of only approximately 0.8%. Thus, the single-distance phase retrieval method can visualize even

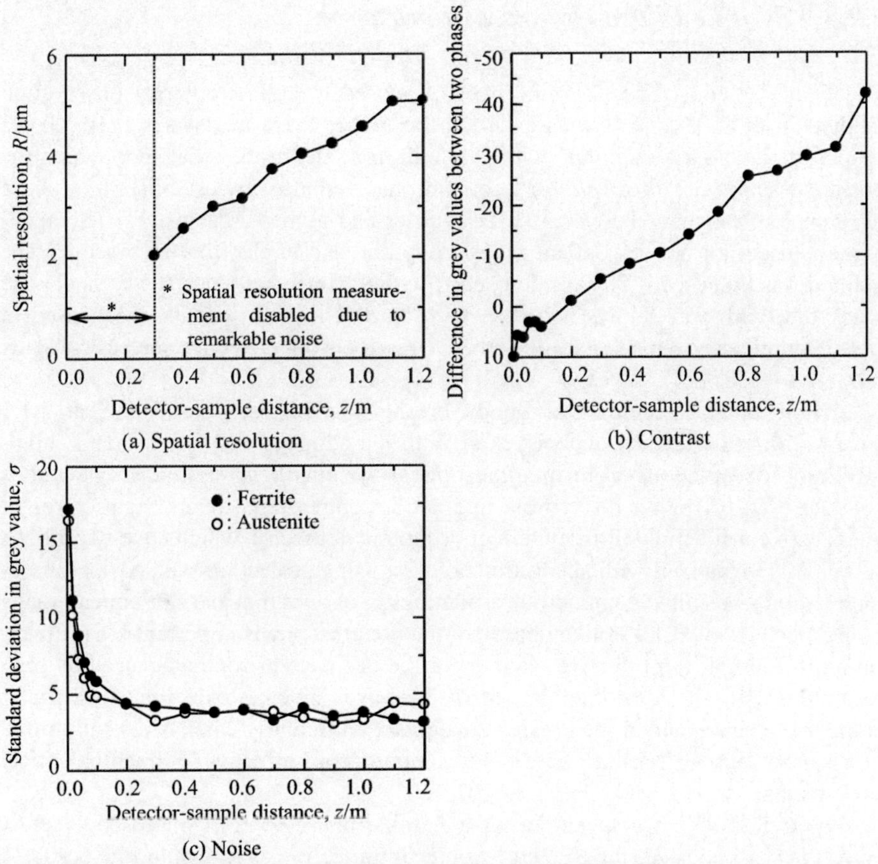

Fig. 5.24 Measurements of spatial resolution, contrast in both phases, and noise in 3D-reconstructed images of two-phase stainless steel composed of ferrite and austenite taken at SPring-8. The spatial resolution is determined by full width at half maximum (FWHM) of the differentiation of the gray value transition in the interface between the two phases [32]

slight differences in density due to different alloying element concentrations, differences in crystallographic structures, and void presence using a simple experiment, making this an extremely effective imaging method for understanding the internal structure of materials or components.

Mayo et al. have applied the single-distance phase retrieval method to laboratory-scale CT as well. They generated X-rays by directing electron beams onto a scanning electron microscope (SEM) equipped with a field-emission gun at a tantalum film to observe the polymer spheres [33]. Paganin's phase retrieval method was applied to the obtained image, and internal observations were successfully conducted using phase contrast [33]. It is believed that this method will be important for industrial X-ray CT scanning in the future.

5.2.2 *Zernike Phase-Contrast Microscope*

The Dutch scientist Fritz Zernike, who proposed the phase-contrast observation method in the 1930s [34], was awarded the Nobel Prize in Physics in 1953. He proposed placing a *phase plate* with a hole in its center at the focal spot behind the objective lens (*back focal plane*) to obtain phase contrast by establishing a phase difference of only π/2 between the zero-order and higher-order waves. This technique is used not only in optical microscopy but also in electron microscopy. The method was first used in X-ray microscopy with the introduction of the Fresnel zone plate, first with soft X-rays and then with hard X-rays. It has also been used in imaging optical systems that utilize other X-ray focusing elements, such as complex refractive lenses [35].

Imaging optical systems featuring the Fresnel zone plates discussed in Sect. 5.1.1 have been used widely in recent years. With the adoption of these systems, high-resolution observations from imaging optical systems have significantly reduced contrast even for rather discernible internal structures (e.g., internal structures or voids with considerably different effective atomic numbers), which have so far been sufficiently visualized with absorption contrast using projection-type X-ray tomography; this has resulted in numerous difficulties in observation and subsequent image processing errors [10]. Zernike phase-contrast microscopy is important for compensating in a simple yet effective manner for the decrease in contrast caused by high resolution [10]. The zero-order and diffracted waves are easy to distinguish spatially at the back focal plane of the Fresnel zone plate, particularly when Köhler illumination is used, making Zernike phase-contrast microscopy effective for imaging optical systems that use Fresnel zone plates [10].

Figure 5.25 shows a schematic of a Zernike phase-contrast microscope setup. Here, we shall consider the usage of Köhler illumination as shown in Fig. 5.10b. A ring-like spot is formed at the back focal plane of the Fresnel zone plate. A phase plate

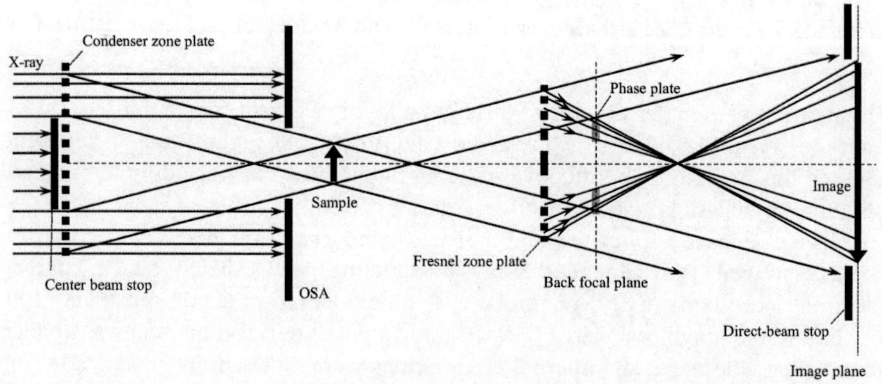

Fig. 5.25 Schematic of a phase contrast tomography setup where a Zernike phase plate was combined with an imaging optical system that uses Köhler illumination and a Fresnel zone plate

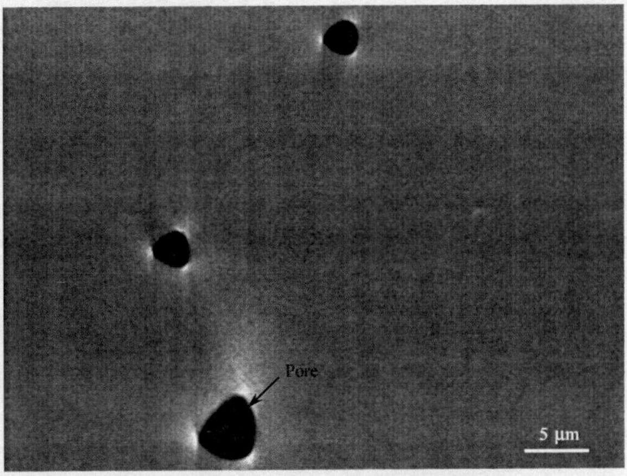

Fig. 5.26 Example of an image obtained with phase contrast tomography using a Zernike phase contrast microscope. Observations of pores in Al–Zn–Mg alloys made using an imaging optical system with an X-ray energy of 20 keV at the SPring-8 BL20XU. Artifacts referred to as halos can be seen near the interface of the internal structure

that corresponds to this shape is placed, and phase modulation is conducted for the zero-order wave only, advancing (*negative contrast*) or delaying (*positive contrast*) the zero-order wave phase by $\lambda/4$ relative to the diffracted wave phase. Advancing the zero-order wave phase by $\lambda/4$ relative to the diffracted wave phase conversely delays the diffracted wave by $\lambda/4$. The delayed portion becomes brighter in the transmission image, which is referred to as "*bright contrast*." In this sense, there is also *dark contrast*. For these cases, larger phase differences up to a certain point result in a larger contrast. Furthermore, absorption contrast imaging can be conducted as-is when the phase plate is shifted away from the optical axis. Incidentally, adjustments to the phase-plate position in the direction of the optical axis are not particularly severe.

The proportional relationship between the contrast obtained by a Zernike phase-contrast microscope and the phase change is limited to cases of small phase changes [10]. Large phase changes can conversely result in decreased contrast, no contrast, or even the inversion of light and dark. Furthermore, as shown in Fig. 5.26, "halo" artifacts that highlight edges can form near the interfaces of the internal structure [10], making observations of the microstructure near the interface difficult. Therefore, this is a point of caution to be remembered when assessing the images obtained with a Zernike phase-contrast microscope.

Finally, an actual example of a phase plate is shown in Fig. 5.27. The phase plate used with an X-ray energy of 8 keV by Takeuchi et al. was made by NTT Advanced Technology Co. and has both a pattern with a zone width of 4 μm etched into tantalum of thickness 0.96 μm and an inverted pattern, which are used for negative and positive contrast, respectively [10]. The phase plate thickness in this case is adjusted according to the X-ray energy.

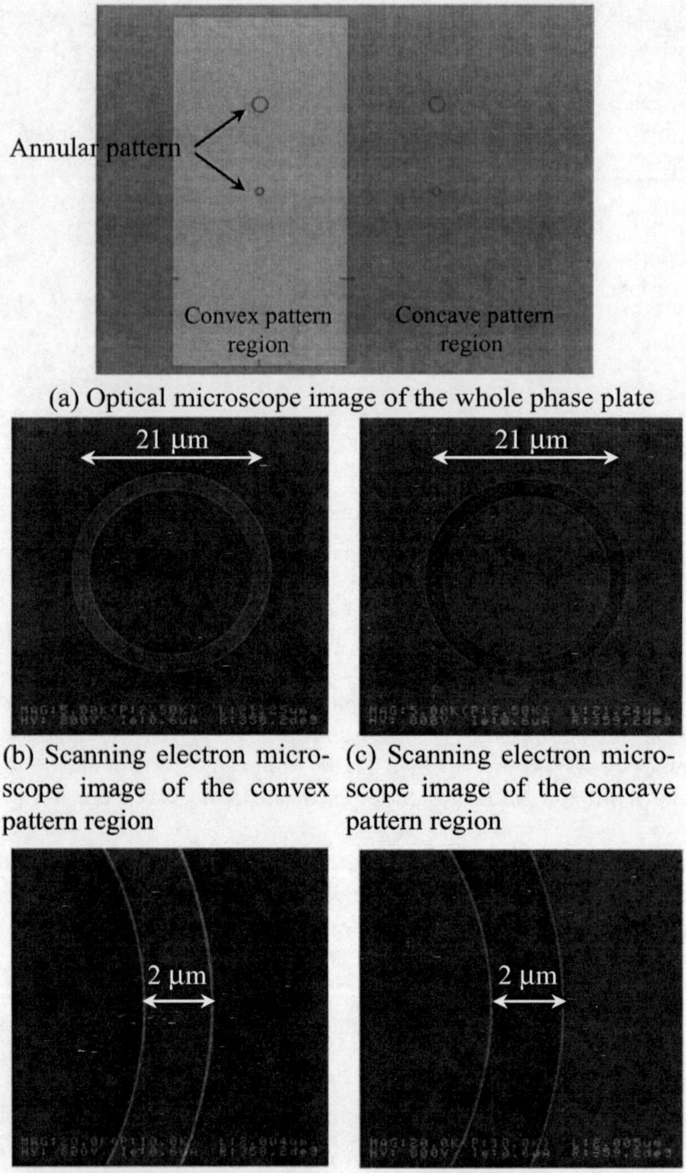

(a) Optical microscope image of the whole phase plate

(b) Scanning electron microscope image of the convex pattern region

(c) Scanning electron microscope image of the concave pattern region

(d) Magnified view of (b)

(e) Magnified view of (c)

Fig. 5.27 Photograph of a phase plate for Zernike phase contrast microscopy (courtesy of Seiichi Itabashi from NTT Advanced Technology Co.)

5.2.3 *Methods Using Interferometers*

(1) Crystal Interferometer

Figure 5.28 shows examples of a phase-contrast imaging method using a *crystal interferometer* [36]. X-ray phase-contrast imaging using a *Bonse–Hart interferometer* was first reported by Momose in 1995 [37]. Figure 5.28a shows a classic example of a

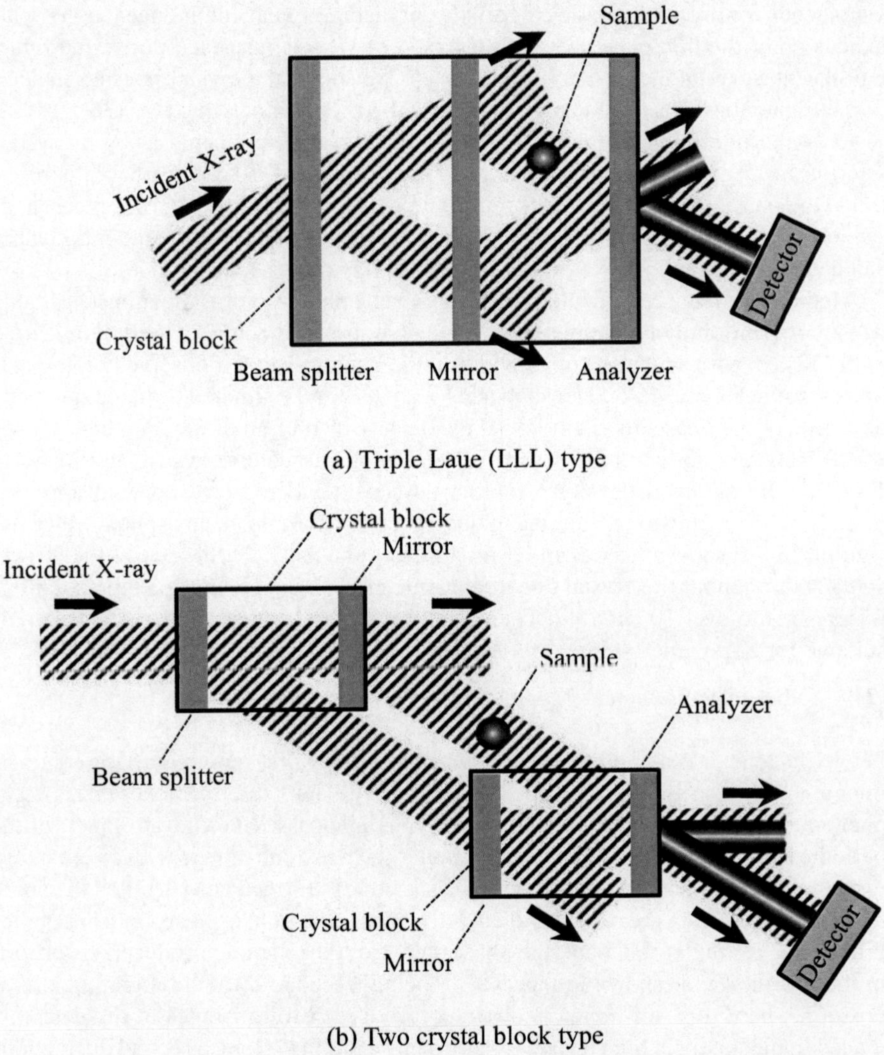

Fig. 5.28 Schematic of phase contrast imaging due to two types of interferometers using crystals [36]

Bonse–Hart interferometer comprising a beam splitter, mirror, and analyzer aligned at equal intervals. First, a beam splitter is used to divide the incident X-ray beam into two beams based on Bragg diffraction. This beam splitter also acts as a monochromator. The beam splitter, mirror, and analyzer consist of three thin plates made from an ingot of a single silicon crystal. Therefore, even the thin plate at the back of the beam splitter automatically satisfies the Bragg condition without any adjustments. Finally, two beams with the same optical path length interfere with one another at the analyzer position. Setting the sample on one optical path will enable observations of the interference fringes at the detector position. The setup in Fig. 5.28a is disadvantageous for use with biological samples in that their heat can induce the thermal expansion of the thin plate and that their field of view is restricted due to the limitation of the crystal block size [36]. Figure 5.28b shows the crystal interferometers that mitigate these disadvantages by splitting the crystal block into two [36].

Only the interference figure is determined from measurements using a crystal interferometer; the phase distribution itself is not obtained. To obtain phase distributions, the phase plate must be inserted in the optical path of the interferometer and several interference figures must be obtained from which phase shift amounts can be calculated. This method is referred to as the "*fringe scanning technique*."

Methods utilizing crystal interferometers enable phase-contrast imaging at a high-density resolution. For example, Yoneyama et al. from Hitachi Co. used monochromatic X-rays with an energy of 17.8 keV at the Photon Factory, observed biological objects using the interferometer shown in Fig. 5.28b, and estimated from the standard deviation of the phase distribution that the density resolution of the imaging system was 0.3 mg/cm^3 [38]. As the density of the biological material was approximately 1 g/cm^3, the obtained density resolution is considered to correspond to approximately 0.03%. However, this technique requires multiple beam splits, which is significantly disadvantageous in terms of X-ray intensity. Furthermore, the X-ray spreads throughout the crystal due to dynamic effects; hence, the spatial resolution is degraded to over 10 μm [39]. Therefore, this method is generally not considered suitable for X-ray micro-/nano-tomography.

(2) Talbot Interferometer

Peter Cloetens, a Belgian beamline scientist at ESRF, first reported phase-contrast imaging based on the *Talbot effect* in 1997 [39]. The "Talbot effect" refers to a phenomenon that occurs when spatially coherent light passes through an object with a periodic structure (such as a diffraction grating), where a self-image with an intensity distribution corresponding to that periodic structure is formed at a specified distance d_T downstream from the original object [40]. When a sample is placed in front of the diffraction grating at this time, the phase shift from the sample produces distortion in the self-image, as shown in Fig. 5.29. Placing a second diffraction grating at the position where this self-image is observed produces Moire fringes in the detector placed immediately behind the second diffraction grating (Here, a second diffraction grating with the same periodicity as the first is typically used.) A high-resolution detector is not required, as the fringe interval of the Moire fringes is larger than

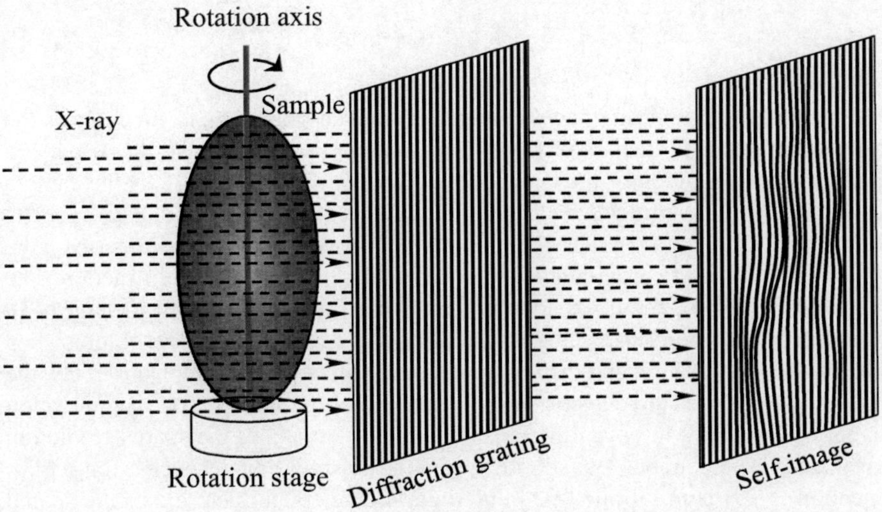

Fig. 5.29 Schematic of phase contrast tomography using a Talbot interferometer. Self-image formation and its deformation due to the insertion of a sample can be seen due to the Talbot effect

the periodicity of the diffraction fringe. The term d_T is the *Talbot distance* and is expressed as follows:

$$d_T = \frac{2mp^2}{\lambda} \tag{5.28}$$

Here, p is the periodicity of the diffraction grating and m is a natural number. The interference figure can be observed using this method. This approach is similar to the crystal interferometer approach in that additional methods are required to obtain the phase distribution.

A spatial coherence length that exceeds the periodic length of the diffraction grating is generally necessary for coherence in phase-contrast imaging using the aforementioned Talbot interferometer [40]. Incidentally, as was already estimated using Eq. (2.35) in Chap. 2, the spatial coherence length when the X-ray energy at the standard SPring-8 undulator light source is 10 keV is approximately 390 μm. Consequently, Talbot interferometers are primarily used in synchrotron radiation facilities. However, there have been reports of experiments involving their use with microfocus tubes [41].

Cone-beam magnification has high utility in phase-contrast imaging for obtaining images with high spatial resolution without using a detector with high spatial resolution. In cases where a cone beam from a microfocus tube is used and the diffraction grating interval is sufficiently narrow, spatial resolution can be expressed as follows when considering a system [42]:

$$\sigma_{sys}^2 \approx \left(1 - \frac{1}{M}\right)^2 \sigma_{src}^2 + \frac{1}{M^2}\sigma_{det}^2 + \frac{1}{2}\sqrt{\frac{\lambda L}{M}} \tag{5.29}$$

Here, σ_{sys}^2, σ_{src}^2, and σ_{det}^2 are the variances of the point spread function of the system (explained in Sect. 7.5.2), the X-ray source distribution, and the dispersion in the spatial resolution of the detector, respectively. Furthermore, L is the camera length, M is the magnification, and R_{so} is the X-ray source–sample distance, with the three variables relating to each other as $M = (R_{\text{so}} + L)/R_{\text{so}}$. The third term on the right-hand side of the aforementioned equation is related to X-ray diffraction. The spatial resolution is generally said to be $2\sigma_{\text{sys}}$ when the Fresnel number is large [43]. As such, the focal point size of the X-ray source stipulates the spatial resolution when the X-ray energy is high and the observation magnification is large. Conversely, the spatial resolution of the detector is influential when the observation magnification is relatively small. Moreover, the Fresnel zone size stipulates the spatial resolution when the Fresnel number is small (i.e., when the transmission distance is long) [43]. Equation (5.29) is not limited to Talbot interferometers but determines the spatial resolution of systems for X-ray propagation-based phase-contrast imaging methods as a whole [42].

Methods using the Talbot interferometer are suited for synchrotron radiation; however, the Talbot–Lau X-ray interferometer shown in Fig. 5.30 is also suited for use with X-ray tubes. It was proposed by Pfeiffer et al. in 2006 [44]. In addition to the two diffraction gratings used with the Talbot interferometer shown in Fig. 5.30a, a third diffraction grating, which creates a hypothetical beam source array, is inserted near the X-ray source, as shown in Fig. 5.30b. The periodicity of the third diffraction grating must be sufficiently reduced in this case to create partially coherent X-rays. A subsequent key point, shown in Fig. 5.31, is to determine the periodicity for the first and second diffraction gratings where the self-images created by the X-rays passing through the adjacent slits in the third diffraction grating are phase-shifted from each other by a period and are superimposed. At this point, $p_1 = p_3\ l_1/l_2$ is satisfied in Fig. 5.31. The focal point size of the X-ray source can reach the millimeter-order if a Talbot–Lau X-ray interferometer is used. Pfeiffer et al. used a radiation source with an effective focal point size of 0.8 × 0.4 mm and observed polymer spheres using phase contrast [44]. However, a trade-off between increasing focal point size and decreasing spatial resolution was unavoidable, as shown in Fig. 5.31.

(3) Prism

Reports on X-ray prism interferometers using synchrotron radiation were published by Lang and Makepeace of the University of Bristol in the U.K. in 1999 [45] and by Suzuki et al. of SPring-8 in 2002 [46]. Both used an 11–15 keV monochromatic X-ray; however, Lang et al. used a diamond prism whereas Suzuki et al. used an acrylic prism for experiments. As described in Chap. 2, the δ/β value of the complex refractive index in light elements such as carbon is approximately 1000 for a wide range of X-ray energies. As such, an X-ray prism that can conduct refraction without absorbing virtually any hard X-rays can be obtained by creating a prismatic column from light

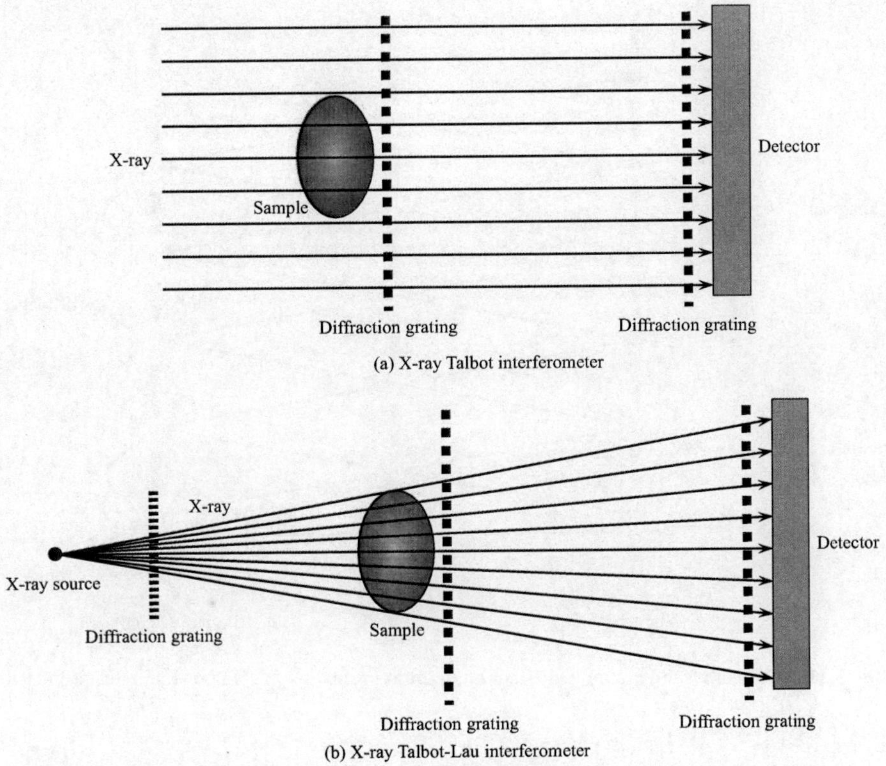

Fig. 5.30 Schematic of a Talbot interferometer and a Talbot-Lau interferometer

materials such as polymers and polishing its surface. As shown in Fig. 5.32, X-rays at a glancing angle θ_1 incident onto a prism with a vertical angle θ_2 form wavefronts inclined by a deflection angle θ_3. This structure can be used as an interferometer by refracting only half of the X-ray beam with the prism and combining the two beams at the image surface, as shown in Fig. 5.32. The deflection angle is expressed as follows when the vertical angle $\theta_2 = \pi/2$ [46]:

$$\theta_3 = \frac{\delta}{\tan\theta_1} \tag{5.30}$$

Here, the refraction angles of 13.5 μrad and 23.2 μrad can be obtained for prisms made of acrylic and diamond material, respectively, when θ_1 is set as 5° and the X-ray energy is set as 15 keV. Thus, beam shifts of 135 μm and 230 μm, respectively, are generated in the horizontal direction of the beam at a position 10 m downstream from the prism. In other words, phase-contrast imaging for a field of view of the approximate size discussed is possible even when X-ray absorption at the prism is

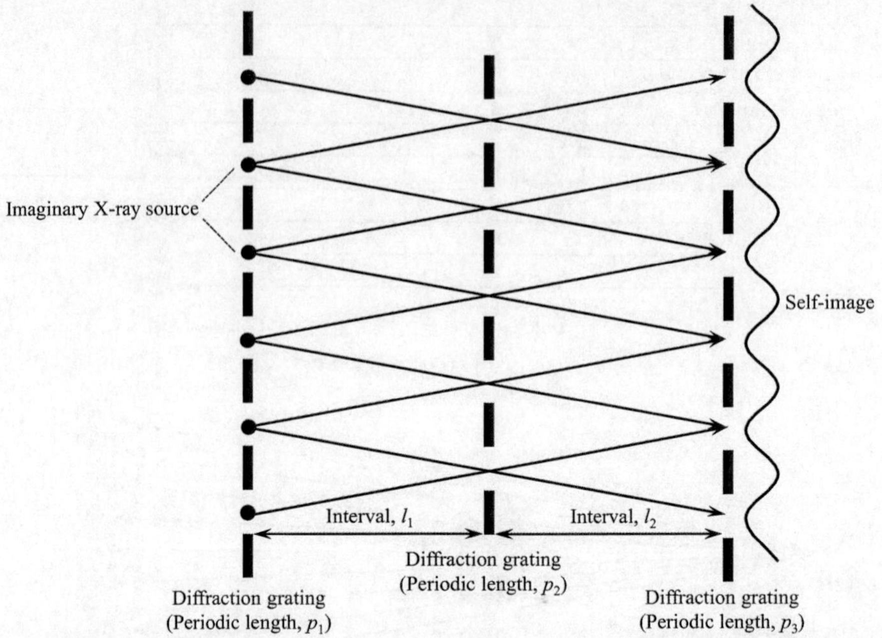

Fig. 5.31 Schematic showing three diffraction gratings arrange with a Talbot-Lau interferometer

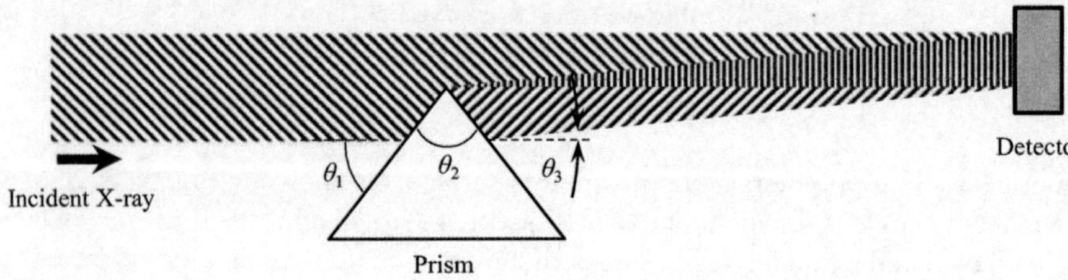

Fig. 5.32 Schematic of a phase contrast imaging method using an X-ray prism. An X-ray with a glancing angle θ_1 incident to a prism with a vertical angle θ_2 propagates after being refracted by θ_3, overlaps with the X-rays, which continued in a straight line to the side of the prism at the detector position, and forms interference fringes

considered. Furthermore, the glancing angle is larger than the critical angle of total reflection discussed in Sect. 2.2.1 (1) [47]. In this range, as per Eq. (5.30), the smaller the glancing angle, the larger is the deflection angle.

As this method requires sufficient spatial coherence, monochromaticity, and a relatively long prism–detector distance, its applications are limited to synchrotron radiation. The prism surface needs to be smooth and mirrored. However, Suzuki et al. used low-cost visible-light prisms sold at hardware stores [47] and determined that no special technologies or expensive materials are required for preparing the prism. They also reported potential applications for X-ray holography or spatial coherence measurements using X-ray prisms [47].

5.2.4 *X-ray Holography*

Holography—invented in 1948 by Hungarian scientist Dennis Gabor, who won the Nobel Prize for Physics [48]—also has a wide array of applications in the X-ray field. At ESRF in particular, the *holotomography* techniques reported by Cloetens in 1999 are employed by a wide range of users.

X-ray holography requires the use of coherent X-ray technology and a synchrotron radiation facility. As shown in Fig. 5.33, X-ray holography involves the acquisition of *holograms* that record X-ray strength distributions due to interference by inducing interference between the X-rays scattered from the sample (*object waves*) and a *reference wave*. The sample absorption/phase distribution can be determined by conducting these measurements at various camera lengths.

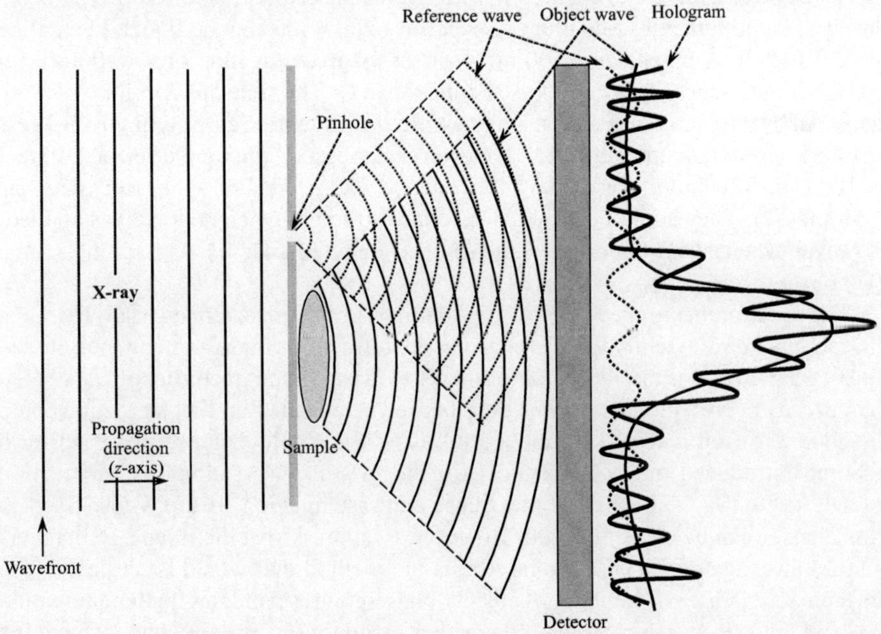

Fig. 5.33 Schematic of simple X-ray holography where a Fresnel zone plate was not used. The parallel X-ray beam is incident to the object and the X-ray wavefront changes following transmission through the object. A hologram is obtained through the interference of this with the reference waves from the pinhole at the image surface

In "*X-ray fluorescence holography*," holograms are produced from the emission of X-rays with energies higher than the absorption edge of the element to be measured and fluorescent X-rays are generated from a specific element in the material. Fluorescent X-rays are emitted in an isotropic manner as spherical waves from the atom, producing two X-rays—one proceeding without dispersion, and one dispersed by the surrounding atoms—that interfere with each other at the detector position. The surrounding crystal structure of the atoms that generated the fluorescent X-rays is reconstructed after the detector is scanned in the vicinity of the sample and a hologram is obtained. Fluorescent X-ray holography also has the advantage of not requiring coherent X-rays. Details on this topic are specialized and beyond the scope of this text; therefore, an overview by Hayashi et al. is provided here as a reference, although further details are omitted [49].

Holotomography has been applied to the observation of structures made of iron, steel, aluminum, or other structures. The cases introduced here are all experimental examples conducted at ESRF. In 2004, Borbély et al. used an X-ray with the energy of 20.5 keV at the ID19 beamline and conducted observations of metal matrix composites where alumina particles were dispersed in an AA6061 aluminum alloy [50]. The pixel size in this case is 1.9 μm and the spatial resolution is estimated to be 4 μm. Transmission images with the two camera length standards of 12 mm and 100 mm were obtained in this experiment [50]. In 2012, Tolnai et al. observed Mg_2Si particles in a cast aluminum alloy and aluminide particles in iron using the ID22NI beamline in ESRF [51]. A pixel size of 60 nm was set for magnification observations with a cone beam using a Kirkpatrick–Baez mirror [51]. The transmission images were acquired by adjusting the camera lengths to four different levels ranging from 29.68 to 44.61 mm [50]. In the same year, Landron et al. observed martenite/ferrite DP steel at ID22NI [52]. In this case, the X-ray energy was 29 keV and the pixel size was 100 nm [52]. They could observe a dual-phase structure when no load was applied; however, when a load was applied, the fringes remained after reconstruction, and the two phases were not observable [52].

The two aforementioned cases of aluminum observation feature internal structures that should be sufficiently visualized using the refraction contrast imaging discussed in Sect. 2.2.2 (2). Furthermore, for in situ observation of the fracturing of DP steel, it has already been demonstrated that dual-phase visualization and segmentation could be achieved even with a load applied by using the single-distance phase retrieval method introduced in Sect. 5.2.1 (2) for the in situ observations of the fracture of DP steels [31, 32]. The various phase-contrast imaging methods discussed in this chapter can be currently used. However, to address specific issues, researchers should investigate on a case-by-case basis the method that would be easiest to use in their research environment: whether a phase-contrast imaging technique would be best (and, if so which type), or whether standard absorption contrast imaging or refraction contrast imaging methods should be applied instead. Particularly in the case of DP steel, even if a high-sensitivity phase-contrast imaging method with high-density resolution capacity is used, simpler techniques may be more effective depending on the observation conditions. Sufficient attention must be paid to this matter.

5.3 Fast Tomography

Here, we provide an overview of *fast tomography*, where the imaging speed is increased and a 3D image is obtained in a short period. The “fast” in fast tomography generally does not refer to the reconstruction process but rather the speed at which transmission image sets are obtained. The primary use of fast tomography in industrial X-ray CT scanners is perhaps for a 100% inspection of production defects in factory production lines or for achieving a higher throughput via optimizing and reducing the time dedicated to the evaluation and analysis of internal structures and defects. X-ray tomography is used at synchrotron radiation facilities for the in situ observations of various important phenomena in scientific research. Figure 5.34 shows representative examples of dynamic phenomena for which 3D visualization is conducted across various scientific research disciplines.

The development of fast X-ray tomography since the first report on fast tomography was published in 2003 is shown in conjunction with the spatial resolution index of pixel size in Fig. 5.35. This is the result of combining the original publication search results in a reference database (SCOPUS) and graphing them alongside the publications at hand. Considering that the vertical axis is logarithmic, the development of fast tomography is accelerating, and its progress is evident even today. Thus, with fast tomography, we can assess and analyze in 3D various phenomena that could not so far be visualized (e.g., metal solidification process, crystallized phase formation and growth behavior during solidification).

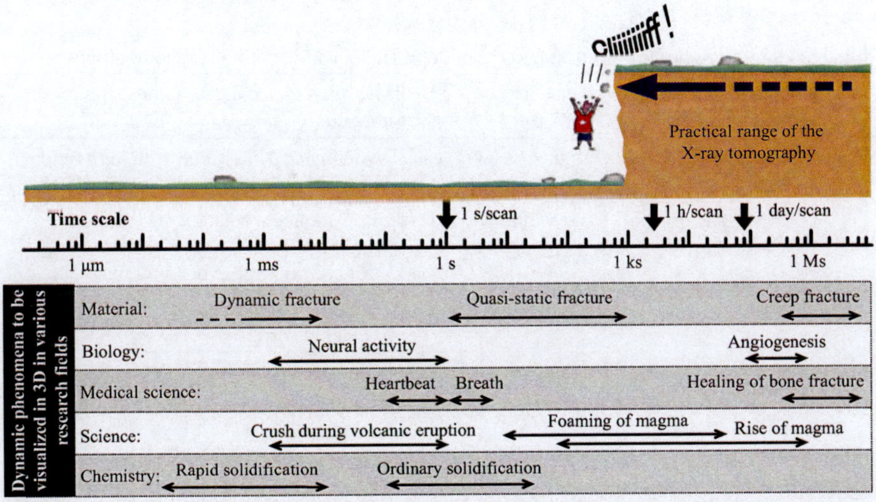

Fig. 5.34 Schematic of the representative dynamic phenomena to be visualized in 3D for each scientific research discipline. Barriers to increasing the speed of imaging in X-ray tomography are derived from either the X-ray source, rotation stage, detector, data acquisition method, or a combination of these factors, forming a limit (i.e. the cliff in the figure). Measures that resolve these issues enable the visualization of the above-mentioned phenomena

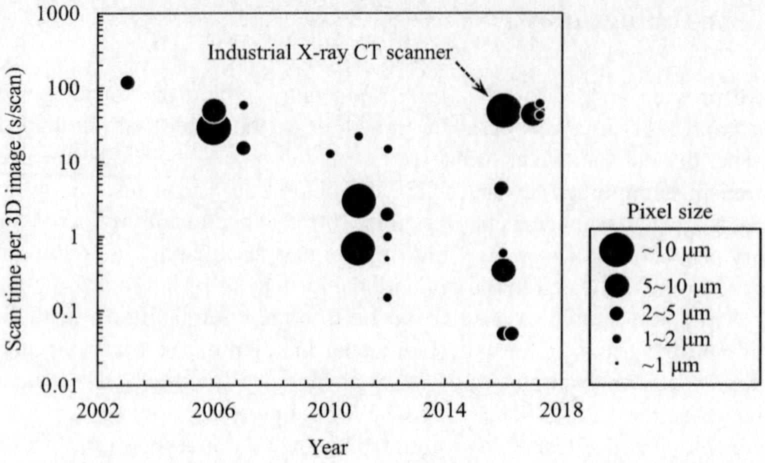

Fig. 5.35 Summary of the original publications relating to fast tomography. The amount of time needed for acquiring a single 3D image is plotted in the form of annual changes. The pixel size is represented by the size of each plot point. Except for one case, all studies used synchrotron radiation

Table 5.2 lists the various measures taken in the X-ray source, detector, and sample rotation stage to increase the speed of X-ray tomography; it also lists the implementation examples, alongside its references. Readers should refer to these while observing the publication year.

Table 5.2 Summary of the various measures undertaken to achieve fast X-ray tomography

Constituting devices and their characteristics			Measures to accelerate tomographic imaging
X-ray source	X-ray tube	Output	Adopting an X-ray tube with high output and small effective focal spot size [53]
	Synchrotron	Band width	Quasi-monochromaticity (Undulator) [54–56]. White X-ray (Wiggler) [57, 58]. White X-ray (Super bending magnet [59])
Rotation stage		Imaging sequence	Continuous rotation [54, 55, 58, 53, 60, 61] Helical scan
		Number of revolution	High speed rotation (600 rpm [59, 60])
Detector		Type	CMOS camera instead of CCD [54, 55, 62, 60, 61]
		Detection method	Photon counting detector [53]
		Binning process	Reasonable utilization of binning [55, 63–65]

5.3.1 Synchrotron Radiation Facilities

As previously discussed in Sect. 4.1.5 (5), a flux of 10^{13} [photons/s/mm^2] is obtained when a Si (111) plane is used in the monochromator at the first experiment hutch of the BL20XU beamline in SPring-8. The flux, without using a monochromator and after the radiation from the undulator is formed with two mirrors, is 10^{15} [photons/s/mm^2] at an X-ray energy of 12.4 keV, which becomes two orders of magnitude higher relative to that when a monochromator is used [54]. The bandwidth in this case is approximately 2%, and the beam becomes a quasi-monochromatic beam [54]. Menke et al. [55] and Shuai et al. [56] conducted experiments using quasi-monochromatic beams from undulators at the Diamond Light Source in the U.K. Lame et al. and the present author's group have conducted fast tomography experiments using white X-ray beams from a wiggler at ESRF [57, 58]. Experiments by the author's group tuned a monochromatic X-ray from a wiggler at the ID15A beamline in ESRF to 60 keV and obtained a flux of 2×10^{14} [photons/s/mm^2] [58]. Furthermore, Zefreh assessed afterglow influences in experiments using white X-ray beams from a *superbending magnet* in TOMCAT at the SLS synchrotron radiation facility in Switzerland [59]. A superbending magnet is a bending magnet with a strong magnetic field. TOMCAT in SLS uses 2.9 T bending magnets to obtain a flux of 10^{13} [photons/s/0.1% bw] immediately upon exiting the bending magnet across a wide energy range of 15–22 keV [62]. Incidentally, the bending magnet at SPring-8 is 0.68 T.

The quantitative aspect when measuring the linear absorption coefficient distribution is sacrificed to some extent when using quasi-monochromatic beams from an undulator. Furthermore, the imaging objective must be the visualization of the shape of the inner structure when using white X-ray beams from wigglers or superbending magnet beams. However, the use of quasi-monochromatic or white X-ray beams has a significant impact on the increased speed of X-ray tomography. Furthermore, large differences in beam size, X-ray energy, and its distributions are produced according to differences in the light source [54]. For example, the divergence angle of the beam is large in the case of the wiggler and the beam does not need to be expanded with mirrors as with the undulator [54]. Meanwhile, as shown in Fig. 4.44, the fundamental wave of the undulator has high brilliance and a peak at a relatively low X-ray energy compared with the wiggler. As such, the use of the wiggler is more efficient in cases of non-light metals such as iron. However, synchrotron radiation from wigglers often includes low-energy components as shown in Fig. 4.44; hence, artifacts referred to as beam hardening (discussed later) are likely to be produced [54]. Monochromators must be used with the understanding that large flux decreases will occur when there are observation-related issues.

Zefreh et al. assessed the influence of afterglow in scintillators during the aforementioned TOMCAT experiments using LuAG (Ce) [59]. Moreover, even among the representative scintillator materials and their characteristics summarized in Table 4.14, this material has a relatively short decay time of 58 ns. According to this study, afterglow with a maximum intensity of approximately 0.1% was observed for 150 ms after the end of illumination when the scintillator thickness was 300 μm.

Setting this thickness to 20 μm or 100 μm results in a decrease in the maximum intensity to 0.1% 25 ms after the illumination ends. Hence, it is understood that 100 μm is the optimal thickness for LuAG (Ce) used in fast tomography, considering light emission intensity as well.

5.3.2 Industrial X-ray CT Scanners

As explained previously in Sect. 4.1.2 (5), the brilliance of X-ray tubes can be determined by dividing the maximum output by the focal point size (surface area). Therefore, it is important to select a tube with a large output and a small effective focal point size [53]. Thus, Fig. 4.26 now shows that there is a correlation to some extent between the effective focal point size of the X-ray tube and the X-ray output. However, there is a considerable amount of variation between the two, particularly in X-ray tubes with a small focal point size. For example, there is a 44-fold difference between the maximum and minimum outputs in tubes with a focal spot size of approximately 4 μm. As listed in Table 4.3, this is likely due to the differences in the cooling mechanisms of the X-ray tubes.

5.3.3 Technical Elements Besides the X-ray Source

As shown in Fig. 5.36, several research projects use an *on-the-fly data acquisition method* instead of a *step-and-shoot data acquisition method* in sample rotation stages with regard to the synchronization of the rotation stage and detector, and the imaging step sequence [53–55, 58]. The latter method has had a wider use even in cases where the fast tomography technique is not particularly advocated. We can consider the relevant reasons quantitatively. For example, with the field-of-view width of 1 mm and 1800 projections/180° imaging conditions frequently used by the author, the amount of blur generated by the continuous rotation of the sample rotation stage between two adjacent transmission images would reach a maximum of only 0.87 μm at the outermost circumference and 0.5 μm at the outer edges of a prismatic column sample. However, the spatial resolution of 3D images in these cases is approximately 1 μm, and there is virtually no influence on the spatial resolution even with this amount of blur.

Incidentally, the experiments conducted by the author in reference [58] used on-the-fly data acquisition with the sample rotation stage and the tension testing machine set up on top for 1024 × 2048 pixel transmission images (voxel size approximately $(1.6\ \mu m)^3$, spatial resolution 4.1 μm), which numbered 450 in total [58]. The imaging speed was 22.5 s/scan [58]. Interrupting the material test and fixing the displacement for each image would result in unloading due to the relaxation phenomena. Considering these factors from a mechanical perspective, the validity of measuring the J-integral, which is a fracture mechanics parameter and is based on deformation theory,

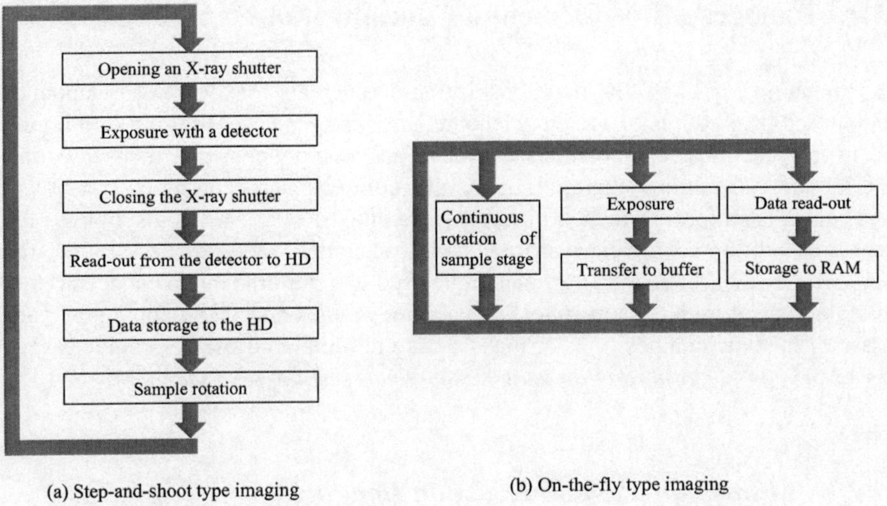

Fig. 5.36 Example of an imaging step sequence relating to a rotation stage. Comparison of step-and-shoot type imaging and on-the-fly type imaging of the sample rotation stage

cannot be guaranteed when the material is unloaded. To counteract this issue, this study succeeded in directly measuring the J-integral from consecutively measured 3D images. Stress-field singularities for the tips of static and propagating cracks, which could not be measured using the conventional approach of repeated load- or displacement-fixed observation (with partial unloading), have been identified with this new method [58].

Next, the CCD cameras used as detectors have recently been replaced with sCMOS cameras [54, 55, 62]. As shown in Sect. 4.4.2 (3) and (4), characteristics such as high quantum efficiency, low noise, and high dynamic range have been achieved even with CMOS cameras being used as sCMOS cameras. These fundamental characteristics, in combination with technologies such as global shutters, have confirmed that the sCMOS camera is an important component for fast imaging. Furthermore, Kumpová reported the application of fast tomography in industrial X-ray CT scanners in 2016 [53]. This study combined high-speed approaches (e.g., the aforementioned on-the-fly data acquisition methods with the sample rotation stage and high-output microfocus X-ray tubes) with an X-ray photon-counting detector that used cadmium telluride as a sensor. Similar to the medical-grade CT scanners in Fig. 3.3b, GE also sells industrial fast X-ray CT scanners in which a gantry that integrates an X-ray tube and a 64-line sensor camera rotates/slides in a helical manner around the sample [66]. This enables scans within as short as 15 s.

5.4 Tomography of Elemental Concentrations

As shown in Eq. (2.9), the mass absorption coefficients for alloys or compounds composed of mixtures of various elements are determined by the rule of mixture, using the chemical compositions of alloys and compounds, and their fractions. Furthermore, structural materials generally comprise three or more constituent elements. Therefore, even if a material is assumed to have no voids and the 3D linear absorption coefficient distributions are accurately reconstructed from X-ray tomography measurements, they cannot be used to determine the 3D concentration distributions of each constituent element. Here, we introduce 3D mapping methods for the chemical compositions of the structure of a material using techniques such as *absorption-edge subtraction imaging* and *fluorescent X-ray tomography*.

5.4.1 Absorption-Edge Subtraction Imaging

Findings on techniques involving medical-grade X-ray CT scanners, which uses transmission images taken at two different X-ray energy ranges to enhance the contrast of specific structures, were published in the 1970s [67]. Images were consecutively captured by changing the tube voltage during this period. There have been publications in the 1980s on techniques that rapidly change the tube voltage within a single scan: in the 1990s, on detectors that combined two light-emitting scintillator layers that scintillate at different X-ray energy ranges; in the 2000s, on X-ray CT scanners that installed two X-ray tubes with a high or low X-ray energy range; and in the 2010s, on energy-discriminating photon-counting detectors [68]. These applied the influence of a tube voltage and the target material on the X-ray energy distributions shown in Fig. 4.2 and Table 4.2, and the measurements of X-ray energy spectra due to the photon-counting detector discussed in Sect. 4.4.5 (3). These techniques are also important in industrial X-ray CT scanners and X-ray tomography using synchrotron radiation and have been used in practice in numerous applications.

As shown in Fig. 2.4, the linear absorption coefficient jumps near the absorption edge—for example, at approximately 7 keV in iron. As discussed in Sect. 2.1.2, the predominant absorption mechanism gradually transitions in X-ray energy ranges outside this region, with both linear regions at low energy and flat regions at high energy. The linear absorption coefficients in these regions only show gradual changes in X-ray energy. Thus, taking transmission images with high and low X-ray energies on either side of the absorption edge of a specified element and then obtaining the difference enables the 3D mapping of the concentrations of that element. However, this requires the precise registration of the two images before pixel subtraction and the correction of the pixel values. Furthermore, the use of a monochromatic beam at a synchrotron radiation facility assures the quantitative characteristics of the elemental concentrations; detectability (approximately 0.5 mass%) and reproducibility equivalent to SEM-EDX can be obtained. Hence, absorption-edge subtraction imaging

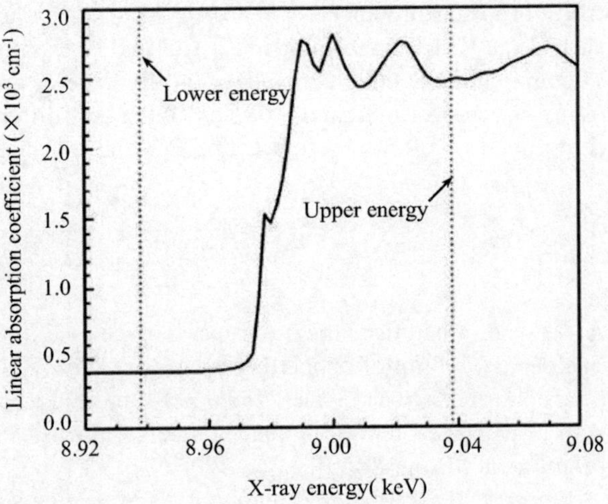

Fig. 5.37 X-ray absorption spectrum near the K absorption edge of copper; copper with a purity of 99.99% was used and measurements were made with an X-ray beam size of 8 × 0.6 mm [69]

has been used for assessing the microstructure of and defects in the interior of a material, spatial heterogeneities in elemental concentrations, and their influences on mechanical characteristics. Generally, elements having an absorption edge within the X-ray energy range that can be obtained in the synchrotron radiation facility can be analyzed. The K absorption edge ranges from vanadium (5.46 keV) to lanthanum (38.92 keV). Using the L-shell absorption edge facilitates the analysis of elements with even larger atomic numbers.

Here, we consider an example of absorption-edge subtraction imaging of copper (which has a K absorption edge at 8.98 keV [69]) in an aluminum base alloy comprising N components. Figure 5.37 shows the X-ray absorption spectrum near the K absorption edge of copper [69]. A fine structure referred to as the "X-ray absorption fine structure" can be observed from the vicinity of the absorption edge to the high-energy region. Imaging is conducted at X-ray energies that are as close to the absorption edge as possible while avoiding these fine structures to ensure the reproducibility of the measurements. The example in Fig. 5.37 uses an X-ray energy of ±0.1 keV from the absorption edge. Here, the mass absorption coefficient μ_m^E at an X-ray energy E is obtained by the reconstruction of a set of transmission images from absorption contrast imaging under suitable conditions and is expressed as follows:

$$\mu_m^{8.88} = x_{Cu}\mu_{m,Cu}^{8.88} + \sum_{n=1}^{N-1} x_i \mu_{m,i}^{8.88} \tag{5.31}$$

$$\mu_m^{9.08} = x_{Cu}\mu_{m,Cu}^{9.08} + \sum_{n=1}^{N-1} x_i \mu_{m,i}^{9.08} \tag{5.32}$$

Here, $\mu_{m,Cu}^E$ and $\mu_{m,i}^E$ are the mass absorption coefficients of copper and the other elements, respectively, at an X-ray energy E; x_{Cu} and x_i are their respective mass fractions. The difference between $\mu_{m,i}^{8.88}$ and $\mu_{m,i}^{9.08}$ can be corrected with equations such as the Victoreen formula in Eq. (2.10). For example, the absorption coefficients of aluminum and silicon each have a difference of approximately 3% between the X-ray energies of 8.88 and 9.08 keV. After this difference is corrected, x_{Cu} can be determined as follows by subtracting Eq. (5.31) from Eq. (5.32):

$$x_{Cu} = \frac{\Delta\mu}{\left(\mu_{m,Cu}^{9.08} - \mu_{m,Cu}^{8.88}\right)\rho_{alloy}} \tag{5.33}$$

$\Delta\mu$ is determined from the difference at each pixel. ρ_{alloy} is the density of the alloy. The detection limit of copper concentration and measurement reproducibility have been confirmed to be 0.5 mass% and ±0.1 mass%, respectively, when measurements with a spatial resolution of approximately 1.3 μm were conducted at the BL20XU beamline in SPring-8 [69].

Figure 5.38 shows the 3D mapping of copper concentrations within an Al–5% Cu alloy interior using these methods [69]. Coarse crystallized particles formed during aluminum casting and heterogeneous copper distributions due to solidifying segregation can be observed; it is apparent that they have been considerably homogenized through heat treatment. The alloy element concentrations after solution treatment

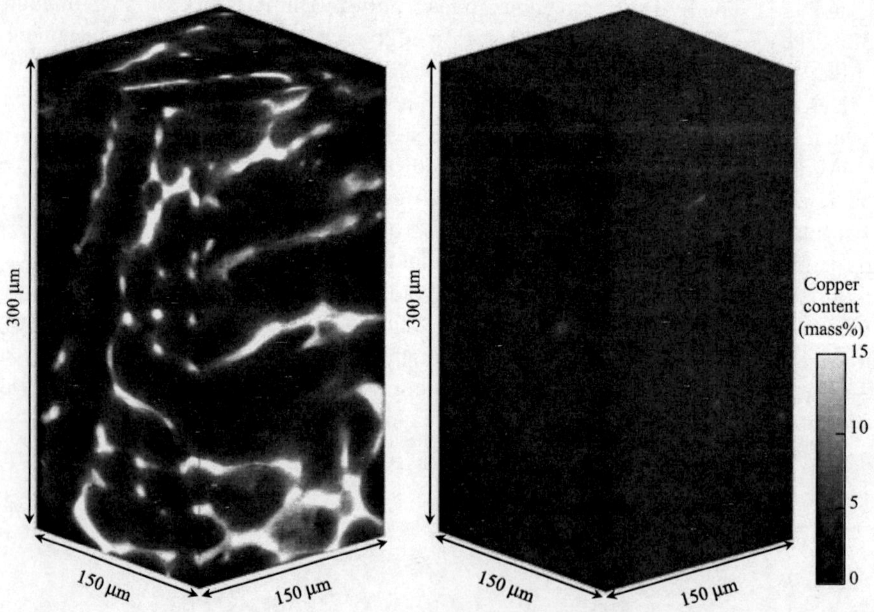

(a) As-cast state (b) After solution treatment at 813 K for 10.8 ks

Fig. 5.38 3D mapping of copper concentration in Al-5% Cu alloy using K absorption-edge subtraction imaging; comparison of as-cast and as-solution-treated samples generally appear homogeneous in standard chemical analyses. However, from the mapping in Fig. 5.38b, it is evident that some degree of heterogeneity remained in the copper concentrations even after sufficient heat treatment. This is related to the mechanical properties and their variations in structural materials. Moreover, these findings are considered important for their practical use. Other reports have shown the applications of these methods for the strength assessments of porous materials. Significant alloy element segregation occurs in porous materials relative to standard structural materials due to rapid solidification during the fabrication process. The relationship between zinc segregation on the cell wall of porous materials and the crack propagation path during fracture was clarified using these techniques [70].

Incidentally, absorption-edge subtraction imaging trials utilizing industrial X-ray CT scanners, in which an X-ray photon-counting detector was used with cadmium telluride or cadmium zinc telluride as sensors, have also been reported. Egan et al. combined a HEXITEC X-ray photon-counting detector with a microfocus tube in a commercially available X-ray CT scanner to conduct absorption-edge subtraction imaging [71]. The HEXITEC detector was developed in 2006 by a consortium formed by EPSRC in the U.K. It is capable of measurements at an energy resolution of 1 keV at peak *full width at half maximum* (*FWHM*) for X-rays with an energy in the range 3–200 keV [71]. Egan et al. visualized the 3D distribution of palladium (absorption edge of 24.35 keV) in an alumina catalyst and the spatial distribution of gold (absorption edge of 80.725 keV) and lead (88.005 keV) in ore collected from gold-bearing hydrothermal veins. The region near the K absorption edge was measured with a 0.24 keV step for palladium imaging. The pixel size for 3D imaging in these cases is coarse at 53–65 μm; however, it was concluded that imaging can be conducted with a pixel size of approximately 5 μm for smaller samples.

5.4.2 XANES Tomography

The spectroscopy approach referred to as "*X-ray absorption near-edge structure*," abbreviated as *XANES*, provides enhanced measurements and analyses compared with absorption-edge subtraction imaging and utilizes the absorption spectrum observed near the absorption edge, such as that shown in Fig. 5.37. It is virtually identical to NEXAFS for organic molecules. Measurements use a monochromatic X-ray to take consecutive transmission images near the absorption edge of a given element A while changing the X-ray energy. This results in an X-ray absorption spectrum near the absorption edge such as that shown in Fig. 5.37 for every pixel. Next, the fractions of chemical species that include element A for each pixel can be obtained by fitting previously known absorption near-edge structure data of the compound, including element A. A 2D chemical species mapping is consecutively obtained while rotating the sample, and a 3D mapping of the chemical species fractions can be obtained through reconstructing the 2D mapping in 3D.

For example, Meirer et al. applied XANES tomography to compounds containing nickel [72]. These compounds were scanned near the nickel absorption edge with a step of 0.5 eV at the synchrotron radiation facility, SSRL. The authors obtained a 3D mapping of the fractions of metallic nickel and nickel oxide within a field of view of 15 μm × 15 μm × 15 μm. Their experiments involved an imaging optical system that used a condenser zone and Fresnel zone plates, achieving a spatial resolution of several dozen nanometers. The algebraic reconstruction methods introduced in Sect. 3.3.1 were used in the image reconstruction. As this method alters the X-ray energy, accurate registration of the multiple transmission images that are obtained is essential. Detailed data analysis must be conducted after data retrieval. Energy changes in the experiments of Meirer et al. resulted in minor changes in the magnification of the obtained images; therefore, a high degree of matching between images by considering these changes is required.

5.4.3 Fluorescent X-ray Tomography

In another approach called "fluorescent X-ray tomography," X-rays are irradiated on a sample; the sample is then rotated, and the excited fluorescent X-rays are sequentially imaged. The 3D spatial distribution of the trace elements within the sample is reconstructed from the obtained image set. This approach was first reported by Boisseau and Grodzins [73]. Experiments by Boisseau et al. focused on the iron and titanium distributions in bees at the synchrotron radiation facility, NSLS. Immediately afterward in 1989, Cesareo et al. reported research using X-ray tubes [74]. Fluorescent X-ray tomography has several advantages, such as a detection limit considerably lower than that of absorption-edge subtraction imaging and the ability to measure multiple elements simultaneously. The potential of fluorescent X-ray tomography application is dependent on the intensity of the generated fluorescent X-rays. As already shown in Eq. (2.15), this intensity is dependent on the fluorescence yield. Therefore, this method often involves the mapping of elements such as chlorine (atomic number 17), and heavier elements.

Figure 5.39 shows a schematic of an experimental setup at BL37XU of SPring-8 by Ohigashi et al. of JASRI. This setup utilized a scanning-type fluorescence X-ray microscope where a Kirkpatrick–Baez mirror-focused beam was scanned. With the linear polarization characteristics of the radiation (i.e., when it was polarized within the trajectory), the Compton scattering strength was minimized in the direction orthogonal to the incident X-ray, and the intensity ratio between the fluorescent X-ray and the background scattering X-ray was maximized. Therefore, we can observe in Fig. 5.39 that the detector is positioned in the direction orthogonal to the incident X-ray.

Figure 5.40 shows the measurement of a 3D distribution of iodine within marine plankton using the aforementioned setup by Ohigashi et al. [75]. The observed plankton had a body length of approximately 500 μm. The incident X-ray energy, in this case, was set to 10 keV, and simultaneous measurements of iodine, chlorine, potassium, calcium, iron, and zinc were conducted. Measurements involved the use of a helical scan (1 μm translation/360° rotation); scans were made in 3-μm steps

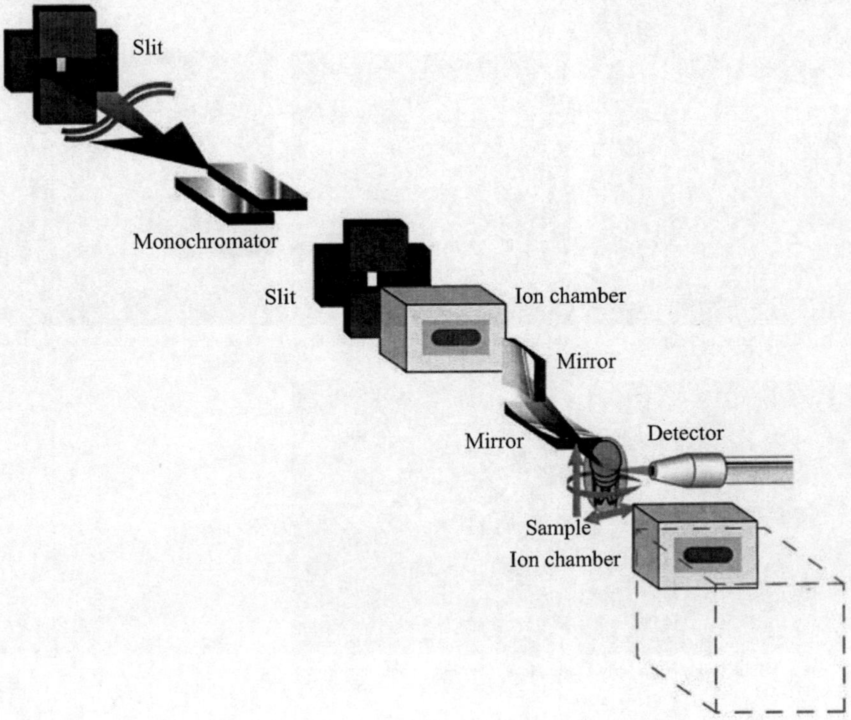

Fig. 5.39 Schematic of a fluorescent X-ray tomography setup (courtesy of Takuji Ohigashi of the Institute for Molecular Science)

for a total of 125 steps in the horizontal direction and in 0.1-μm steps in the vertical direction; then, the sample was rotated 180° in steps of 2.4°. A total of 150 projections were taken for the transmission images. Consequently, this imaging required 70.6 h. Thus, from a practical standpoint, an extremely extended measurement time is the trade-off for superior spatial resolution and detectability.

An issue with fluorescent X-ray tomography is that the incident X-ray and fluorescent X-ray are both absorbed by the sample itself. For example, the true magnitude correlations between high-concentration and low-concentration regions in a sample interior may even be inverted during measurement, depending on the positional relationship within the horizontal sample cross-section of the two regions. Ohigashi et al. reported research focused not only on the visualization of elemental distributions but also on the correction of incident and fluorescent X-ray self-absorption to conduct accurate measurements of elemental concentrations [76].

Other studies have used X-ray photon-counting detectors with cadmium telluride as a sensor for industrial X-ray CT scanners [77]. Fluorescent X-ray tomography can conduct X-ray energy spectrum measurements under low illumination, exploiting the characteristics of photon-counting detectors.

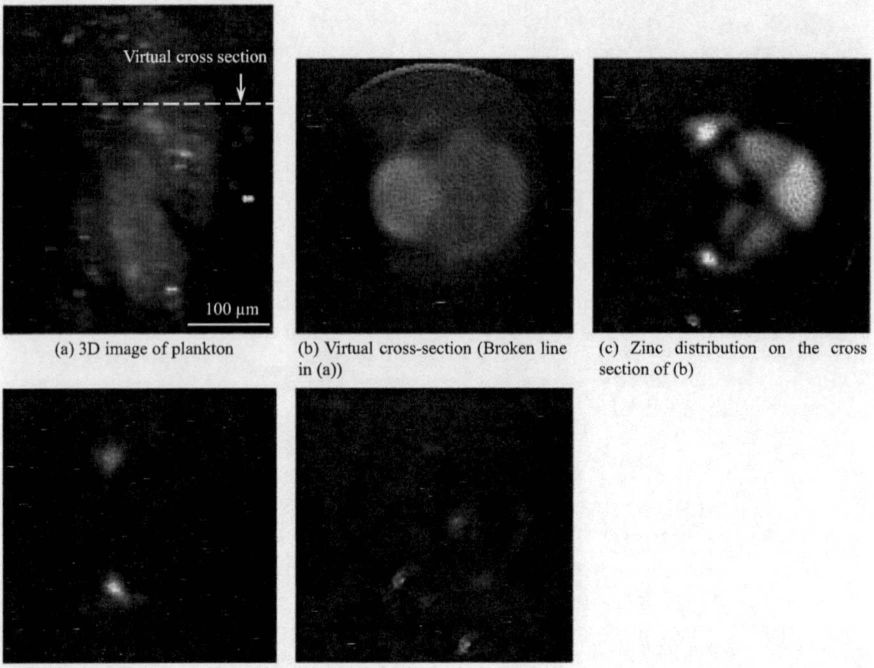

(a) 3D image of plankton

(b) Virtual cross-section (Broken line in (a))

(c) Zinc distribution on the cross section of (b)

(d) Iodine distribution on the cross section of (b)

(e) Iron distribution on the cross section of (b)

Fig. 5.40 Implementation example of fluorescent X-ray tomography; 3D iodine distribution within marine plankton (courtesy of Takuji Ohigashi of the Institute for Molecular Science)

5.5 Polycrystalline Tomography

In this section, we introduce methods that facilitate the imaging of crystallographic grains in polycrystalline structures, the measurement of their respective crystallographic orientations, and the mapping of local crystallographic orientations within each crystallographic grain. The thickness of the grain boundary, which is the interface between adjacent crystallographic grains, is generally considered to be a few layers of atoms at most—in other words, in the order of 1 nm. As previously discussed, the spatial resolution level for 3D imaging using an X-ray imaging optical system is approximately 2 orders of magnitude larger than this value. Hence, grain boundaries cannot be directly imaged with X-ray tomography. Furthermore, the grain itself usually cannot be visualized either due to the X-ray phase and absorption changes or due to differences in the crystallographic orientation of the crystallographic grains. However, the *liquid metal wetting technique* discussed in Sect. 5.5.1 is a simple method that can be applied when only the shape of the grain boundary is to be imaged. Meanwhile, techniques that apply X-ray diffraction or those that combine X-ray tomography and X-ray diffraction must be used for cases where the crystallographic orientation is to be measured, in addition to the grain shape. Several methods have been proposed in actual practice for these cases. However, each of these methods requires a detailed understanding of X-ray diffraction and complex algorithms, and thus, they are outside the scope of this study. Therefore, references will be provided here and only an overview of these topics will be presented.

5.5.1 Liquid Metal Wetting Technique

"*Gallium-enhanced microscopy*" involves diffusing gallium so that it penetrates the grain boundaries of aluminum and conducting observations using a scanning electron microscope. In 2000, Ludwig et al. at ESRF used the large difference in linear absorption coefficients between aluminum and gallium to conduct 3D observations of the grain boundaries using X-ray tomography [78].

The combinations of metals whose grain boundaries are to be observed and the metals used for the decoration of the grain boundaries can be selected from combinations of solid and liquid metals; this produces the phenomenon of liquid-metal embrittlement. In other words, the negative phenomenon of embrittlement, which induces metal fracture, is interrupted immediately before fracture and is instead utilized to positive effect in imaging via grain boundary decoration.

An explanation is provided here using the example of the most common application of gallium-enhanced microscopy, i.e., the visualization of aluminum grain boundaries. Gallium is a solid metal under room temperature and atmospheric pressure, but it has a low melting point of 29.8 °C. Placing melted gallium in contact with solid aluminum results in the rapid penetration of the aluminum grain boundaries by the gallium. Grain boundary misorientation dependencies and anisotropy within the grain boundaries are known to be present for this grain boundary diffusion. For example, reports have indicated that gallium diffuses at speeds of 8.8 μm/s along the grain boundary and 0.3–0.7 nm/s from the grain boundary to the grain interior [79]. There is an approximate 10,000-fold difference between the two speeds.

Figure 5.41 shows the results of visualization experiments conducted by the author's group at BL47XU in SPring-8. The dendrites and silicon particles in the aluminum cast alloy were effectively observed, and the spatial positional relationships between the particles and the grain boundary were well understood. The experimental procedure involved applying gallium to the sample surface; subsequently, the surface was scratched and the oxidized films were partially removed. Gallium diffused and penetrated into the grain boundaries during the few minutes at which a temperature above the melting point of gallium was maintained. Finally, a 3D image of the grain boundary such as in Fig. 5.41 could be obtained by removing the excess gallium attached on the surface with adhesive tape to prevent metal artifacts [80]. Examples of placing aluminum and gallium in contact with each other in sodium hydroxide to diffuse the gallium stably have been presented. The liquid metal wetting technique can be used as a simple 3D analysis method of polycrystalline structures, and for the assessment of relationships between various damage/fracture phenomena and polycrystalline structures.

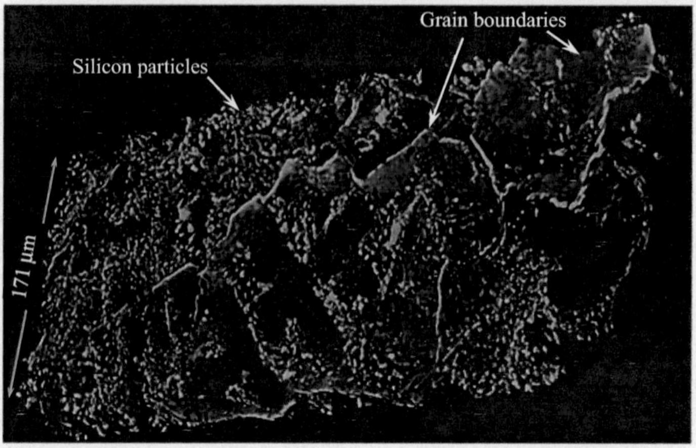

Fig. 5.41 Visualization of the crystallographic grain boundaries in Al-7% Si alloy using gallium. Granular eutectic silicon particles are also shown in the figure. Silicon particles are present in the gaps of the dendrite arms and the presence of the crystallographic grain boundary encompassing the dendrite can be seen

Furthermore, this is a general-use technique than can be implemented with industrial X-ray CT scanners equipped with a microfocus tube and having a sufficiently high spatial resolution.

5.5.2 *Diffraction Contrast Tomography*

Diffraction contrast tomography (*DCT*) non-destructively determines the overall crystallographic grain shape and orientation using relatively simple measurements and short-duration analyses. This method was proposed in 2007 by Ludwig et al. at ESRF [81]. Figure 5.42 shows a schematic of the experimental setup [82]. The basic procedure involves irradiation with an X-ray beam that has a width that can sufficiently cover the horizontal width of the sample and the placement of a detector with a larger field of view than that in standard X-ray tomography. In experiments on hydrogen embrittlement in aluminum alloys conducted by the author's group at SPring-8, a 2,046 × 2,046-pixel CMOS camera was used for imaging within a 6.3 mm × 6.3 mm range [83]. Each crystallographic orientation satisfies the Bragg conditions individually and diffracts the X-rays when the sample is rotated 180° or 360° while being irradiated with X-rays. As shown in Fig. 5.42, the transmission image of the crystallographic grains that diffract the X-ray becomes darker (*extinction*); the crystallographic grain image is projected in the direction in which the X-ray was diffracted. These results are successively recorded as images. Higher-accuracy measurements become possible by rotating the sample 360° and utilizing a *Friedel pair* [82]. As shown in Fig. 5.43, a Friedel pair refers to a set of diffrac-

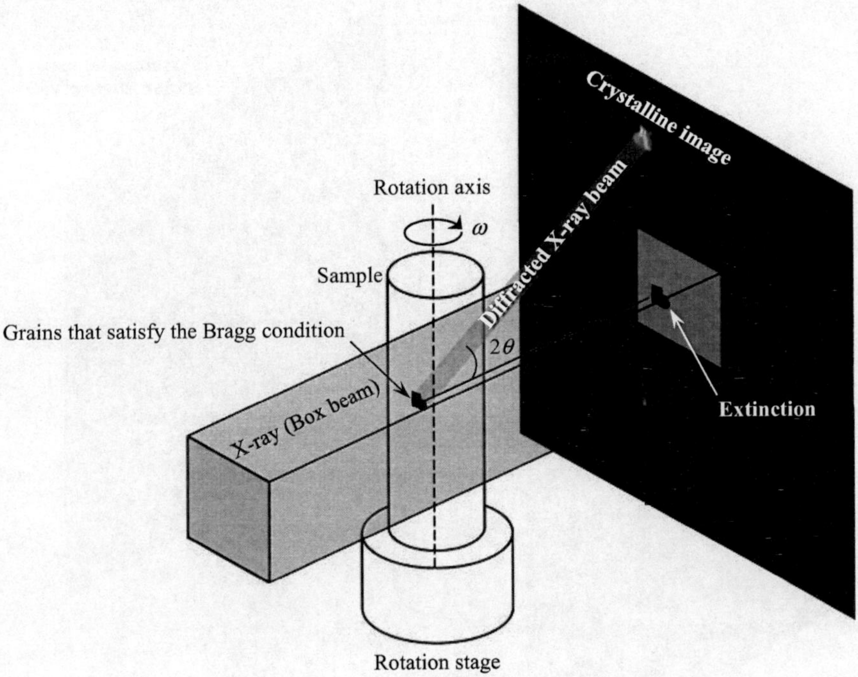

Fig. 5.42 Schematic of an experimental setup of diffraction contrast tomography

tion images measured at 180° from the sample rotation angle, from the (h, k, l) and $(\bar{h}, \bar{k}, \bar{l})$ planes of a single crystallographic grain. When the X-ray diffraction measurements conducted during sample rotation are expressed as in Fig. 5.43, the Friedel pair has parallel and reverse-direction *scattering vectors* (the scattering X-ray wavenumber vectors with the incident X-ray wavenumber vectors subtracted) and diffraction beams. Furthermore, the original crystallographic grain is positioned on the line segment that connects the two diffraction images. With this approach, the crystallographic grain and diffraction image can be successfully matched by removing erroneous data due to factors such as *double diffraction*. Furthermore, crystallographic grain pairs that exhibit consistency in geometric shape, crystallographic orientation, and positional information within the sample are labeled as Friedel pairs based on identical crystallographic grains. Several diffraction images that are sufficient for conducting the 3D reconstruction of the crystallographic grain image can be obtained even after removing those whose diffraction images overlapped with other crystallographic grains, those that diffracted outside the detector imaging range, and those with low contrast. The algebraic reconstruction method introduced in Sect. 3.3.1 is used for the 3D reconstruction of each crystallographic grain [82].

Figure 5.44 shows an example of a 3D visualization of Al-Zn-Mg alloy crystallographic grains conducted by Hirayama et al. at the author's laboratory at BL20XU of

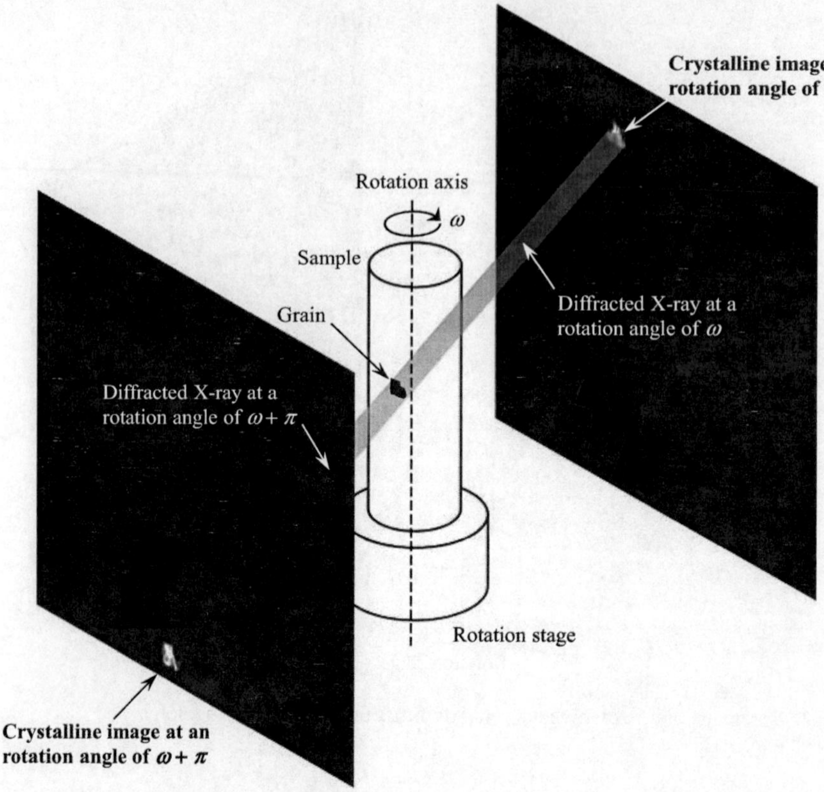

Fig. 5.43 Schematic of a Friedel pair. For a crystallographic grain that satisfies the diffraction conditions at a given angle, a diffraction image is obtained from the same crystallographic grain by rotating the sample 180°. Next, rotating the entire sample 180° and stacking with the first diffraction image established a coordinate system where the beam and detector are hypothetically rotated while the sample is fixed. The crystallographic grain that diffracted the X-ray is positioned on the line segment that links the two crystalline images. Furthermore, the diffraction angle can be determined without knowing the crystallographic grain position in the sample

SPring-8. Although there are some overlaps and gaps between the crystallographic grains adjacent to the grain boundaries, it is evident that the 3D shapes of the grains are accurately reconstructed overall.

In 2013, Ludwig et al. achieved DCT at the laboratory level using microfocus tubes [84]. Products that include modules and analysis software for DCT can now be used with commercially available industrial X-ray scanners. McDonald et al. conducted DCT experiments on titanium alloys using an industrial X-ray scanner and concluded that the 3D imaging of crystal grains up to a grain diameter of approximately 40 μm was possible [85].

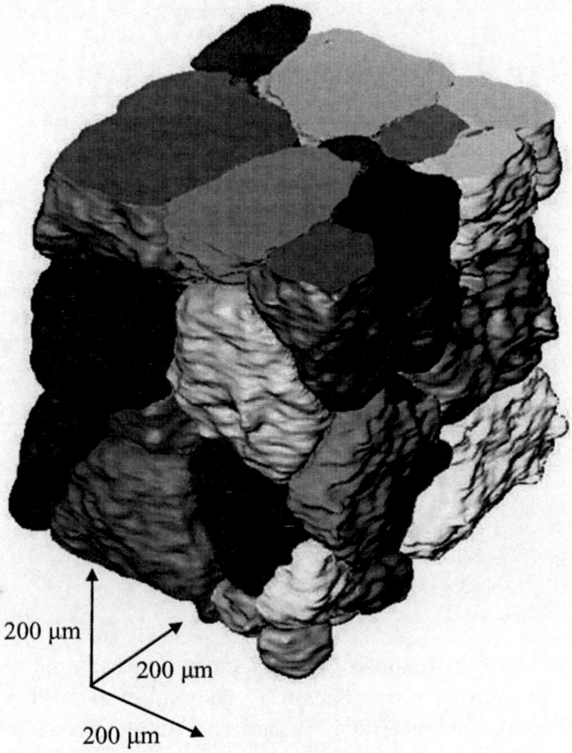

Fig. 5.44 Example of a 3D image of a polycrystalline structure obtained from diffraction contrast tomography

5.5.3 *3D-XRD*

3D-XRD is a non-destructive polycrystalline structure visualization technique based only on X-ray diffraction and was developed by Poulsen et al. of Riso National Laboratory in Denmark [86]. As shown in Fig. 5.45, the approach involves setting up either two detectors with different camera lengths or a single camera with camera lengths varied across several positions. A fan beam collimated to a thickness of 5–10 μm is applied to a rotating sample, and the diffraction spots are recorded with a two-dimensional detector. The diffraction spots are matched with each position within the crystallographic grain after measurement, and the crystallographic orientations are mapped in 3D in real space. Regions with the same crystallographic orientation are identified as crystal grains; the boundaries of regions with different crystallographic orientations are identified as grain boundaries to visualize the shapes of crystallographic grains. The 3D-XRD method has been applied to studies on the recrystallization behavior of metal materials, crystallographic grain growth behavior at high temperatures, and crystal lattice rotations during deformation [86]. This method is referred to as "*high-energy X-ray diffraction microscopy*" (*HEDM*) at the synchrotron radiation facility, APS; technical development and applied research to analyze the deformation and damage in polycrystalline materials have been conducted there as well [87].

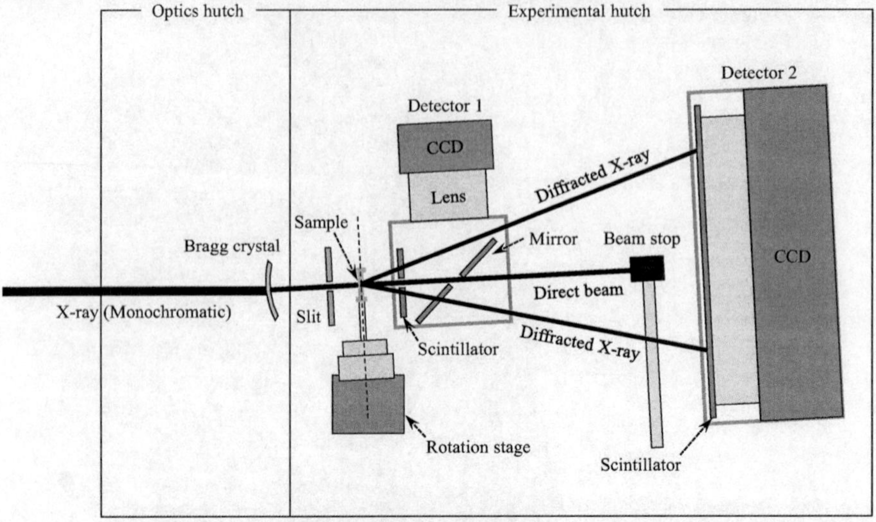

Fig. 5.45 Schematic of an experimental setup for 3D-XRD. This shows an example of experiments conducted by Kobayashi et al. at the ID11 beamline in ESRF

Figure 5.46 shows an implementation example of the scanning-type 3D-XRD method using a microbeam by Hayashi et al. of Toyota Central R&D Labs [88]. The observed material is a standard steel material (SPCC steel). The experiment was conducted at BL33XU (Toyota beamline) in SPring-8 and scanning was conducted by focusing a 50 keV monochromatic X-ray with a Kirkpatrick–Baez mirror. These experiments demonstrate that crystallographic grains as small as 5 μm can be determined.

The 3D-XRD method is outside the range of topics discussed in this text, as it is an indirect imaging method based on X-ray diffraction experiments; interested readers should refer to Paulsen's work [86].

5.5.4 *Diffraction-Amalgamated Grain-Boundary Tracking (DAGT)*

The *diffraction-amalgamated grain-boundary tracking* (*DAGT*) method was developed by the author's group to analyze deformation and fracture crystallographically. DCT and 3D-XRD use X-ray diffraction phenomena; hence, the plastic deformation of the sample simultaneously results in a diffraction spot spread within the detector plane and the sample rotation direction (ω direction in Fig. 5.42), and decreased diffracted X-ray intensity. Therefore, it is difficult or even impossible to measure crystallographic orientation or determine the 3D shape of crystallographic grains during plastic deformation. The DAGT method is based on the *grain-boundary tracking*

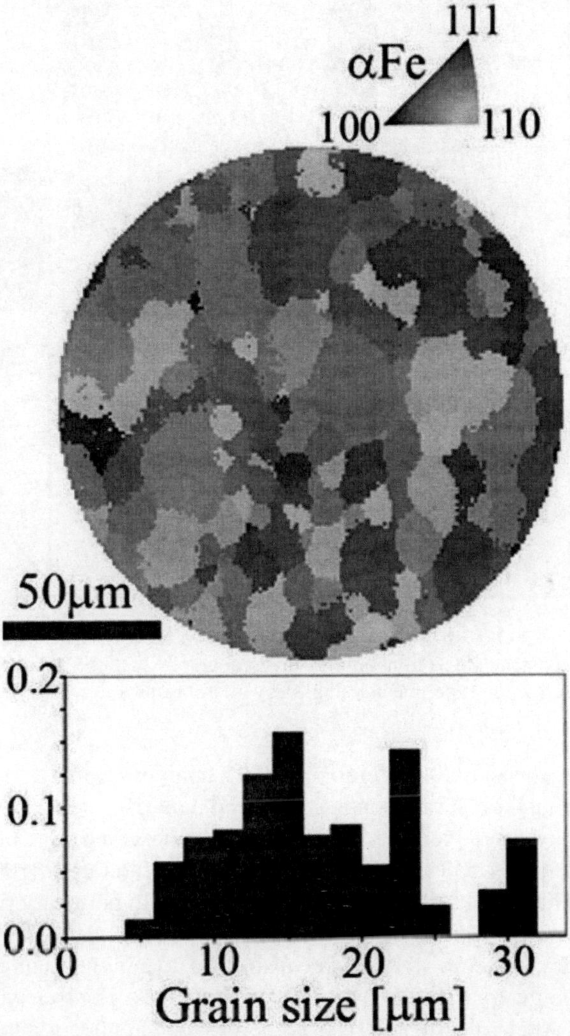

Fig. 5.46 Crystallographic orientation mapping and crystallographic grain diameter histogram for a steel material based on scanning-type 3D–XRD (courtesy of Yujiro Hayashi of Toyota Central R&D Labs) [88]

technique (GBT) [89], in which X-ray tomography is combined with the liquid metal wetting technique introduced in Sect. 5.5.1. X-ray diffraction measurements using an X-ray microbeam are combined with this approach to develop a technique that facilitates the elucidation of morphological changes in crystallographic grains with a high accuracy, even during the plastic deformation of the sample [90].

Figure 5.47 shows a schematic of experiments with the DAGT method. First, a pencil beam (5–10 μm) is scanned while rotating the sample 180° and an X-ray diffraction image is obtained. Next, the sample is deformed/fractured while the process is consecutively imaged in 3D using X-ray tomography. Subsequently, the grain boundary is decorated with liquid metal, and 3D images are obtained once

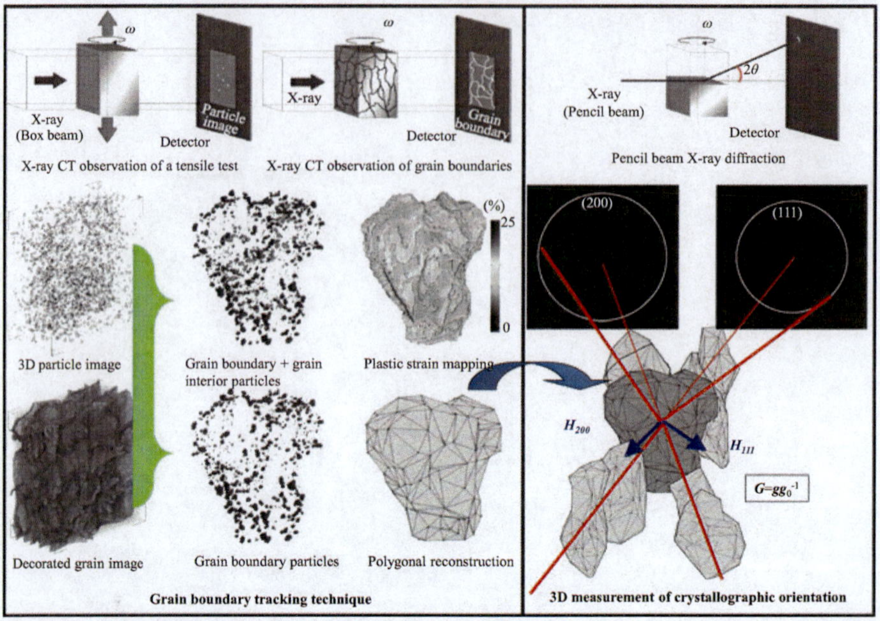

Fig. 5.47 Schematic of the DAGT method process, which combines the grain boundary tracking technique (left) (combining X-ray tomography with the liquid-metal wetting technique) and X-ray diffraction experiments using X-ray microbeams

again using X-ray tomography. The pair of 3D images obtained before and after grain boundary decoration is referenced, and particles that appear on the grain boundary planes are identified among all those observed. The number of observable particles within a 600 μm square observation area is up to 100,000 for an aluminum alloy. The physical displacement of all the grain boundary particles is traced back in time via the consecutively obtained 3D images. The 3D morphology of the crystallographic grains can be reconstructed as polyhedrons by linking the adjacent grain boundary particles using triangular planes. The trajectories of all the grain boundary particles can be traced from before load application to immediately before fracture; therefore, the manner in which the 3D morphology of all the crystallographic grains changed during the deformation/fracture of the material can also be observed. Finally, the various crystallographic grains and X-ray diffraction spots are associated, and the crystallographic orientation distribution of the interior of each crystallographic grain is calculated. This method can trace the physical displacement of all particles, including those within grains; hence, the 3D distribution of plastic strain in the interior of the material and its changes can also be visualized simultaneously.

Figure 5.48 shows the 3D reconstruction of the crystallographic grain morphology and the crystallographic orientation measurement results for 23 crystallographic grains present in a given region of a material interior [90]. The DAGT method can map the 3D plastic strain distributions of each crystal grain interior according to each

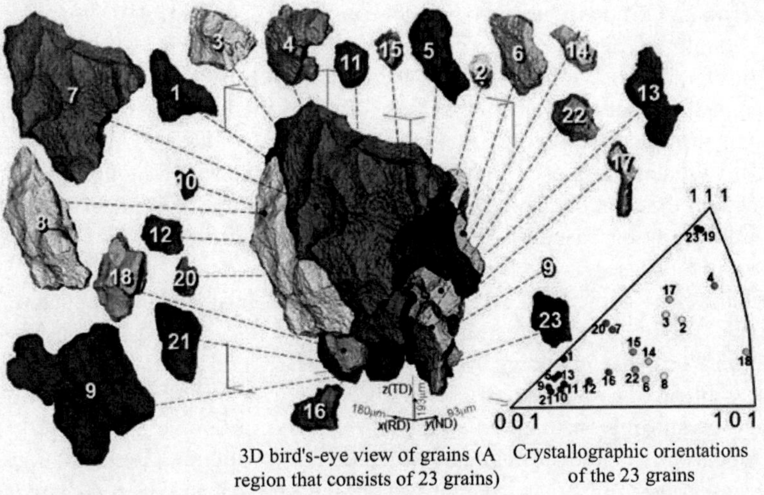

Fig. 5.48 Visualization example of crystallographic grains based on the DAGT method [90]. 3D display of the crystallographic grains in an Al-3% Cu alloy (left) and its corresponding crystallographic orientation distribution (right: inverse pole figure). The original figure is in color; please refer to the figure in the original publication for details

strain component. Therefore, the local Taylor factor and Schmid factor are calculated for each crystallographic orientation, and the local deformation behavior can be studied. An extremely heterogeneous plastic strain distribution can be observed even in a single crystallographic grain interior in the applied example shown in Fig. 5.47. A type of coordinated deformation behavior across the polycrystalline material, wherein plastic deformation appears to propagate across multiple crystallographic grains, can be observed here as well.

5.6 Other Tomography Techniques

There are other 3D imaging techniques than those discussed here that utilize the image reconstruction techniques introduced in Sect. 3.3. Representative examples include *X-ray small-angle scattering* tomography. Schroer et al. first reported this technique in 2006 [91]. Orientations of nanoscale structures such as particles in a material can be expressed in 3D using *X-ray small-angle scattering tomography*. When applying this technique to isotropic samples such as colloids that cause isotropic scattering, 3D reconstruction can be conducted by simply integrating the scattering intensities. However, specialized measurement techniques and reconstruction methods (e.g., data retrieval when the sample is rotated across multiple rotational axes) are necessary when analyzing microstructures with anisotropic properties, such as orientation.

Schaff et al. first proposed a method for this purpose in 2015, successfully measuring the 3D orientation distribution of collagen fibers in teeth during experiments at the synchrotron radiation facility SLS in Switzerland [92].

Chen et al. combined a Talbot–Lau X-ray interferometer, a phase-contrast imaging technique introduced in Sect. 5.2.3 (2) suited for X-ray tubes, with a rotating anode-type X-ray tube and proposed an X-ray small-angle scattering tomography measurement method that can be conducted with industrial X-ray CT scanners [93]. Cone-beam reconstruction based on Feldkamp's algorithm introduced in Sect. 3.3.4 (2) was used in this case.

Furthermore, tomography using *X-ray topography* has been implemented. X-ray topography facilitates the imaging of lattice defects (e.g., dislocation) using X-ray diffraction. The basic procedure involves the use of a single crystal material, in which the X-ray diffraction intensity differences between the crystal lattice defects and the complete crystal region of the material are utilized. Examples of 3D versions of X-ray topography include research by Mukaide et al., which implemented *step scanning section topography* using synchrotron radiation [94], and the *topo-tomography* method proposed by Ludwig [95]. These methods are limited to the analyses of silicon and ceramics and are unable to visualize dislocation distributions in polycrystalline materials in practical use, such as steels or aluminum alloys. References are included on this topic, although a more in-depth discussion is omitted. A thorough overview by Kajiwara from SPring-8 is referenced for readers interested in further information [96].

High-resolution 3D imaging techniques besides the imaging X-ray micro-/nano-tomography introduced in Sect. 5.1 include X-ray tomography based on *coherent diffractive imaging* (*CDI*) [97, 98]. As shown in Fig. 5.49a, this method involves the irradiation of coherent X-rays onto a sample and the measurement of diffraction intensity patterns at a distance away from the sample. This diffraction intensity pattern can be expressed by a convolution of the Fourier transform of the sample

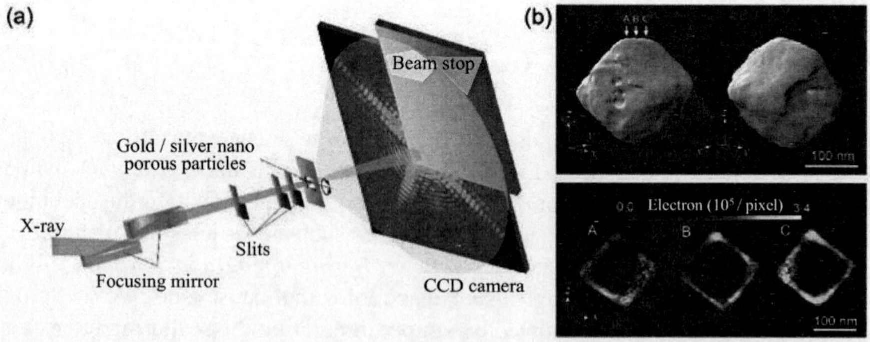

Fig. 5.49 **a** Overview of tomography based on coherent diffraction imaging. **b** Observation image of gold/silver nano-porous particles using coherent diffraction imaging-based tomography. The top image in (**b**) is the particle surface image and the bottom image in (**b**) is the cross-section image (courtesy of Yukio Takahashi of Osaka University)

function and the irradiation function, which expresses the wavefront of the incident X-ray. A sample image, as shown in Fig. 5.49b, can then be obtained by conducting phase retrieval calculations on the obtained data. This enables 3D imaging of the internal structure of a material at an ultra-high spatial resolution of up to 10 nm without using an imaging optical system, such as a Fresnel zone plate. Takahashi et al. from Osaka University have actively pursued this line of research in Japan [97, 98]. This approach can be considered a lens-less high-spatial-resolution 3D imaging technique and is on the cutting edge of technology in terms of spatial resolution among 3D imaging techniques using X-rays. It has been considered particularly useful for samples composed of light elements, such as biological tissue. Synchrotron radiation facilities are essential when implementing this technique, and synchrotron radiation sources that have high brilliance and low emittance are particularly suited to it.

References

1. Y. Suzuki, H. Toda, *Advanced Tomographic Methods in Materials Research and Engineering*, ed. by John Banhart (Oxford University Press, 2008), (Sect. 7.1)
2. H. Takenaka, T. Koyama, H. Takano, K. Kagoshima, Optics **42**, 289–295 (2013)
3. A. Snigirev, I. Snigireva, C. R. Physique **9**, 507–516 (2008)
4. H. Toda, SPring-8 priority partner user proposal/utilization proposal experiment report (proposal number 2017AA0076), experimental proposal title, "4D imaging techniques of structural materials and further advancement of its surrounding analysis techniques", (2017), https://user.spring8.or.jp/apps/experimentreport/detail/21890/ja
5. NTT Advanced Technology Corporation, https://www.ntt-at.com/product/x-ray_FZP/. Accessed Mar 2018
6. Applied Nanotools Inc., https://www.appliednt.com/x-ray-zone-plates/. Accessed Mar 2018
7. V. De Andrade, J. Thieme, P. Northrup, Y. Yao, A. Lanzirotti, P. Eng, Q. Shen, Nucl. Instrum. Methods Phys Res Sect A Accel. Spectrom. Detect. Assoc. Equip. **649**, 46–48 (2011)
8. J. Kirz, J. Opt. Soc. Am. **64**, 301–309 (1974), https://doi.org/10.1364/josa.64.000301.125
9. A. Takeuchi, K. Uesugi, Y. Suzuki, S. Itabashi, M. Oda, J. Synchrotron Radiat. **24**, 586–594 (2017)
10. A. Takeuchi, Y. Suzuki, K. Uesugi, Development of a homogeneous field of view Zernike phase contrast X-ray microscope (J-GLOBAL ID: 200902254194013128), in *KEK Proceedings* (2008), pp. 35–38
11. Y. Suzuki, A. Takeuchi, K. Uesugi, M. Hoshino, in *AIP Conference Proceedings*, vol. 160 (2011), pp. 1365, https://doi.org/10.1063/1.3625329
12. ZEISS Xradia 810 Ultra, https://www.zeiss.co.jp/microscopy/products/x-ray-microscopy/xradia-810-ultra.html#inpagetabs_da30-6. Accessed Mar 2018
13. T. Uruga, M. Nomura, Introduction to radiation beamline optics, ed. by H. Ohashi, K. Hirano (Japanese Society for Synchrotron Radiation Research, 2008)
14. The Center for X-Ray Optics (CXRO), Lawrence Berkeley National Laboratory (LBNL), X-ray interactions with matter calculator, https://henke.lbl.gov/optical_constants/. Accessed Mar 2018
15. K. Yamauchi, H. Mimura, Y. Mori, Surface Sci **22**, 152–159 (2001), https://doi.org/10.1380/jsssj.22.152
16. S. Aoki, A. Takeuchi, J. Jpn. Soc. Synchrotron Radiat. Res. **10**, 194–199 (1997)
17. A. Snigirev, V. Kohn, I. Snigireva, B. Lengeler, Nature **384**, 49–51 (1996)

18. B. Lengeler, C.G. Schroer, M. Richwin, J. Tummler, M. Drakopoulos, A. Snigirev, I. Snigireva, Appl. Phys. Lett. **74**, 3924–3926 (1999)
19. C.G. Schroer, J. Meyer, M. Kuhlmann, B. Benner, T.F. Gunsler, B. Lengeler, C. Rau, T. Weitkamp, A. Snigirev, I. Snigireva, Appl. Phys. Lett. **81**, 1527–1529 (2002)
20. Y.I. Dudchik, F.F. Komarov, M.A. Piestrup, C.K. Gary, H. Park, J.T. Cremer, Spectrochim. Acta Part B **62**, 598–602 (2007)
21. S. Bajt, M. Prasciolu, H. Fleckenstein, M. Domarацký, H.N. Chapman, A.J. Morgan, O. Yefanov, M. Messerschmidt, Y.D. Kevin, T. Murray, V. Mariani, M. Kuhn, S. Aplin, K. Pande, P. Villanueva-Perez, K. Stachnik, J.P.J. Chen, A. Andrejczuk, A. Meents, A. Burkhardt, D. Pennicard, X. Huang, H. Yan, E. Nazaretski, Y.S. Chu, C.E. Hamm, Light Sci. Appl. **7**, 17162 (2018), https://doi.org/10.1038/lsa.2017.162
22. H.C. Kang, H. Yan, R.P. Winarski, M.V. Holt, J. Maser, C. Liu, R. Conley, S. Vogt, A.T. Macrander, G.B. Stephenso, Appl. Phys. Lett. **92**, 221114 (2008)
23. M.R. Teague, J. Opt. Soc. Am. **73**, 1434–1441 (1983)
24. K. Ishizuka, Microscopy **40**, 188–192 (2005)
25. A. Burvall, U. Lundstr¨om, P.A.C. Takman, D.H. Larsson, H.M. Hertz, Opt. Express **19**, 10359 (2011)
26. Y. Suzuki, N. Yagi, K. Uesugi, J. Synchrotron Radiat. **9**, 160–165 (2002)
27. D. Paganin, S.C. Mayo, T.E. Gureyev, P.R. Miller, S.W. Wilkins, J. Microsc. **206**, 33–40 (2002)
28. T. Weitkamp, D. Haas, D. Wegrzynek, A. Rak, J. Synchrotron Radiat. **18**, 617–629 (2011)
29. ImageJ Website, https://imagej.nih.gov/ij/plugins/ankaphase/. Accessed Apr 2018
30. T.E. Gureyev, A.W. Stevenson, D.M. Paganin, T. Weitkamp, A. Snigirev, I. Snigireva, S.W. Wilkins, J. Synchrotron Radiat. **9**, 148–153 (2004)
31. H. Toda, A. Takijiri, M. Azuma, S. Yabu, K. Hayashi, D. Seo, M. Kobayashi, K. Hirayama, A. Takeuchi, K. Uesugi, Acta Mater. **126**, 401–412 (2017)
32. H. Toda, F. Tomizato, R. Harasaki, D. Seo, M. Kobayashi, A. Takeuchi, K. Uesugi, ISIJ Int. **56**, 883–892 (2016)
33. S.C. Mayo, T.J. Davis, T.E. Gureyev, P.R. Miller, D. Paganin, A. Pogany, A.W. Stevenson, S.W. Wilkins, Opt. Express **11**, 2289–2302 (2003)
34. F. Zernike, Zeitschrift für technische Physik **16**, 454–457 (1935)
35. Y. Kohmura, K. Okada, A. Takeuchi, H. Takano, Y. Suzuki, T. Ishikawa, T. Ohigashi, H. Yokosuka, Nucl. Instrum. Methods Phys. Res. A **467–468**, 881–883 (2001)
36. A. Momose, Opt. Express **11**, 2303–2314 (2003)
37. A. Momose, Nucl. Instrum. Methods Phys. Res. Sect. A Accel. Spectrom. Detect. Assoc. Equip. **352**, 622–628 (1995)
38. A. Yoneyama, A. Nambu, K. Ueda, S. Yamada, S. Takeya, K. Hyodo, T. Takeda, in *11th International Conference on Synchrotron Radiation Instrumentation (SRI 2012)*, Journal of Physics: Conference Series, vol. 425 (2013), pp. 192007
39. P. Cloetens, J.P. Guigay, C. De Martino, J. Baruchel, Opt. Lett. **22**, 1059–1061 (1997)
40. A. Momose, J.Jpn. Soc. Synchrotron Res. **23**, 382–392 (2010)
41. M. Engelhardt, J. Baumann, M. Schuster, C. Kottler, F. Pfeiffer, O. Bunk, C. David, Appl. Phys. Lett. **90**, 224101 (2007)
42. S.W. Wilkins, Y.I. Nesterets, T.E. Gureyev, S.C. Mayo, A. Pogany, A.W. Stevenson, Philosoph. Trans. R. Soc. A Math. Phys. Eng. Sci. **372**, 20130021 (2013), https://doi.org/10.1098/rsta.2013.0021
43. E. Gureyev, Y.I. Nesterets, A.W. Stevenson, P.R. Miller, A. Pogany, S.W. Wilkins, Opt. Express **16**, 3223–3241 (2008)
44. F. Pfeiffer, T. Weitkamp, O. Bunk, C. David, Nat. Phys. **2**, 258–261 (2006)
45. A.R. Lang, A.P.W. Makepeace, J. Synchrotron Radiat. **6**, 59–61 (1999)
46. Y. Suzuki, Jpn. J. Appl. Phys. **41**, L1019 (2002)
47. Y. Suzuki, J. Jpn. Soc. Synchrotron Res. **18**, 75–83 (2005)
48. D. Gabor, Nature **161**, 777–778 (1948)
49. K. Hayashi, Y. Takahashi, E. Matsubara, Mater. Jpn. **40**, 801–807 (2001)

50. A. Borbély, F.F. Csikor, S. Zabler, P. Cloetens, H. Biermann, Mater. Sci. Eng. A **367**, 40–50 (2004)
51. D. Tolnaia, G. Requena, P. Cloetens, J. Lendvai, H.P. Degischera, Mater. Sci. Eng. A **550**, 214–221 (2012)
52. C. Landron, E. Maire, J. Adrien, H. Suhonen, P. Cloetensc, O. Bouaziz, Scripta Mater. **66**, 1077–1080 (2012)
53. I. Kumpová, M. Vopálenský, T. Fíla, D. Kytýř, D. Vavřík, M. Pichotka, J. Jakůbek, V. Veselý, in *Proceedings of 2016 IEEE Nuclear Science Symposium, Medical Imaging Conference and Room-Temperature Semiconductor Detector Workshop (NSS/MIC/RTSD)* (2016), https://doi.org/10.1109/nssmic.2016.8069950
54. K. Uesugi, T. Sera, N. Yagi, J. Synchrotron Radiat. **13**, 403–407 (2006)
55. H.P. Menke, M.G. Andrew, J. Vila-Comamala, C. Rau, M.J. Blunta, B. Bijeljic, J. Vis. Exp. **120**(e53763), 1–10 (2017)
56. S. Shuai, E. Guo. A.B. Phillion, M.D. Callaghan, T. Jing, P.D. Lee, Acta Mater. **118**, 260–269 (2016)
57. O. Lame, D. Bellet, M. Di Michiel, D. Bouvard, Nucl. Instrum. Methods Phys. Res. B **200**, 287–294 (2003)
58. H. Toda, E. Maire, S. Yamauchi, H. Tsuruta, T. Hiramatsu, M. Kobayashi, Acta Mater. **59**, 1995–2008 (2011)
59. K.Z. Zefreh, in *Proceedings of 2016 IEEE Nuclear Science Symposium, Medical Imaging Conference and Room-Temperature Semiconductor Detector Workshop (NSS/MIC/RTSD)* (2016), https://doi.org/10.1109/nssmic.2016.8069830
60. K.J. Dobson, S.B. Coban, S.A. McDonald, J.N. Walsh, R.C. Atwood, P.J. Withers, Solid Earth **7**, 1059–1073 (2016)
61. R. Daudin, S. Terzi, P. Lhuissier, J. Tamayo, M. Scheel, N.H. Babu, D.G. Eskin, L. Salvo, Acta Mater. **125**, 303–310 (2017)
62. R. Mokso, F. Marone, M. Stampanoni, in *AIP Conference Proceedings* 1234, vol. 87, no. 1234 (2010), pp. 87–90, https://doi.org/10.1063/1.3463356
63. E. Maire, V. Carmona, J. Courbon, W. Ludwig, Acta Mater. **55**, 6806–6815 (2007)
64. N. Limodin, L. Salvo, M. Suéry, M.D. Michiel, Acta Mater. **55**, 3177–3191 (2007)
65. N. Limodin, L. Salvo, E. Boller, M. Sue´ry, M. Felberbaum, S. Gailliégue, K. Madi, Acta Mater. **57**, 2300–2310 (2009)
66. GE Website, https://www.gemeasurement.com/inspection-ndt/radiography-and-computed-tomography/speedscan-ct-64. Accessed May 2018
67. R.A. Rutherford, B.R. Pullan, I. Isherwood, Neuroradiology **11**, 23–28 (1976)
68. W.A. Kalender, *Computed Tomography: Fundamentals, System Technology, Image Quality, Applications*, 3rd edn. (Bellingham, WA, Wiley Interscience, 2011), p. 316
69. H. Toda, T. Nishimura, K. Uesugi, Y. Suzuki, M. Kobayashi, Acta Mater. **58**, 2014–2025 (2010)
70. Q. Zhang, H. Toda, Y. Takami, Y. Suzuki, K. Uesugi, M. Kobayashi, Philos. Mag. **90**, 1853–1871 (2010)
71. C.K. Egan, S.D.M. Jacques, M.D. Wilson, M.C. Veale, P. Seller, A.M. Beale, R.A.D. Pattrick, P.J. Withers, R.J. Cernik, Sci. Rep. **5**, 15979 (2015), https://doi.org/10.1038/srep15979
72. F. Meirer, J. Cabana, Y. Liu, A. Mehta, J.C. Andrews, P. Pianetta, J. Synchrotron Radiat. **18**, 773–781 (2011)
73. P. Boisseau, L. Grodzins, Hyperfine Interact. **33**, 283–292 (1987)
74. R. Cesareo, S. Mascarenhas, Nucl. Instrum. Methods Phys. Res. Sect. A Accel. Spectrom. Detect. Assoc. Equip. **277**, 669–672 (1989)
75. T. Ohigashi, Y. Terada, A. Takeuchi, K. Uesugi, K. Kubokawa, J. Phys. Conf. Ser. **186**, 012093 (2009)
76. T. Ohigashi, N. Watanabe, H. Yokosuka, S. Aoki, in *AIP Conference Proceedings*, vol. 705 (2004), pp. 1352–1355
77. C. Yoon, Y. Kim, W. Lee, IEEE Trans. Nucl. Sci. **63**, 1844–1853 (2016)
78. W. Ludwig, D. Bellet, Mater. Sci. Eng. A **281**, 198–203 (2000)
79. W. Ludwig, E. Pereiro-López, D. Bellet, Acta Mater. **53**, 151–162 (2005)

80. M. Kobayashi, H. Toda. K. Uesugi, T. Ohgaki, T. Kobayashi, Y. Takayama, B.G. Ahn, Philos. Mag. **86**, 4351–4366 (2006)
81. W. Ludwig, E.M. Lauridsen, S. Schmidt, H.F. Poulsen, J. Baruchel, J. Appl. Crystallogr. **40**, 905–911 (2007)
82. W. Ludwig, P. Reischig, A. King, M. Herbig, E.M. Lauridsen, G. Johnson, T.J. Marrow, J.Y. Buffière, Rev. Sci. Instrum. **80**, 033905 (2009), https://doi.org/10.1063/1.3100200
83. K. Hirayama, Y. Sek, T. Suzuki, H. Toda, K. Uesugi, A. Takeuchi, Material, special issue "In-Situ X-Ray Tomographic Study of Materials," (2018), under review
84. A. King, P. Reischig, J. Adrien, W. Ludwig, J. Appl. Crystallogr. **46**, 1734–1740 (2013)
85. S.A. McDonald, P. Reischig, C. Holzner, E.M. Lauridsen, P.J. Withers, A.P. Merkle, M. Feser, Sci. Rep. **5**, 14665 (2015), https://doi.org/10.1038/srep14665
86. H.F. Poulsen, *Three-Dimensional X-Ray Diffraction Microscopy, Mapping Polycrystals and their Dynamics* (Springer Tracts in Modern Physics, Springer, Berlin, 2004)
87. U. Lienert, S.F. Li, C.M. Hefferan, J. Lind, R.M. Suter, J.V. Bernier, N.R. Barton, M.C. Brandes, M.J. Mills, M.P. Miller, B. Jakobsen, W. Pantleon, Advanced materials analysis, part II research summary. JOM **63**, 70–77 (2011)
88. Y. Hayashi, Y. Hirose, Y. Seno, T. Yoshida, SPring-8/SACLA user information, **22**, 8–13 (2017)
89. H. Toda, Y. Ohkawa, T. Kamiko, T. Naganuma, K. Uesugi, A. Takeuchi, Y. Suzuki, M. Kobayashi, Acta Mater. **61**, 5535–5548 (2013)
90. H. Toda, T. Kamiko, Y. Tanabe, M. Kobayashi, D.J. Leclere, K. Uesugi, A. Takeuchi, K. Hirayama, Acta Mater. **107**, 310–324 (2016)
91. C.G. Schroer, M. Kuhlmann, S.V. Roth, R. Gehrke, N. Stribeck, A. Almendarez-Camarillo, B. Lengeler, Appl. Phys. Lett. **88**, 164102 (2006)
92. F. Schaff, A. Malecki, G. Potdevin, E. Eggl, P.B. Noël, T. Baum, E.G. Garcia, J.S. Bauer, F. Pfeiffer, Nature **527**, 353–356 (2015)
93. G.-H. Chen, N. Bevins, J. Zambelli, Z. Qi, Opt. Express **18**, 12960–12970 (2010)
94. T. Mukaide, K. Kajiwara, T. Noma, K. Takada, J. Synchrotron Radiat. **13**, 484–488 (2006)
95. W. Ludwig, P. Cloetens, J. Härtwig, J. Baruchel, B. Hamelin, P. Bastie, J. Appl. Crystallogr. **34**, 602–607 (2001)
96. K. Kajiwara, S. Iida, T. Mukaide, S. Kawada, J. Crystallogr. Soc. Jpn. 12–17 (2012)
97. A. Suzuki, S. Furutaku, K. Shimomura, K. Yamauchi, Y. Kohmura, T. Ishikawa, Y. Takahashi, Phys. Rev. Lett. **112**, 053903 (2014)
98. Y. Takahashi, N. Zettsu, Y. Nishino, R. Tsutsumi, E. Matsubara, T. Ishikawa, K. Yamauchi, Nano Lett. **10**, 1922–1926 (2010)

Chapter 6
X-Ray CT Scanners and Application Examples

The various devices and technical components that compose an X-ray CT scanner were comprehensively discussed in Chap. 4. Furthermore, applied technologies conducted with industrial X-ray CT scanners were discussed in Chap. 5. It is likely exceptionally rare that standard users of X-ray CT scanners would examine and select the various constituent devices upon understanding the various technical components of X-ray tomography and assemble an industrial X-ray CT scanner by themselves. Instead, users are more likely to select commercially available X-ray CT scanners and introduce them to their affiliated institution, contract measurements to public research institutions (e.g. universities) or private measurement contractors, or rent the devices from these institutions.

Commercially available X-ray CT scanners are rapidly advancing, and higher-performance/higher-functionality products are released almost every year. As such, it may appear that providing an overview of the specifications of commercially available X-ray CT scanners may not be appropriate for a technical book that focuses on fundamental technologies as presented here. However, X-ray CT scanner performance cannot be entirely discussed with just the various constituent devices and technical components discussed in Chap. 4. In other words, the methods in which the various component technologies that were not fully discussed in Chaps. 4 and 5 are combined and the techniques that have integrated the various component technologies, careful consideration and innovations, and advanced user-friendly software more often than not stipulate the strengths, versatility, and characteristic applications of the X-ray CT scanner system. Therefore, with the understanding that this book will inevitably become obsolete, the present chapter introduces the latest X-ray CT scanners that are commercially available at the time that this book was written, as well as superior examples of their application. These X-ray CT scanners may appear somewhat outdated with time. However, the technologies and innovations incorporated in these scanners will certainly be effectively applied in subsequent X-ray CT scanners or future developments will be based on them.

H. Toda, *X-Ray CT*,
https://doi.org/10.1007/978-981-16-0590-1_6

In writing this chapter, many of the manufacturers of these X-ray CT scanners were directly approached for information regarding the characteristics of the products from each company and their notable application examples. Support was also provided by the radiation department (department head at the time of writing this book: Masahito Natsuhara, Shimadzu Co.) of the Japanese Inspection Instrument Manufacturers' Association (JIMA), a business group of nondestructive inspection instrument manufacturers in Japan.

6.1 General-Use X-Ray CT Scanners for Industrial Use

Table 6.1 displays specifications of representative industrial X-ray CT scanners. Photographs of industrial X-ray CT scanners with high versatility among the scanners listed in Table 6.1 are shown in Fig. 6.1. Manufacturers have provided detailed specifications, which are normally not included even in catalogs, for this book. Combined with the descriptions of the various constituent devices shown in Chap. 4, we can develop a deeper understanding of industrial X-ray CT scanners.

The 14 types listed in Table 6.1 cover the representative tube voltage range of industrial X-ray CT scanners that use X-ray tubes. Many of these use tungsten as a reflection-type target. However, the cooling methods for tubes vary according to their output. Air-cooling is sufficient for low-output radiation sources but water-cooling and oil-cooling become necessary with higher output. Furthermore, X-ray tubes include sealed and open variations. The open type allows for filaments to be exchanged when they are broken, but the sealed type needs to have the entire tube replaced. In recent years, some manufacturers have provided maintenance services in which the sealed tube is temporarily collected, the filament exchanged, and the tube returned. Materials with a small atomic number (e.g. beryllium) are often used for the radiation window of the X-ray source and the window material can avoid issues caused by the absorption of low-energy X-rays. Furthermore, aluminum is frequently used as a window material for devices that require high-energy X-rays.

Figure 6.2 is a visualization example of fiber-reinforced plastic (FRP) using the X-ray CT scanner (Shimadzu Co. inspeXio SMX225CT FPD HR) shown in Table 6.1 with a tube voltage of 225 kV. This X-ray CT scanner is equipped with a large-scale, high-resolution flat-panel detector (corresponding to 14 million pixels). The tube voltage is relatively high at 225 kV, but low-energy X-rays can be used due to the specialized carbon plate used in the radiation window of the X-ray source. This enables the clear visualization of carbon fiber- or glass fiber-reinforced plastic composites (CFRP and GFRP, respectively) or non-woven fibers. This also is due to the high dynamic range of the incorporated flat panel detector. Comparisons of Fig. 6.2a and b show how much the dynamic range difference (14 bit vs. 16 bit) in a detector plays a role in image quality. These types of fiber orientation measurements in CFRP or GFRP can be easily conducted with optional software. Figure 6.2b is an example of a carbon fiber-reinforced thermoplastic (CFRTP). Industrial X-ray CT scanners can provide a measurement technology environment, which controls the

Table 6.1 Detailed specifications of representative X-ray CT scanners

Manufacturer		SHIMADZU		NIKON		OMRON
Type		inspeXio SMX-90CT Plus	inspeXio SMX225CT FPD HR	XT H 225 ST	XT H 450	VT-X750
X-ray source	Type	Sealed tube	Open tube	Open tube	Open tube	Sealed tube
	Output (W)	−10	−135	225	450	−39
	Tube voltage (kV)	−90	−225	225	450	−130
	Tube current (mA)	0.11	0.6	0–2	0–2	−0.3
	Focal spot size (μm)	5	−4	3	80	
	Maximum beam diameter (μm)	–	7	–	–	–
	Target material	Tungsten	Tungsten	Tungsten, etc	Tungsten	Tungsten
	Target type	Reflective target	Reflective target (cylindrical)	Reflective target	Reflective target	–
	Target angle (°)	45	45	–	–	–
	Emission angle (°)	40	40	–	–	45
	Filter	Various materials	Various materials	Cu, etc	Cu	Various materials
	Window material	Be	Special carbon	Be	Al	Be
	Tube cooling system	Air cooling	Water cooling	Water cooling	Oil cooling	Air cooling

(continued)

Table 6.1 (continued)

Manufacturer		SHIMADZU		NIKON		OMRON
Type		inspeXio SMX-90CT Plus	inspeXio SMX225CT FPD HR	XT H 225 ST	XT H 450	VT-X750
Stage	Load bearing capacity (kg)	2	12	50	100	4 (Transportation conveyor)
	Inclination of a rotation axis	No	Possible (optional)	Possible	No	No
Detector	Type	FPD	FPD	FPD	FPD/CLDA	FPD
	Scintillator	CMOS	Amorphous	–	–	–
	Number of pixels	1000 × 1000	3000 × 3000	2000 × 2000/4000 × 4000	2000 × 2000/4000 × 4000	–
	Pixel size (μm)	50 × 50	139 × 139	200/100	200/100/400	–
	Dynamic range	12 bit	16 bit	–	–	14 bit
Total	Max. spatial resolution (μm)	10	4	–	–	6/pixel
	Offset scan	Possible	Possible	–	–	No
	Volume rendering software	Attached	Attached	–	–	Attached
	Image analysis software	VGStudio	VGStudio	VGStudio	VGStudio	Attached
	Dimensions (mm)	830 × 601 × 587	2170 × 1350 × 1857	2414 × 1275 × 2202	3616 × 1828 × 2249	1550 × 1925 × 1645
	Weight (kg)	250	3100	4200	14,000	2970

(continued)

Table 6.1 (continued)

RX SOLUTIONS		ZEISS		Hitachi	
DeskTom130	EasyTom150	ZEISS Xradia 810 Ultra	ZEISS Xradia 520 Versa	HiXCT-1 M	HiXCT-9 M
Sealed tube	Sealed tube	Open tube	Sealed tube	Compact linear accelerator	Compact linear accelerator
–40	–75	875	10	–	–
–130	–150	35 (Effective value: 5.4 keV)	160	950	9000
–0.3	–0.5	25	0.075	–	–
5	5	75	–	2000	1600
–	–	75	–	–	–
Tungsten	Tungsten	Chromium	Tungsten	Tungsten	Tungsten
Reflective target	Reflective target	Rotating anode	Transmissive target	Transmissive target	Transmissive target
–	–	–	0	–	–
–	–	–	–	45	30
Various materials	Various materials	–	Various materials	No	No
Be	Be	Be	Diamond	–	–
Air cooling	Air cooling	Water cooling	Air cooling	Air cooling	Water cooling
2	30	1	25	100	100
Possible (optional)	Possible (optional)	–	–	–	–
FPD	FPD	FZP + optical lens + CCD	Optical lens + CCD (FPD)	Line sensor	Line sensor
CsI or Gadox	CsI or Gadox	– (Patented technology)	– (Patented technology)	Si semiconductor	Si semiconductor
1920 × 1536	1920 × 1536	1024 × 1024	2048 × 2048 (CCD)	750–	750–
127 × 127	127 × 127	–	–	–	–

(continued)

Table 6.1 (continued)

RX SOLUTIONS		ZEISS		Hitachi	
DeskTom130	EasyTom150	ZEISS Xradia 810 Ultra	ZEISS Xradia 520 Versa	HiXCT-1 M	HiXCT-9 M
16 bit	16 bit	16 bit	16 bit	16 bit	16 bit
4	5	0.05	0.7	200	200
Possible	Possible	–	Possible	Possible	Possible
Attached	Attached	Attached	Attached	–	–
VGStudio, VGStudioMAX	VGStudio, VGStudioMAX	ORS Dragonfly Pro (optional)	ORS Dragonfly Pro (optional)	VGStudio	VGStudio
1250 × 800 × 1800	2100 × 1100 × 2000	2180 × 1200 × 2170	2170 × 1190 × 2090	830 × 601 × 587	6000 × 6000 × 6000
700	2000	2600	2468	60,000–80,000 with radiation shielding	15,000 without radiation shielding

JED	RF	Yamaha motor
CTH200 FPD	NAOMi-CT	YSi-X TypeHD
Sealed tube	Sealed tube	Sealed tube
−150	−500	−39
−200	50–100	−130
−0.75	2–10	−0.3
20	500	−16
–	–	−50
Tungsten	Tungsten	Tungsten
Reflective target	Reflective target	Reflective target
–	5	45
40	–	100

(continued)

Table 6.1 (continued)

JED	RF	Yamaha motor
Various materials	Cu	Various materials
Special glass	–	Be
Air cooling	Oil/air cooling	Air cooling
10	10	2 (Transportation conveyor)
No	No	No
FPD	FPD	FPD
Amorphous	CSI	–
1025 × 1025	1232 × 1216	–
200 × 200	100	–
16 bit	12 bit	14 bit
–	5 LP/mm	7
Possible	Possible	No
Attached	Attached	Attached
Attached	Attached	Attached
1550 × 1092 × 1413	623 × 338 × 297.5	1710 × 1883 × 1705
1860	50	2900

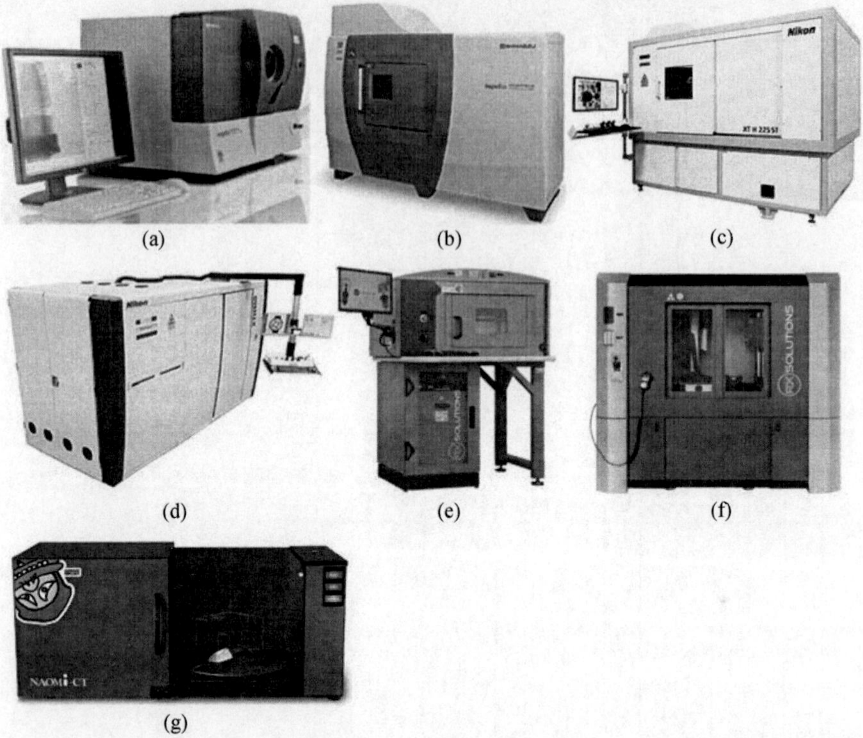

Fig. 6.1 Photographs of commercially available X-ray CT scanners described in Table 6.1; **a** Shimadzu Co. inspeXio SMX-90CT Plus model from Table 6.1 and **b** inspeXio SMX225CT FPD HR model (both courtesy of Masahito Natsuhara of Shimadzu Co.); **c** and **d** are the Nikon XT 225 ST model and XT H 450 model, respectively, from Table 6.1 (both courtesy of Satoshi Kazama of Nikon); and **e** and **f** are RX SOLUTIONS DeskTom 130 model and EasyTom 150 model from Table 6.1 (both courtesy of Wataru Adachi of Seikoh Giken). **g** RF Co. NAOMi-CT model (courtesy of Kazumi Kodaira of RF Co.)

important factor of fiber misalignment when handling CFRTP. As shown in Fig. 6.3, other applications of this X-ray CT scanner include the deformation analysis of lithium-ion batteries and high-spatial-resolution/high-contrast observations of separators. High spatial resolution and high contrast can be obtained in this way. Additionally, fast tomography with a maximum speed of 33 s/scan and fast image reconstruction in as fast as 5 s has become possible with detector improvements and software updates.

Figure 6.4 shows imaging conducted on pearls. This was conducted using the RX SOLUTIONS device (DeskTom130) shown in Table 6.1. A pearl is a light substance with inorganic materials like calcium carbonate and calcium oxide as its principal components. Figure 6.4 shows that the pearl core and pearl layers, as well as its boundary layers, were clearly visualized. We can observe that the pearl core/pearl layer and boundary layer, which have minimal compositional differences, are clearly

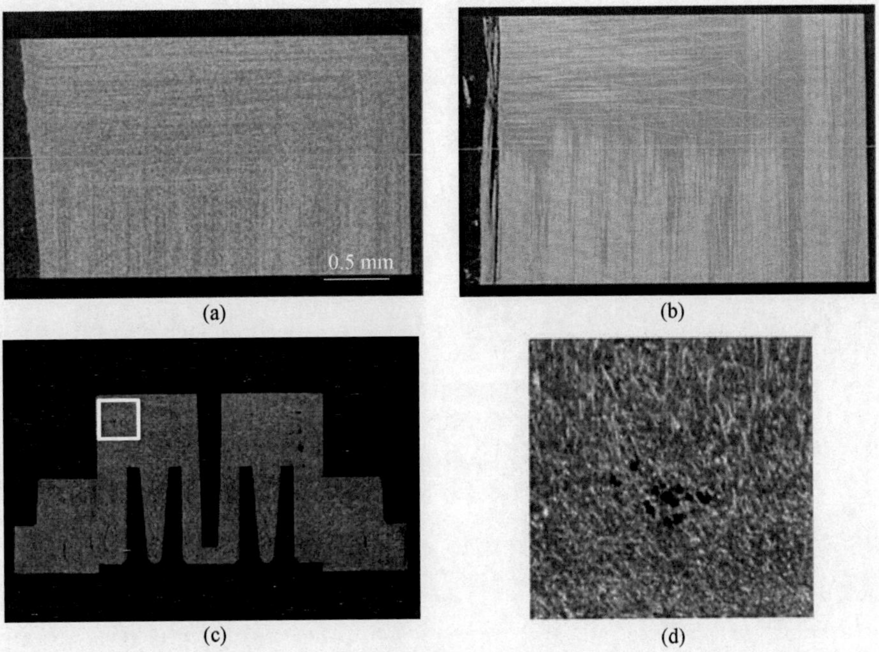

Fig. 6.2 **a** Cross-section image when visualizing CFRP using an X-ray CT scanner with a 14-bit flat panel detector; **b** is a visualization example using an X-ray CT scanner with a 16-bit flat panel detector described in Table 6.1 (Shimadzu Co. inspeXio SMX225CT FPD HR model); **c** visualization of GFRP using the same device; and **d** magnification of the enclosed area in **c** (both courtesy of Masahito Natsuhara of Shimadzu Co.)

distinguished. This visualization demonstrates how the boundary layer is broken and how this results in the deformation of the pearl surface.

Many X-ray CT scanners can handle imaging regions of interest that are considerably smaller than the sample diameter. This enables the local/high-magnification observation of large samples, which do not fit within the field of view within a detector, such as electric circuit boards. Figure 6.3 is one such observation example. Furthermore, Fig. 6.5 shows an observation of a soldered section of an electronic circuit board using the RX SOLUTIONS machine (DeskTom130) shown in Table 6.1. The existence of pores in the solder, which are likely to result in fracture, can be confirmed using the microfocus tube.

Many of the devices in Table 6.1 can handle offset scans. As described in Sect. 3.3.5 (1), offset scans can be used to magnify the sample size by integer multiples of the X-ray beam width. In this case, the included software can automatically stitch together multiple transmission images due to the high-accuracy offset. However, care must be taken concerning the penetrative power of the X-ray.

Incidentally, it can be observed from Table 6.1 that many X-ray CT scanners use a microfocus X-ray tube. The need for high spatial resolution is also firmly rooted in industrial X-ray CT scanners. There have been products that have two tubes with

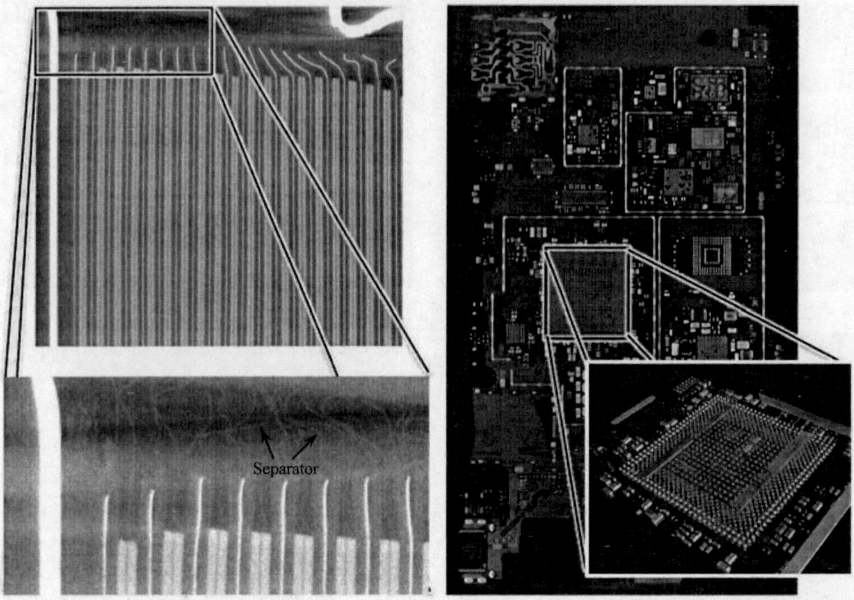

(a) (b)

Fig. 6.3 **a** Example of a high-resolution/high-contrast observation of lithium ion battery deformation and the separators and **b** an observation of the electronic circuit board of a smartphone and a high-magnification observation of the enclosed CPU. Each are visualization examples using the same X-ray CT scanner (Shimadzu Co. inspeXio SMX225CT FPD HR model) from Fig. 6.2 (courtesy of Masahito Natsuhara from Shimadzu Co.)

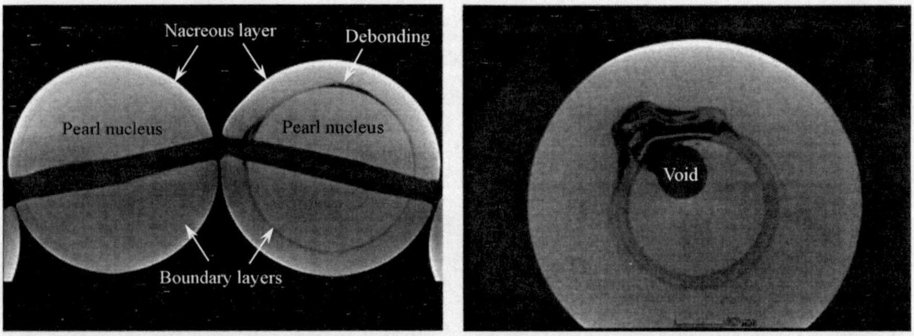

(a) (b)

Fig. 6.4 Interior structure of a pearl observed using an RX SOLUTIONS device (DeskTom130) in Table 6.1. The pearl on the right in **a** and the pearl in **b** show damaged boundary layers between the pearl core and pearl layers (courtesy of Wataru Adachi from Seikoh Giken)

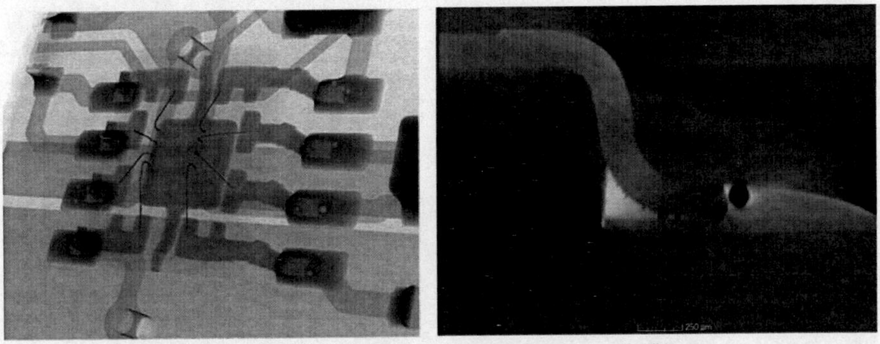

(a) (b)

Fig. 6.5 **a** Transmission image of an electronic circuit board after the completion of soldering; **b** is a virtual cross-section observed from the side in a 3D image of a soldering section. The device used is the RX SOLUTIONS DeskTom130 model shown in Table 6.1 (courtesy of Wataru Adachi of Seikoh Giken)

varying focal spot size/output values installed, also referred to as multifocus. Its applications are innumerable, from the currently-expanding field of layered manufacturing to assessment of the aforementioned CFRP and GFRP fiber orientation, to various assessments in archaeology. For example, high-spatial-resolution imaging with low-energy X-rays in the archaeological disciplines can be used to determine whether an ancient tree was a coniferous or broadleaf tree and can even be used to analyze what the tree species was. The focal spot size of X-ray tubes and maximum spatial resolution of X-ray CT scanners are shown in Table 6.1. Their current state is such that the former is generally used as an index for spatial resolution. However, it must be noted that the focal spot size of X-ray tubes is but one of the factors that regulate the spatial resolution of X-ray CT scanners. Data provided directly from the manufacturer are shown here, and there is no follow-up on how the maximum spatial resolution of the device is measured or assessed. Those who have read this book would likely understand what spatial resolution is and how this is measured. It is recommended that the cross-sectional images of a sample that approximates the desired 3D image and imaging conditions be obtained and evaluated independently.

Figure 6.6 shows an observation example of an archaeological artifact excavated from the ancient ruins of Antikythera in Greece using an X-ray CT scanner (Nikon XT H 450 model) with a 450 kV tube voltage that enables the use of X-rays with high energies even among those in Table 6.1. This device is unique in that it is capable of having an X-ray tube with a relatively small focal point size despite having a tube voltage of 450 kV. For this reason, a high spatial resolution is achieved despite the high X-ray energy and output, where a maximum spatial resolution of approximately 200 μm is achieved. Usage of this X-ray CT scanner has enabled the visualizations of detailed patterns that were buried under rust on the surface of relatively large artifacts. These observations clarified that this archaeological artifact with a previously-unknown use was the world's oldest astronomical calendar. This

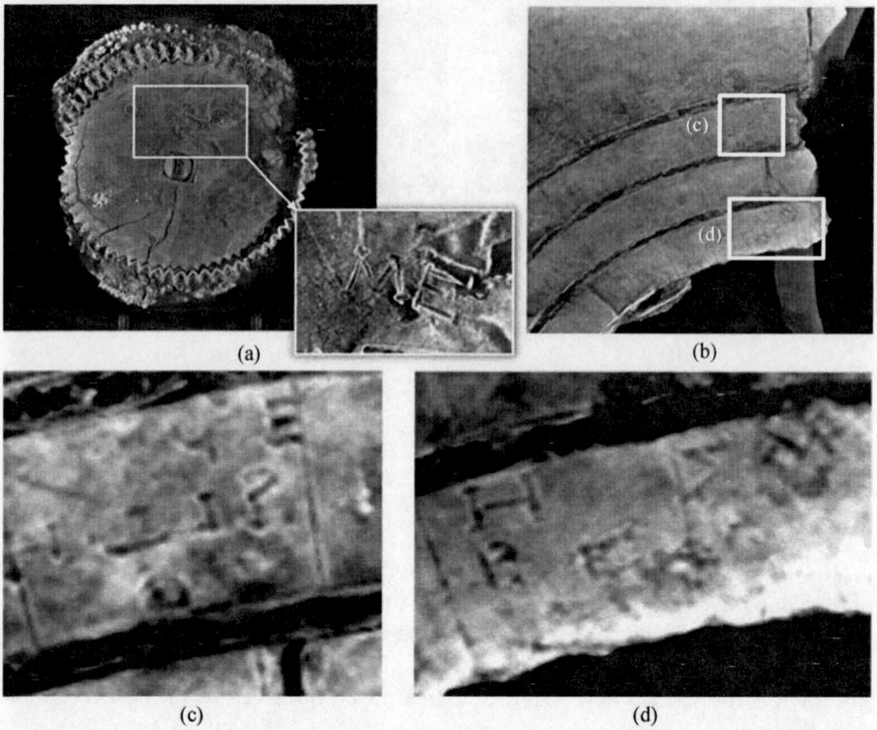

Fig. 6.6 **a** Full view of the excavated artifact from the ancient ruins of Antikythera in Greece; **b** shows a section of this; **c** and **d** are further magnifications of the section in (**b**); and (**c**) and (**d**) show patterns that indicate that this artifact was a calendar. These were visualized using a 450 kV tube voltage X-ray CT scanner (Nikon XT H 450 model) described in Table 6.1 (courtesy of Satoshi Kazama of Nikon)

is considered a good example of how industrial X-ray CT has provided valuable academic contributions.

Figure 6.7 shows an application example of an X-ray CT scanner with a slightly smaller tube voltage of 300 kV. This device has a microfocus tube whose distance between the X-ray focal spot and X-ray irradiation window is shorter at roughly 5 mm and whose automatic changes in focal spot dimensions, as per the cathode power of the X-ray tube, achieve a balance between X-ray penetrative power and relatively high spatial resolution. Figure 6.7 shows 3D images of gray cast iron and ductile cast iron materials where a specimen measuring □ approximately 2 mm was extracted. The graphite portions of each of the materials are extracted and displayed here. The thin and long flaky graphite is visualized with sufficient spatial resolution.

Many of the devices in Table 6.1 can handle inclinations in the sample rotation stage. This enables the implementation of the laminography discussed in Sect. 3.3.5 (3). In practice, many X-ray CT scanners with software that can handle laminography are commercially available, enabling imaging of plate-like samples that do not fit

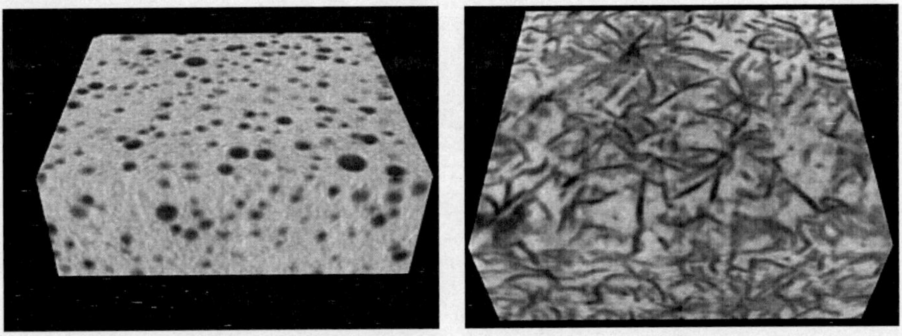

(a) (b)

Fig. 6.7 Graphite extracted from **a** spherical graphite cast iron and **b** flaky graphite cast iron. The microscale flaky graphite is visible. A 300 kV tube voltage X-ray CT scanner (Toshiba IT & Control Systems TOSCANER-33000μFD-ZII model) was used for observation (courtesy of Junichi Iwasawa of Toshiba IT & Control Systems)

within the X-ray beam range in standard imaging methods, such as electronic circuit boards, aluminum or steel body panels for automobiles, and fiber metal laminate thin plates for airplanes.

As shown in the detector row in Table 6.1, many products use a flat-panel detector. Image intensifiers typically offer the advantages of relatively low cost, high efficiency, and a high frame rate. However, flat-panel detectors have the advantages of a high dynamic range, few distortions such as with image intensifiers, and a large effective field of view. Combined with recent price reductions, flat-panel detectors are increasingly used. Furthermore, the 450-kV tube voltage device (Nikon XT H 450 model) in Table 6.1 also has a model with a line sensor camera. As discussed in Sect. 4.4.2 (1), the X-ray beam is formed into a fan beam using the collimator when using the line sensor camera to have the X-ray illuminate the sample only within the field-of-view of the line sensor camera. Forward scattering of high-energy X-rays in the case of a parallel beam or cone beam results in the emission of scattered X-rays into the detector pixel from the whole sample; this reduces contrast and spatial resolution. The use of a fan beam can effectively prevent this.

Figure 6.8 is a schematic that shows how image quality changes when either a line sensor camera or flat-panel detector is used. Imaging with a line sensor camera requires vertical scanning of the sample across 500–2000 layers; consequently, the imaging time necessarily becomes longer. However, the influence of scattering X-rays coming from the sample regions in the vertical directions is no longer present and high-quality images with low noise levels can be obtained. However, the scattered X-rays coming from adjacent regions in the longitudinal direction of the line sensor cannot be avoided, therefore, adjustments in the sample-detector interval are necessary.

Figure 6.9 shows an observation example of an aluminum die-casting cylinder block using a line sensor camera. X-ray scattering at high X-ray energies from cone beam-type X-ray CT scanners can make the linear profile sections swollen in a

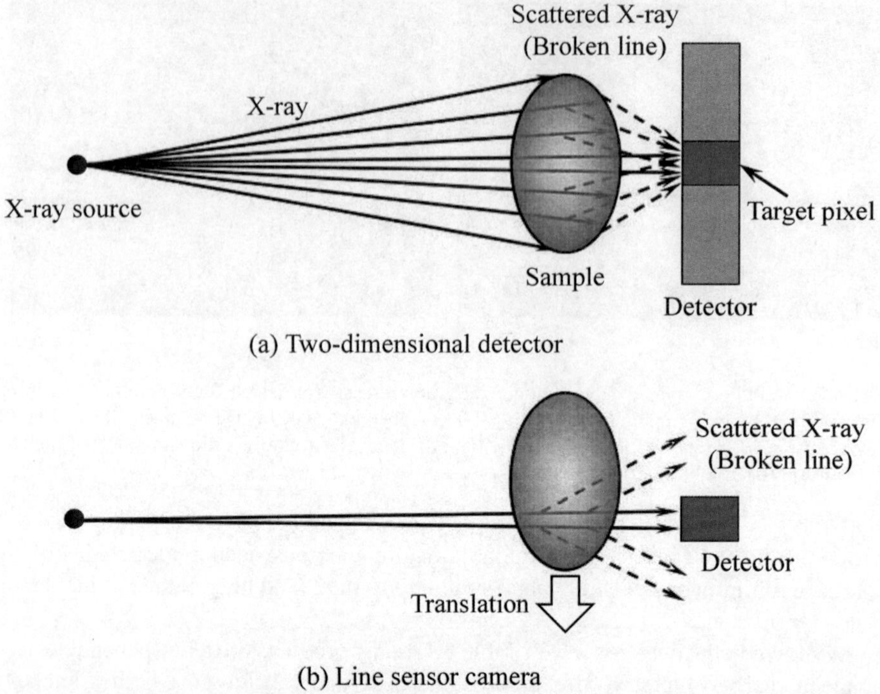

Fig. 6.8 Schematic showing differences in image quality between a line sensor camera device and a two-dimensional detector device (e.g. flat panel detector) when imaging with high-energy X-rays

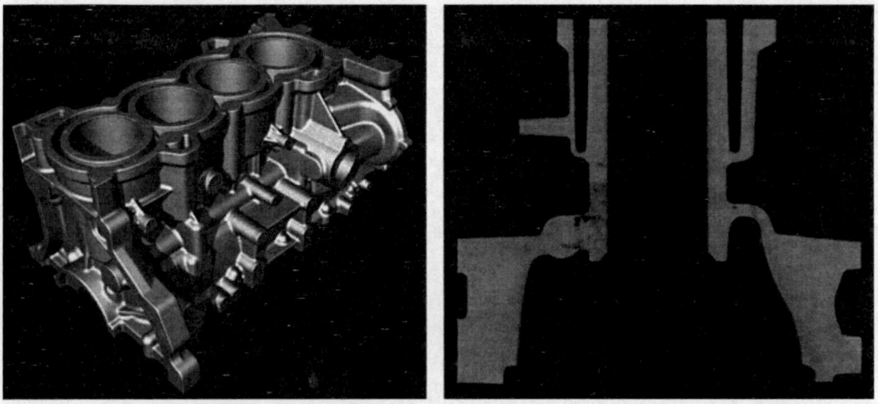

Fig. 6.9 Visualization example of an aluminum die-casting cylinder block; **a** is the total image and **b** is a virtual cross-section, which visualizes manufacturing defects such as internal cavities. A 450 kV tube voltage X-ray CT scanner (Toshiba IT & Control Systems TOSCNANER-24500twin model) was used for observations (courtesy of Junichi Iwasawa of Toshiba IT & Control Systems)

barrel-like shape. Line sensor cameras can ensure that geometrically accurate 3D images with no distortion can be obtained. Two rows of line sensor cameras are set up in the X-ray CT scanner and a third cross-section image can also be measured by supplementing the space between these two rows. Furthermore, increasing the scintillator thickness used for the line sensor camera can enable the acquisition of approximately 300 mm of transmission for aluminum with a sufficient S/N ratio. For this reason, aluminum products such as those shown in Fig. 6.9 can detect drill holes measuring φ0.3 mm in the interior of areas with an outer diameter of 100 mm. This is due to the utilization of an X-ray tube with a high target power tolerance for the focal point dimensions.

Among the devices listed in Table 6.1, there are some low-cost devices in the range of tens of thousands of USD made possible by technology transfers from the diffusion of dentistry X-ray CT scanners and the appropriation of their components. Figure 6.1g is one such example. Its characteristics include a lightweight main body at 50 kg, a size that can be placed on a desktop, and its compatibility with household power sources while being moved around on a cart. As shown in Fig. 6.10, application examples include various advances in disciplines where prohibitively expensive industrial X-ray CT scanners could not be frequently used, such as fundamental research on food product development, sports applications, and the confirmation of sport glove fit.

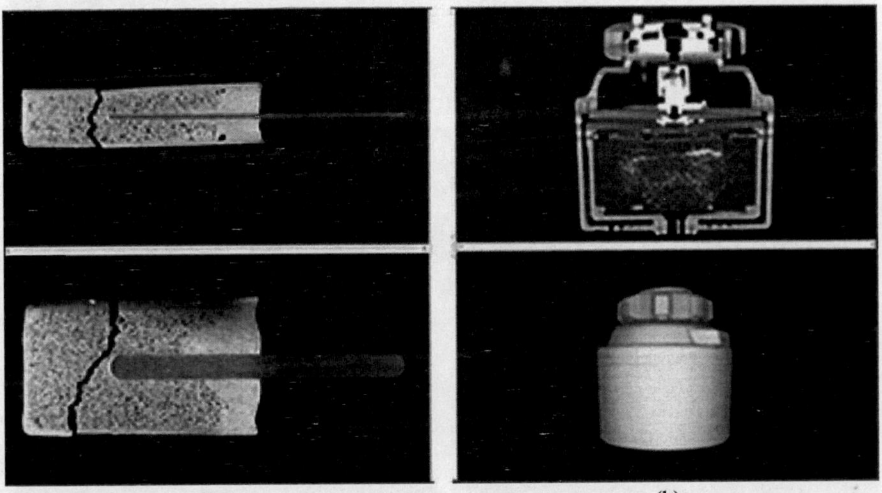

(a) (b)

Fig. 6.10 **a** Distribution of bubbles within a bar of ice cream. **b** is the observation of the internal structure of a water purifier. Miniature light-weight devices (RF Co. NAOMi-CT model) were used for both observations (courtesy of Kazumi Kodaira of RF Co.)

6.2 High-Energy Industrial X-ray CT Scanners

X-ray CT with X-ray energies in the MeV range has been developed and manufactured since the 1980s. Compact electron accelerators have been used as their radiation source. Devices that use radiation sources with energies of 950 kV, 3 MV, 6 MV, and 9 MV can be utilized as industrial products. Figure 6.11 shows an example of a commercially available product. These product specifications are recorded in Table 6.1. Devices with 12-MV radiation sources had been produced in the past but

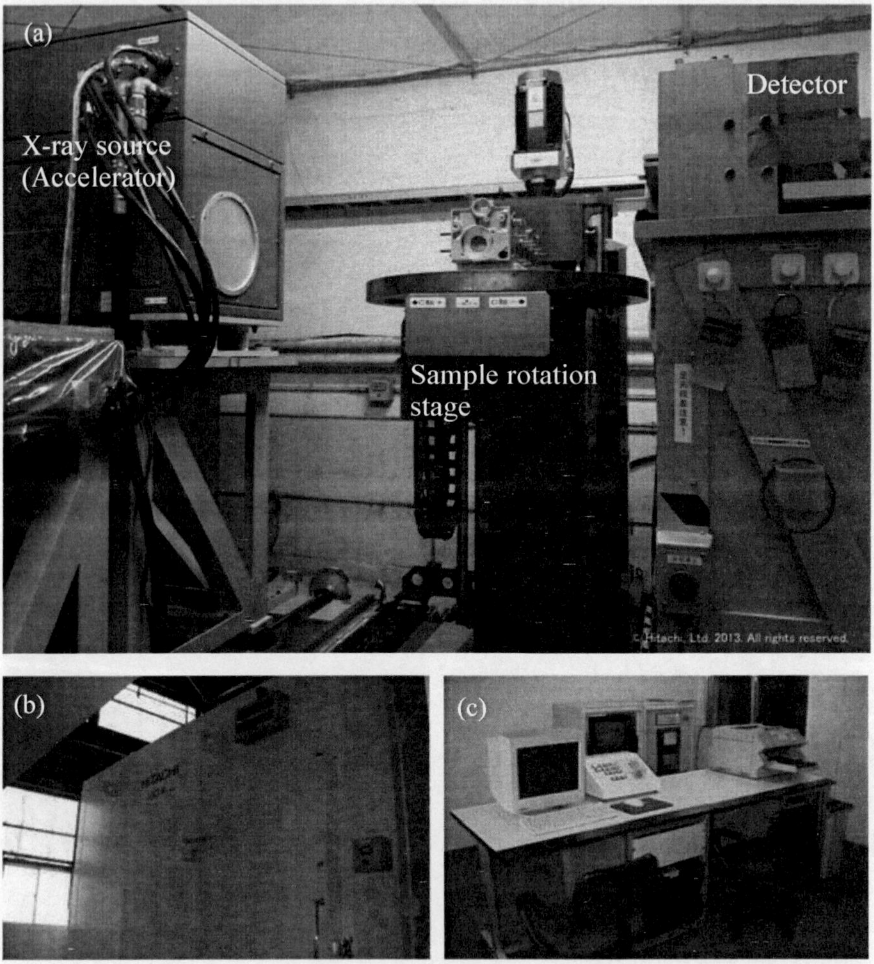

Fig. 6.11 Example of a high-energy industrial-use X-ray CT scanner that is commercially available and described in Table 6.1. Photographs of the **a** Hitachi Co. HiXCT-9M model device and the HiXCT-1M, **b** shielding container, and **c** control panel/operating device (courtesy of Katsutoshi Sato of Hitachi Co.)

currently no longer exist. X-rays with MeV-level energies have a high penetrating power of materials, so decreased sensitivity in standard X-ray detectors is unavoidable. For this reason, line sensor detectors or detection circuits that are specialized for X-rays of MeV-level energies have been developed and implemented. This enables the acquisition of images with minimal artifacts due to low noise and a high dynamic range. The shape of the X-ray beam is a fan beam. For example, X-ray CT scanners with a drive voltage of 9 MV can observe steel and aluminum alloy materials with thicknesses up to 32 cm and 96 cm, respectively. The field of view is φ600–800 mm, the height 500–1000 mm, and a sample rotation stage with a load capacity of 100 kg is used. Proper shielding is necessary to prevent the leakage of high-energy X-rays in high-energy X-ray CT scanners out of the device. For example, the total weight of a 950 kV device can be up to 60–80 t when set up in a shielding container made of steel plates. Furthermore, the 9 MV machine can no longer be shielded with steel plates and must be housed in a reinforced-concrete shielding structure with 2 m thick concrete.

In addition to the so-called third-generation CT with a rotate/rotate system, measurement systems include the second-generation translate/rotate system and a new system referred to as the double-rotate system [2]. Fast imaging is characteristic of the rotate/rotate system, with speeds up to 10 s/slice. Meanwhile, the translate/rotate system is suited for imaging with a wide field of view but has a lengthy imaging time. Furthermore, the double-rotate system combines fast imaging and high-spatial-resolution imaging, requiring an imaging time that is approximately four times longer than the rotate/rotate system but with a spatial resolution that is superior by a factor of 1.5.

Figure 6.12 shows differences in image quality between scan systems. Innovations in the measurement system enable the clean visualization of difficult-to-detect

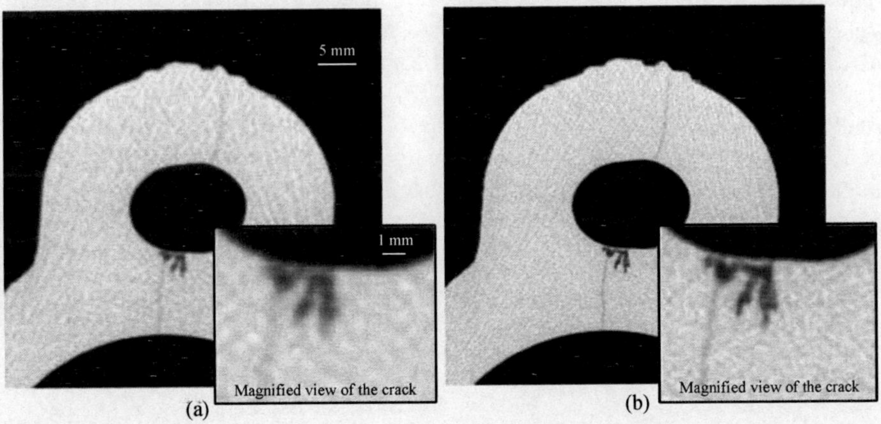

Fig. 6.12 Observation example of a cavity and crack in a cast iron component using a high-energy industrial-use X-ray CT scanner. Image **a** was obtained using the rotate/rotate system (pixel size 0.4 mm, 1500 × 1500 pixels, scan time 15 s) and **b** was obtained with the double-rotate system (pixel size 0.2 mm, 3000 × 3000 pixels, scan time 50 s). These were imaged using the Hitachi Co. HiXCT-9M-SP model (courtesy of Katsutoshi Sato of Hitachi Co.)

(a) (b)

Fig. 6.13 Observation example of an automobile component (alternator) interior using a high-energy industrialuse X-ray CT scanner; **a** is a 3D observation of the inner coil by making the external aluminum components semi-transparent and **b** is the segmentation of the constituent components, where each component is shown in a disassembled manner to show the assembly conditions. These were imaged using the Hitachi Co. HiXCT-9M-SP model (courtesy of Katsutoshi Sato of Hitachi Co.)

cracks with small openings in brittle cast iron parts. Furthermore, Fig. 6.13 shows a 3D imaging example of an automobile part. The noise and artifacts are sufficiently reduced while using high-energy X-rays, enabling not only the visualization of large-scale automobile part interiors but also assessments based on the segmentation and extraction/discrimination of its constituent components.

Fraunhofer EZRT operates an XXL-CT X-ray CT scanner, which can scan larger-scale products as is. This is shown in Fig. 6.14. XXL-CT is a device capable of conducting 3D imaging of automobiles, cargo containers, airplane fuselages, and large-scale ship engines as is. A rotation stage, which has a width of 3.2 m, height of 5 m, and load capacity of 10 t, is combined with 9 MeV high-energy X-rays to conduct 3D imaging with a focal spot size of approximately 3 mm. Automobiles are positioned upright on the rotation stage when conducting 3D imaging. It is anticipated that the applications of these types of CT scanners for large-scale structures will increase in the future.

Fig. 6.14 X-ray CT scanner comprising an XXL-CT that can scan large automobile-scale products as-is, located in Fraunhofer EZRT in Germany (courtesy of Thomas Kessler of the Fraunhofer Laboratory)

6.3 High-Resolution Industrial X-Ray CT Scanners

Industrial X-ray CT scanners that seek high spatial resolutions using standard microfocus tubes are introduced in Sect. 6.1. Industrial X-ray CT scanners that use X-ray focusing elements used in imaging-type X-ray micro-/nanotomography discussed in Sect. 5.1 are commercially available.

Figure 6.15 shows a visualization example of active materials in an electrode sheet of a commercially available secondary lithium-ion battery using an industrial X-ray CT scanner equipped with an imaging optical system that uses a Fresnel zone plate (introduced in Sect. 5.1.1) and an illumination system that uses a condenser zone plate. The internal 3D structure and voids between the particles can be clearly observed in Fig. 6.15. Observations and quantitative assessments of the opened and closed micropores are conducted from these images, as well as observations of internal structures closely associated with the performance and deterioration of the secondary lithium-ion battery including the debonding of the grain boundary between the primary particles accompanying the storage and discharge of lithium ions during lithium-ion battery charging and discharging. The device in Fig. 6.16, also recorded in Table 6.1, was used for this observation. This device uses a Fresnel zone plate with an outermost zone width of 35 nm and a nominal maximum spatial resolution of 50 nm. Observations in Fig. 6.15 show that the sample has a diameter of approximately 64 μm and a spatial resolution of 150 nm during imaging. The manufacturer has referred to this as a nano-resolution X-ray microscope and interpolates 3D imaging in the intermediate region between the micron-level spatial resolution

Fig. 6.15 Example where an electrode sheet was extracted from a commercially-available secondary lithium ion battery and where high-resolution visualizations of the 3D structure of active materials in an electrode (left) and inter-particle voids (right) were conducted. Imaging was conducted with a Carl Zeiss ZEISS Xradia 810 Ultra model (Fig. 6.16) (courtesy of Nobuhiro Hayami, Benjamin Hornberger, Stephen T. Kelly, and Hrishikesh Bale of Carl Zeiss Co.)

of microfocus X-ray CT and the nano-order region of electron microscopes. This device produces 3.5 keV monochromatic X-rays by using an approximately 0.9 kW high-brilliance X-ray source. According to the device manufacturer, chrome is used for the X-ray tube target and monochromatic characteristics with an energy resolution of approximately $\Delta E/E = 1.7 \times 10^{-3}$ are obtained through the combination of $K_{\alpha 1}/K_{\alpha 2}$ X-rays and a Fresnel zone plate. This enables an image optical system, which uses a Fresnel zone plate, to be achieved in commercially available industrial X-ray CT scanners. A phase plate is inserted in the optical path to set up a Zernike phase contrast microscope (discussed in Sect. 5.2.2). A stage with an in situ observation device introduced in Sect. 4.5 can be used for the sample rotation stage and in situ observations of tension, compression, and nano-indentations can be made. There are considerable limits with regards to sample size, however, this has been applied in a wide range of disciplines, including material development (e.g. polymers, ceramics, and batteries), natural resource exploration (e.g. oil and minerals), and bioengineering.

Fig. 6.16 Example of a commercially-available high-resolution industrial-use X-ray CT scanner described in Table 6.1. This is the Carl Zeiss ZEISS Xradia 810 Ultra model (courtesy of Nobuhiro Hayami of Carl Zeiss Co.)

6.4 High-Functionality Industrial X-Ray CT Scanners

The previous section detailed how considerably high spatial resolutions could be obtained in absorption contrast tomography using industrial X-ray CT scanners equipped with a microfocus X-ray tube. However, there are occasionally instances with various observation subjects where the contrast is low and observations cannot be made or where there are problems with regards to quantitative assessment. Of the polycrystalline structure tomography methods introduced in Sect. 5.5, the diffraction contrast tomography in Sect. 5.5.2 can be implemented at the laboratory level by using products that incorporate modules and analysis software specific to industrial X-ray CT scanners.

Figure 6.17 shows a 3D visualization example of a crystallographic structure of polycrystalline iron. A clean 3D visualization is obtained without gaps or overlaps in the crystallographic grains, which have a crystallographic grain diameter of approximately 100 μm. This enables not only the accurate determination of the 3D

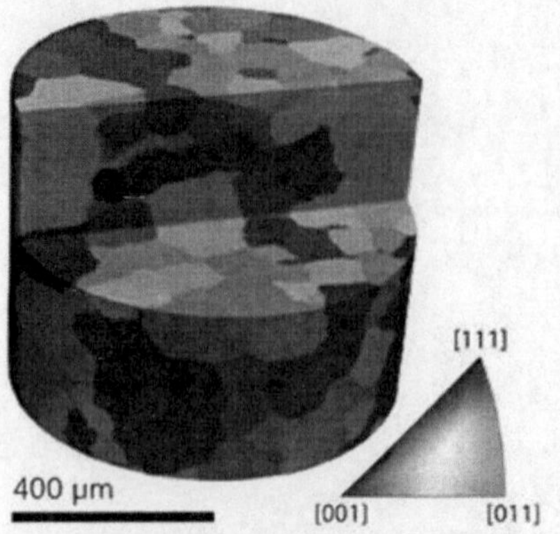

(a) 3D polycrystalline structure and corresponding inverse pole figure

(b) Halved 3D image

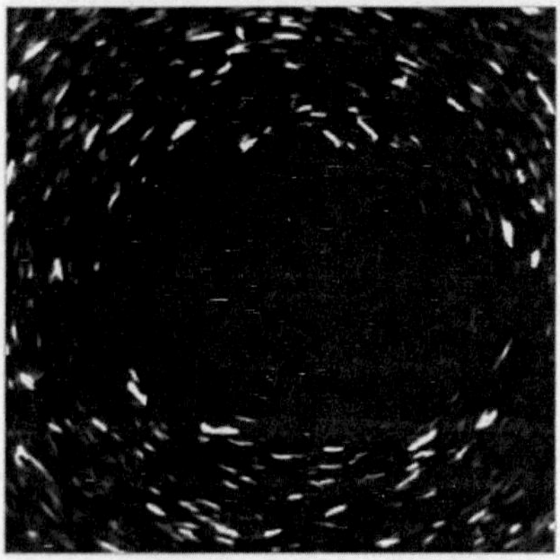

(c) Diffraction pattern of grains

Fig. 6.17 **a** and **b** are 3D images of the polycrystal structure of polycrystalline iron. The colors of each crystallographic grain indicate the crystal orientations expressed in the inverse pole figure included in **a**. The Carl Zeiss ZEISS Xradia 520 Versa model (Table 6.1 and Fig. 6.18 for reference) was used for imaging (courtesy of Nobuhiro Hayami of Carl Zeiss Co.)

morphology of the crystallographic structure but also crystallographic assessments of various physical phenomena (e.g. recrystallization/growth behavior, relationships between crystallographic structure and various mechanical properties/corrosion characteristics).

Figure 6.18 shows a photograph of the industrial X-ray CT scanner used for imaging in Fig. 6.17. The basic structure of the device is the same as that of an industrial X-ray CT scanner with a standard microfocus X-ray tube. This device adopts a scintillator and a detector that uses an objective lens from an optical microscope in a manner similar to synchrotron radiation X-ray tomography. In other words, the device combines geometric X-ray magnification projection using a microfocus X-ray

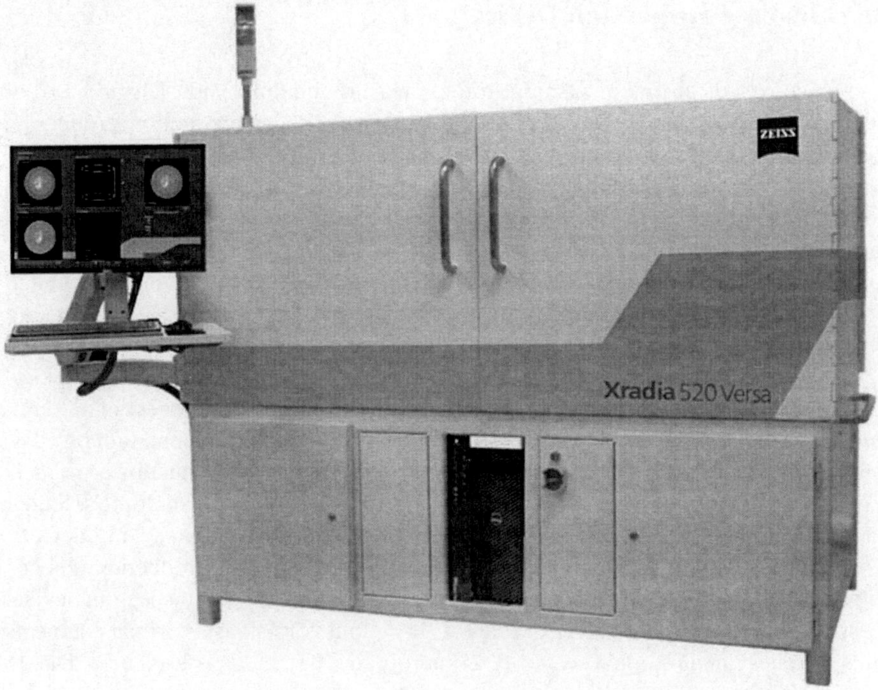

Fig. 6.18 Example of a commercially available high-functionality industrial-use X-ray CT scanner described in Table 6.1. This is the Carl Zeiss ZEISS Xradia 520 Versa model (courtesy of Nobuhiro Hayami of Carl Zeiss Co.)

tube with an optical image converted to visible light. A geometric magnification in the order of 100:1 is usually necessary to achieve a spatial resolution below 10 μm with micro-tomography using a microfocus X-ray tube. Meanwhile, this device can control the geometric magnification based on the cone beam optical system to a factor of 1–10 and achieves a larger working distance by magnifying after visible light conversion. This enables the implementation of material tests that use relatively large testing devices and in situ observations under heating or electrification even among the in situ observation devices introduced in Sect. 4.5. An aperture plate and beam stop can be installed during diffraction contrast tomography to detect only diffracted light while shielding 0th-order light. An X-ray diffraction pattern as is shown in Fig. 6.17c is then obtained while rotating the sample. This allows for the acquisition of geometric information such as crystal coordinates, orientation, size, and morphology of crystallographic grains with a minimum grain size of around 40 μm and an angular resolution of the crystallographic orientation of 0.5°. Standard absorption contrast tomography imaging can also be conducted in the same sample; therefore, the relationships between the crystallographic structure obtained from diffraction contrast tomography and the cracking obtained from absorption contrast tomography can be directly assessed as well.

6.5 In-Line Inspection Device

The objectives of in-line inspections are to measure the dimensions/shape of products, components, and substrates that move down a mass-production production line, identify the presence of defects, and conduct quality control. If these can be conducted for the inspection of all products, the delivery of defective products can be prevented, significantly impacting product reliability. In addition to human-based inspections, in-line inspections until now have involved external appearance inspections with visible-light cameras, transmission image inspections using X-rays, and robot-based measurements. Line sensor cameras introduced in Sect. 4.4.2 (1) are widely used for in-line inspections based on transmission images.

Since the start of the twenty-first century, X-ray CT scanners and similar technologies have been used for in-line inspections during the production process of automobile parts, semiconductors, and electric circuits. The wafer implementation process, which falls under the post-processing step of the semiconductor production process, is the primary subject for these types of technology. Furthermore, the multiple soldered parts in electric circuits (e.g. solder wetting defects, through-hole packing defects, or insufficient soldering amounts) are frequently the source of production defects. Figure 6.19 is an example of such a fast X-ray CT-type automatic inspection device. Its specifications are described in Table 6.1. Its output is small as its primary subjects are substrates; an air-cooled sealed-type microfocus X-ray tube is used here. Due to

Fig. 6.19 Photograph of an in-line inspectionuse X-ray CT scanner described in Table 6.1; Omron VT-X750 model (courtesy of Nobuharu Sugita of Omron)

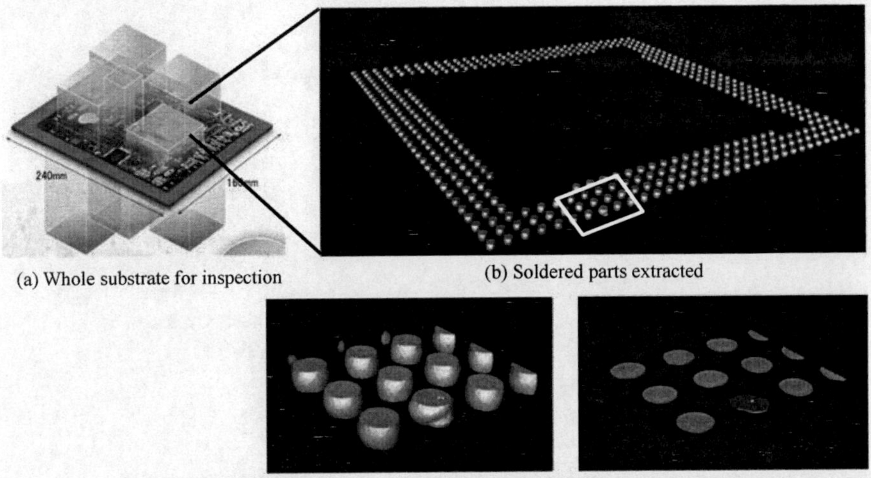

(a) Whole substrate for inspection (b) Soldered parts extracted

(c) Magnified views of the solders highlighted by a white rectangle shown in (b)

Fig. 6.20 Inspection example of a soldered section of the electric circuit using a commercially available in-line inspection X-ray CT scanner (Omron VT-X750 model) shown in Table 6.1 (courtesy of Nobuharu Sugita of Omron)

its high spatial resolution, this single unit is capable of conducting not only mass-production inspections but also analyses of defective products diverted as a result of mass-production inspections. This device is unique in that it applies a specialized measurement system referred to as parallel CT [1]. This involves the turning of the camera and sample in a circular trajectory that is parallel to the substrate instead of a rotation stage. This enables fast tomography wherein a single 3D image can be obtained in 4 s. Although the device in Fig. 6.19 is for in-line inspection, there is also a high-spatial-resolution (0.3 μm/pixel) device for off-line inspection. Figure 6.20 shows the verification of soldering junction strength using the device in Fig. 6.19. A single defective part was identified from several hundred soldered sections, in which pore-like defects were visible in its interior.

Figure 6.21 also shows an inspection device that is specialized for the in-line inspection of electric circuits, with a structure that combines X-ray tomography, X-ray transmission inspections, visible-light/infrared microscopy observations, and laser distance measurements [3]. The in-line setup of this device in an electric circuit production line enables X-ray tomography observations at a spatial resolution that is high enough for detailed soldering junction inspections while simultaneously identifying marks due to visible/infrared light and X-ray observation reference plane measurements using a laser range finder. This device has the characteristic of being able to conduct fast imaging that can keep up with the production line cycle time despite taking such a high number of measurements simultaneously. Figure 6.22 is a summary of the information obtained from each of these inspections. This device is capable not only of detecting various production defects such as the presence of voids in a heat sink or back fillet of leads, but also various defective product ejections

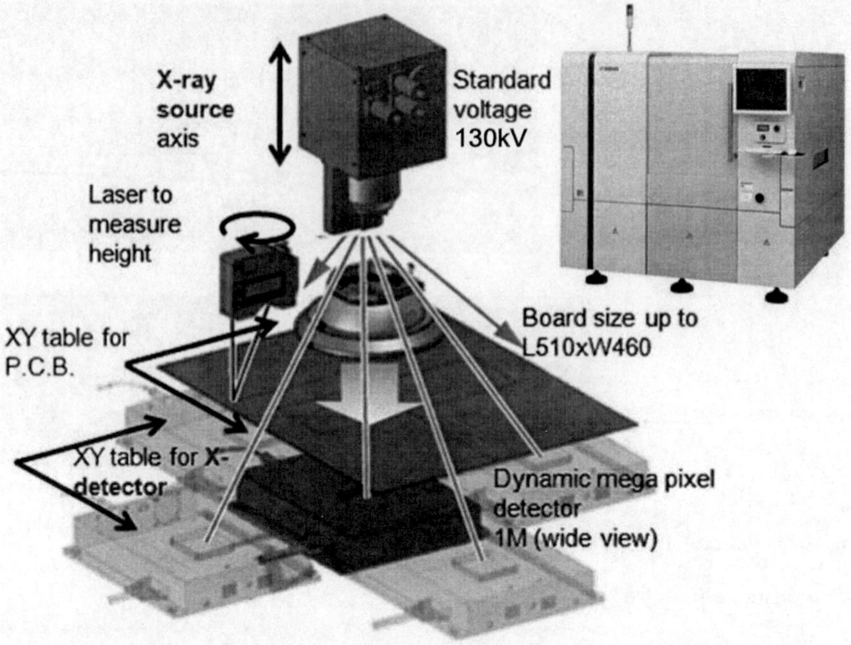

Fig. 6.21 Photograph (top right) of a commercially available in-line inspection-type X-ray CT scanner described in Table 6.1 and an internal layout where visible-light inspection, laser displacement sensor, and an X-ray imaging system co-exist; Yamaha Motor Company Tsi-X model (courtesy of Akira Kakuta of Yamaha Motor Company)

and traceability assurances such as serial number identification and the detection of product assembly errors, all simultaneously and autonomously. Furthermore, despite using a sealed tube, the device is innovative regarding its maintenance as the X-ray tube filament is exchanged and rebuilt when the filament is broken.

Meanwhile, Fig. 6.23 shows an X-ray CT scanner that is built for imaging with standard X-ray tomography. Its specifications are listed in Table 6.1. This device uses a high-output X-ray tube and is used for the complete part-by-part inspection of products like aluminum alloy components. Despite conducting standard X-ray tomography, it is capable of conducting fast imaging at speeds of 1 min/scan, which correspond to the production lines of automobile components. Figure 6.24 shows an observation example where this device was used to observe internal defects (e.g. cavities) in the interior of an aluminum die-casting product. This device can be set up even in harsh usage environments that would be harsh for precision instruments due to temperature or dust, like in aluminum die-casting lines. Inspections are conducted from when the casting product enters the inside of the device until the product quality is determined, as this is synchronized with the production cycle of die-cast machines. In addition to manual inspections by staff members, there are also

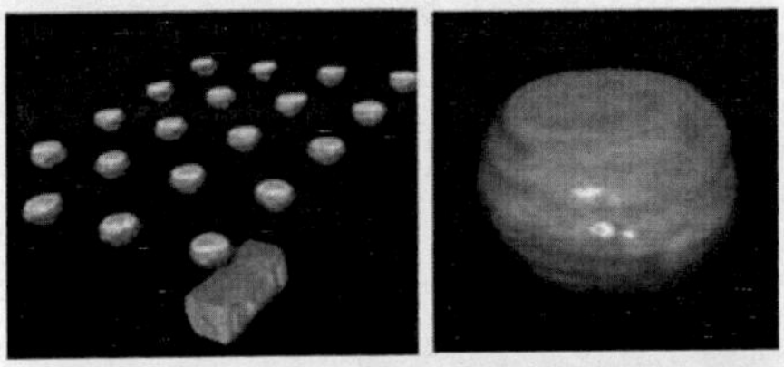

(a) High-speed sliced surface inspection using X-ray tomography

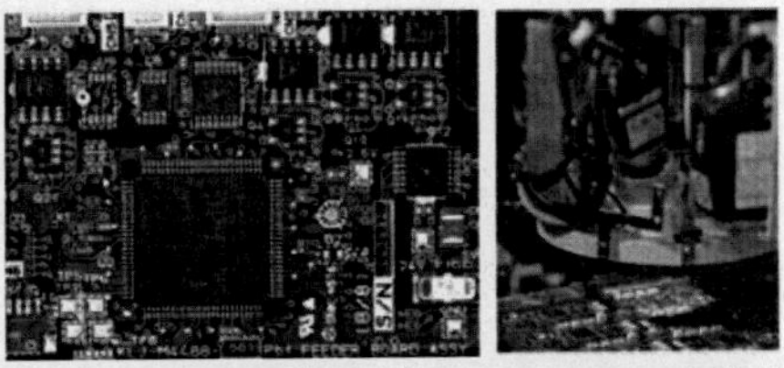

(b) High-speed optical microscope image obtained simultaneously with X-ray tomography images

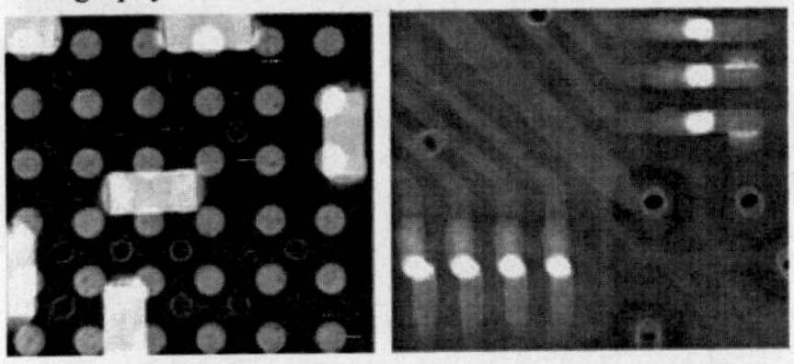

(c) X-ray radiographs obtained simultaneously with the optical inspection

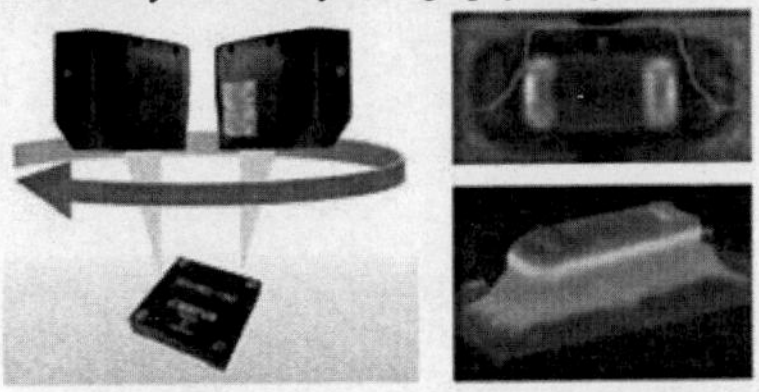

(d) Laser height position measurement

Fig. 6.22 Inspection example of an electric circuit using a commercially available in-line inspection-type X-ray CT scanner (Yamaha Motor Company Ysi-X model) described in Table 6.1. X-ray tomography, X-ray transmission, optical microscopy observations, infrared light observations, and laser inspections can be conducted simultaneously (courtesy of Akira Kakuta of Yamaha Motor Company)

Fig. 6.23 Photograph of a commercially available in-line inspection-type X-ray CT scanner described in Table 6.1; JED CTH200FPD model (courtesy of Osamu Kinoshita of JED, Co., Ltd.)

(a) Test piece for inspection

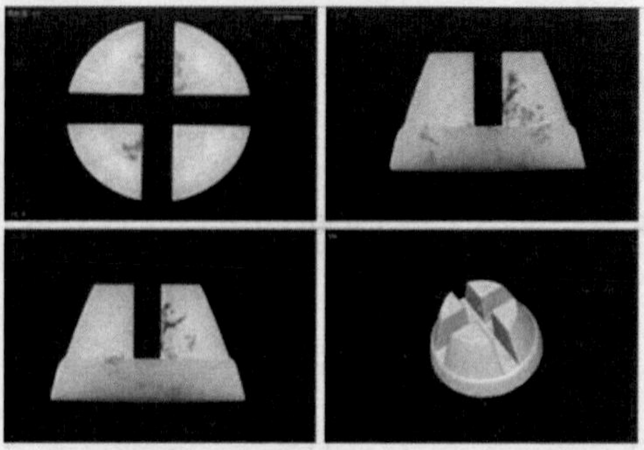

(b) Inspection of blow holes using a 3D image and virtual cross sections

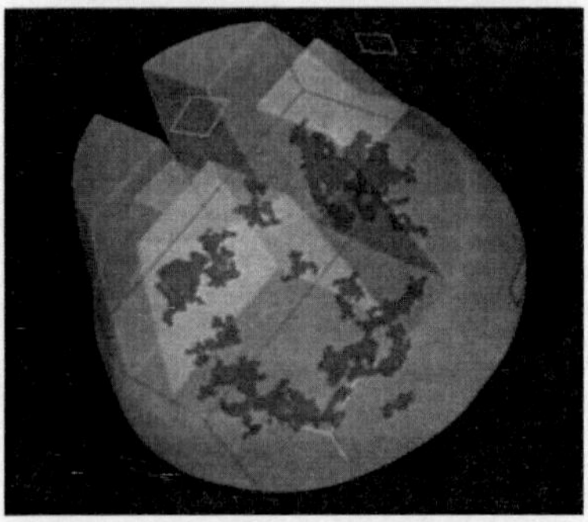

(c) 3D rendered blow holes

Fig. 6.24 Inspection example of an aluminum die-cast component using a commercially available in-line inspection-type X-ray CT scanner (JED CTH200FPD model) described in Table 6.1 (courtesy of Osamu Kinoshita of JED, Co., Ltd.)

completely automated inspections conducted in coordination with industrial robots. Product inspections can immediately and automatically determine the quality of all products while automatically saving and recording various numerical data, which can be used for subsequent defect dimension measurements.

6.6 X-Ray Tomography Using Synchrotron Radiation

The current state of the technology used at the imaging beamlines in SPring-8 and the Photon Factory, as well as recent application examples, is introduced here.

6.6.1 Projection-Type X-Ray Tomography

Projection-type X-ray tomography can be implemented at SPring-8 primarily through the two beamlines of BL20XU and BL20B2. The former is a beamline that uses an undulator and is suited for high-resolution work, whereas the latter is a beamline that uses bending magnets and is suited for imaging with a wide field of view. Both beamlines primarily use a 2048 × 2048-pixel Hamamatsu Photonics sCMOS camera measuring 6.5 μm × 6.5 μm. The former is often combined with a BM-3 beam monitor that combines a scintillator with a visible-light lens and mirror. The latter frequently uses a BM-2. Combinations of different beam monitors allow for changes in the effective pixel size. As can be observed later in Fig. 7.15, the effective spatial resolution for both cases is exactly double that of the pixel size (approximately 0.5 μm for projection-type X-ray tomography in BL20XU) [23]. In other words, the spatial resolution is limited by the Nyquist frequency, which is based on the sampling theorem (discussed in Sect. 7.5.1 (2)).

Figure 6.25 shows an imaging example using BL20B2 [4]. Beamlines that use bending magnets are more suited for metals even if they are porous with over 90% porosity. The effective pixel size, in this case, is 2.7 μm, the spatial resolution is 5.4 μm, and both the field of view width and height are 5.4 mm. Clean visualizations of not only the complex and irregular structure of the porous metal cell wall but also of the interior micropores and precipitated hydrogen was obtained. With the support of the 3D image-based simulations (discussed in Sect. 8.6), this research has elucidated the deformation/fracture behavior of porous metals.

BL20XU is used when further spatial resolution is required. Figure 6.26 shows a visualization of creep fracture behavior in steel materials at high temperatures. The effective pixel size, in this case, was 0.5 μm, the spatial resolution was approximately 1.0 μm, and both the field of view width and height were approximately 1 mm. Exposure to high temperature after heat-resistant steel with 9% chrome was welded created defects referred to as type IV voids in the heat-affected zone. This was observed at 650 °C under a load of 70 MPa after 17,220 h. Defects (creep voids) occurred at the specified grain boundary of the prior austenite and the formation of voids, which had not been recognized until now, were observed in the crystallographic grains. Micro-scale voids were extremely clearly captured and the various 3D/4D analysis methods in Chaps. 8 and 9 can be used.

Figure 6.27 shows the observations of the inner pyramidal layer in the cortex of the frontal lobe of the human brain using BL20XU [5]. Absorption contrast with an X-ray energy of 12 keV was used for observation. The effective pixel size, in this case, was 0.51 μm and the effective spatial resolution was 1.2 μm. The top area of Fig. 6.27 corresponds to the brain surface. These observations clearly show the pyramidal neuron, which is the primary excitatory neuron that exists in the cerebral cortex and hippocampus [5].

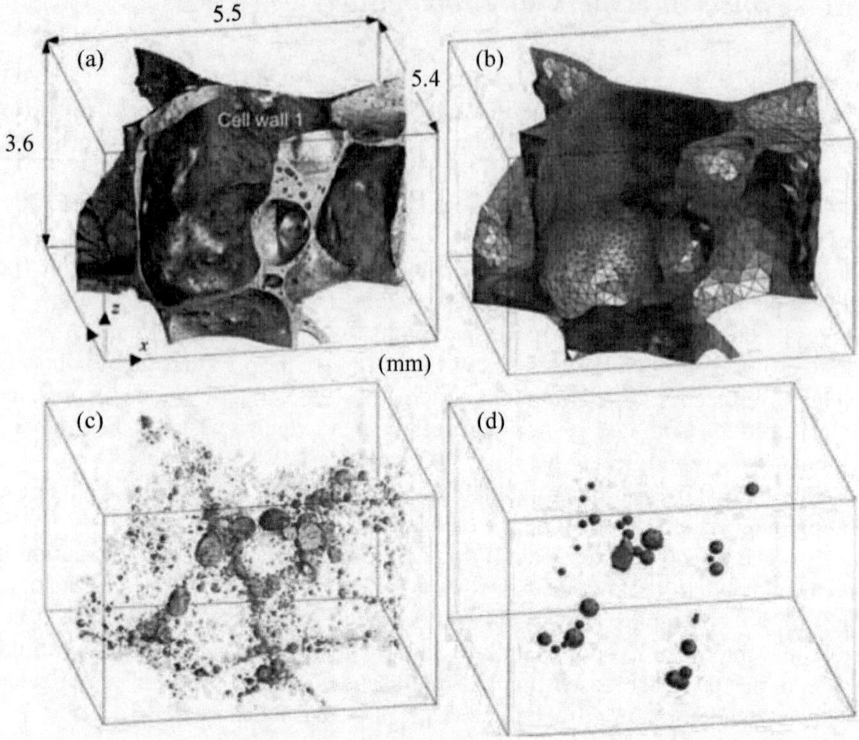

Fig. 6.25 Observation example using BL20B2 in SPring-8. The X-ray energy at BL20B2 was set at 20 keV and 3D imaging was conducted at experimental hutch 1 with an effective pixel size of 2.73 μm [4]. The observation subject was porous aluminum; **a** and **c** show the porous metal cell wall and the interior pores; **b** and **d** show the surface extractions for image-based simulations (STL files)

6.6.2 Imaging-Type X-Ray Tomography

Experimental devices for imaging-type X-ray tomography have been made available to users since the mid-2000s at a number of imaging beamlines in SPring-8. Throughout the 10 years following this period, imaging with relatively low X-ray energies of 8–10 keV has been conducted for various applications. Figure 6.28 shows the observation results of precipitates in aluminum conducted during this early stage [6]. Precipitates in Al–Ag alloys have been observed at high spatial resolution in this research. Imaging-type X-ray tomography experiments were exclusively conducted at BL47XU at the time, with the illumination system comprising a parallel beam passing through a diffusion plate; Köhler illumination had not yet been used. The effective pixel size, in this case, was 88 nm and an effective spatial resolution approximately double this was obtained. However, the sample size was 56.8 μm (H) ×

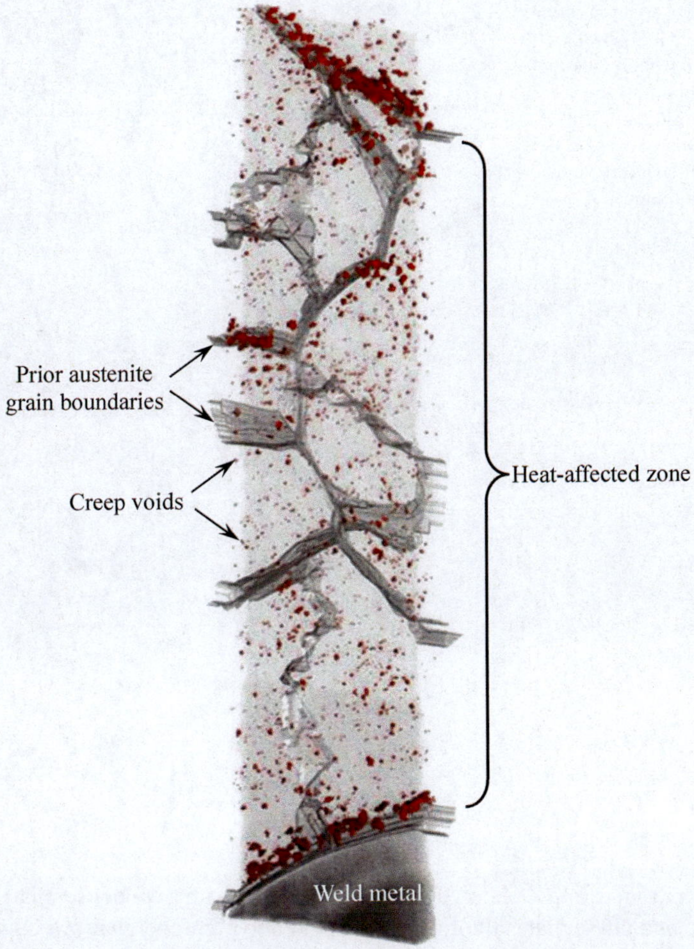

Fig. 6.26 Observation example using BL20XU at SPring-8. The X-ray energy was set at 37.7 keV and 3D imaging was conducted at experimental hutch 1 with an effective pixel size of 0.5 μm. The observation subjects are the type IV voids formed by creep of the welded material of the high-chrome heat-resistant steel

35.3 μm (V) and thinner than the width of a hair due to the X-ray energy limitations of 9.8 keV, as there was considerable difficulty in sample preparation and handling.

Subsequent research and development on imaging-type X-ray tomography at SPring-8 has continued in earnest, with 3D imaging that uses 20-keV and 30-keV X-ray energies achieved in 2017 and 2018, respectively. Currently, imaging-type X-ray tomography can be conducted at BL47XU and BL37XU, with BL20XU at the center. The mainstream setup is as shown in Fig. 5.25, which is phase-contrast imaging-type X-ray tomography that combines Köhler illumination, a Fresnel zone

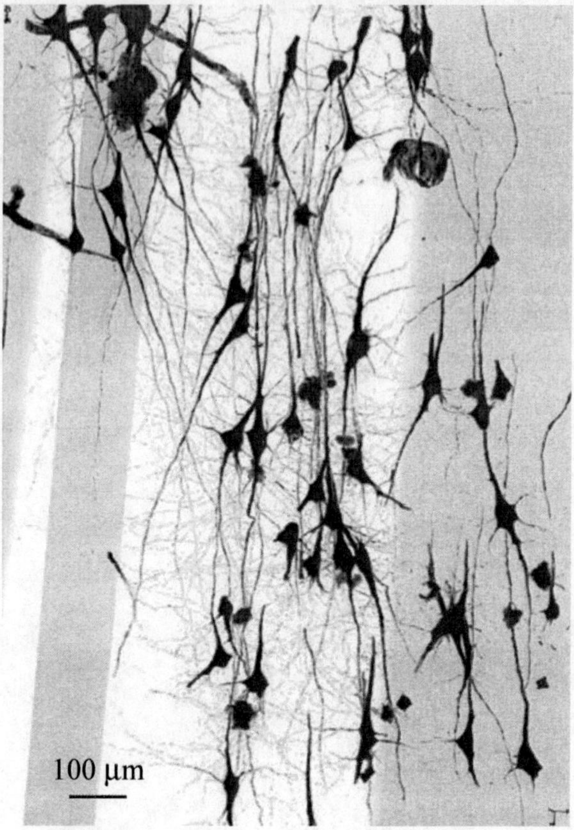

Fig. 6.27 Observation of the inner pyramidal layer of the cortex of the frontal lobe in the human brain using BL20XU at SPring-8 [5]. The X-ray energy was set at 12 keV (courtesy of Ryuta Mizutani of Tokai University)

plate, and a Zernike phase plate. SPring-8 is equipped with a condenser plate, Fresnel zone plate, and phase plate, all of which are suited to high-energy usage but are difficult to micromachine. The realization of the apodization Fresnel zone plate (discussed in Sect. 5.1.1 (3)) in particular is key to achieving high-energy imaging-type X-ray tomography. BL20XU can now switch between projection and imaging modes at a touch. It may appear like there are no major differences between this facility and the imaging-type X-ray tomography of other facilities with regard to spatial resolution. However, as discussed in Sect. 7.5.3 (1), other reports on spatial resolution measurements have not frequently accurately identified and measured single cycles in the periodic structure of the test object from the image data and instead displayed half of this as the spatial resolution. Sufficient care must be taken when comparing performance based on reports from different institutions.

Figure 6.29 shows an observation example of nanovoids that formed during hydrogen embrittlement in an aluminum alloy. Combinations of imaging-type X-ray tomography, projection-type X-ray tomography, and HAADF-STEM have enabled observations of the formation of high-density nanovoids across multiple scales, from several nm to 10 μm [7]. In addition to these types of direct observations, nanovoid

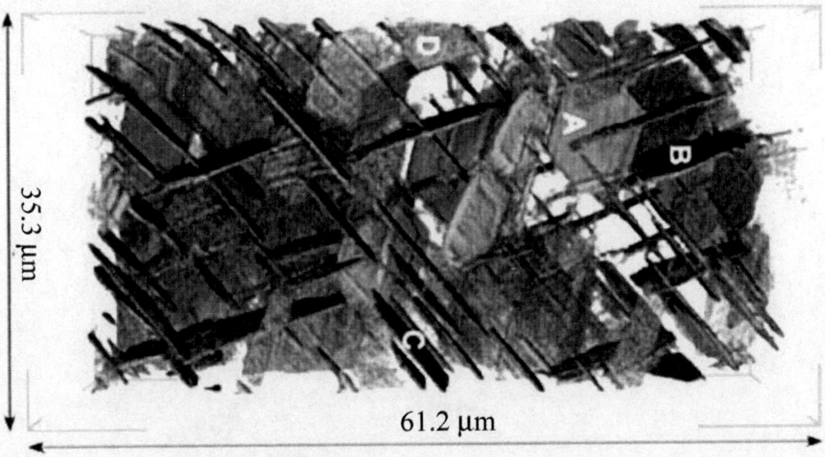

Fig. 6.28 Observation example using BL47XU at SPring-8 [6]. The X-ray energy was set at 9.8 keV and 3D imaging was conducted with an effective pixel size of 88 nm. Observation subjects are aging precipitates in over-aged Al-Ag alloy material

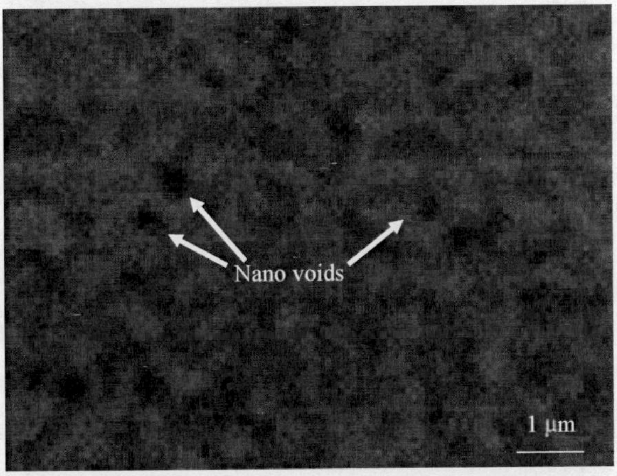

Fig. 6.29 Observation of nanovoids formed during hydrogen embrittlement of A7150 aluminum alloys using BL20XU at SPring-8. The X-ray energy was set at 20 keV and 3D imaging was conducted with an effective pixel size of approximately 60 nm

formation and growth can be perceived as the generation of tensile strains under hydrostatic pressure, as shown later in Fig. 9.13 [7].

Figure 6.30 shows imaging-type X-ray tomography-based observations of the cortex of the frontal lobe in a human brain, observed in Fig. 6.27 with projection-type X-ray tomography (separate sample). The experiment, in this case, was conducted using BL37XU, the X-ray energy was 8 keV, and the effective spatial resolution was 100 nm. Absorption contrast was used in the imaging method. The pyramidal neurons in the cortex of the frontal lobe in the human brain and the surrounding neurites can be clearly observed. Furthermore, the spinous process can be observed in the neurite.

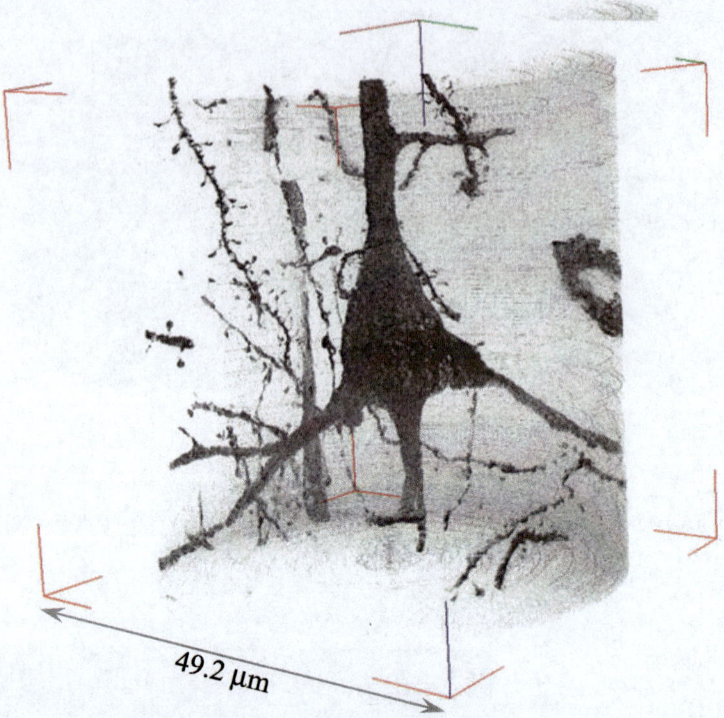

Fig. 6.30 Imaging X-ray tomography observations at BL37XU at SPring-8 of the inner pyramidal layer in the cortex of the front lobe in the human brain, which was observed with projection X-ray tomography in Fig. 6.27 [5]. The X-ray energy was set at 8 keV (courtesy of Ryuta Mizutani of Tokai University)

The spinous process is formed or destroyed depending on neural activity in the brain and the observations of its number and morphology are important for understanding the mechanisms of neural pathway formation.

6.6.3 Phase-Contrast Tomography

Imaging beamlines at SPring-8 currently requires either a combination of projection-type X-ray tomography and the single-distance phase retrieval method (discussed in Sect. 5.2.1 (2)) or imaging-type X-ray tomography that uses a Zernike phase-contrast microscope (discussed in Sect. 6.6.2). First, Fig. 6.31 shows an observation example of the former with DP steel using methods employed by Paganin [8]. This is a DP steel that comprises a dual-phase structure of martensite and ferrite, where observations of deformation, damage, and fracture behavior under tensile loads are observed in a material where the martensite has a volume fraction, which barely forms a 3D

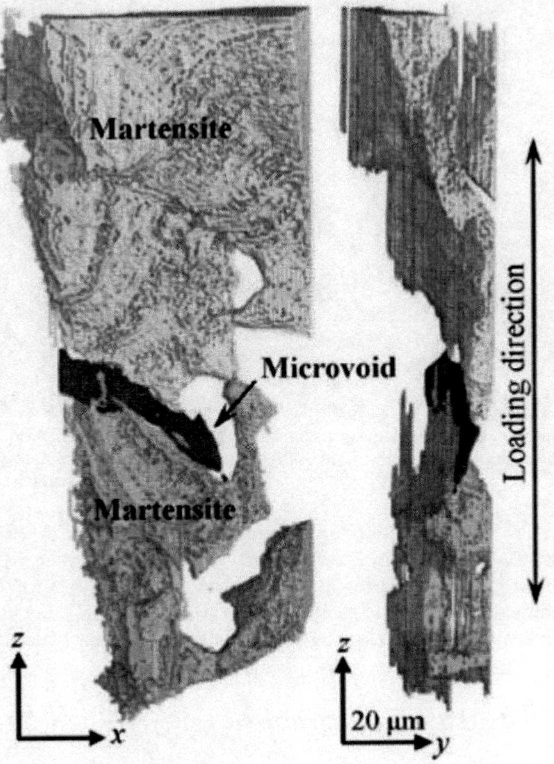

Fig. 6.31 In-situ observation example of martensite phase damaging behavior in DP steel using BL20XU at SPring-8 [8]. The X-ray energy was set at 37.7 keV and 3D imaging was conducted with an effective pixel size of 0.5 μm. The gray areas in the figure are the martensite phases and the black areas are crack-like voids where the martensite phase was fractured. The original figure is in color; please refer to the original publication for details

network. The carbon concentration in the martensite phase is approximately 0.33%, whereas the carbon concentration in ferrite is low at approximately 0.001%. As such, the martensite and ferrite phases have densities of 7.87 and 7.76 g/cm^3, with a difference of approximately 1.4% between them. The imaging in Fig. 6.31 uses this density difference. Discrimination between the two phases in ferrite-pearlite steel, which has a density difference of approximately 0.4%, is possible with this method [9].

Figure 6.32 shows a visualization of a retained austenite phase in TRIP steel using a Zernike phase-contrast microscope. In Fig. 6.32a, where phase retrieval of the projection-type X-ray tomography was conducted using Paganin's methods, the presence of the austenite phase could be visualized; however, its morphology was unclear. Meanwhile, imaging-type X-ray tomography using a Zernike phase-contrast microscope enabled the visualization of the morphology of the austenite phase with sufficient spatial resolution and contrast. This research captured the gradual changes in the austenite phase. It is anticipated that findings valuable for microstructural design will be obtained in the future.

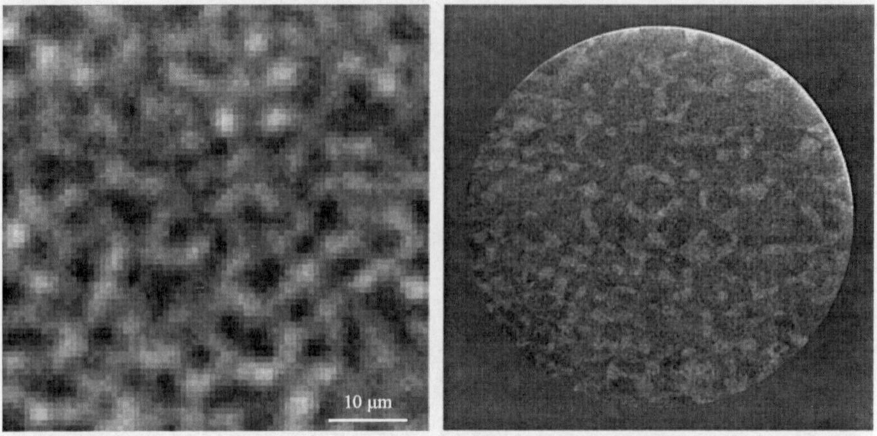

(a) Projection-type X-ray tomography image that has been phase-retrieved by means of Paganin's method

(b) Sample in (a) observed using imaging-type X-ray microtomography technique combined with Zernike phase-contrast optics

Fig. 6.32 Observation example of the austenite phase in TRIP steel using BL20XU at SPring-8 [8]. The X-ray energy is set at 20 keV and in-situ observations of the deformation-induced austenite transformation while the specimen is under tension are shown. The effective pixel size in **b** is 60 nm. Phase retrieval with Paganin's method after projection X-ray tomography imaging with an effective pixel size of 0.5 μm is shown for comparison in **a**. Both are shown with the same magnification

6.6.4 Fast Tomography

Examples conducted by the author at ESRF using white X-rays will be introduced here for fast tomography. Figure 6.33 shows in situ observations of fracture toughness tests conducted on an A2024 aluminum alloy [10]. Changes in the distribution of the driving force for crack propagation can be observed. Fixing the displacement each time X-ray tomography imaging is conducted results in unloading during in situ observations of metal material fracture due to relaxation phenomena. Elasto-plastic analysis applications for the obtained 3D images cannot be validated for experiments, which experience unloading. This research took single scans every 22.5 s, which was, at the time, an extremely accelerated setup, through the combination of white light, a camera capable of high-speed readout, and an imaging process that minimized time loss. The research conducted in situ observations of crack propagation during fracture toughness experiments conducted at a fixed cross-head speed without any interruptions and analyzed the strain field of the crack tip. These methods were used to reduce imaging time for a single 3D image by 1/50th of the original duration while maintaining superior spatial resolution when a monochromatic X-ray was used. The obtained images visualized not only cracks but also microstructures such as particles. As shown in Fig. 6.33, this enabled the clear visualization of crack tip morphology corresponding to stress-field singularities of the crack tip (i.e. crack-tip blunting/sharpening), as well as the transitions and spatial spread of stress-field singularities accompanying crack propagation.

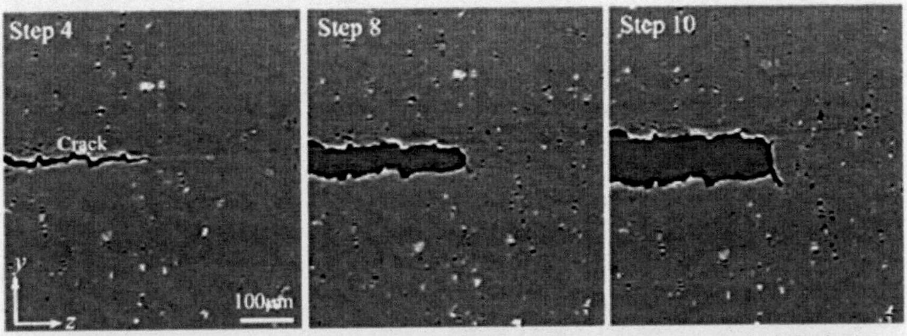

(a) Cross-sections at $x = 59$ μm (An example of remarkable crack-tip blunting)

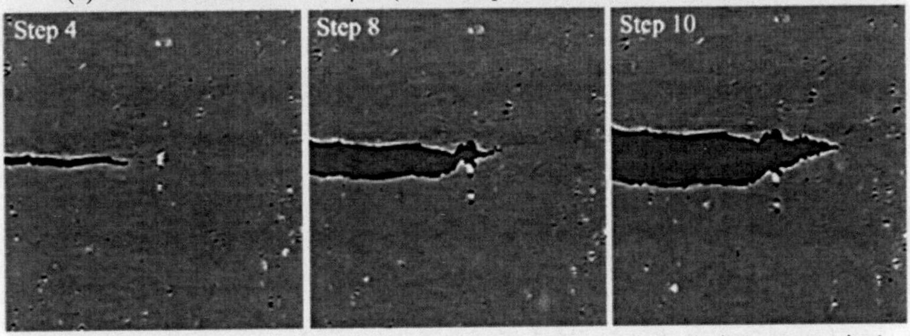

(b) Cross-sections at $x = 577$ μm (An example of sharp crack-tip during propagation)

Fig. 6.33 In-situ observations of fracture toughness experiments of a fatigue pre-cracked A2024 aluminum alloy using the ID15A beamline at ESRF [10]. Blunting (top row) and sharpening (bottom row) of the propagating crack tip can be observed by continuously conducting fast tomography observations while applying tensile loads without any unloading

6.6.5 *Tomography of Elemental Concentrations*

Figure 6.34 shows the application of absorption-edge subtraction imaging introduced in Sect. 5.4.1 on an Al–Si–Cu ternary alloy after a high-temperature solution-treatment, as well as the quantification of the spatial distribution of copper concentrations in the microstructure and its relationship with fracture [11]. A 3D image was obtained at the X-ray energies above and below the K-absorption edge of copper (upper side: 9.038 keV, lower side: 8.938 keV), on which pixel subtraction was conducted. The Al–Si–Cu ternary alloy exhibited the highest strength at 807 K as a solution-treated sample when the solution treatment temperature was varied between 773 and 824 K and both a tensile strength improvement of over 12% and reduction in solution treatment time of approximately 1/5th were achieved relative to the standardized heat treatment conditions (T6 treatment) [11]. This temperature corresponds to a range above the ternary eutectic point and significant increases in copper concentration, pore formation/growth, and matrix copper concentration were observed due to local melting such as in Fig. 6.34. This research shows that pores formed from eutectic melting deteriorate material characteristics and act as preferential pathways

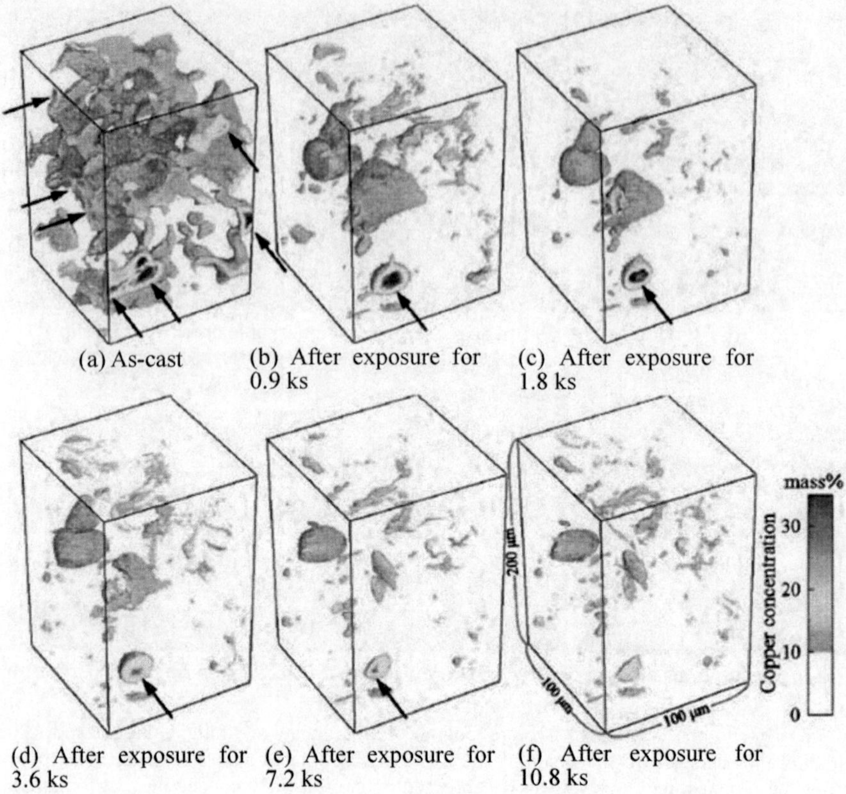

Fig. 6.34 Changes in the microstructure during solution treatment at a temperature of 807 K, where the pure ternary system Al–Si–Cu alloy has the highest strength when solution treatment temperatures varied between 773 and 824 K. Absorption-edge subtraction imaging was used for 3D mapping of copper concentration distributions [11]. The original figure is in color; please refer to the figure of the original publication for details

for crack propagation [11]. However, it was also demonstrated that the beneficial effects of increased precipitate due to increased copper concentration outweigh the negative effects of local melting when heat treatment is applied for a short period even if above the ternary eutectic point and that this results in the improvement of the material characteristics [11].

Other applications of absorption-edge subtraction imaging involve the visualization of alloy element distributions in the cell walls of cast porous metals [12] and the mapping of tungsten concentration distributions in steels [13].

An imaging-type XAFS tomography setup can be utilized at the PF-AR NW2A beamline in the Photon Factory; Fig. 6.35 shows this setup. A Carl Zeiss X-ray microscope Xradia Ultra was introduced in this beamline and X-ray energies of 5–11 keV can be used. Figure 6.36 shows observations of Yb–Si–O particles prepared

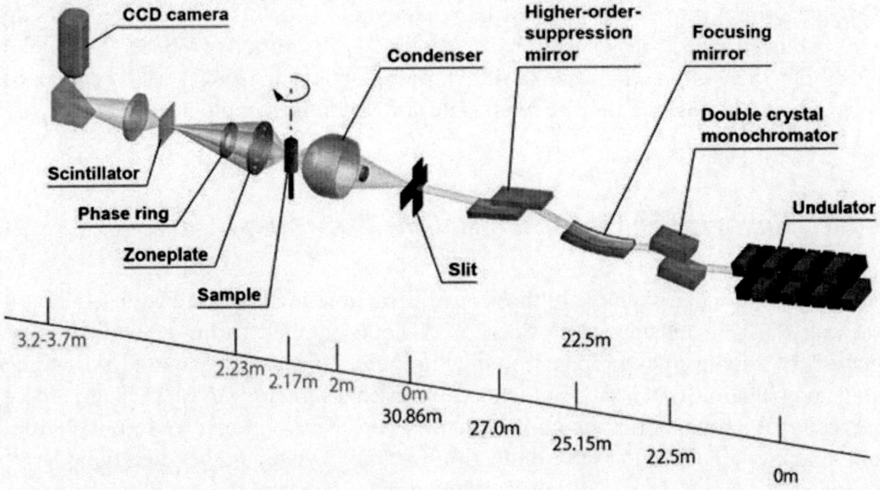

Fig. 6.35 Imaging XAFS tomography setup, which can be used at the Photon Factory (PF-AR NW2A beamline) (courtesy of Masao Kimura of KEK)

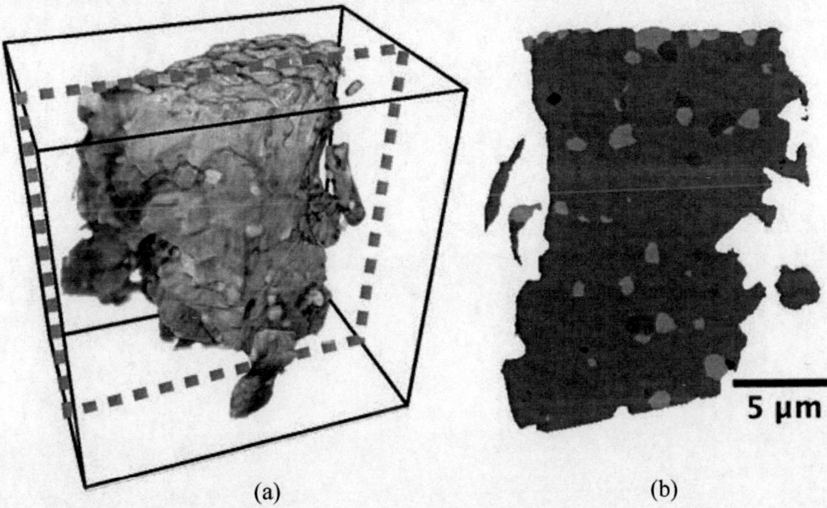

Fig. 6.36 X-ray microscopy observation results where an $Yb_2Si_2O_7$ oxidized plate was heated at high temperatures. **a** 3D image at an X-ray energy of 8.98 keV. **b** Results where the chemical conditions of Yb were differentiated using measurements that combine X-ray CT and X-ray absorption spectroscopy (XAFS-CT) (corresponds to the cross-section of the dotted lines in **a**). The reference has a color figure; please refer to this figure for details (courtesy of Masao Kimura of KEK) [14]

by pulverizing that measure 10–20 μm in diameter [14]. X-ray tomography imaging was conducted at 36 X-ray energy levels near the L_{III} absorption of Yb (8.9–9.06 keV) and a 3D image with a pixel size of 48.8 nm was obtained. This research succeeded in cleanly distinguishing the chemical states of Yb in the sample interior [14].

6.6.6 *Tomography for Polycrystalline Structures*

Figure 6.37 shows an example of the X-ray diffraction-amalgamated grain-boundary tracking (DAGT) introduced in Sect. 5.5.4 being applied to the crystallographic analysis of fatigue crack propagation in aluminum alloys for airplanes [15]. Fatigue crack propagation is heavily influenced by crystal structure even in stage IIb of fatigue crack propagation, which is generally considered insensitive to microstructure; cracks were shown to have deflections or to be twisted in the grain boundaries where large differences in crystallographic orientation were observed. Furthermore, acceleration and deceleration due to the crack closure patches at the crystallographic grain boundary or crack tip vicinity were microscopically observed. A particularly

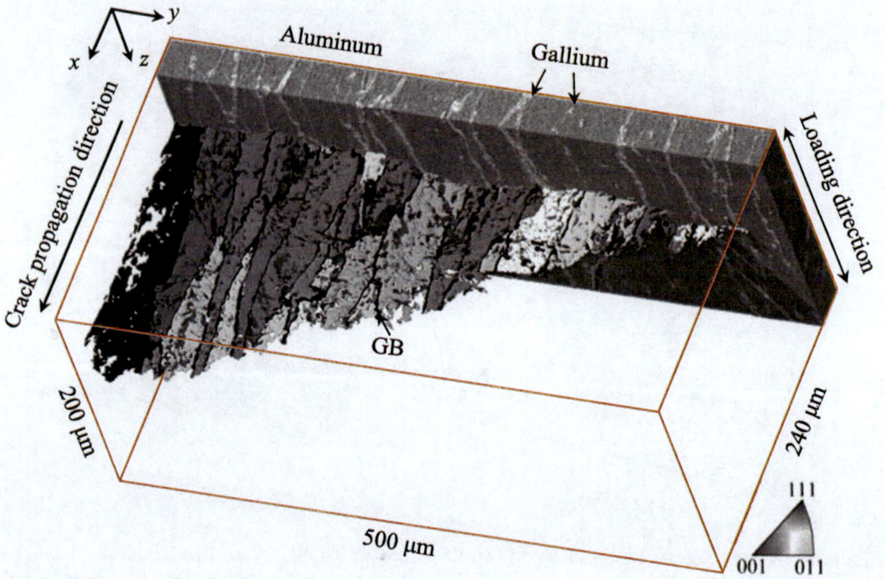

Fig. 6.37 Relationship between fatigue crack propagation and crystallographic structure in an A7075 aluminum alloy [15]; research example that combines 3D imaging of a crystallographic structure using DAGT and continuous observations of crack propagation behavior. The influence of crystallographic structure can be seen in regions where stable crack growth is observed. The original figure is in color; please refer to the figure in the original publication for details

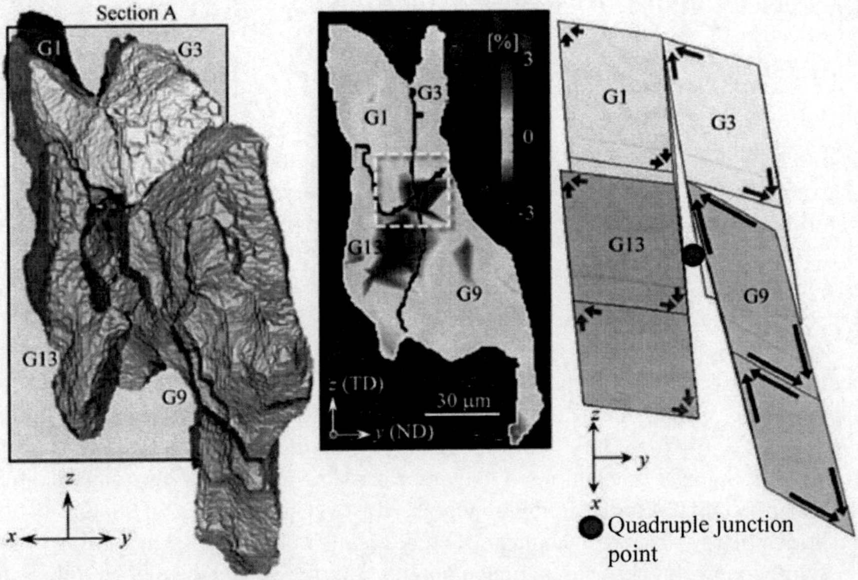

Fig. 6.38 Visualization example of crystallographic grains during tensile deformation using DAGT [16]; 3D display of the four crystallographic grains that surround the grain boundary quadruple point within an Al-3% Cu alloy (left) and a 3D strain mapping shown with a virtual cross-section (see Sect. 9.2.3 for reference). Right: four unit boxes selected from the crystallographic grain where the strains measured by DAGT are applied and where the displacements are emphasized by a factor of 5. The formation of voids due to hydrostatic tension near the grain boundary quadruple point is simulated. The original figure is in color; please refer to the figure in the original publication for details

high crack propagation rate was observed when propagating along a specified crystal orientation.

Figure 6.38 shows an example of DAGT applied to the analysis of damage during tensile deformation [16]. Hydrostatic strain distribution was measured at the quadruple junction point of the grain boundaries. The deformation in the vicinity of the grain boundary was substituted with unit boxes for higher visibility, demonstrating that there was particularly large shear deformation present at crystallographic grain 9 (G9) and the deformations of the other three crystallographic grains were relatively coordinated [15]. The hydrostatic tension generated by this mismatch accelerates void formation and propagation and acts as the cause of macro-scale fracture [15].

6.7 Selection of Devices and Conditions

6.7.1 Device Selection

As observed until now, the spatial resolution of X-ray tomography devices is a little <1 μm at most with projection-type X-ray tomography and is equivalent to a high-spatial-resolution optical microscope. Furthermore, imaging-type X-ray tomography has a spatial resolution that is worse than the scanning electron microscopes, which are widely used for surface observations, by two orders of magnitude. For this reason, it is sometimes the case that internal structures or microstructures close to the effective spatial resolution are observed with X-ray tomography. Furthermore, scanning electron microscopes, optical microscopes, and stereoscopic microscopes have the characteristic of easily being able to change their magnification during observation. Meanwhile, X-ray tomography cannot change its magnification or has serious limitations in doing so. For example, it may appear at first glance that the magnification in a cone-beam CT can be freely changed. However, increasing the magnification and attempting to observe a smaller structure results in the sample protruding out of the field-of-view. In these cases, image quality deterioration is inevitable, unless the specialized image reconstruction techniques as in Sect. 3.3.5 (2) are used. A factor that must be controlled in these instances is the *trade-off between spatial resolution and the field-of-view*. For example, the various X-ray CT scanners summarized in Table 6.1 frequently have a two-dimensional detector with 1000–4000 pixels. As shown with the solid line in Fig. 6.39, the spatial resolution is stipulated by the sample size divided by the pixel number, multiplying this by two, when a device that assures high spatial resolution is used and when the Nyquist frequency based on the sampling theorem discussed in Sect. 7.5.1 (2) constrains the spatial resolution. When the sample size doubles, in this case, the spatial resolution worsens by a factor of two. However, the spatial resolution levels off, as shown in the dotted line in Fig. 6.39 when there are some influences from the X-ray source, sample rotation stage, or detector.

Figure 6.40 shows a schematic of these types of X-ray tomography constraints. Scanning electron microscopes and optical microscopes have a wide range of visualization, so they have many opportunities to favorably conduct observations without any particular considerations. Meanwhile, the chemical composition, density, and size of internal structures/microstructures, which can be visualized, vary according to the performance of the X-ray CT scanner itself, as well as various imaging conditions such as X-ray energy, sample size, and X-ray brilliance.

For these reasons, X-ray tomography does not show much promise as a versatile observation method. First, one must abandon the notion of "the greater also serves for the lesser," instead narrowing down the method to the most important observation subject, determining the optimal X-ray energy, spatial resolution, and imaging techniques based on the fundamental knowledge provided in this book, as well as closely examining the devices, imaging conditions, and samples to be used. Figure 6.41 shows an overview of this process. Conventional surface observation devices were

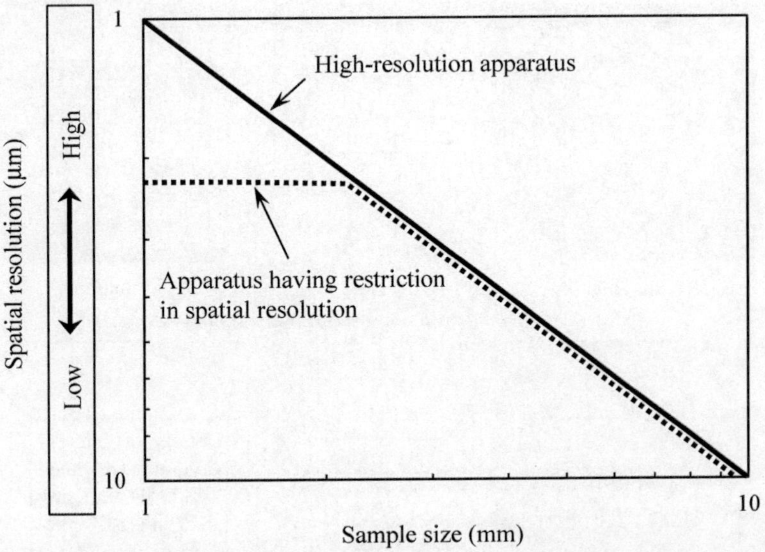

Fig. 6.39 Trade-off between spatial resolution and sample size. Two cases where the spatial resolution is constrained across the entire sample size range due to the Nyquist frequency based on the sampling theorem and where spatial resolution limitations exist due to the device are shown schematically. A detector pixel number of 2000 pixels and a rotation step in the sample rotation stage whose angular increments correspond to this are assumed

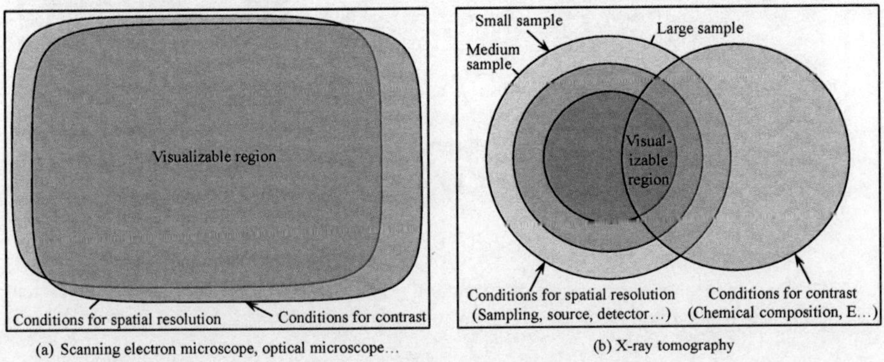

Fig. 6.40 Comparisons between the conventional scientific/engineering observation method of scanning electron microscopy or optical microscopy, as well as X-ray tomography. The rectangular outer frame stipulates the whole. Furthermore, the figures in the frame are constrained by spatial resolution and contrast. The ability to visualize (inside the diagrams) or not visualize (outside the diagrams) is determined based on these conditions. Areas where the diagrams overlap are the regions where the samples are observed in terms of both spatial resolution and contrast. In the case of X-ray tomography in (**b**), the sample size plays a large role in the visualizable range (internal structures, microstructures and their size)

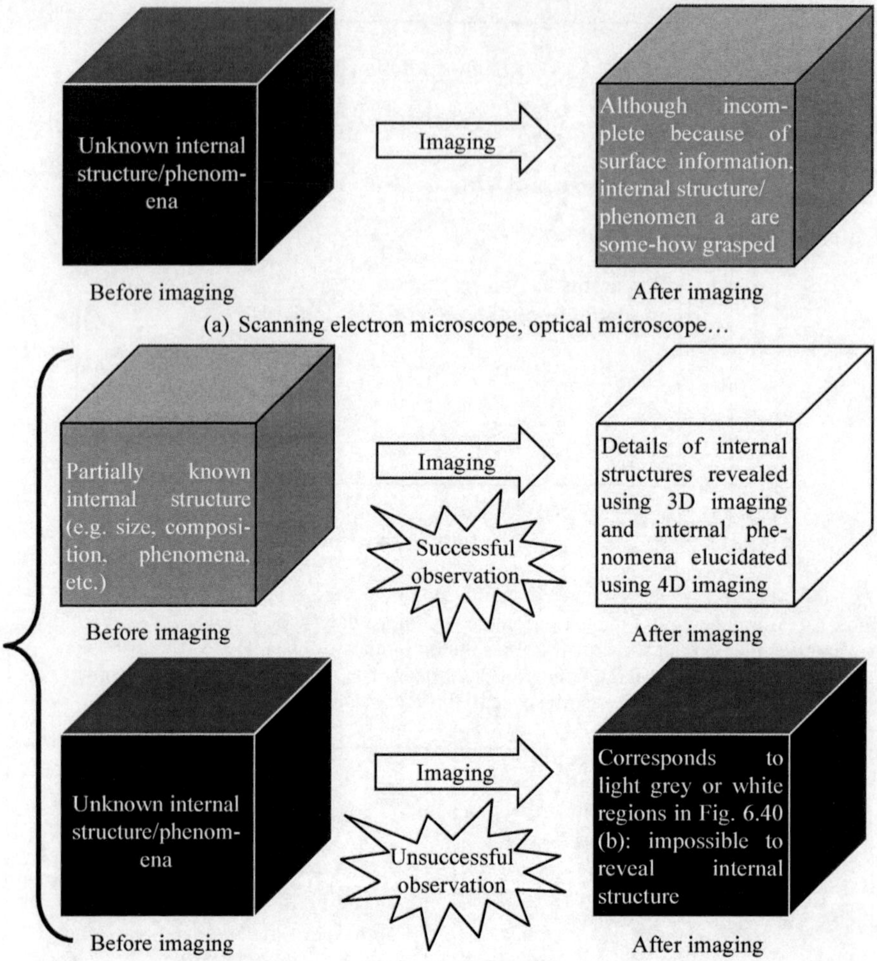

Fig. 6.41 Schematic showing necessary information (black box-gray box) for conventional scientific/engineering observation methods and X-ray tomography prior to observation, and observation results (black box-white box)

expected to transform a black box object into a white box object; in other words, the expectation is that at least some kind of information can be obtained even if the sample is unknown by some form of observation. However, what is actually obtained is only superficial information and assessments are conducted by averaging the observable information as individual structures cannot be assessed. In this sense, the endpoint in Fig. 6.41a is set as a gray box because an accurate determination/understanding of the internal structure or phenomenon cannot be made. By contrast, X-ray tomography is thought to take a gray box sample, where the interior is known to some degree and convert it completely into a white box. Researchers occasionally state, "there was a sample (black box) that I wanted to observe but couldn't effectively do so." This is oftentimes the case when visualization is technically possible but where results yield no significant findings due to insufficient advance considerations on the sample, actual device, or imaging conditions.

However, discoveries have been made in samples such as in Fig. 6.41a, where the internal structure was previously unknown, leading to significant findings. A representative example of this is the discovery of high-density pores in aluminum reported in Ref. [17]. This occurred when high-density hydrogen micropores with diameters less than 10 μm were coincidentally discovered during synchrotron radiation experiments conducted in 2001 to observe fatigue cracks in aluminum [18]. These types of pores could not be observed with conventional cross-section observations as they were filled with abrasive powder during sample preparation processes like cutting and polishing. There have been many subsequent reports indicating that aluminum alloy strength, fatigue, and high-temperature defects are controlled by these micropores [19], resulting in contributions not only to metal damage/fracture but also to hydrogen embrittlement and stress-corrosion cracking research, as well as new material development.

Until now, the key to observation success in X-ray tomography has been in selecting the "right tool" for the job. However, as presented in this chapter, the functionality of industrial X-ray CT scanners is consistently increasing. Some scanners are equipped with multiple X-ray sources or detectors and devices that can significantly change X-ray energies. That being said, there are limits to the range over, which a single device can cover. Furthermore, industrial X-ray CT scanners are expensive, and it is prohibitively difficult to prepare different devices for each observation subject. We anticipate further active approaches in these cases, such as applications in coordination with the various X ray CT scanners, which are being introduced among industrial technology centers throughout the country and the usage of synchrotron radiation facility setups. There may be a high barrier to entry with regards to synchrotron radiation facilities for those who have no experience using them. However, many of these facilities can be used for virtually no cost and high-quality and rich information can be obtained with a few hours' worth of experiments, which could otherwise never be obtained with industrial X-ray CT scanners. As the structure and size of the obtained data are the same, there is no need for concern on how to handle the data. Industrial X-ray CT scanner users have experience with 3D imaging and are versed to some extent with 3D image assessment and processing. As such, the barrier to entry for synchrotron radiation facilities is rather low, once accustomed to their particularities.

6.7.2 Realities of 3D Imaging

The discussion below focuses on projection-type X-ray tomography and considers actual X-ray tomography measurements. The factors to first keep in mind with regards to visualizations are as follows:

(1) Is the sample, component, or product observable from a size standpoint?
(2) Can X-rays sufficiently transmit through the sample size?
(3) Is the spatial resolution sufficient for observing the internal structure and microstructure of interest?
(4) Can a sufficient contrast be obtained for observing the internal structure and microstructure of interest?

Spatial resolution is first constrained due to the trade-offs between spatial resolution and the field-of-view (discussed in Sect. 6.7.1) if condition (1) among these is prioritized. However, the possibility exists that condition (3) may not be satisfied due to this aspect. Thus, the first step is to somewhat characterize the internal structure or texture and identify the X-ray CT scanners and imaging conditions (e.g. number of projections) that can satisfy condition (3). This requires a more or less *a priori approach* , such as using stereoscopic microscopes, optical microscopes, or scanning electron microscopes in a supplemental manner, using reference or bibliographic information. Next, assessments using the information learned in Chaps. 2 and 5 and discussed in Chap. 7 are to be conducted to determine whether conditions (2) and (4) can be satisfied. Except for in-line inspections, observations after cutting the sample should be considered a possibility for satisfying condition (1). One of the greatest benefits of X-ray tomography is its ability to conduct internal or external observations without any chemical or physical pre-processing (e.g. sample cutting or polishing). However, instead of unilaterally insisting on this, efforts should be made to cut the sample and stitch a 3D image after multiple scans of each component, accurately determining its internal structure in this manner.

6.7.3 Sample Size and X-Ray Energy Selection

According to Grodzins, the incident X-ray intensity I_0 necessary for transmission image projection when photon noise is predominant is expressed as follows [20]:

$$I_0 wLt < \frac{DBe^{\mu D}}{w(w\mu)^2\left(\frac{\sigma}{\mu}\right)^2} \tag{6.1}$$

Here, the left-hand side of the equation represents the incident photon number, which is obtained by the cross-sectional area wL (w is the pixel size and L is the beam transmission length) of the beam that penetrates the sample multiplied by the exposure time t.

Furthermore, D is the diameter of the cylindrical sample, μ the average linear absorption coefficient, B a constant that depends on the reconstruction algorithm (approximately 2) [20], and σ is the standard deviation of the image signal (i.e. photon noise).

Assuming that the linear absorption coefficient distribution within the sample is homogeneous, the above-mentioned equation has a minimum value when $\mu D = 2$ [20] or $\mu\ D = 2.22$ [21]. Under these conditions, a constant S/N ratio results in minimal exposure time and a constant exposure time results in a maximal S/N ratio. According to Eq. (6.1), this corresponds to a transmissivity I/I_0 of 11–14%. As shown in Eq. (2.10), μ is dependent on λ, so the X-ray energy should be adjusted so that the transmissivity is at this level. For example, X-ray energy of 14–16 keV corresponds to a transmissivity of 11–14% for a pure aluminum sample with a diameter of around 1 mm. Elements like copper, zinc, and iron, which have atomic numbers considerably larger than aluminum, typically comprise several to 10% of aluminum alloys. For this reason, the author typically uses an X-ray energy of 20 keV for imaging aluminum alloys. Furthermore, the optimal μD becomes rather small for cases where detector-based noise has a large influence, requiring a larger transmissivity of approximately 20–30%. Recent simple analyses by Vopálenský have indicated optimal values of $\mu D = 1$ and transmissivity of $I/I_0 = 37\%$ [22]. Although these analyses have been conducted under bold assumptions, we believe they approximate experimental realities.

References

1. H. Fukami, T. Morita, G. Nakamura, Y. Horii, H. Noda, M. Sakai, N. Arakawa, H. Hayashi, K. Yoshida, K. Murakami, Patent publication, X-ray inspection method, X-ray inspection device and X-ray inspection program, application number 2009002814
2. K. Sato, S. Izumi, Patent publication, Computed tomography imaging device and computed tomography imaging method, patent number 3653992
3. Y. Ito (applicant, Yamaha Motor Company), Patent release publication, patent publication 2013-142678, released July 22, 2013
4. H. Toda, M. Takata, T. Ohgaki, M. Kobayashi, T. Kobayashi, K. Uesugi, K. Makii, Y. Aruga, Adv. Eng. Mater. **8**, 459–467 (2006)
5. R. Mizutani, R. Saiga, S. Takekoshi, M. Arai, A. Takeuchi, Y. Suzuki, Microscopy Today **23**(5), 12–17 (2015)
6. H. Toda, K. Uesugi, A. Takeuchi, K. Minami, M. Kobayashi, T. Kobayashi, Appl. Phys. Lett. **89**, 143112 (2006)
7. H. Su, H. Toda, R. Masunaga, K. Shimizu, H. Gao, K. Sasaki, Md.S. Bhuiyan, K. Uesugi, A. Takeuchi, Y. Watanabe, Acta Mater. **159**, 332–343 (2018)
8. H. Toda, A. Takijiri, M. Azuma, S. Yabu, K. Hayashi, D. Seo, M. Kobayashi, K. Hirayama, A. Takeuchi, K. Uesugi, Acta Mater. **126**, 401–412 (2017)
9. H. Toda, D. Seo, Section 3.4, "Fracture behavior analysis in a two-phase steel using phase contrast imaging". Flex. Strong Steel Mater. NTS (2016), 185–198
10. H. Toda, E. Maire, S. Yamauchi, H. Tsuruta, T. Hiramatsu, M. Kobayashi, Acta Mater. **59**, 1995–2008 (2011)
11. H. Toda, T. Nishimura, K. Uesugi, Y. Suzuki, M. Kobayashi, Acta Mater. **58**, 2014–2025 (2010)
12. Q. Zhang, H. Toda, Y. Takami, Y. Suzuki, K. Uesugi, M. Kobayashi, Phil. Mag. **90**, 1853–1871 (2010)
13. M. Kobayashi, H. Toda, A. Takijiri, A. Takeuchi, Y. Suzuki, K. Uesugi, ISIJ Int. **54**, 141–147 (2014)
14. Y. Takeichi, T. Watanabe, Y. Niwa, S. Kitaoka, M. Kimura, Microsc. Microanal. **24**, 484 (2018)

15. H. Li, H. Toda, K. Uesugi, A. Takeuchi, Y. Suzuki, M. Kobayashi, Mater. Trans. **56**, 424–428 (2015)
16. H. Toda, T. Kamiko, Y. Tanabe, M. Kobayashi, D.J. Leclere, K. Uesugi, A. Takeuchi, K. Hirayama, Acta Mater. **107**, 310–324 (2016)
17. H. Toda, T. Hidaka, M. Kobayashi, K. Uesugi, A. Takeuchi, K. Horikawa, Acta Mater. **57**, 2277–2290 (2009)
18. H. Toda, I. Sinclair, J.-Y. Buffière, E. Maire, K.H. Khor, P. Gregson, T. Kobayashi, Acta Mater. **52**, 1305–1317 (2004)
19. H. Toda, T. Inamori, K. Horikawa, K. Uesugi, A. Takeuchi, Y. Suzuki, and M. Kobayashi: Materials Transactions, 54(201), 2195–2201
20. L. Grodzins, Nucl. Instrum. Methods Phys. Res. **206**, 541–545 (1983)
21. W. Graeff, K. Engelke, *Handbook of Synchrotron Radiation*, vol. 4 (Elsevier, Amsterdam, 1991), pp. 361–405
22. M. Vopálenský, D. Vavřik, and I. Kumpová: Proceedings of the 7th Conference on Industrial Computed Tomography, Leuven, Belgium (iCT 2017), (2017), 7 pages
23. H. Toda, M. Kobayashi, T. Kubo, K. Moizumi, D. Sugiyama, Y. Yamamoto, T. Harada, K. Hayashi, Y. Hangai, Y. Murakami, Journal of Japan Institute of Light Metals **63**, 343–349 (2013)

Chapter 7
Fundamentals of 3D Images

Until now, we have covered the fundamentals of X-rays and then, sequentially learned about image reconstruction, various component devices, various applied technologies, and the realities of 3D imaging. Now, we finally envision the actual process of conducting 3D imaging. Applying what we have learned until now, we select a suitable device in the case of an industrial X-ray CT and set up suitable component devices in the case of synchrotron radiation X-ray tomography. Imaging conditions should be firmly established based on solid foundational knowledge and guidelines. If the device used is one of the latest industrial X-ray CT scanners, then user-friendly software may automatically determine imaging conditions to some extent. However, the ability to independently determine imaging success by looking at the obtained 3D image and pursuing more favorable conditions can make a significant difference in whether the imaging in X-ray tomography was successful.

X-ray tomography has been the traditional choice for imaging complex internal or external structures, items with parts having varying degrees of X-ray absorption, or items with complex regions or microstructures. It is difficult in such cases to obtain images with optimal image quality using imaging conditions that are consistently the same, pre-determined, or completely automatic. Furthermore, X-ray tomography generates unique artifacts in the obtained 3D image. Understanding this is also an essential component. Furthermore, it is often the case with industrial X-ray CT scanners or synchrotron radiation X-ray tomography that various material or sample sizes must be handled with a single device. In contrast, the subjects of medical X-ray CT scanners are restricted to human bodies and all hospitals seek to diagnose illnesses in the same organ with the same image quality standard. For this reason, image quality assessments that are unique to the device are important for medical X-ray CT scanners. However, the image quality assessments of each 3D image obtained when various material and sample sizes are used are essential with industrial X-ray CT scanners or synchrotron radiation X-ray tomography. It is through suitable image quality assessments that researchers and engineers can determine whether the 3D image sufficiently reflects the internal structure to be observed or where the observable limits apply to the size and contrast.

H. Toda, *X-Ray CT*,
https://doi.org/10.1007/978-981-16-0590-1_7

The objective of this chapter is to develop a deeper understanding of 3D imaging, conducting a simple examination of the images obtained to gain a general understanding of the advantages and disadvantages of the approach and of the defects that can occur during the process. We will also devote space in this chapter to a more detailed analysis of 3D imaging, as well as a quantitative understanding of its pros and cons.

7.1 3D Image Structure

The 3D picture element that corresponds to the *pixel* that comprises a 2D image is referred to as a *voxel*. A voxel is not necessarily a cube. Until the miniaturization of line sensor camera elements, it was typical for medical X-ray CT scanners to have *anisotropic voxels* (Fig. 7.1b for reference) whose measurement pitch in the human body axis direction (i.e. slice thickness) was long. Industrial X-ray CT scanners and synchrotron radiation X-ray tomography typically generate 3D images comprising cubic *isotropic voxels* (Fig. 7.1a for reference) using 2D detectors. In this case, the voxel size of the 3D image is stipulated by the effective pixel size (so-called pixel size) determined by the vertical/horizontal pixel size of the detector and the magnification before and after visible light conversion. Cross-sections that are orthogonal to the human body axis are often used for diagnosis in medical X-ray CT scanners, but the cross-sections for the three intersecting directions are in most cases equally important

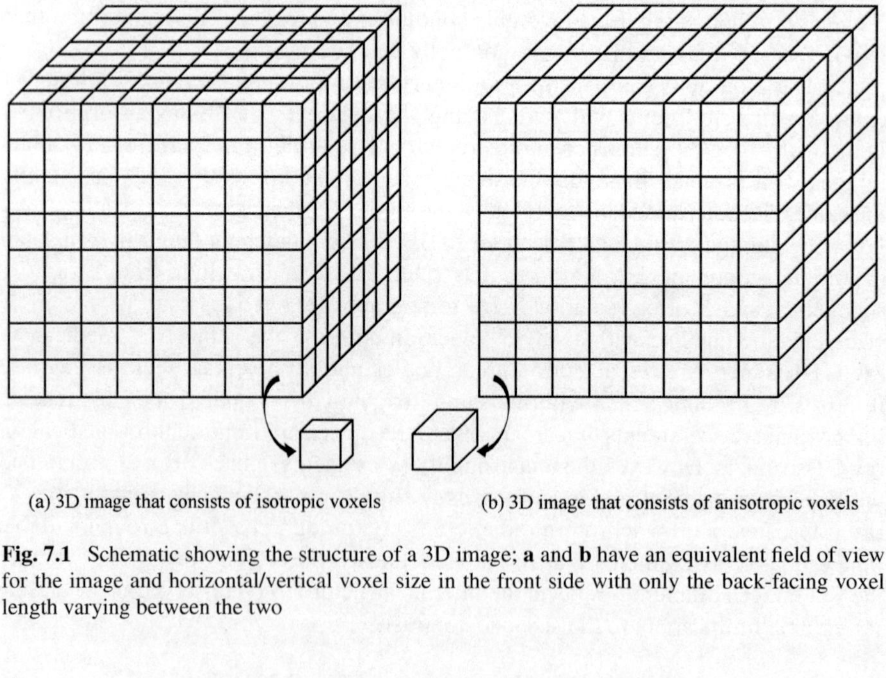

(a) 3D image that consists of isotropic voxels (b) 3D image that consists of anisotropic voxels

Fig. 7.1 Schematic showing the structure of a 3D image; **a** and **b** have an equivalent field of view for the image and horizontal/vertical voxel size in the front side with only the back-facing voxel length varying between the two in industrial X-ray scanners and synchrotron radiation X-ray tomography. For these reasons, isotropic voxels have become necessary. The areas shown by images in all directions are assessed using *multi-planar reconstructions* (*MPR*) shown in Fig. 7.2.

Image file formats include stacks of 2D images in general formats, like TIFF, BMP, or JPEG, or DICOM, which is a standard format used in medical imaging.

Images obtained by X-ray tomography primarily use 16-bit ($2^{16} = 65{,}536$ gradations), 12-bit ($2^{12} = 4{,}096$ gradations), and 8-bit ($2^{8} = 256$ gradations) bit-depth parameters. For example, the reconstructed linear absorption coefficient distributions in absorption contrast tomography are expressed in a *grayscale* such as 8-bit.

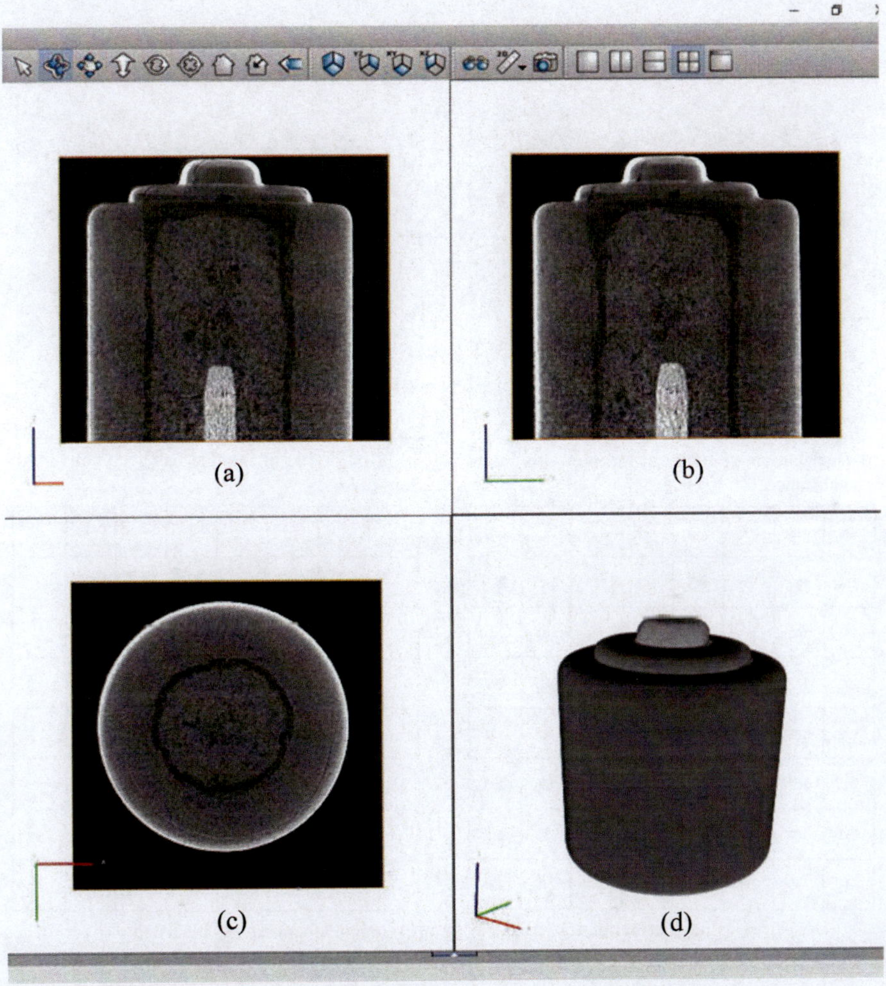

Fig. 7.2 Example of an MPR display of a 3D image of a battery; **a** *x*–*z* cross section, **b** *y*–*z* cross section, **c** *x*–*y* cross section, and **d** 3D image

Furthermore, distributions of the amounts measured by methods like phase-contrast tomography and various applied imaging methods are displayed. Figure 7.3 images a sample whose linear absorption coefficient gradually changes with distance from the center and which is shown with an 8-bit grayscale ranging from white (gray value of 255) to black (gray value of 0). Furthermore, Fig. 7.3d shows a *binarization* where a given *gray value* was set as a *threshold value*. This is, so to speak, a 1-bit black and white image that is used for structure size and morphology assessments. A decreased bit count of a 3D image reduces the amount of information but also reduces the amount of data, making handling easier. For example, an 8-bit 2,048 × 2,048-pixel image requires 4 MB of data but this doubled to 8 MB with a 16-bit

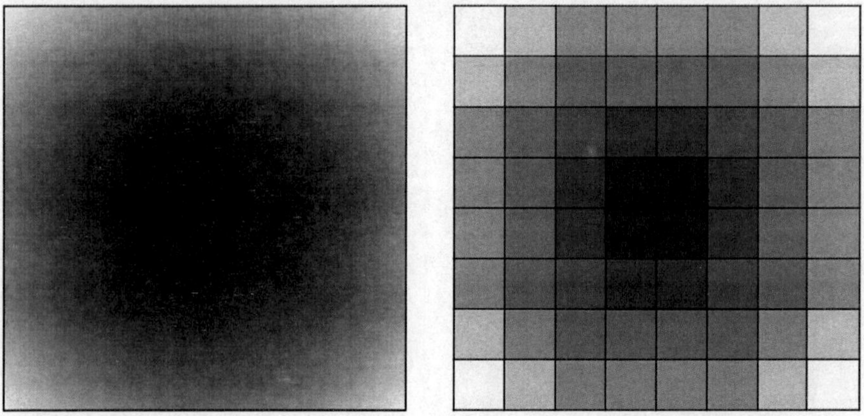

(a) Distribution of linear absorption coefficient in a substance

(b) 8-bit digital image that has been converted from (a)

234	215	205	192	192	205	215	234
215	187	168	143	143	168	187	215
205	168	80	35	35	80	168	205
192	143	35	10	10	35	143	192
192	143	35	10	10	35	143	192
205	168	80	35	35	80	168	205
215	187	168	143	143	168	187	215
234	199	215	192	192	205	215	234

(c) Grey value distribution for (b)

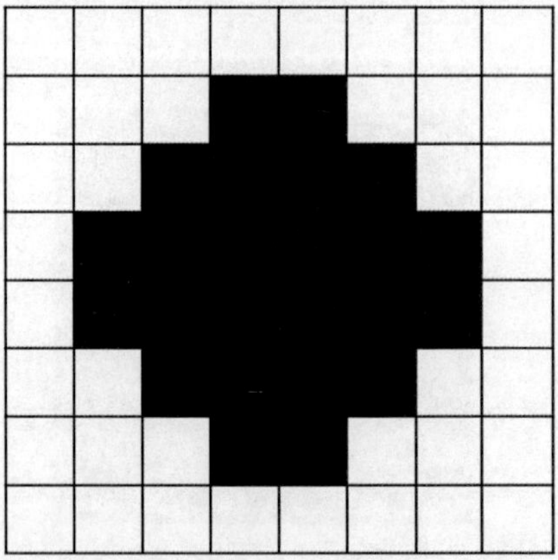

(d) Image obtained by binarizing (b) with a threshold value of 150

Fig. 7.3 Pixel display of a sample with a linear absorption coefficient gradient and schematics, which show its gray value distribution **a** is the distribution of linear absorption coefficient in a substance; **b** is an 8-bit digital image that has been converted from **a**; **c** is grey value distribution for **b**; and **d** is an image obtained by binarizing **b** with a threshold value of 150 image. Sufficient results can be obtained with an 8-bit depth if the imaging objective is only to display the distributions of numerous structures. Meanwhile, images with a larger bit depth (e.g. 16-bit) should be used when it is necessary to express differences in pixel values over 100 times larger, when there are many structures to be distinguished, or when a given structure has a gray value distribution and needs to be visualized. Thus, this should be determined on a case-by-case basis. In addition to these black-and-white or *binary images*, 3D images can also display multiple scalar quantities or vector values. For example, color images with intensities for the three RGB (red, green, blue) components can be displayed as well.

Chapter 4 discussed various types of noise such as photon noise that is unavoidable with X-ray tomography. Many artifacts are also unavoidable (discussed later). For these reasons, the grayscale range of 3D images as captured often has default settings in image reconstruction software that cover an unnecessarily-wide range of linear absorption coefficients, such as setting the maximum and minimum values of the linear absorption coefficients as the upper and lower edges, respectively. With this in mind, this must be changed to a grayscale range that corresponds to the linear absorption coefficients of multiple internal structures to be observed, as shown in Fig. 7.4. Furthermore, the bit count should be simultaneously reduced as necessary, as shown in Fig. 7.4. For example, a clear 3D image can be obtained when there are multiple structures present in a sample by setting the structure A with the lowest gray value as black (0 for an 8-bit gray value) and structure B with the highest gray value as white (255 for an 8-bit gray value). However, visualizing the gray value distributions of structures A or B requires having the observed gray value distribution fit inside an 8-bit gray value range. For this reason, it is not uncommon to use different gray value ranges for each structure to be analyzed. In either case, the basic principle is to look at the 3D image and assess it while constantly checking which gray values correspond with each of the internal structures that can be determined.

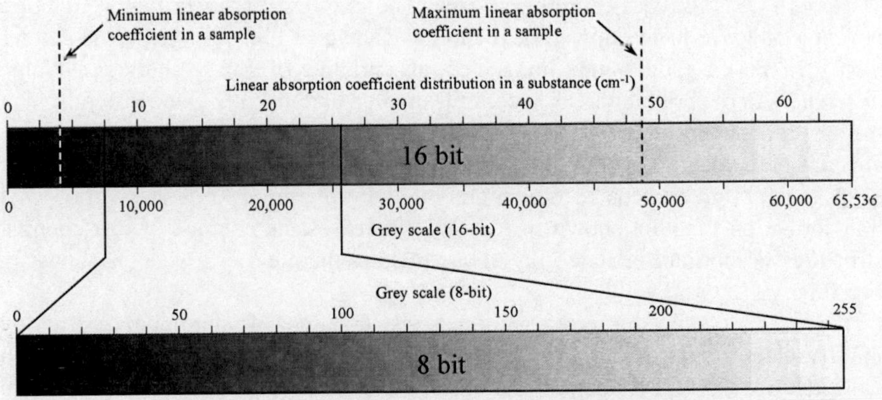

Fig. 7.4 Transformation from a 16-bit to 8-bit image, and the simultaneous change in gray-scale range

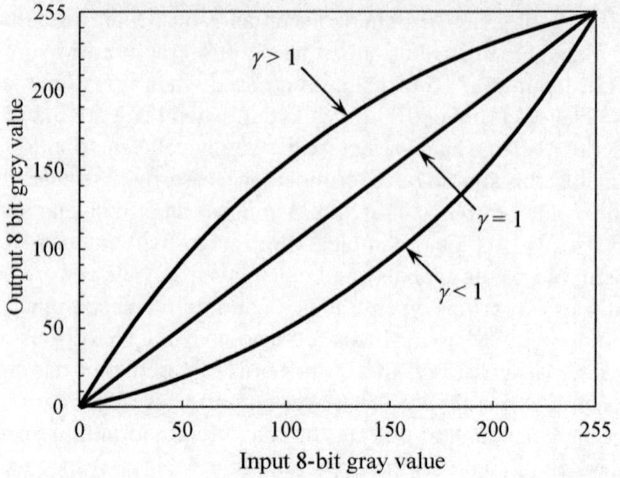

Fig. 7.5 Gray-level transformation function used for the gamma correction. Gray-level transformations for three different gamma values are shown

Incidentally, the transformation of the grayscale range usually follows a linear relationship. However, non-linear *gray-level transformation functions* such as those shown in Figs. 7.5 and 7.6 can also be used as necessary. Representative examples include the *gamma correction* shown below. The relationship between the input gray value g_n and output gray value g_{out} is as follows:

$$g_{out} = g_{max}\left(\frac{g_{in}}{g_{max}}\right)^{\frac{1}{\gamma}} \tag{7.1}$$

Here, g_{max} is the maximum gray value. The gray-level transformation function becomes concave in the upward direction as shown in Fig. 7.5 when γ is greater than 1. In this case, the output image becomes brighter overall, with the gray value differences in the brighter areas becoming smaller, and the gray value differences in the darker areas becoming greater. The gray-level transformation function becomes concave in the downward direction when γ is less than 1 and the gray value differences in the brighter areas become highlighted. Furthermore, an S-shape gray-level transformation function shown in Fig. 7.6a increases the contrast of low-contrast structures in the intermediate gray areas, making them easier to see, as shown in Fig. 7.6c.

Figure 7.7a shows a cross-sectional visualization of aluminum cast alloy microstructures using X-ray tomography. This figure is an 8-bit image that has been converted from a 16-bit image. Furthermore, Fig. 7.7b is the gray value histogram after 8-bit conversion. This sample includes gray-colored pores dispersed in the aluminum matrix (hydrogen interior: gray value of 0), intermetallic compounds that include iron (gray value of 255), and silicon particles whose gray values are close to

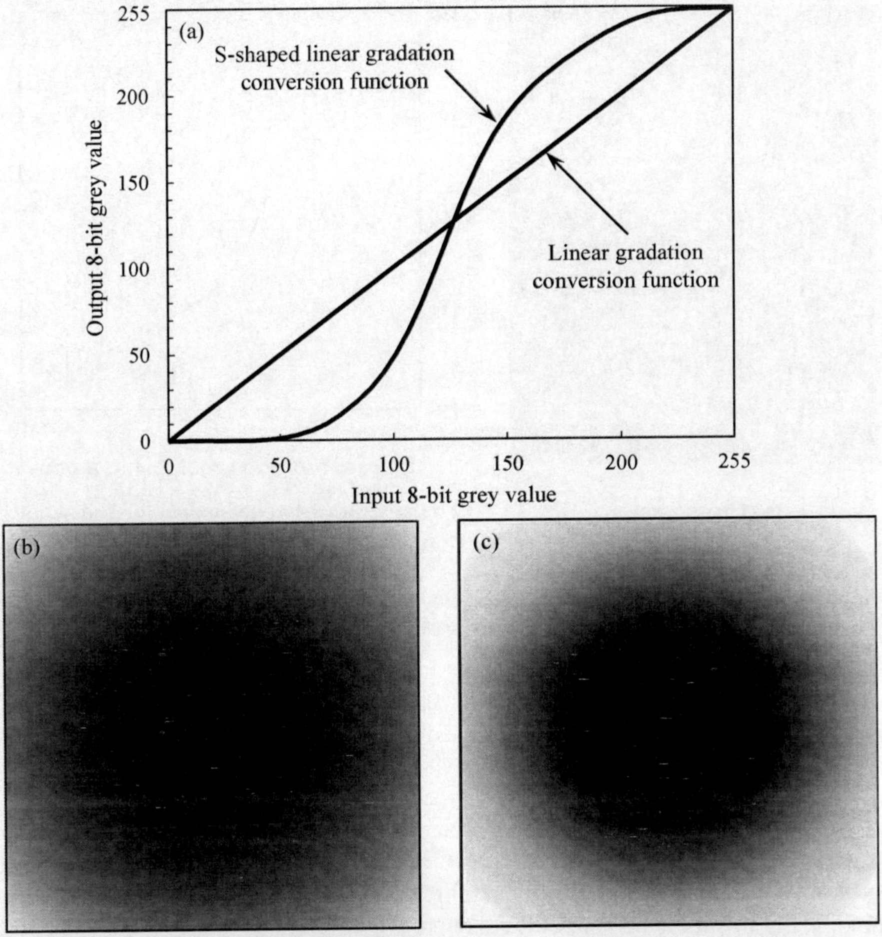

Fig. 7.6 Transformation of the image shown in Fig. 7.3. **a** using a non-linear gray-level transformation function; **b** is the original image and **c** is the image after transformation

aluminum. The pores can be clearly distinguished in the gray value histogram, but the other phases are not visible as peaks in the gray value histogram. It is extremely important to look at the obtained imaged in conjunction with the histogram, identify where the structures of interest fall on the gray value histogram, and pre-emptively investigate how to express those structures within the 3D image.

Changing the linear absorption coefficient range covered in a 3D image obtained by X-ray tomography and performing filtering operations on it are conducted routinely. Care must also be taken when handling these types of images. Careful consideration must be afforded to the cutting/pasting of images and processing only one section of an image, as well as to comprehensive secondary processing procedures such as nonlinear gray-level transformations. For example, if conducting a

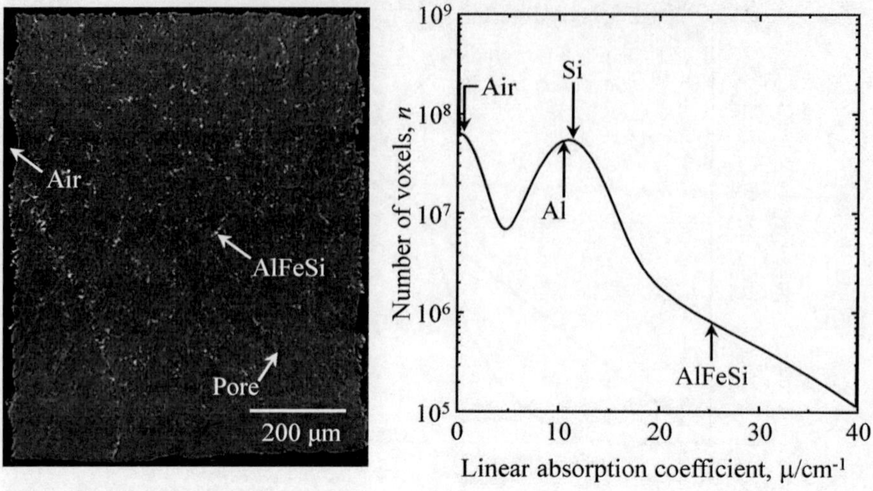

(a) Cross-section

(b) Histogram of pixel value for the whole 3D image

Fig. 7.7 **a** Cross-section image of Al-7% Si aluminum cast alloy using X-ray tomography. The imaging beamline BL20XU at SPring-8 was used and imaging was conducted with an X-ray energy of 20 keV; **b** histogram of gray value following 8bit conversion

gray-level transformation, the function used should be clearly stated, and the purpose and procedures used in this processing should be verifiable by a third party. The reason for this is that unexpected data processing may be perceived as a breach of research ethics by a researcher and of data manipulation by an engineer. Processing which changes the results derived from an image is not in the least acceptable. This strikes at the core of the trust and reliability that form the basis of science and technology. Scientists should apply the foundational knowledge learned from this book and proceed carefully and humbly with the analysis of 3D images, so that they may not, in their haste, whether intentionally or otherwise, swear black is white in the interpretation of the data.

7.2 Examination of 3D Images

The quality of 3D images is discussed before delving into image examination methods. The term "image quality" is not defined even in standards such as the Japanese Industrial Standards (JIS B7442: 2013), which stipulate the specialist terminology relating to industrial X-ray CT scanners [1]. Definitions relating to this term are also not included in the medical radiation terminology list (updated edition in 2012) relating to medical X-ray tomography; the first edition was published far earlier in 1991 in Japan [2]. The medical image engineering handbook issued in

2012 states, "the definition of image quality is ambiguous, and has been an ongoing issue of investigation." [3] The perspective of some is that "image quality = resolution" but this is not entirely correct. The most general understanding of image quality is one that involves considerations of spatial resolution, detectability, noise, contrast, and artifacts. Each of these individual factors should be quantitatively assessed with methods that can be reproduced using objective measures. This chapter will look at these factors in order. Using the term "image quality" to express these elements, these individual factors will be comprehensively assessed with some degree of subjectivity, considering the objectives and significance of the 3D images used in research and development.

After measuring with X-ray tomography, image reconstruction of either a single cross-section or the entire data is conducted using one of the image reconstruction methods discussed in Chap. 3. The first step is to visually examine the quality of the 3D image, checking whether it requires re-measurement or whether there are any artifact-related issues (discussed later in Sect. 7.6). If the imaging is deemed to have been unsuccessful due to some unexpected reason, this is investigated and the measurement must be repeated. As detailed in Sect. 3.2.1, either investigating the sinogram or re-confirming the obtained transmission image stack as a fast-forwarded video is effective at this stage. The latter can likely be completed with built-in software in the device if it is an industrial X-ray CT scanner. Furthermore, if imaging was conducted using X-ray tomography at a synchrotron radiation facility, it can be completed via the image-processing features of the software that controls the CCD or sCMOS camera and acquires the images [4] or via software released by the beamline in the synchrotron radiation facility [5]. Similar assessments can be conducted with general image processing software when these cannot be used. For example, the image processing software ImageJ can easily conduct the confirmation of transmission images if the images are imported in the form of a TIFF-image stack [6]. Figure 7.8 shows an example of this confirmation process.

Other simple methods for checking for image quality deterioration involve comparisons between 0° and 180° transmission images. The 0° and 180° images will be the same when the range for transmission image acquisition is 0° to 180°. The only differences are that the image is flipped horizontally with regard to the rotational axis and the imaged time is shifted by almost exactly a single X-ray tomography imaging duration. The former can be used for the assessment of system alignment, such as the inclination between the rotation and detector axes and the horizontal shift in the rotation axis from the center of the detector. The latter can also be used to determine the amount of sample drift and deformation that occurred during an imaging period. A schematic of this is shown in Fig. 7.9. Slow sample movements due to factors such as the thermal expansion, plastic deformation, and creep deformation of the sample or sample holder are difficult to detect from visual inspections of the sinogram or transmission image video. However, this can be clearly yet simply determined with comparisons between the 0° and 180° transmission images.

A summary of the representative issues that occur during imaging is shown in Table 7.1. The primary methods, which can investigate the causes of these issues are also shown in Table 7.1. These methods should be used to troubleshoot these issues.

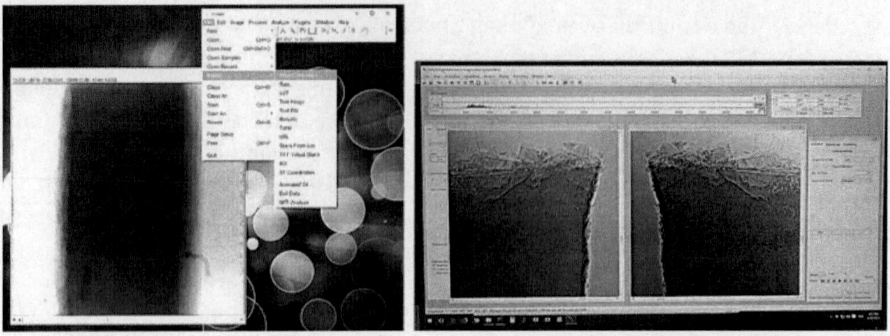

(a) Checking a transmission image using ImageJ

(b) Confirmation of transmission images in a synchrotron radiation facility

Fig. 7.8 **a** Screenshot of a display where transmission images are confirmed using a fast-forward video in ImageJ; **b** Shows an experiment image at the beamline BL20XU in SPring-8. Comparisons between 0° and 180° transmission images are conducted using the Hamamatsu Photonics software HiPic following measurements with X-ray tomography

7.3 Noise

As detailed in Sect. 4.4.1, noise is caused by various factors during X-ray imaging. Noise is unavoidable with 3D images measured by X-rays. The quantitative determination of noise is first necessary for its effective handling during X-ray tomography measurements, image processing, and image analysis.

7.3.1 Standard Deviation

The simplest way to assess noise is to calculate the *standard deviation* of the gray values in the 3D image. This is also referred to as *RMS granularity*. The standard deviation *SD* in a 3D image with *n* pixels is expressed as follows:

$$SD = \sqrt{\frac{1}{n}\sum\nolimits_{i=1}^{n}(g_i - \bar{g})^2} \tag{7.2}$$

Here, g_i is the gray value of the *i*th pixel and $\bar{g}$ is the average gray value of all pixels. The normalized version of this, referred to as the *normalized standard deviation NSD*, is also used for noise assessment.

$$NSD = \frac{1}{\bar{g}}\sqrt{\frac{1}{n}\sum\nolimits_{i=1}^{n}(g_i - \bar{g})^2} \tag{7.3}$$

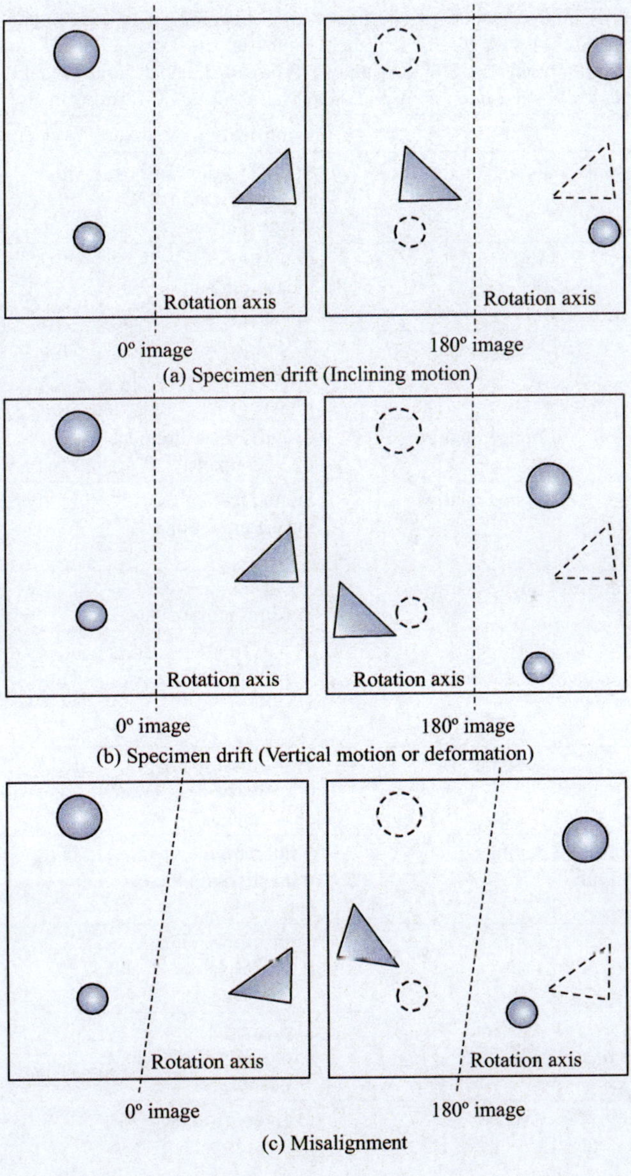

Fig. 7.9 Schematic showing the determination of the causes for decreased image quality using comparisons of 0° and 180° transmission images; **a** and **b** show deformation or vertical/horizontal movement of the sample or its supporting holder; **c** shows a case of misalignment between the rotation axis of the sample rotation stage and the detector. Normal conditions are when the rotation angle is vertical and when the same structures in the 0° and 180° images are symmetrical with respect to the rotation axis

Table 7.1 Representative issues that occur during imaging, their types, and methods for investigating its causes. Issues that arise in both industrial-use X-ray CT (ICT) and synchrotron radiation X-ray tomography (SRCT) are summarized. SRCT and ICT in the table each refer to synchrotron radiation X-ray tomography and industrial-use X-ray CT, respectively

Trouble	Type	Confirmation method	Associated technique
Shortage of spatial resolution	Image quality	Spatial resolution measurement in a 3D image	All
High noise level	Image quality	S/N ratio measurement in a 3D image	All
Shortage in flux	Image quality	S/N ratio measurement in a 3D image	All
Poor contrast	Image quality	Contrast measurement in a 3D image	All
Eccentricity and surface runout of a rotation stage	Image quality	Spatial resolution measurement in a 3D image	Mainly SRCT
Intensity fluctuation of an Xray beam	Image quality	Sinogram, movie of transmission images, etc.	Mainly SRCT
Nonuniformity and speckle noise of an X-ray beam	Image quality	Transmission image	Mainly SRCT
Misalignment	Artifact	Transmission image (Comparison of 0° and 180° images)	SRCT
Missing wedge of information (Angular limitation)	Artifact	Sinogram, movie of transmission images, etc.	Mainly SRCT
Sample drift	Artifact	Transmission image (Comparison of 0° and 180° images)	All
Sample deformation/change	Artifact	Transmission image (Comparison of 0° and 180° images)	All
Protrusion of a sample from a field of view	Artifact	3D reconstructed image	All
Beam hardening	Artifact	Reconstructed image (Virtual cross section)	Mainly ICT
Ring artifact	Artifact	Reconstructed image (Virtual cross section)	All
Metal artifact	Artifact	Reconstructed image (Virtual cross section)	All

(continued)

Table 7.1 (continued)

Trouble	Type	Confirmation method	Associated technique
Scattering artifact	Artifact/image quality	S/N ratio measurement in a 3D image, etc.	All
Undersampling	Artifact	Reconstructed image (Virtual cross section)	All
Corn beam artifact	Artifact	Reconstructed image (Virtual cross section)	ICT
Deviation of a rotation axis	Artifact	Reconstructed image (Virtual cross section)	All
Deflection artifact	Artifact	Reconstructed image (Virtual cross section)	Mainly SRCT

The standard deviations of multiple 3D images with variable exposure times are measured as an example. Images that have too short of an exposure time will inevitably have a large amount of noise. The standard deviations of the images in Fig. 7.10a–d are written in the figures. The amount of noise can be effectively expressed by using the standard deviation. However, we can also easily imagine that assessments with only standard deviation would not be possible if the gray value distributions in areas other than the noise are different. Considering the two types of samples shown in Fig. 7.11, the 3D image of sample A has lower noise but larger fluctuations in the gray value of the internal structure, while the structure of sample B is relatively homogeneous but has a poor image S/N ratio. The standard deviations of the gray values in both cases are 13. As the average gray values for each are also 48.5, the normalized standard deviations are both the same (i.e., 0.27). In these cases, assessments using the noise power spectrum discussed in the next section are necessary.

7.3.2 Noise Power Spectrum

The *noise power spectrum* (*NPS*) was previously referred to as the *Weiner spectrum* (*WS*) but has been currently standardized to the former by International Electrotechnical Commission (IEC) stipulations [7].

A 2D Fourier transform $F(u, v)$ of the image $g(x, y)$ whose noise is to be assessed is defined by Eq. (3.5). The mathematical definition of the noise power spectrum *NPS* shown in two dimensions is set as the square of the absolute value of the Fourier transform, as shown in the following equation [8].

$$NPS(u, v) = \lim_{X,Y\to\infty} \frac{1}{XY}\left\langle\left|\int_{-X/2}^{X/2}\int_{-Y/2}^{Y/2} g(x, y)e^{-2\pi i(xu+yv)}dxdy\right|^2\right\rangle \tag{7.4}$$

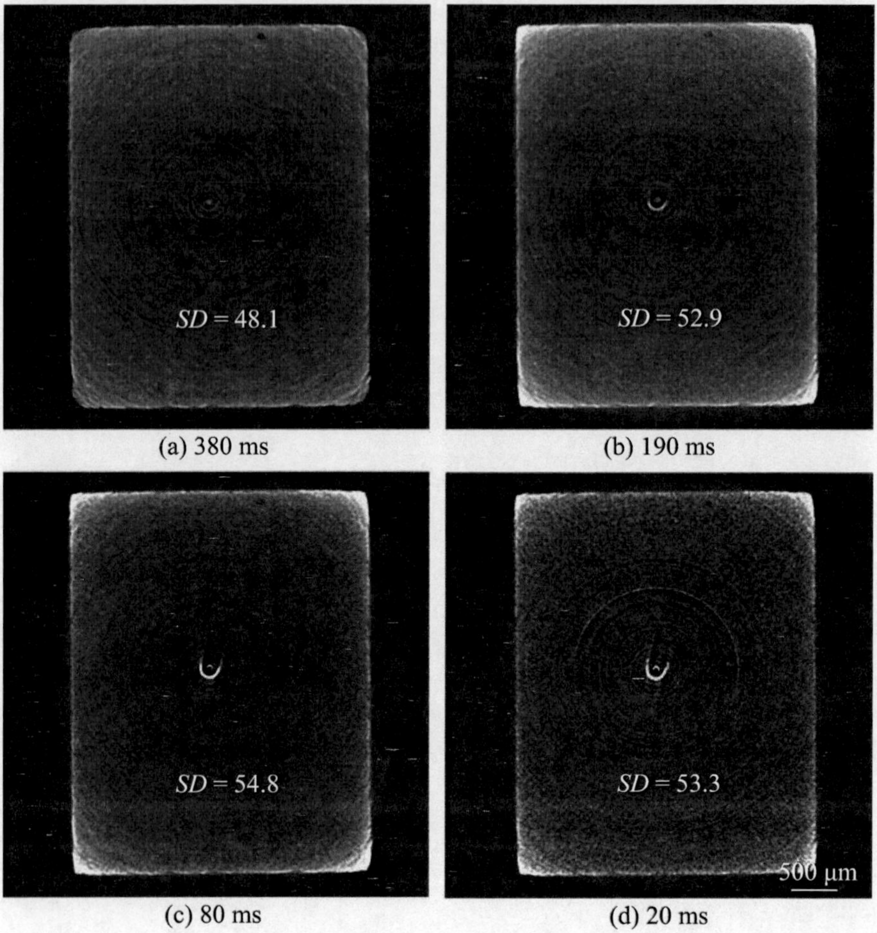

Fig. 7.10 Measurement results of noise due to standard deviation values produced when the exposure time in 3D imaging of the same sample was changed to induce different noise levels. Imaging conducted with the same device and imaging conditions are described in Sect. 7.6. **a**, **b**, **c** and **d** are 380, 190, 80 and 20 ms in exposure time, respectively

Here, X and Y are the sample sizes in the x and y directions. As such, the pixel sizes in the x and y directions are set as Δx and Δy and the pixel numbers are set as N_x and N_y, which yield $X = \Delta x N_x$ and $Y = \Delta y N_y$. The angle brackets in Eq. (7.4) signify the average of the infinite set of images. Furthermore, the noise power spectrum is a Fourier transform of an auto-correlation function. Equation (7.4) is effective only for cases of a Gaussian distribution where the average is 0; there are no images with an infinitely large size in reality [8]. As such, calculating the noise power spectrum in the region of interest within an actual image for a two-dimensional case yields the following expression:

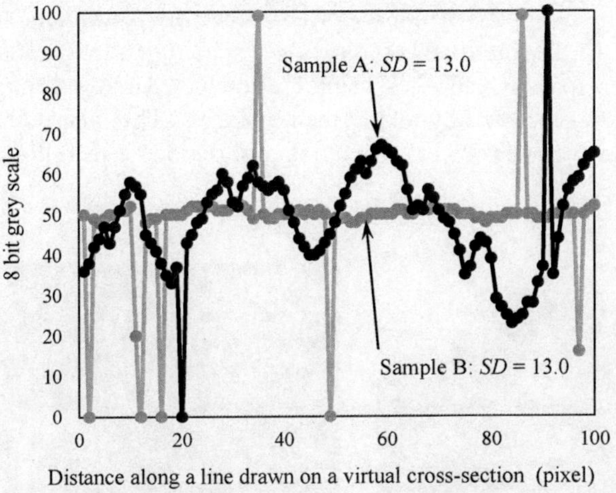

Fig. 7.11 Gray value distributions of two samples whose internal structure and microstructure are entirely different. A line segment is drawn across a given cross-section in a 3D image and the gray value histogram along that line segment is plotted. Sample A has large changes in X-ray absorption of the internal structure, but the image S/N ratio is favorable. Meanwhile, sample B has a relatively homogeneous internal structure, but the image has high amounts of noise

$$NPS_{2D}(u, v) = \frac{\Delta x \Delta y}{N_x N_y} \left\langle \left| DFT_{2D}\left\{ g(x, y) - \overline{g(x, y)} \right\} \right|^2 \right\rangle \tag{7.5}$$

Here, DFT refers to the discrete Fourier transform, which also appeared in Chap. 3. The above-mentioned equation applies not only to the (x, y) cross-section but the (y, z) and (z, x) cross-sections as well. Furthermore, the 3D noise power spectrum of a given region of interest is as follows:

$$NPS_{3D}(u, v, w) = \frac{\Delta x \Delta y \Delta z}{N_x' N_y' N_z'} \left\langle \left| DFT_{3D}\left\{ g(x, y, z) - \overline{g(x, y, z)} \right\} \right|^2 \right\rangle \tag{7.6}$$

Here, $\overline{g(x, y)}$ and $\overline{g(x, y, z)}$ are the average values and are subtracted in both equations for image trend elimination. Furthermore, N_x', N_y', and N_z' are the sizes of the region of interest (pixel number). The 1D noise power spectrum typically takes the average of *NPS* in the same spatial frequency (circumference of each concentric circle) of the 2D noise power spectrum to determine the spatial frequency distribution of the *NPS*.

$$NPS_{1D}(\omega) = \int_0^{2\pi} NPS_{2D}(\omega, \theta) d\theta \tag{7.7}$$

Here, the angular wave frequency is set as ω, and the Cartesian coordinates were transformed to polar coordinates (ω, θ).

Measurement examples of noise power spectra with different dimensions are shown in Fig. 7.12. The intuitive 1D noise power spectrum is frequently used. Regardless of the dimension in which assessments are conducted, multiple regions of interest must be taken and assessed by taking their averages. The measurement error e_{NPS} when taking the average *NPS* in an *m* number of regions is as follows [9].

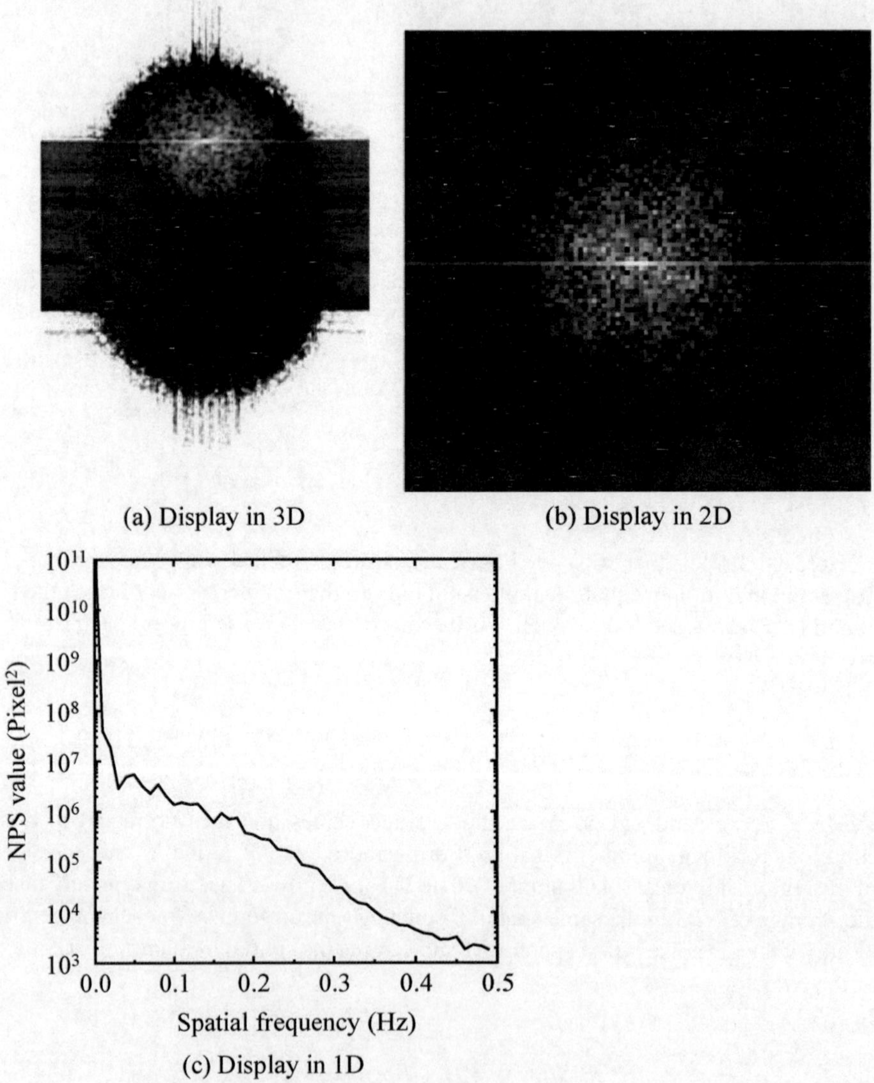

Fig. 7.12 Noise power spectrum of the 3D image shown in Fig. 7.10 **a** depicted in 1, 2, and 3 dimensions. **a** and **b** are displays in 3D and 2D, respectively. The 1D data in **c** (1D) is the data in **b** averaged in the circumferential direction

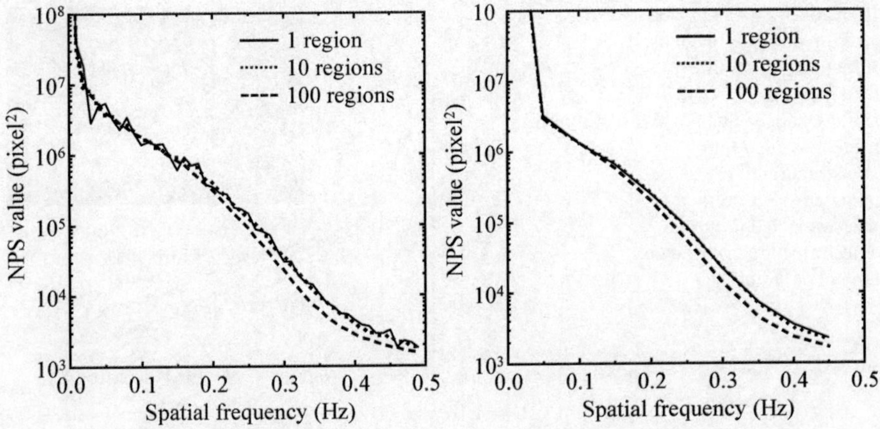

(a) Raw data (b) After applying moving average

Fig. 7.13 The 1D noise power spectrum in Fig. 7.12 where measurements were made with varying sampling region numbers; **a** is the raw data and **b** is the data after a 5-section moving average was applied

$$e_{NPS} = \sqrt{\frac{1}{m} NPS} \tag{7.8}$$

Thus, a considerably large amount of sampling must be conducted to minimize errors. For example, the error drops to 1/10 after taking the averages of 100 regions of data. Figure 7.13 shows a 1D noise power spectrum with changes to the sampling number. Multiple regions of interest in a single image, must be measured, and several image sets must be prepared for a single sample to minimize error. Figure 7.13 shows the result of applying a 5-section moving average on data. These types of data processing are occasionally conducted to capture overall trends. Even measurements in a single region are capable of giving a picture of overall trends using the moving average, but they are inaccurate compared with measurements using 100 regions.

Figure 7.14 shows a 1D noise power spectrum of an image that was reconstructed with filtered back projection following the application of various reconstruction filters on projection data of aluminum-copper alloys in Figs. 3.27 and 3.28. The vertical and horizontal axes of Fig. 7.13 each show the extent of noise and spatial frequency. Although this image was measured using synchrotron radiation, noise ranging from coarse noise with low spatial frequency to fine noise with high spatial frequency is distributed in the image. Among these, filters with high spatial resolution capabilities like the Ram-Lak filter and the Shepp-Logan filter result in increased noise across a wide-ranging spatial frequency range. Meanwhile, filters with relatively low spatial resolution functionality such as the Hamming filter, Hann filter, and Parzen filter show considerably low noise.

The above-mentioned *NPS* calculations are relatively straightforward when using package software such as ImageJ or MATLAB. ImageJ also has a plug-in for NPS

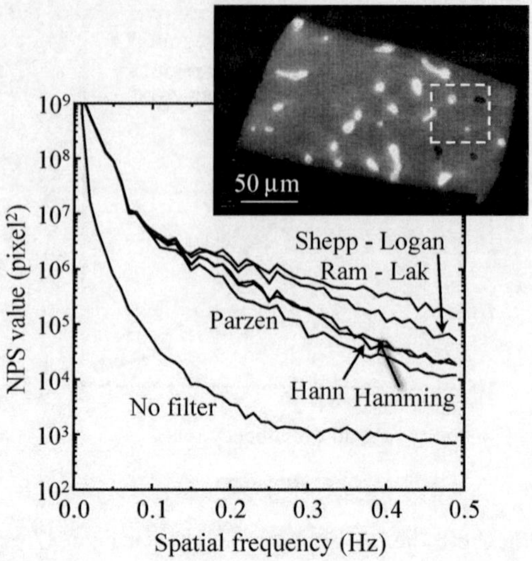

Fig. 7.14 1D noise power spectrum of a 2D image used for reconstruction filter assessments in Figs. 3.27 and 3.28 in Chap. 3. The noise differences due to the reconstruction filter can be compared. Measurements were conducted in the white-framed region shown in the photograph

calculations [10]. Simple calculations can still be conducted even if one does not have much experience using this type of analysis software if existing software such as CT image calculation programs released by the Japanese Society for CT Technology is used [10]. Software from the Japanese Society for CT Technology also includes MTF calculation functionalities (discussed later). These include trial versions for non-members mentioned in the reference, which allow for many more functionalities once becoming a member of the society [11].

7.4 Contrast

7.4.1 Basic Concepts

As mentioned in the introduction, device characteristics must be assessed in medical X-ray CT scanners through the image quality of 3D images. A standardized reference referred to as a phantom is used in these cases [12]. Measuring these allows for the assessment of image quality (including contrast) to determine performance changes in the device over time, daily. With regard to contrast, a term referred to as contrast resolution is stipulated in the Japanese Industrial Standards, which includes a list of terms for industrial X-ray CT devices. According to this, contrast resolution refers to the following: "the difference in industrial CT values (average) between a given section and its background which are at the limit of detection, taking into account the noise in a concentration resolution image" [1]. Noise has a large influence when the contrast between the background and observation subject is low. This is because the X-ray sources, detectors, and reconstruction functions appear similar in all the hospitals, particularly in medical X-ray CT scanners. Specifically, assessments are conducted using the standard deviation and noise power spectrum discussed in the previous section, as well as the *contrast-to-noise ratio*. In this manner, contrast is assessed as an index that quantitatively describes the visual impressions when looking at the same section of a human body. There are multiple definitions for the contrast-to-noise ratio; please refer to the works cited for further information [13].

In contrast, it is important to conduct image quality assessments on all 3D images obtained using a diverse set of materials and sample sizes with industrial X-ray CT scanners and synchrotron radiation X-ray tomography. Suitable image quality assessments allow researchers to determine whether the internal structure to be observed is sufficiently reflected in the obtained 3D image and where the observable limits are in terms of size or contrast.

Image quality assessments handle images that vary greatly by X-ray source, detector, reconstruction method, and their various associated conditions. The basic principles for obtaining contrast in these cases not only involve differences in linear absorption coefficients obtained during absorption contrast tomography but the various methods discussed in Sect. 2.2 or Chap. 5. With this in mind, it has been the author's experience that assessments of noise, artifacts, and contrast should be conducted independently for industrial X-ray CT scanners and synchrotron radiation X-ray tomography. Furthermore, unlike with medical X-ray CT scanners, where visual assessments are the fundamental component, industrial X-ray CT scanners and synchrotron radiation X-ray tomography often seek volume rendering based on segmentation (discussed in Chap. 8) and quantitative assessment and measurement on morphology and size. Thus, the ultimate measure for contrast quality following the application of suitable filters (discussed in Chap. 8) involves whether the various phases and regions in the sample interior can be accurately segmented while removing noise and artifact effects. It is vital when applying industrial X-ray CT scanners and synchrotron radiation X-ray tomography for cutting and polishing samples to first conduct two-dimensional observations; these observations can be obtained using devices capable of easily acquiring a wealth of information on factors such as chemical composition at a higher spatial resolution (such as scanning electron microscopes). The existence and details of the structure to be observed should be understood as thoroughly as possible beforehand, as diverse samples will be observed with the limited spatial resolution and contrast of X-ray tomography.

7.4.2 Quantitative Assessment

There are several standard measures of contrast. One of these, referred to as the *Weber contrast* C_W, is as shown below:

$$C_W = \frac{g_f - g_m}{g_m} \tag{7.9}$$

Here, g_f and g_m refer to the gray values for the region/phase and the background, respectively, of the observation subject. The Weber contrast is suitable when a small region is dispersed in a homogeneous background. Furthermore, the *Michelson contrast* C_M is as shown below:

$$C_M = \frac{g_h - g_l}{g_h + g_l} \tag{7.10}$$

Here, g_h and g_l refer to the two regions of the observation subject, with the former having the high gray value and the latter having a lower gray value. Michelson contrast is suitable for cases where there are dark and light phases. Furthermore, Eq. (7.2) is used as RMS contrast when the image patterns are random and granular. Care must be taken in each case so that the gray value of the observation subject structure, rather than the noise, is used. The contrast measurement that is most suitable for the characteristics of the spatial distribution of the structure of interest in the sample should be selected among the many measurement methods available.

The Japanese Industrial Standards that define the terminology for industrial X-ray CT device state that Michelson contrast should be calculated based on the CT value of air (according to JIS, this refers to a gray value of an arbitrary scale) [1]. Care must be taken in these cases as this is not accurately calculated when the air or observation subject structure is set outside of the scale during the grayscale or bit depth conversions shown in Fig. 7.4.

7.5 Spatial Resolution

7.5.1 Fundamental Aspects

(1) Spatial Resolution as a Device and 3D Image Spatial Resolution

Spatial resolution is an index that relates whether two points in proximity to one another can be distinguished as two different points; it is stipulated as the minimal distance in which this distinction occurs [1]. Often, the spatial resolution of X-ray tomography devices is inferior to two-dimensional visualization devices commonly used in laboratories (e.g. scanning electron microscopes or transmission electron microscopes). For this reason, sometimes sample sizes and internal structures that are close to the effective spatial resolution are assessed in X-ray tomography. Thus, it is essential that the effective spatial resolutions obtained in the 3D imaging conducted by the reader are correctly determined.

As summarized in Table 7.2, the factor with the lowest accuracy among the diverse set of factors in X-ray tomography (e.g. X-ray source, sample rotation stage, X-ray focusing element in the case of the imaging optical system, and detector) acts as the rate-limiting element that dictates the maximum spatial resolution achievable by

Table 7.2 Major elements that can potentially constrain the spatial resolution in industrial-use X-ray CT scanners and synchrotron radiation X-ray tomography. The *sign refers to factors relating to the Nyquist frequency based on the sampling theorem

Elements	Factors
X-ray tube, etc.	Effective focal spot size
	Transmissivity (X-ray energy)
Imaging optical system Enlarged projection	Spatial resolution of an X-ray focusing device
	Magnification
	Illumination system
Sample rotation stage	Eccentricity and surface runout of a rotation stage
	Sample drift
Detector Fiber optics Optical lens	Effective pixel size*
	Various noise
	Scintillator thickness
	Scintillator stopping power
	Diffraction limit of visible light
Sample	Drift due to insufficient sample fixing
	Material deformation/transformation
Imaging condition	Flux (Exposure time)
	Rotation step*
Reconstruction	Reconstruction filter
	Incomplete dataset in cone beam reconstruction
	Filtering effects during phase retrieval process
Others	Various artifact
	Blurring due to Fresnel diffraction

an X-ray CT scanner. Additionally, this can be confirmed in Eq. (5.29) and (7.30). The spatial resolution is also greatly affected by the sample, imaging conditions, and various types of noise and artifacts. Intrinsically, the spatial resolution, which expresses performance as a system, should be assessed independently of the noise, artifacts, and imaging conditions, whose magnitude and type vary with different conditions. However, noise and artifacts can be said to be unavoidable with X-ray tomography. Furthermore, these can have an influence not only on image quality but also visualization success depending on the sample or imaging conditions. For these reasons, the spatial resolution should be calculated using the 3D images obtained under the imaging conditions and environment that can realistically be achieved; moreover, a device-/sample-/imaging condition-specific local spatial resolution in a given region of interest in the sample interior should be assessed.

There are multiple measures relating to spatial resolution in medical and industrial X-ray tomography and there may be some variation or even at times confusion in their definitions as in, for example, terms such as resolution, resolving power, spatial resolution, sharpness, or detectability. Some expressions specify the resolution and resolving power for images and spatial resolution for device specifications. However, for the previously mentioned reasons, this book systematically defines spatial resolution as the spatial resolution that can be measured for 3D images, which is the outcome of X-ray tomography. Furthermore, the term "detectability" must be accurately defined as a separate measure from spatial resolution.

Of the various conditions in Table 7.2, multiple factors relating to the devices have been discussed up until now in Chaps. 4 and 5. Unfortunately, however, the spatial resolution of industrial X-ray CT scanners is frequently not presented as the maximum spatial resolution in catalogs, etc. as part of comprehensive specifications based on the factors listed in Table 7.2. Often, the focal point size of the X-ray tube or the effective pixel size is shown instead of the spatial resolution or sometimes as the spatial resolution itself. As can be observed in the examples shown below, these often produce significant errors.

Figure 7.15 shows valuable data where the sample size varied in a Russian matryoshka-doll-like pattern, as shown in Fig. 7.15a, b, and where the sample was scanned at the same position and compared among two types of industrial X-ray CT scanners and Synchrotron radiation X-ray tomography (high-resolution imaging beamline BL20XU at SPring-8) [14]. The dashed line in Fig. 7.15c is the relationship between the pixel size and the spatial resolution, which is based on the sampling

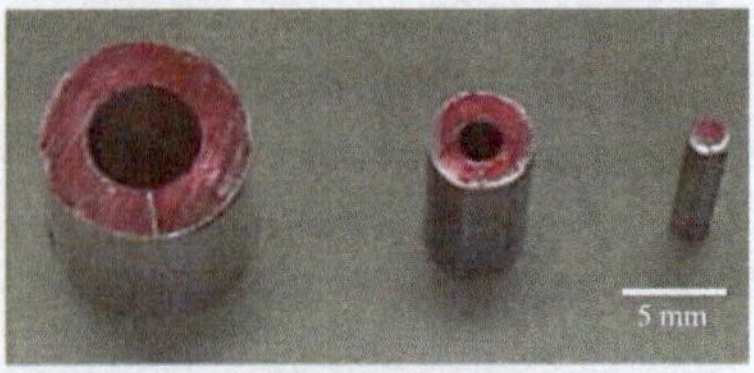

(a) A nest of test jigs for spatial resolution measurement. The right cylinder was used for the measurement.

(b) After assembling three jigs shown in (a)

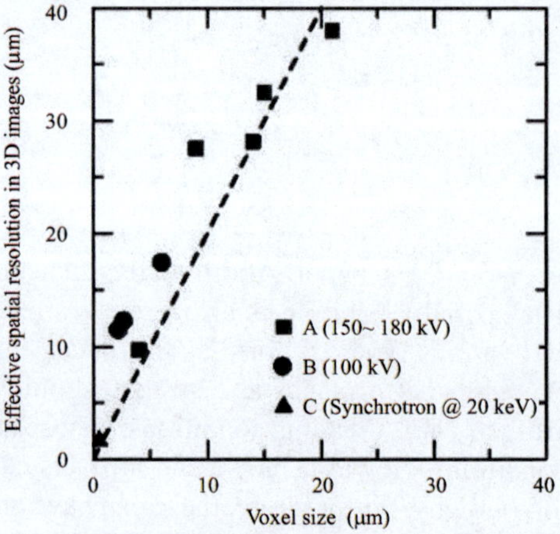

(c) Spatial resolution values of three X-ray CT scanners measured using the jigs shown in (a)

Fig. 7.15 The relationships between 3D pixel size and effective spatial resolution were measured and compared in nested test pieces shown in **a** and **b** using two types of industrial X-ray CT scanners (devices A and B) and the synchrotron radiation tomography instrument BL20XU at SPring-8 (device C) [14]. **c** shows the spatial resolution values of three X-ray CT scanners measured using the jigs shown in **a**. The dashed line in the figure expresses the sampling theorem

theorem discussed in the following section. Changing the pixel size in industrial X-ray CT scanners A and B changes the spatial resolution for each device in which the focal spot size of the X-ray tube does not constrain the spatial resolution. The measurement points for both industrial X-ray CT scanner A and synchrotron radiation X-ray tomography are mostly above the dashed line. This indicates that the spatial resolution is constrained by the Nyquist frequency based on the sampling theorem discussed in the following section. Meanwhile, the spatial resolution of industrial X-ray CT scanner B is lower than the level stipulated by the sampling theorem by a factor of 1.5–2.5. This signifies that a factor other than those marked with * in Table 7.2 controls the effective spatial resolution. Identifying the factor which constrains the spatial resolution in these cases and implementing suitable measures can result in considerable improvements to spatial resolution, which correspond to this difference.

Figure 7.16 uses the three types of devices in Fig. 7.15 and shows observations/comparisons of the same sample at the same positions while changing the pixel size. A more detailed structure is visualized as devices with higher spatial resolution

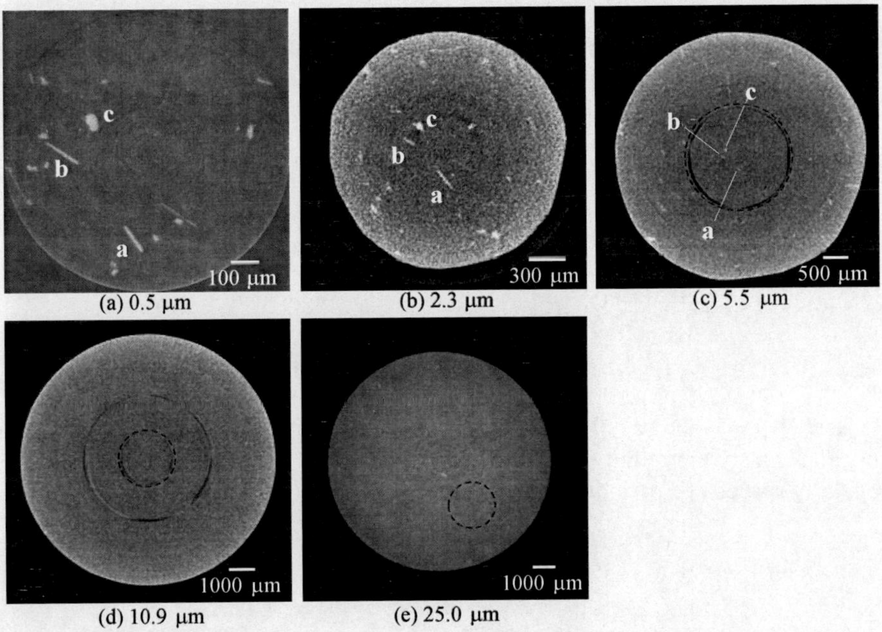

Fig. 7.16 Nested test pieces shown in the left side of Fig. 7.15 scanned at the same positions and with different voxel sizes using the two types of industrial X-ray CT scanners (devices A and B) and the synchrotron radiation tomography instrument BL20XU at SPring-8 (device C), referred to in Fig. 7.15 [14]. Device C was used for the pixel size of 0.5 µm in **a**; device B for 2.3, 5.5, and 10.9 µm in **b**, **c** and **d**, respectively; and device A for 25 µm in **e**. The **a** and **b** in the figure refer to the acicular Al_3Ti particles in the aluminum, and **c** refer to the TiB_2 particles. The dashed circles in (**c**–**e**) designate the same region in the sample shown in (**b**). Only (**a**) shows imaging of a region of interest in a portion of the sample shown in (**b**)

capabilities are used and when the pixel size becomes smaller in the same device (Fig. 7.16b–d). Structures that were clearly visible in Fig. 7.16b cannot be visualized at all in Fig. 7.16d, where the pixel size has increased by a factor of approximately 5. It is apparent that the sample size must be minimized to the extent possible and the sample must be placed as close to the X-ray tube as possible to acquire a sufficient spatial resolution. This is a frequently overlooked fact among industrial X-ray CT users.

The Fresnel diffraction of the X-ray and the Rayleigh diffraction limit, after the X-ray was converted to visible light, act as the bottlenecks for spatial resolution when improving the various conditions in Table 7.2 for projection-type X-ray tomography using a parallel beam. These stipulate the physical limits of spatial resolution, therefore, projection X-ray tomography cannot exceed a maximum spatial resolution of slightly less than 1 μm. Even with the example of synchrotron radiation X-ray tomography in Fig. 7.15, the effective spatial resolution is approximately 1 μm when the pixel size is approximately 0.5 μm; it is clear that the physical limits of spatial resolution have been reached.

(2) Sampling Theorem

Sampling in a 3D space in all *x*-, *y*-, and *z*-directions at a short periodicity is necessary for X-ray tomography to achieve a sufficiently small spatial resolution relative to the structure size to be observed in the sample. First, we consider this with a 1D image for simplicity. A 1D image is shown in Fig. 7.17a. The actual physical object is continuous even with observations with magnifications at the atomic level. Digital images are considered discrete expressions of continuous physical objects. Mapping a continuous function with a discrete function is referred to as *sampling*. The process of obtaining a digital image from a detector is multiplying a sampling function $s\ (x)$ by a primitive function $f\ (x)$, as shown in the following equation:

$$f_s(x) = f(x)s(x) \tag{7.11}$$

Here, $f_s\ (x)$ is the sampled image and is as shown in Fig. 7.17e. The *sampling function* is also referred to as the *comb function*, taking the form of delta functions regularly spaced at a periodicity $\Delta\ x$, as shown in Fig. 7.17c.

$$s(x) = \sum_{n=-\infty}^{\infty} \delta(x - n\Delta x) \tag{7.12}$$

The pitch of this sequence of functions stipulates the spatial resolution. As shown in Fig. 7.17b, the Fourier transform of the primitive function $f\ (x)$ either has the spatial frequency components included in the original image below ν_c or is bandwidth limited. This ν_c is referred to as the *cutoff frequency*. Furthermore, the comb function is used for the Fourier transform of the sampling function $S\ (x)$, as shown in Fig. 7.17d:

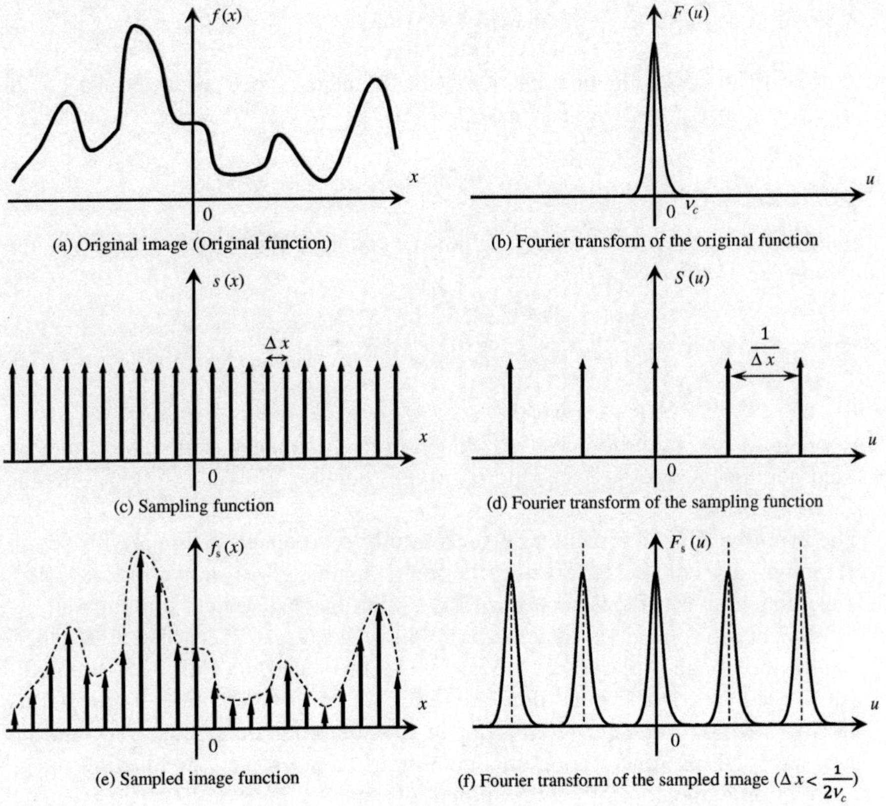

Fig. 7.17 Schematic of spectra in real space and frequency space when imaging a 1D image with a detector. The sampling function is an example of a delta function stipulated by the sampling pitch Δx determined by the pixel size of the detector; $\Delta x < \frac{1}{2\nu_c}$ (over-sampling) is satisfied in this case and the original image can be accurately reconstructed. **a** an original image; **b** Fourier transform of **a**; **c** sampling function; **d** Fourier transform of **c**; **e** sampled image function; and **f** Fourier transform of **e** ($\Delta x < 1/2\nu_c$)

$$S(x) = \frac{1}{\Delta x} \sum_{n=-\infty}^{\infty} \delta\left(u - \frac{n}{\Delta x}\right) \tag{7.13}$$

The multiplication operation on the sampling function in Eq. (7.11) is equivalent to convoluting the sampling function in the frequency space.

$$F_s(u) = F(u) * S(u) = \frac{1}{\Delta x} \sum_{n=-\infty}^{\infty} F\left(u - \frac{n}{\Delta x}\right) \tag{7.14}$$

In other words, $F(u)$, obtained by applying a Fourier transform on the primitive function as shown in Fig. 7.17f, can be periodically repeated by sampling it. The following rectangular function is used to extract the spectra of the original image from this.

$$H(u) = rect(u\Delta x) \tag{7.15}$$

A filtering process using this function in the frequency space produces an accurate Fourier transform of the original image.

$$F(u) = H(u)F_s(u) \tag{7.16}$$

Finally, an inverse Fourier transform of the equation shown above results in the following expression:

$$f(x) = \sum_{n=-\infty}^{\infty} f(n\Delta x)\mathrm{sinc}\left\{\frac{\pi}{\Delta x}(x - n\Delta x)\right\} \tag{7.17}$$

In other words, multiplying a *sinc* function to the images sampled at a period Δx and adding these reproduces the primitive function. This is referred to as the *sampling theorem.*

The primitive function cannot be successfully reproduced in Fig. 7.18b due to overlapping adjacent spectra even if filtering is conducted as shown in Eq. (7.16). This phenomenon is referred to as *aliasing*. Conditions where the sampling number is insufficient are referred to as *undersampling*. Aliasing artifacts (discussed later) appear on the image in these cases. Meanwhile, the spectra in Fig. 7.18c are sufficiently spaced apart; these conditions are referred to as *oversampling*. Figure 7.18a, where the *Nyquist frequency* f_N is half of the sampling frequency $1/\Delta x$ and $f_N = \nu_c$ is satisfied, presents the optimal conditions for accurately conducting image reconstruction while controlling the amount of data.

$$f_N = \frac{1}{2\Delta x} \tag{7.18}$$

Therefore, the original image can be accurately restored when the maximum frequency component ν_c of the original image is known by sampling at an interval finer than double this value (i.e., $2\nu_c$). There are usually detailed multiscale structures (e.g. material microstructures) in a sample. As such, Eq. (7.16) can be stated in real terms as follows: "the spatial resolution necessary for imaging the internal structure to be observed should be recognized and measurements should be taken at a pitch that is half this value."

Figure 7.19 shows a 2D case. The concepts here are the same for this 2D case. In other words, the measurement pitches Δx and Δy for the two x- and y- directions are determined by the pixel size of the detector. In this case, Eq. (7.17) is expressed as follows:

$$f(x, y) = \sum_{n=-\infty}^{\infty} \sum_{m=-\infty}^{\infty} f(n\Delta x, m\Delta y)\mathrm{sinc}\left\{\frac{\pi}{\Delta x}(x - n\Delta x)\right\}\mathrm{sinc}\left\{\frac{\pi}{\Delta y}(y - m\Delta y)\right\} \tag{7.19}$$

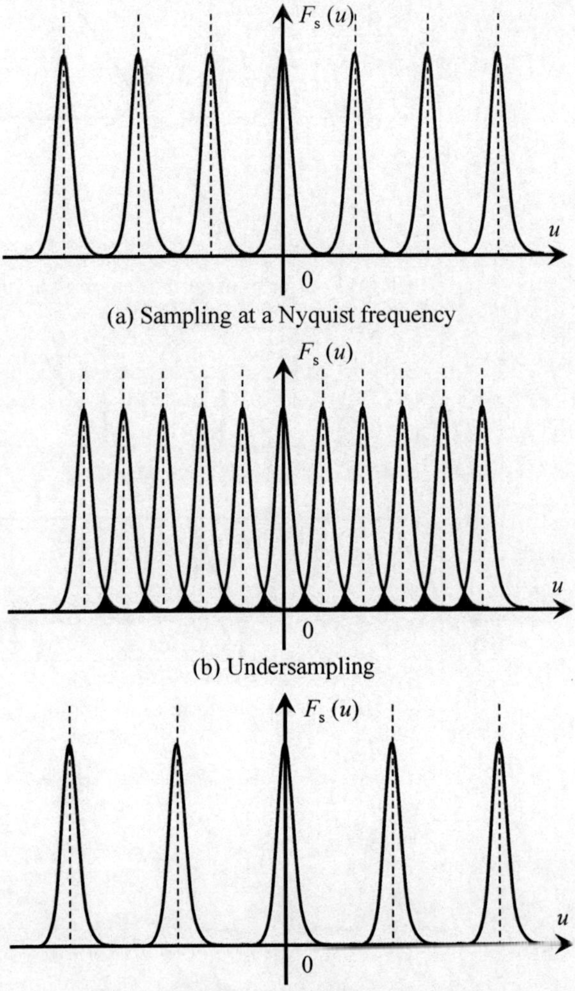

Fig. 7.18 A 1D image obtained from a detector and normalized, after which a Fourier transform was applied; the schematic shows three cases with different sampling frequencies. **a** is sampling at a Nyquist frequency; **b** is undersampling; and **c** is oversampling

As shown in Fig. 3.42a in Chap. 3, the rotation axis of the sample rotation stage is placed during actual X-ray tomography measurements so that it is perpendicular to the incident direction of the X-ray. The 2D detector is used to conduct continuous imaging of transmission images multiple times while rotating the sample by 180°, with this rotation axis at the center. Thus, with regard to the cylindrical coordinates (r, φ, z) corresponding to the rotation axis of the sample rotation stage, the effective pixel size of the detector must exceed the Nyquist frequency in the r- and z-directions, as well as the rotation step (interval expressed as distance) of the sample rotation stage in the φ direction. For example, the ideal spatial resolution in the φ direction at the edge of the field-of-view when conducting imaging of a total of M transmission images during 180° rotation with a field-of-view width and height of D are $\pi D/M$.

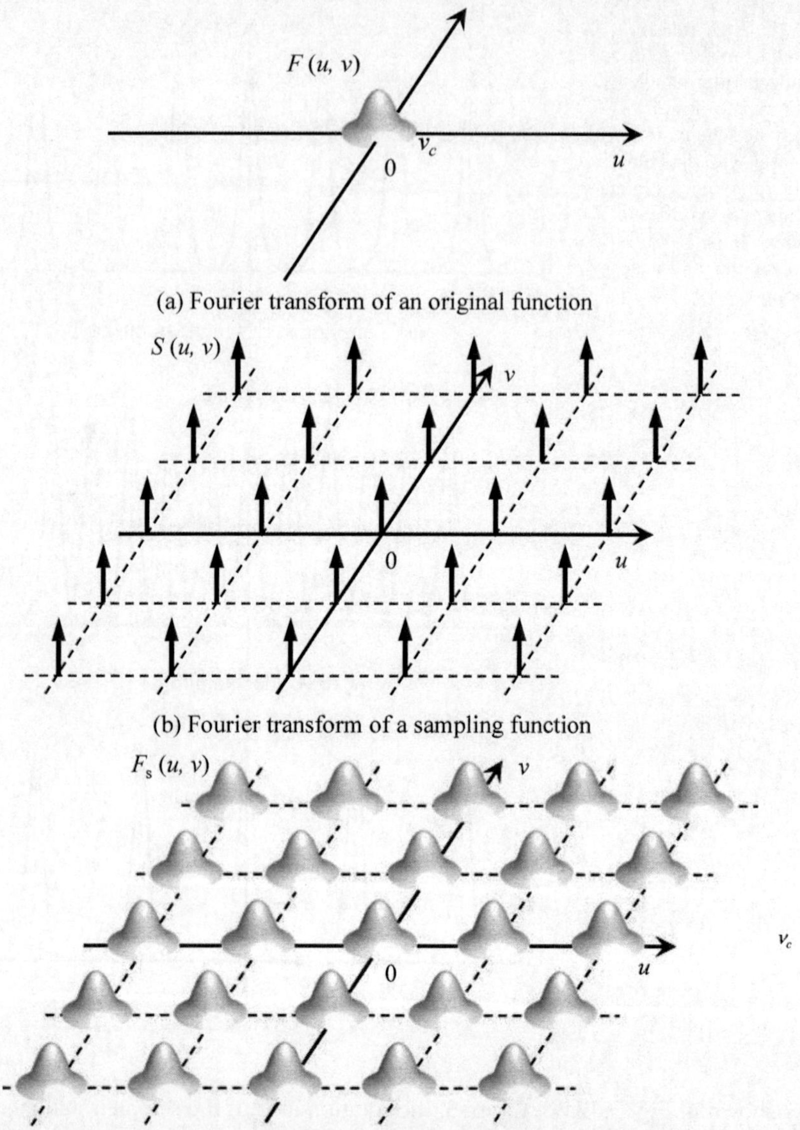

(c) Fourier transform of a sampled 2D image (Δx, $\Delta y < \frac{1}{2v_c}$)

Fig. 7.19 Schematic of spectra in frequency space when imaging was conducted using a 2D detector. The sampling function is a sequence of delta functions stipulated by the sampling pitches Δx and Δy, which are determined by the detector and pixel size; Δx, $\Delta y < \frac{1}{2\nu_c}$ (oversampling) is satisfied in this case and the original image can be accurately reconstructed. **a** is the Fourier transform of an original function; **b** is the Fourier transform of a sampling function; and **c** is the Fourier transform of a sampled 2D image (Δx, $\Delta y < 1/2\nu_c$)

The maximum spatial resolution value in the r- and z-directions when a detector with a pixel number of N in both the horizontal and vertical directions is used for measurement is $2D/N$. The selection of the number of pixels in a detector is not flexible in many cases. Thus, the number of projections becomes $\pi N/2$ if an isotropic spatial resolution is desired. However, the fact that various factors from the light source to the detector overlap and determine the effective spatial resolution has previously been discussed. Furthermore, the sample size is often smaller, to some extent, than the field-of-view size, therefore, the sampling pitch in the r-direction can be coarser than this based on the ratio between sample size and field-of-view size.

(3) Detectability

Spatial resolution and *detectability* are two completely different indices. Although detectability cannot be inferior to spatial resolution, the opposite is possible. In other words, detectability can be a value that is several times smaller than the pixel size, depending on the conditions. Detectability is largely influenced by spatial resolution, imaging methods and conditions, sample structure, and noise.

A schematic of this is shown in Fig. 7.20. Here, we consider the case of the 3D imaging of a material that includes 0.3–0.6-μm diameter particles and pores, as shown in Fig. 7.20a, with a 1-μm spatial resolution device/conditions. When the noise or artifact effects are low, the gray value of the pixels, in which such microstructures are embedded, changes, facilitating the discrimination and binarization of the pixels by employing the binarization shown in Fig. 7.20c. The noise/artifacts cannot be distinguished from the microstructure at this stage. However, as shown in Fig. 7.20d, the capability of detecting microstructures below the spatial resolution can be verified by repeating observations in the same location while growing this structure or by supplementing these with other observations/analysis methods. The detectability, in this case, is 0.27 μm when the maximum threshold was set as an 8-bit gray value of 240. This is 1/3 that of the spatial resolution. Thus, detectability is not only influenced by the structure size but also the microstructure to be detected and the linear absorption coefficient of the matrix. Positioning also plays a role in this as microstructures that are located between pixels are more difficult to detect than structures that are not; in other words, detectability is not homogeneous in a 3D image.

Improvements in microstructure detectability are considered achievable when black and white fringes, generated with refraction contrast imaging (discussed in Sect. 2.2.2 (2)), are utilized as the interfaces can be emphasized and the gray value of the microstructure can be significantly increased or decreased.

Detectability is difficult to directly measure experimentally unless special model samples are prepared. However, as mentioned above, researchers should investigate and be cognizant of the appropriate detectability level at each step of 3D image handling by attaining a deep understanding of the sample, basic imaging principles, internal and microstructure of the sample to be observed, imaging conditions, and the supplemental usage of confirmatory methods. This can minimize the erroneous loss of valuable information that may appear like noise at first glance through processes like filtering.

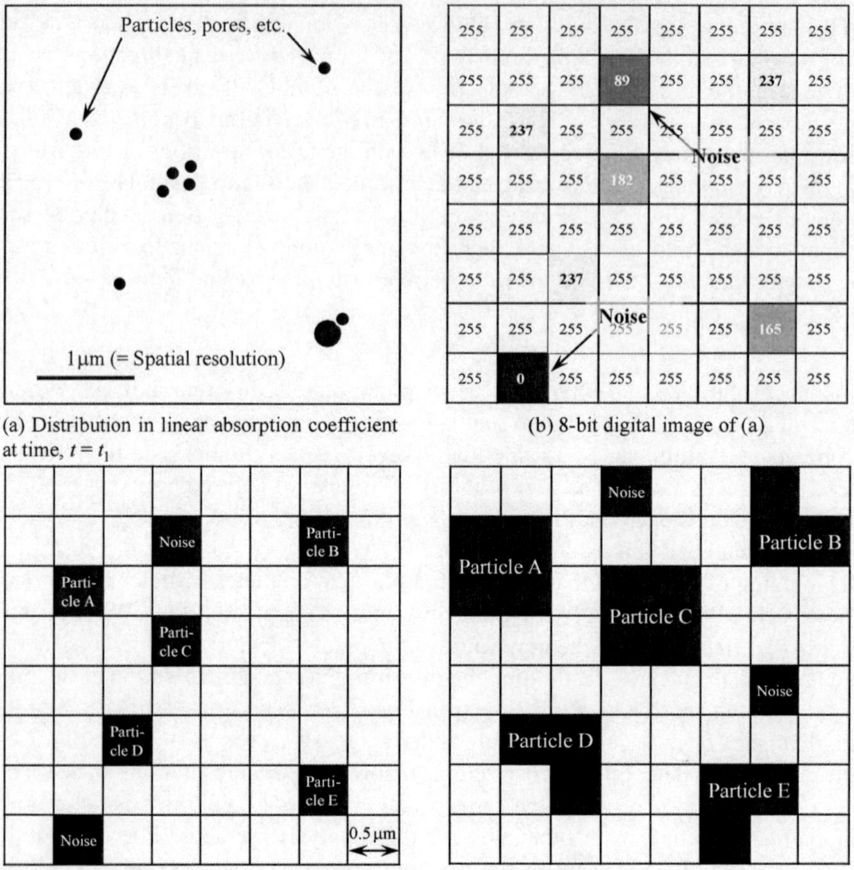

(a) Distribution in linear absorption coefficient at time, $t = t_1$

(b) 8-bit digital image of (a)

(c) After binarizing (b) with a threshold value of 240

(d) Segmented 2D image at time, $t = t_2$ ($t_2 > t_1$)

Fig. 7.20 Schematic where 3D imaging was conducted on a material (gray value of 255) with microscale particles A–E (gray values of 165 to 237 in image **b**) show in **a** using a 1-μm spatial resolution (effective pixel size of 0.5 μm) setup; **b** is the obtained image, **c** is the binarized image, and **d** is a re-observation of the particles, which grew after a given period. Image **b** includes two points of noise, five microstructures, and its aggregates. Image **d** does not show noise in the positions in **b** and the microstructure has grown. For this reason, both structures exceed the spatial resolution and can be clearly identified

7.5.2 *Spatial Resolution Assessment*

(1) PSF, LSF, and ESF

A mathematical expression of the blur in an image taken by an X-ray CT scanner is the convolution of a non-blurred original image with an impulse response. The impulse response for an image is referred to as *the point spread function* (*PSF*). The spatial resolution takes some finite value by adding all of the information relating to spatial resolution, such as the X-ray CT scanner and image reconstruction. Based on this, the PSF indicates how the spread of an infinitesimally small point in the sample is imaged. Thus, the PSF is mathematically expressed as the imaging of a delta function, as shown below:

$$PSF(x, y)R\{\delta(\xi, \eta)\} \tag{7.20}$$

Here, the function R (ξ, η) is the system response and coordinates (x, y) and (ξ, η) are for the detector and sample, respectively. As shown above, the image g (x, y) is expressed with the following convolution integration [15]:

$$g(x, y) \int \int_{-\infty}^{\infty} f(\xi, \eta)PSF(x - \xi, y - \eta)d\xi d\eta \tag{7.21}$$

Figure 7.21a shows a schematic of the PSF. Furthermore, Fig. 7.22a shows the original image with the point-like structure and the image after this that was imaged [16]. A narrower spreading width of the PSF results in less image blur and a superior spatial resolution.

Imaging point structures, which are smaller than the pixel size, are possible, for example, with medical CT scanners by using a phantom where a 0.1-mm diameter metal wire is floated in the air. However, this is not practical for industrial X-ray CT scanners, particularly those that use a microfocus tube with a small effective pixel size. For example, the QRM GmbH sells test objects that float 3-, 10-, and 25-μm diameter tungsten wires in the air for micro-tomography use [17]. However, this can be used for assessments with a spatial resolution of several dozen μm at most.

The PSF could be expressed in 1D if it were isotropic, as shown in Fig. 7.21. This is referred to as the *line spread function* (*LSF*). Figure 7.21b shows a schematic of the LSF and Fig. 7.22b shows an image for this case as well [16]. As was also the

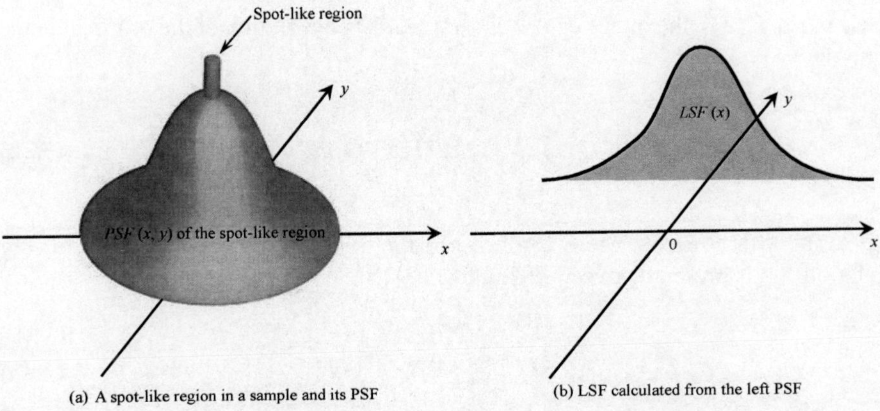

Fig. 7.21 Schematic showing the relationship between the spot-like region in a sample and its PSF and the LSF, which corresponds to its 1D profile. **a** is a spot-like region in a sample and its PSF; and **b** is LSF calculated from the PSF

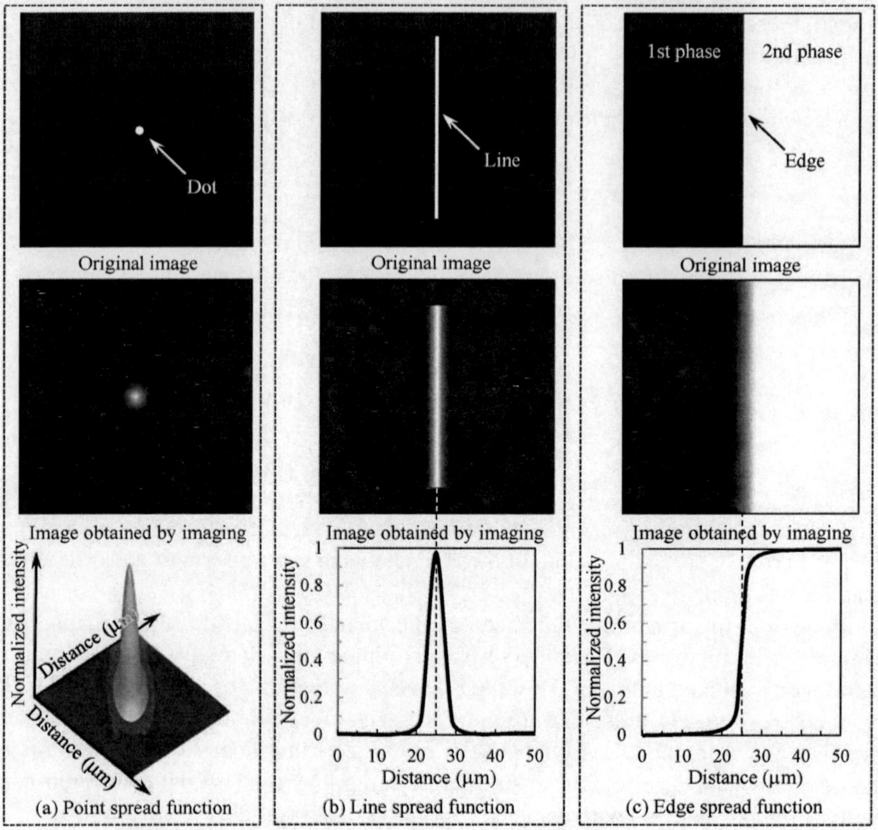

Fig. 7.22 Schematic showing the original image expressed with PSF, LSF, and ESF; the original image blur; and the shapes of each function. **a** is point spread function; **b** is line spread function; and **c** is edge spread function

case in Eq. (7.21), the image can be shown as a 1D convolution of the original image and the LSF [15].

$$g(x) \int_{-\infty}^{\infty} f(\xi) LSF(x-\xi) d\xi \tag{7.22}$$

A slit with a narrow width can be used for LSF measurements. The following relationship is present between PSF and LSF [15].

$$LSF(x) \int_{-\infty}^{\infty} PSF(x, y) dy \tag{7.23}$$

As such, the LSF value at a given position x is equal to the cross-sectional area of PSF in the plane parallel to the y-axis for the same x-coordinates. Furthermore, the relationship between the width of LSF spread and spatial resolution is the same for PSF, which is often two-dimensionally isotropic; the LSF also exhibits bilateral symmetry in this case.

Similarly, the effect where blurring occurs between the interface of the two phases can be assessed using the *edge spread function* (*ESF*). This is shown in Fig. 7.22c [16]. This is mathematically expressed as a step function response where LSF becomes the derivative of ESF [15]:

$$\frac{d}{dx}ESF(x) = LSF(x) \tag{7.24}$$

As shown in Fig. 7.22, the ESF becomes a measure for 1D spatial resolution similar to the LSF. Focusing on practical applications, as some edges are always embedded in the sample, ESF measurements could be termed more versatile than LSF measurements, which require the preparation of a narrow slit.

The PSF, LSF, and ESR are all functions that describe spatial resolution in real space. In terms of practical use, the full widths at half maximum (FWHM) of the PSF and LSF can be used as indices for spatial resolution. Figure 7.23 shows a schematic of this. When the distance between two characteristic points approaches the FWHM, a locally stacked profile comprising the PSF or LSF profiles of the two characteristic points is measured, as shown in Fig. 7.23b. There is still a depression in the center of the curve in Fig. 7.23b, so the two characteristic points can still be distinguished. The two points cannot be distinguished if the distance between these two points is smaller (Fig. 7.23c). As previously mentioned, LSF is determined from the differentiation of ESF, so ESF can also be used for spatial resolution assessments based on FWHM.

(2) MTF

Another standard practice is to convert the PSF, LSF, and ESF discussed in the previous section into the frequency space, assessing spatial resolution with the *modulation transfer function* (*MTF*). An overview of this is provided in this section.

As shown in the following equation, an *optical transfer function* (*OTF*), which is a spatial resolution assessment index in frequency space, can be obtained by applying a 2D Fourier transform on the PSF as shown in Eq. (3.5).

$$OTF(u, v) = F\{PSF(x, y)\} \tag{7.25}$$

Conversely, the PSF is obtained by applying an inverse Fourier transform on the OTF. The OTF is a complex number, and its absolute value is equal to MTF as shown in the following equation. Meanwhile, the phase component of OTF is referred to as the *phase transfer function* (*PTF*).

$$OTF(u, v) = MTF(u, v)e^{-iPTF(u,v)} \tag{7.26}$$

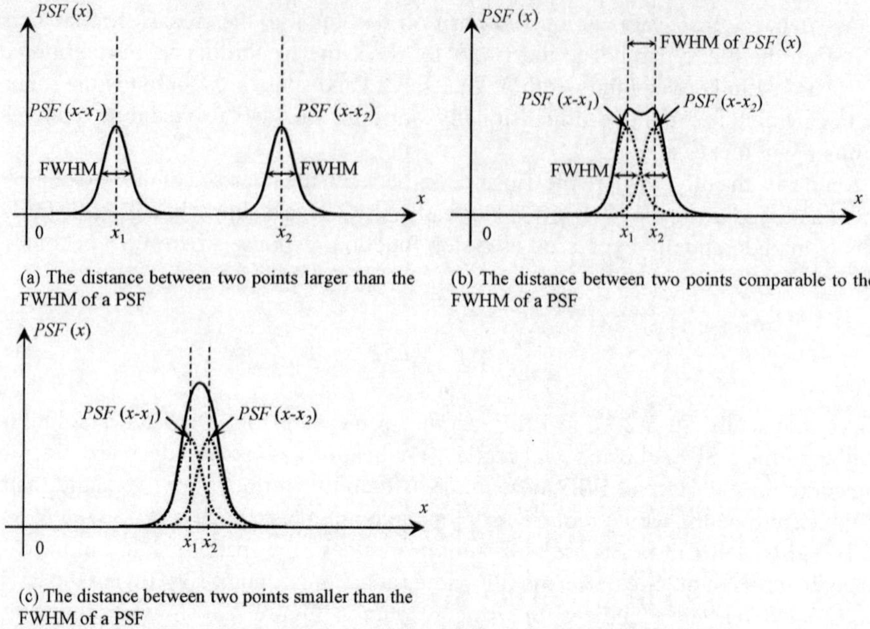

(a) The distance between two points larger than the FWHM of a PSF

(b) The distance between two points comparable to the FWHM of a PSF

(c) The distance between two points smaller than the FWHM of a PSF

Fig. 7.23 Schematic showing the validity of spatial resolution assessments based on PSF, LSF, and ESF. Three PSF cases are shown where the distance between the two points relative to the FWHM of PSF is significantly larger, equal, or narrower. **a** the distance between two points larger than the FWHM of a PSF; **b** the distance between two points comparable to the FWHM of a PSF; and **c** the distance between two points smaller than the FWHM of a PSF

The PSF generally has an isotropic form, so the phase component PTF becomes 0. The MTF is then related to PSF as shown in the following equation:

$$MTF(u, v) = |OTF(u, v)| = |F\{PSF(x, y)\}| \tag{7.27}$$

In this manner, the MTF is a 2D function. However, considering the general ease of visual assessments, this is frequently assessed as a 1D profile that follows a given axis that passes through the origin of the frequency space. The MTF, in this case, is related to LSF and ESF as shown in the following equation:

$$MTF(u) = F\{LSF(x)\} = F\left\{\frac{d}{dx}ESF(x)\right\} \tag{7.28}$$

MTF assessment is enabled by measuring the LSF and ESF. The MTF generally uses the value at the origin of the frequency space *LSF* (0) and is expressed by normalizing *MTF* (*u*) in the range of 0–1.

Figure 7.24 shows an example of an assessment using the MTF. The horizontal axis of Fig. 7.24 is spatial frequency. Changes in the resolution characteristics can

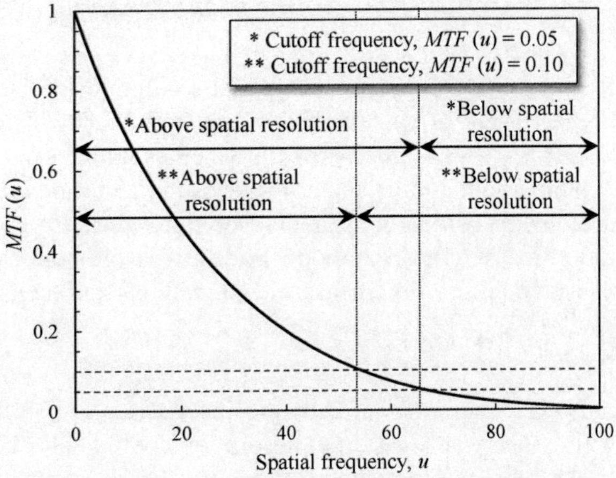

Fig. 7.24 Schematic showing examples of spatial resolution assessments using the MTF. Two analysis examples where the cutoff frequency was changed are shown

be assessed as a function of the spatial frequency using the MTF, which is a function of spatial frequency. In other words, superior resolution characteristics are obtained when these values are closer to $MTF\ (u) = 1$ for a given spatial frequency. As shown in Fig. 7.24, the spatial resolution is often stipulated by the spatial frequency values when the $MTF\ (u)$ value is 0.05, 0.10, or 0.20. These are expressed as 5% MTF, 10% MTF, etc.

The relationship between the ultimate spatial resolution calculated in a 3D image and the various factors that constrain it, which are listed in Table 7.2, was previously discussed in Sect. 7.5.1 (1). Expressing this with MTF yields the following equation [18]:

$$MTF_{total}(u) = MTF_1(u) \cdot MTF_2(u) \cdot \cdots \cdot MTF_k(u) \cdot \cdots \tag{7.29}$$

Here, MTF_{total} is the total MTF, k is the kth factor out of the various factors listed in Table 7.2, and $MTF_k\ (u)$ is the corresponding MTF. However, there is a relationship between MTF and spatial frequency such as that shown in Fig. 7.24 for each factor (e.g. detector, X-ray source). In these cases, $MTF_{\text{total}}\ (u) \leqq MTF_k\ (u)$ is satisfied for all u and v values. Furthermore, the following relationship between the various factors in Table 7.2 and the total FWHM of the LSF is obtained when the spatial resolution is expressed in terms of the FWHM of the LSF [18]:

$$FWHM_{total} = \sqrt{FWHM_1^2 + FWHM_2^2 + \cdots \cdot + FWHM_k^2 + \cdots \cdot} \tag{7.30}$$

Thus, for an FWHM relating to a factor A, which constrains the final spatial resolution ($FWHM_{\text{A}}$), when the FWHM for a given factor B ($FWHM_{\text{B}}$) is such that $FWHM_{\text{B}}/FWHM_{\text{A}} = 0.1$, the influence of FWHM_{B} on the total FWHM is at most 0.5%. In other words, improvements in spatial resolution require the specification of factors that constrain the spatial resolution and the implementation of countermeasures for improving the spatial resolution with those factors. Equation (5.29) should be referenced in conjunction with this.

7.5.3 Spatial Resolution Measurement

Representative examples of spatial resolution measurement methods are outlined below. There are a number of methods besides these and their variations are numerous. This is not a problem when using the same method to assess 3D images obtained with similar materials/measurement methods, but sufficient care must be taken when comparing spatial resolution values obtained with different spatial resolution measurement methods. For this reason, publications and reports must detail the spatial resolution measurement methods and their associated conditions.

(1) 3D Test Object

A 2D test chart is created by placing a series of parallel line regions on a substrate with considerably different absorption coefficients. This is used for the adjustment of the focal point of the microfocus tubes via transmission images and the maintenance/management of industrial X-ray CT scanners. Spatial resolution measurement test specimens (a.k.a. "3D testing charts") are also commercially available. Figure 7.25a shows a *test object* (corresponding to a phantom in medical terms) for spatial resolution measurement use, produced by the Japan Inspection Instruments Manufacturers' Association (JIMA) [19]. Three grooves are created for each of the vertical and horizontal directions on a silicon chip using micromachining techniques for semi-conductors, gold is embedded in these etched grooves, and the silicon chip itself is embedded into a 2-mm diameter acrylic stick. The line widths and line spacing are 3, 4, 5, 6, and 7 μm. Some 3D test objects are produced in Germany; those with a line width of 1 μm come in two variations: one with 10 spatial resolution levels between 1–10 μm, and one with 9 spatial resolution levels at 5, 10, 15, 20, 25, 30, 50, 100, and 150 μm [20]. As these are simply grooves placed in silicon, the JIMA-produced test objects are widely used internationally.

Figure 7.25b shows a test object used by the author for the 3D imaging of steel. Grooves are placed in a 0.5-mm diameter stainless steel wire with focused ion beam (FIB) fabrication and a spatial resolution measurement across 28 levels with line widths ranging from 0.5 to 5.9 μm can be conducted. Figure 7.26 shows the spatial resolution when the test object in Fig. 7.25b is used to visualize steel materials with synchrotron radiation X-ray tomography [21]. In these cases, spatial resolution is assessed in 3D by measuring it in two directions with the test object, as shown in Fig. 7.25b, and in one direction with the edge spread function shown in Sect. 7.5.2 (1). Decreases in spatial resolution appear due to the forward scattering of X-rays when the camera length is short and, particularly, when high-energy X-rays are present. Furthermore, the spatial resolution in the vertical direction is superior when compared to that in the horizontal direction. This is because the front end slit stipulates that the beam sizes in the horizontal and vertical directions are 400 and 15 μm, respectively; therefore, blurring in the horizontal direction cannot be avoided due to the penumbra [21].

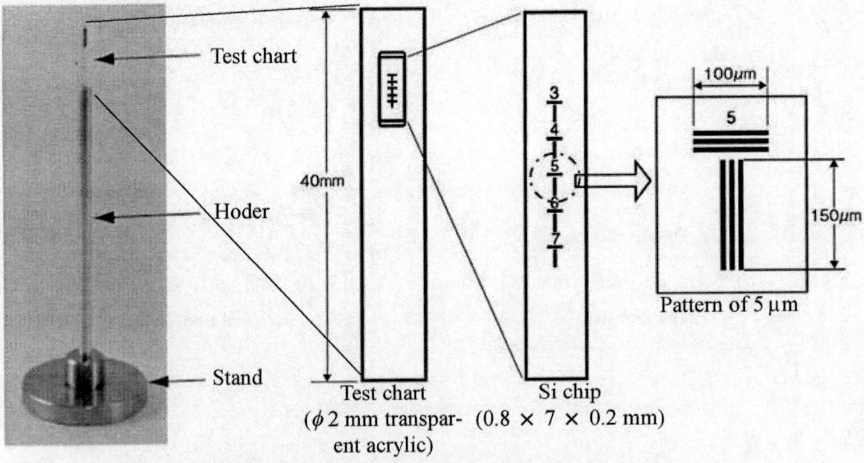

(a) Test chart for X-ray CT provided by JIMA

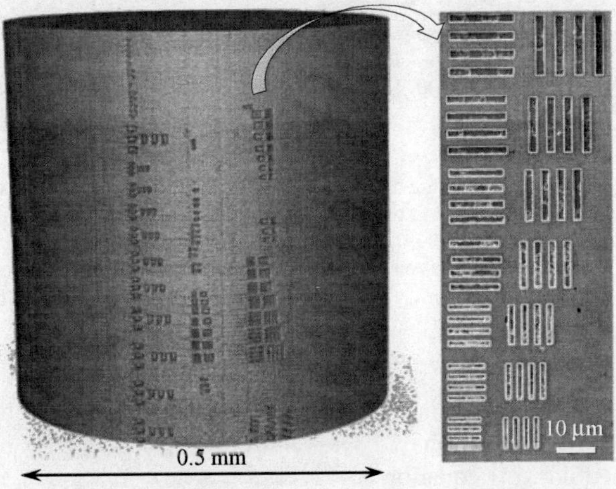

(b) Self-made 3D test chart prepared by the author using FIB

Fig. 7.25 Example of a 3D test object for X-ray CT; **a** is the test object used for spatial resolution measurements produced by the Japanese Inspection Instruments Manufacturers' Association (JIMA) (from the JIMA catalog) with approval from JIMA for publication [19] and **b** is a device made by the author where FIB was used to create grooves on a stainless steel wire surface. The right-hand side shows a magnification of a selected part (3D image from X-ray CT) [21]

The simplest method for measuring spatial resolution while using the test object is to objectively determine the limit at which a contrasted pattern can be identified in a virtual cross-section of a 3D image. However, this can introduce ambiguity due to individual decision standards. For this reason, the spatial resolution is often determined based on the spatial frequency reaching a given MTF value, as discussed

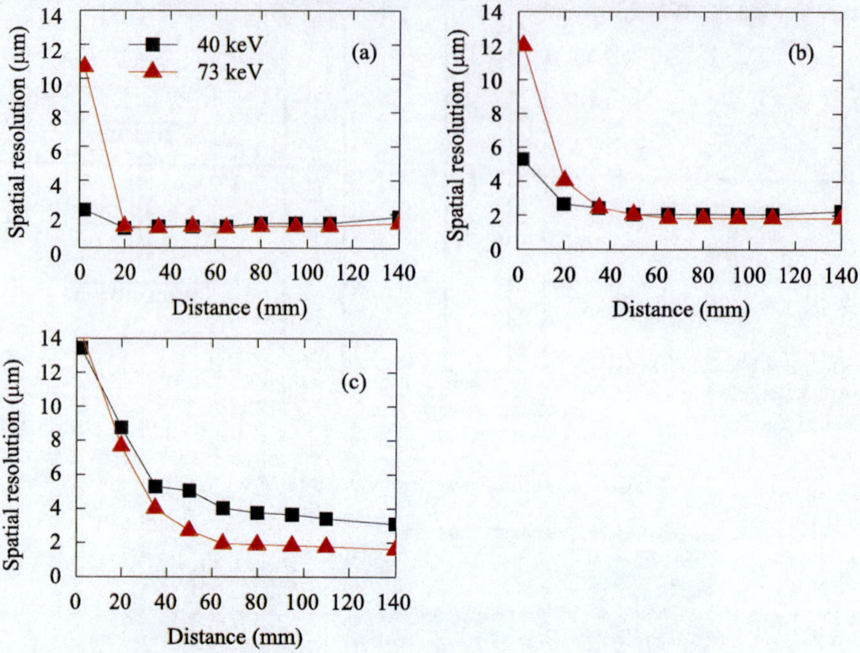

Fig. 7.26 Example where 3D imaging of steel was conducted using X-ray tomography at BL20XU in SPring-8 and where the spatial resolution was measured at two different X-ray energy levels while changing the distance from the sample and detector [21]. The 3D test object of Fig. 7.25 (**b**) was used for the *z*- and -directions (**a** and **b**, respectively). The edge spread function discussed in Sect. 7.5.2 (1) was used for the *r*-direction (**c**) in cylindrical coordinates (*r*, *φ*, *z*) corresponding to the rotation axis of the sample rotation stage

in Sect. 7.5.2 (2). This procedure determines the contrast between the groove and space sections of the test object and determines the input/output contrast ratio [18], as shown in the following equation.

$$MTF(u) = \frac{\frac{g_{\max}(x) - g_{\min}(x)}{g_{\max}(x) + g_{\min}(x)}}{\frac{f_{\max}(x) - f_{\min}(x)}{f_{\max}(x) + f_{\min}(x)}} \tag{7.31}$$

Here, $f_{\max}(x)$ and $f_{\min}(x)$ refer to the maximum and minimum values of the primitive function $f(x)$, respectively, whereas $g_{\max}(x)$ and $g_{\min}(x)$ refer to the maximum and minimum values of the image $g(x)$, respectively. Comparisons between two different devices can be conducted with sufficient reproducibility if the test objects are accurately fabricated. Another advantage is that the spatial resolution measurement can be easily verified from an objective standpoint with regards to the obtained image. Meanwhile, only specified spatial frequencies held by the test object can be measured, so the accuracy of spatial resolution measurements becomes a problem if the device cannot perform in the vicinity of that specified spatial frequency. Furthermore, a single period of the periodic structure of the test object must be accurately identified from the image data and a length that is its integer multiple must be accurately measured [22]. For these reasons, this must be measured by a sampling frequency that satisfies the sampling theorem as shown in Eq. (7.18) [22]. However, it is often the case with spatial frequency near the spatial resolution that there may only be a few pixels' worth of sampling points for each period of the periodic structure of the test object. There have been several reports on methods that increase the measurement accuracy in these cases. Please refer to these as required [22].

(2) Edge Response

Methods discussed in Sect. 7.5.2 (1) for analyzing the edge spread function have the advantage of directly measuring spatial resolution from 3D images of samples to be observed without using specialized devices. These methods are widely used and measure the edge sections of sample surfaces or sample interiors: in other words, the pixel transitions (i.e., edge responses) in the boundary sections, such as between the sample and air. After determining the ESF in this manner, the ESF is differentiated to calculate the LSF, and either its FWHM is determined or a Fourier transform is applied on the LSF to determine the MTF. This MTF-based method is recommended in ASTM E1570-11. The ESF is determined by approximating the pixel value transitions in the obtained image with a sigmoid function as shown in the following equation:

$$ESF(x) = \frac{a}{1 + e^{(-bx+c)}} + f \tag{7.32}$$

Here, a, b, c, and f are coefficients that determine the translation, amplification, and width of the transition zone in the S curve. Figure 7.27 shows a schematic of the ESF where three different spatial resolution levels were obtained by varying the coefficients in the sigmoid function. The differences in spatial resolution expressed as the width of the transition zone can be clearly expressed by the differences in the coefficients of Eq. (7.32). Previous reports, which measured the spatial resolution by using the edge response, have used polynomial approximations, error functions, [23] or the sum of Gaussian functions and exponential functions [24]. The accuracy of the approximations directly influences MTF measurement accuracy, therefore, an optimal function must be selected.

Pixel value transitions, which span the edge, must be measured at a sampling frequency that satisfies the sampling theorem in Eq. (7.18) similar to the test object case when assessing the edge response. However, it is often the case that there is only a few pixels' worth of sampling points (pixels) in the pixel value transition zone across the edge when using high-spatial-resolution devices whose spatial resolutions are stipulated by the sampling pitch based on the sampling theorem. There have been reports of accuracy increases in these cases. A representative example is the inclined edge method. This method is also standardized by ISO12233 [25]. The

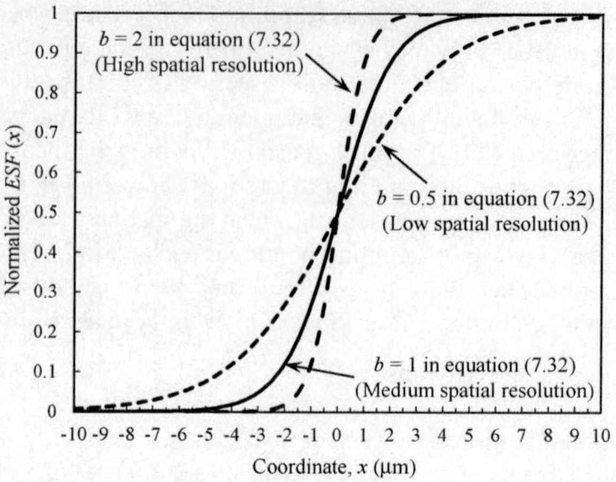

Fig. 7.27 Sigmoid function types. From Eq. (7.32), $a = 1$ and $c = f = 0$, and a normalized ESF was displayed by varying b to reflect high, medium, and low spatial resolutions

inclined edge method measures an edge response on straight edges or two-phase boundaries/surfaces that are inclined only slightly from the pixels of the 2D detector by approximately 1.5–3°. The alignment is slightly shifted for each pixel along the boundary due to this inclination. This is shown schematically in Fig. 7.28. Integrating the data in the direction along the boundary at a sampling pitch considerably smaller than the pixel size of the 3D image along the boundary allows for a high-density measurement of the pixel value profile across the edge and the determination of a synthesized LSF. As mentioned previously, the ESF is often approximated by fitting it to a function whose shape is similar to the edge response. Meanwhile, with ISO12233, space is divided by 1/4 of the pixel size, and discrete data are obtained by averaging the values inside these subpixels.

A smooth surface or interface in the sample interior, which is sufficiently below the spatial resolution, must be used to apply the inclined edge method. If this is not the case, then a test object must be prepared to conduct spatial resolution measurements of physical objects that have a similar smooth edge.

Figure 7.29 shows a 3D image of a high-purity aluminum alloy obtained using SPring-8. The sample-detector distance in Fig. 7.29b was intentionally set as long and the interface between the sample and air (sample surface) was highlighted by the black-and-white fringes created using the refraction contrast imaging. The voxel size of the 3D image in this was 0.5 μm and the effective spatial resolution was approximately 1 μm. Figure 7.30 shows the ESF obtained using Eq. (7.32) to approximate the pixel value transitions at the interface of the image shown in Fig. 7.29; this was then differentiated to calculate the LSF. The center of the figure also shows the LSF calculated by directly differentiating the measured ESF data. The LSF directly determined from the ESF is not bilaterally symmetric. This was primarily thought to be due to the X-ray refraction between the aluminum-air interface. Additionally,

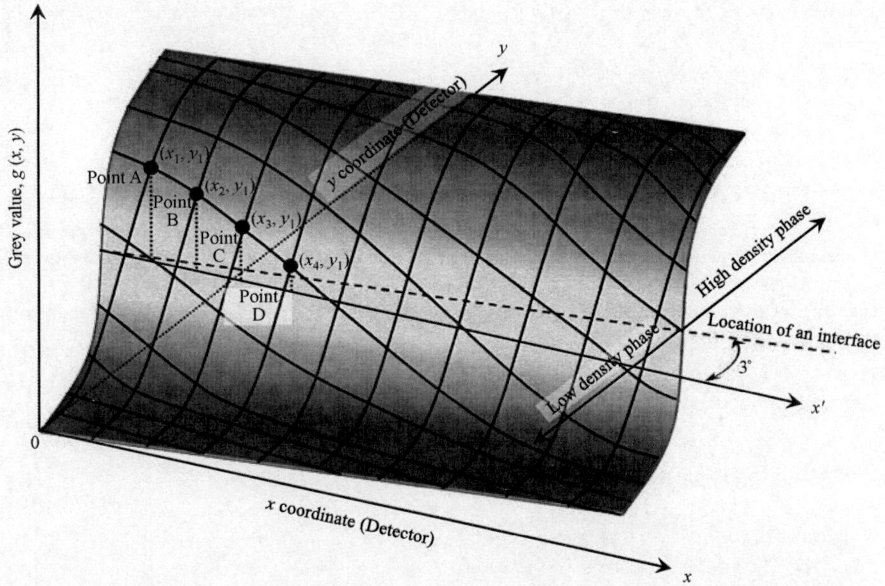

Fig. 7.28 Schematic showing the 2D measurement of pixel values along an interface, which has a slight inclination relative to the high-density and low-density phase edge orientation in terms of the coordinates of the detector. Points A–D in the figure are measurement points, which lie on the x-axis of the same y-coordinate (y_1) ($x_1 < x_2 < x_3 < x_4$). The pixel value gradually decreases from Point A to Point D as this becomes further and further away from the interface; i.e. the edge of the high-density phase, which has a large pixel value with a pitch that is considerably lower than the pixel size

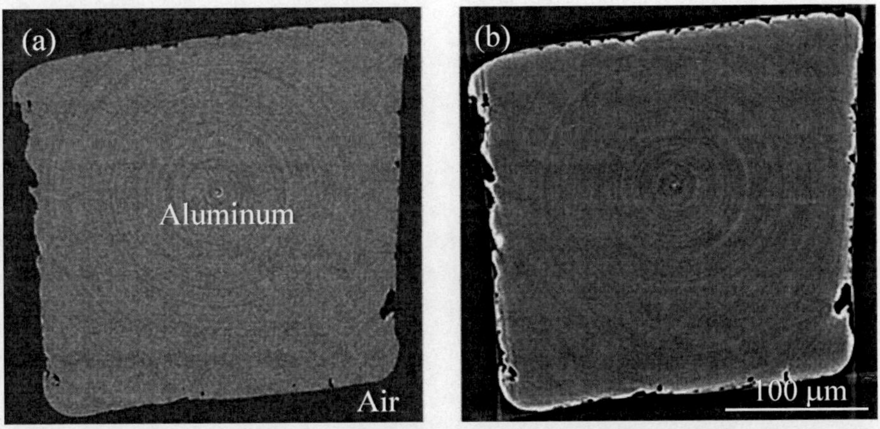

Fig. 7.29 Cross-section image (raw data) of pure aluminum (99.999%) imaged with synchrotron radiation X-ray tomography at the BL20XU beamline in SPring-8. Images were taken with an X-ray energy of 20 keV and reconstruction was based on filtered back projection using the Hann filter; **a** is when the sample-detector distance is 3 mm and **b** is when the distance is 55 mm (courtesy of Kazuyuki Shimizu of Kyushu University)

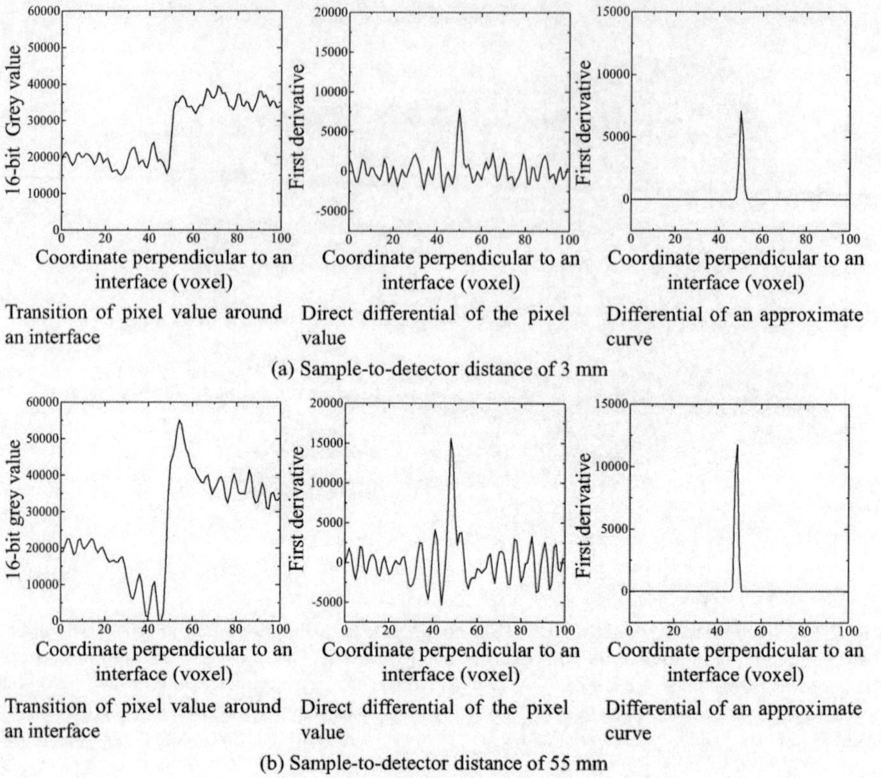

Fig. 7.30 Pixel value transitions were measured in the aluminum-air interface of the 3D images in Fig. 7.29 (left), which were either directly differentiated to obtain the LSF (middle) or the pixel value transitions at the interface were fitted with a sigmoid function, which was then differentiated (right). **a** sample-to-detector distance of 3 mm; and **b** sample-to-detector distance of 55 mm. (courtesy of Kazuyuki Shimizu of Kyushu University)

a similarly asymmetric LSF is generated with beam hardening, which is a representative artifact present in industrial X-ray CT scanners. The asymmetry of LSF is more prominent when the sample-detector distance in Fig. 7.30 is long. Consequently, sufficient care must be taken when measuring different spatial resolution values between the interfaces between dissimilar phases in the sample interior and sample surface, in cases where refraction or artifacts at the matrix–air interface of the sample surface are prominent.

Figure 7.31 shows the MTFs obtained by applying a Fourier transform on the LSF calculated by directly differentiating the ESF and on the LSF obtained by differentiating an approximation curve for the ESF. Figure 7.30 shows an undershoot (i.e., the skirts of the peak have negative peaks on both sides) for the LSF. This results in an overshoot of the MTF in values above 1 in Fig. 7.31. The LSF undershoot is due to the enhancement of the edges; the usage of filter functions during reconstructions, which can produce artifacts near the edges; X-ray refraction at the interface; and

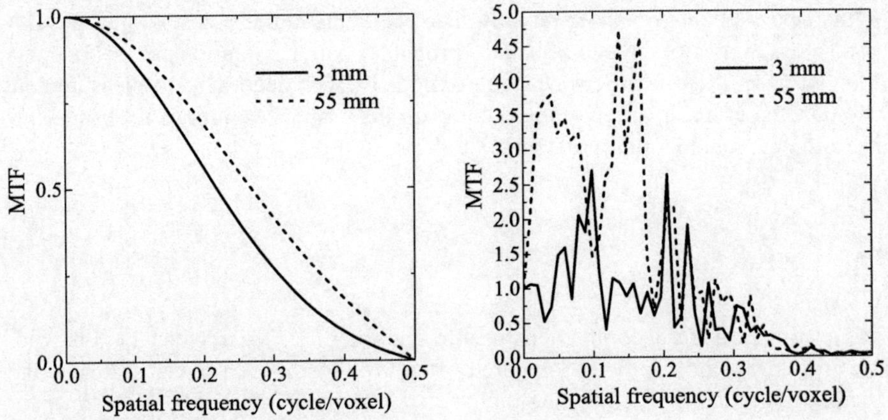

(a) MTF obtained from ESF of the curve approximated with a sigmoid function

(b) MTF obtained from directly measured ESF

Fig. 7.31 MTF obtained by applying a Fourier transform on the LSF in Fig. 7.30. Comparison between the MTF obtained from the LSF after fitting the pixel value transitions in the interface with a sigmoid function and differentiating it in **a**, and the MTF obtained from the LSF after directly differentiating the pixel value transitions in the interface in **b**. (courtesy of Kazuyuki Shimizu of Kyushu University)

beam hardening artifact generation. Furthermore, the fluctuations in the skirts of the LSF appear as sharp peaks in the MTF curve. Furthermore, the skirt range is smooth when fitted with a sigmoid function, so no fluctuations in the curve are observed.

Comparisons of the spatial resolutions determined in this way are shown in Table 7.3. When using the FWHM of the LSF to measure the spatial resolution, a likely spatial resolution is obtained for cases where an approximative curve was obtained with a sigmoid function. Measurements with the MTF yield virtually the same values independent of the large curve fluctuations in the low spatial frequency domains shown in Fig. 7.31, regardless of whether an approximative curve was used,

Table 7.3 Spatial resolutions measured using the data from Fig. 7.30 obtained from the image in Fig. 7.29; Comparisons of spatial resolutions determined from the LSF FWHM for the two sample-detector distance levels and the spatial resolutions stipulated by the spatial frequency value when the *MTF* (*u*) value was 0.05, 0.10, or 0.20 (courtesy of Kazuyuki Shimizu of Kyushu University) (μm)

ESF measurement method	Sample-to detector distance	FWHM	MTF		
			5% MTF	10% MTF	20% MTF
Fitting with a sigmoid function	3 mm	1.1	1.2	1.3	1.5
	55 mm	1.0	1.1	1.1	1.3
Measured value	3 mm	1.2	1.2	1.3	1.3
	55 mm	1.4	1.2	1.3	1.4

or the measured data was used directly. The 5% MTF value is close to the FWHM-based measured values in this case and a probable spatial resolution can be obtained. In either case, there is an error of approximately 50% according to measurement methods, therefore, it is important to compare the spatial resolutions with the same measurement method/measurement conditions.

7.6 Artifacts

This section provides a comprehensive introduction of *artifacts*, which are unavoidable in X-ray tomography, encountered often in industrial X-ray CT scanners and synchrotron radiation X-ray tomography. Here, the industrial X-ray CT scanner for micro-tomography (Skyscan 1072: X-ray tube voltage of 20–100 kV, spot size of <5 μm, 14-bit cooled CCD detector), which can be freely used by the author, was used to for imaging and reconstruction under conditions that intentionally generated artifacts, as shown in the corresponding figures. Artifacts encountered during experiments at SPring-8 are also included here. These aspects should be kept in mind as unavoidable artifacts are handled during X-ray tomography.

7.6.1 Artifacts Due to Interactions Between X-Rays and Objects

(1) Beam Hardening

Figure 7.32 shows an image of aluminum with a thickness of several mm scanned at a tube voltage of 80 kV using the industrial X-ray CT scanner. Figure 7.32a is an image taken without filtering. The voxel value is large in the vicinity of the sample surface, particularly in the corners, where the X-ray path becomes longer. The aluminum structure is completely homogeneous in the cross-section and the changes in color are due to artifacts referred to as beam hardening, which occur when the X-ray is not monochromatic. According to Fig. 7.10, beam hardening is significant when the exposure time is decreased from a suitable value (380 ms).

The generated X-ray energy when using an X-ray tube has a distribution like those seen in Figs. 4.18 and 4.55. X-ray energies can be distributed across a wide range—up to 80 keV, in the case of the X-ray CT scanner used. Furthermore, as shown in Fig. 2.4 and Figs. 2.10, 2.11, 2.12, 2.13 and 2.14, the linear absorption coefficient has a strong X-ray energy dependency. X-rays with low X-ray energy are absorbed immediately below the surface; only those with relatively high energy transmit to the material interior, resulting in changes to the spectrum of the transmission X-ray.

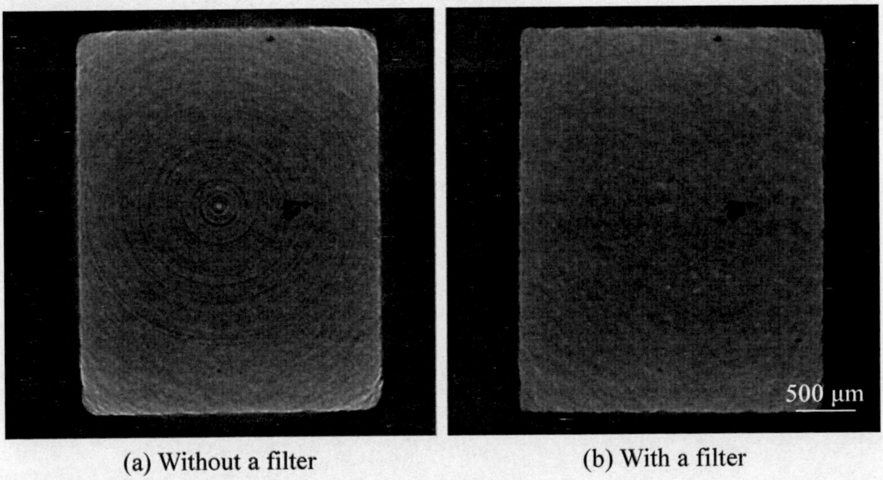

Fig. 7.32 Cross-section images of aluminum specimens scanned with a Skyscan 1072 and subsequently reconstructed in **a**. The number of projections was 1800 images/180°, the exposure time was 380 ms, and the voxel size was 4.6 μm. A 0.5-mm thick aluminum filter was used in **b**

Equation (2.5) shows the transmission X-ray intensity when the incident X-ray intensity and linear absorption coefficients are expressed as a function of X-ray energy. The projection data *p*, in this case, are as follows:

$$p = log\left(\frac{I}{I_0}\right) = -log\left(\int_E I_0(E)e^{-\int_0^L \mu(z,E)dz}dE\right) \tag{7.33}$$

In this manner, projections show a nonlinear relationship to transmission length when the X-ray is not monochromatic.

Beam hardening can be prevented by correcting the projection data with theoretically or experimentally derived correction curves [26] or by placing a filter between the sample and X-ray source. Figure 7.32b shows imaging with a 0.5-mm thick aluminum filter. The placement of the filter shows a considerable reduction in beam hardening. In some devices, the filter is automatically selected, like those in Fig. 4.53. The reduction in flux and worsening S/N ratio must also be considered when a filter is applied.

Even when an X-ray tube is used, a suitable combination of the target material of the X-ray tube and filter material can prevent beam hardening by obtaining quasi-monochromatic light. For example, Jenneson et al. obtained quasi-monochromatic light near the Kα X-ray of molybdenum (17.4 keV) by combining with a molybdenum target, a tube voltage of 50 kV, and a 100-μm thick molybdenum filter [27].

(2) X-Ray Scattering Effects

As shown in Fig. 2.10 through 2.13, the Compton effect is predominant at high X-ray energies, where forward scattering tendencies increase. This has an adverse effect on image quality during X-ray imaging. As detailed in Sect. 6.1, this is also the reason why line sensor cameras can achieve a higher image quality than flat panel detectors in industrial X-ray CT scanners when X-ray tube voltages are high.

As previously shown in Fig. 7.26, the spatial resolution decreases significantly, when the sample-detector distance is small and the X-ray energy is high. Figure 7.33a shows a 3D image in this case. The S/N ratio significantly decreased due to the effects of forward scattering. This not only reduces spatial resolution but also affects linear absorption coefficient measurements. In the author's experience, this effect is prominent when the X-ray energy is over 40 keV when imaging steels with monochromatic X-ray. This can be avoided by setting an appropriate distance between the sample

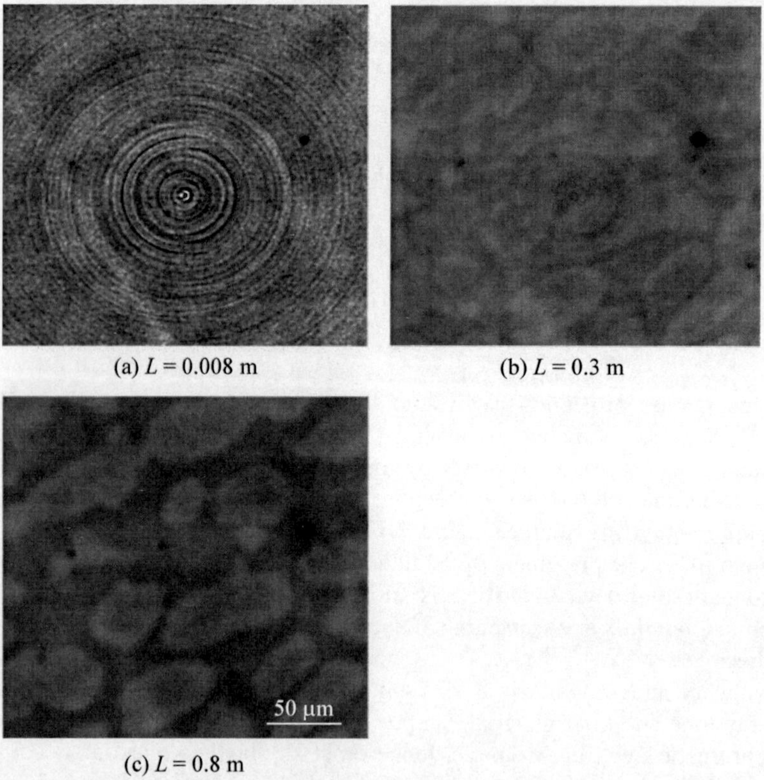

Fig. 7.33 Example where 3D imaging of steel was conducted using X-ray tomography at BL20XU in SPring-8 and where the sample-detector distance L varied during measurement [21]. **a**, **b** and **c** are 0.008, 0.3 and 0.8 m in L, respectively. The sample was a two-phase stainless steel and the X-ray energy used was 37.7 keV

and detector, as shown in Fig. 7.33b, c. However, care must be taken as excessive distances can result in gradually decreasing spatial resolution as shown in Fig. 7.26.

(3) X-Ray Refraction Effects

Section 2.2.2 (2) introduced refraction contrast imaging as an imaging method that utilized the geometric optics phenomenon of X-ray refraction. Refraction phenomena not only have positive effects such as emphasizing the contours of the interface or surface but also the negative effect of generating artifacts. These artifacts are particularly evident during synchrotron radiation X-ray tomography.

Figure 7.34 shows representative examples of this. Figure 7.34a shows artifacts that occur in the vicinity and in front of a crack when the X-ray beam is parallel to the crack surface. The aluminum matrix in the vicinity of the crack varies widely in voxel values and it is difficult to accurately segment the crack and specify the coordinates of the crack tip position. Furthermore, the contours of the crack surface are excessively emphasized due to refraction contrast, as shown in section A. This is a major barrier in studying the fatigue crack closure phenomenon. Similar artifacts can be observed in Fig. 8.25a.

Meanwhile, Fig. 7.34b shows artifacts that occur due to the refraction of X-rays from a flat sample side. The right figure shows white lines that do not exist in reality on the extended lines of the surface. These artifacts are not visible in the left figure, where the sample was placed immediately adjacent to the detector. Fringes that were effectively used for emphasizing contours in refraction contrast imaging are excessively generated in the right figure, where the sample-detector distance is longer. In these cases, the fringes conversely act to inhibit the accurate determination of the sample surface position/shape.

(4) Metal Artifacts

Figure 7.35 shows a cross-section with excessive gallium present on an aluminum surface following a liquid gallium coating applied in the liquid metal wetting technique (discussed in Sect. 5.5.1). As shown in Fig. 2.4, the linear absorption coefficients of gallium and aluminum vary by a factor of several dozen at X-ray energies of over 10 keV. Streaks extend radiatively in the vicinity of the highly-X-ray-absorbing gallium with it at the center. These are also referred to as streak artifacts. However, many other elements cause streaks, so the term *metal artifact*, which occurs with human body/metal implant combinations in medical X-ray CT scanners, is used here.

The mechanism by which metal artifacts occur is not straightforward. In other words, various factors such as artifacts due to beam hardening or photon depletion, scattering, refraction effects, detector characteristics, etc. overlap with the presence of high X-ray absorption phases.

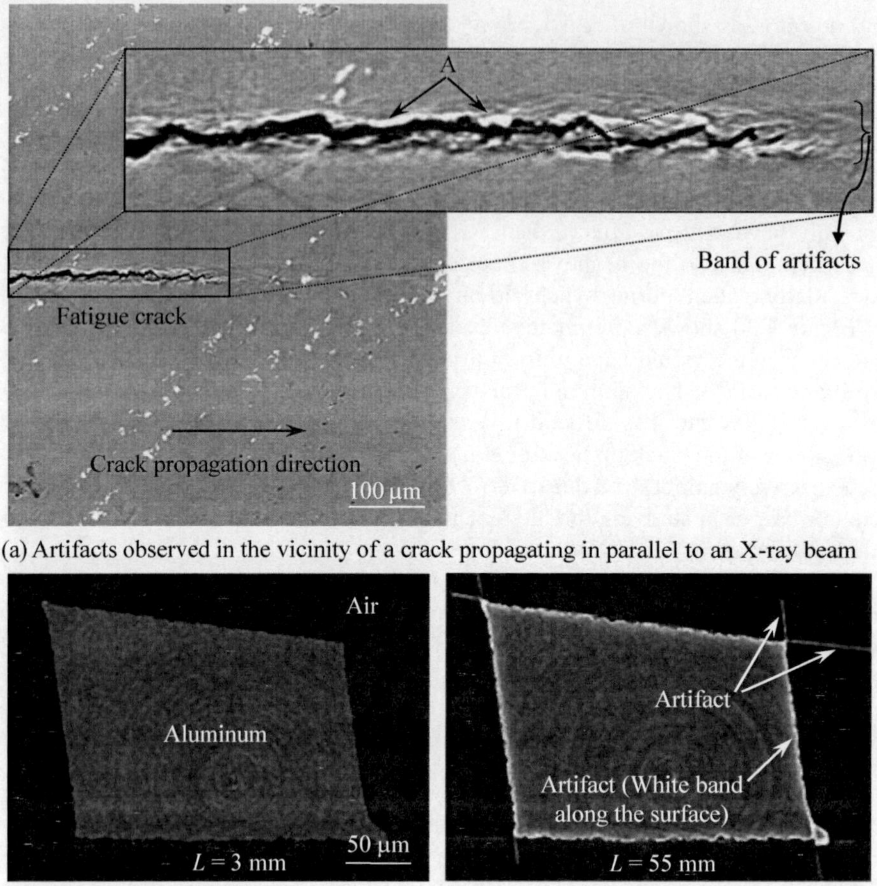

(a) Artifacts observed in the vicinity of a crack propagating in parallel to an X-ray beam

(b) Artifacts stemming from flat sample surface

Fig. 7.34 Examples of artifacts due to X-ray refraction; **a** is the vicinity of a fatigue crack tip in A7075 aluminum alloy and its magnification and **b** is an Al–Cu alloy. Experiments were conducted at BL20XU in SPring-8 and the sample-detector distance was set as 55 mm, except for the left figure in **b**. The X-ray energies were all 20 keV (courtesy of Shohei Yamamoto, who was a graduate of the author's laboratory (currently at NTN Co.); and Kazuyuki Shimizu of Kyushu University)

7.6.2 *Artifacts Due to Device*

(1) Ring Artifacts

Ring-like artifacts were observed in the cross-section images of Figs. 5.23, 7.10, 7.16, 7.32, 7.33, etc., in the x–y plane, where the orthogonal coordinate system was set such that the z-direction was parallel to the sample rotation axis. These are referred to as *ring artifacts*. The center of the ring artifact is the center of rotation. These are popular artifacts in X-ray tomography and occur due to nonlinear responses of

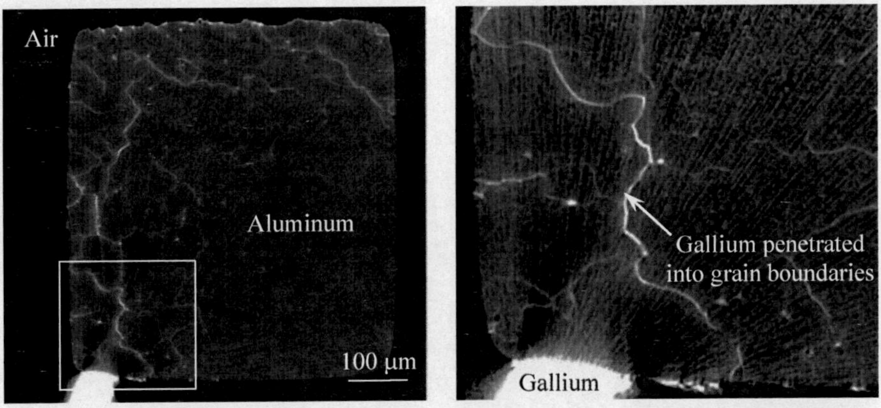

(a) Whole image

(b) Magnified view of the region highlighted with a white rectangle in (a)

Fig. 7.35 Example of metal artifacts. The high X-ray absorbing gallium was affixed to the surface of the aluminum alloy and artifacts were produced near the vicinity of the gallium. Experiments were conducted at BL20XU in SPring-8, the sample-detector distance was 55 mm, and the X-ray energy was set at 20 keV. **a** the whole cross section; and **b** a magnified view of the region highlighted with a white rectangle in **a**

part of the pixels in a detector. These appear even more emphatically when there is electrical/thermal instability in the detector, variation in the X-ray beam, and insufficient exposure time. If the ring artifact occurs near the center of rotation, the sample or the important observation regions in the sample should be offset in the field-of-view if possible.

The standard filtering processes discussed in Sect. 8.1 are not suitable for the removal of ring artifacts. There have been numerous reports since the 2000s on filters designed to remove ring artifacts and these should be used instead. ImageJ also allows for ring artifact removal filters through its Xlib plug-in [28].

(2) Artifacts Due to Misalignment

Figure 7.36 shows misalignment between the rotation axis of the sample rotation stage and the detector axis. For example, insufficient adjustment of the tilt angle in detectors may result in the inclination between the axes of the sample rotation stage and the detector during synchrotron radiation experiments. Furthermore, insufficient stiffness of the sample rotation stage and its vicinity may cause inclinations as it sinks even if the sample weight is below the load capacity of the sample rotation stage. Figure 7.37 shows an image where misalignment occurred. A relatively large misalignment of 9 pixels (0.258°) was present in the top edge of the detector surface between the rotation and detector axes. In these cases, even the form of the sample

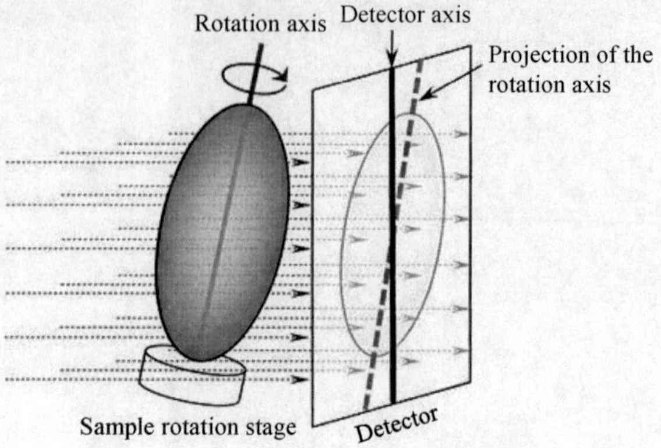

Fig. 7.36 Explanatory figure of when the detector and sample rotation stage axis are misaligned

surface and internal structure could not be accurately measured. Furthermore, the image quality greatly decreased. Quantitative image analysis is difficult in these conditions. If the axis inclination were in the direction shown in Fig. 7.36, this can be corrected at the transmission image stage by rotating the image by an amount equal to the inclination. Corrections are difficult if the inclination is in a parallel direction to the X-ray beam.

(3) Sample Rotation Stage Eccentricity Effects

The positional accuracy of the sample rotation stage (discussed in Sect. 4.3.1) is an important factor that can lead to worsened image quality. Clear artifacts are not generated, creating a further hindrance. An analysis of sample rotation stage eccentricity conducted by the author in the early stages of imaging tomography is introduced in Fig. 7.38 [29]. First, a portion of an aluminum microstructure due to imaging tomography taken with an effective pixel size of 88 nm is shown in Fig. 7.38a. Fine particles with a high X-ray absorption are affixed on the specimen surface and the eccentricity of the sample rotation stage is measured by calculating its trajectory. The eccentricity based on this when the sample was rotated 180° was a maximum of 0.4 pixels with an average of 0.15 pixels. This extent of eccentricity does not influence spatial resolution. The particle image, in this case, is shown in Fig. 7.38a. As shown on the right side of the figures, Fig. 7.38b, c show cases where the eccentricity of the rotation stage was gradual (periodicity of 22.5°) or choppy (periodicity of 0.6°). The transmission images were rotated/translated based on the eccentric trajectory and the reconstructed images obtained for each case were calculated. The eccentricity in each case was ±3 pixels. As can be observed from the figure, the high-frequency case was minimally affected even with the same eccentricity and the particle morphology varied considerably when the eccentric motion was gradual.

(4) Sample Drift

Artifacts can occur during scanning when the sample itself experiences time-dependent deformation or changes. Oftentimes, however, it is due to the sample

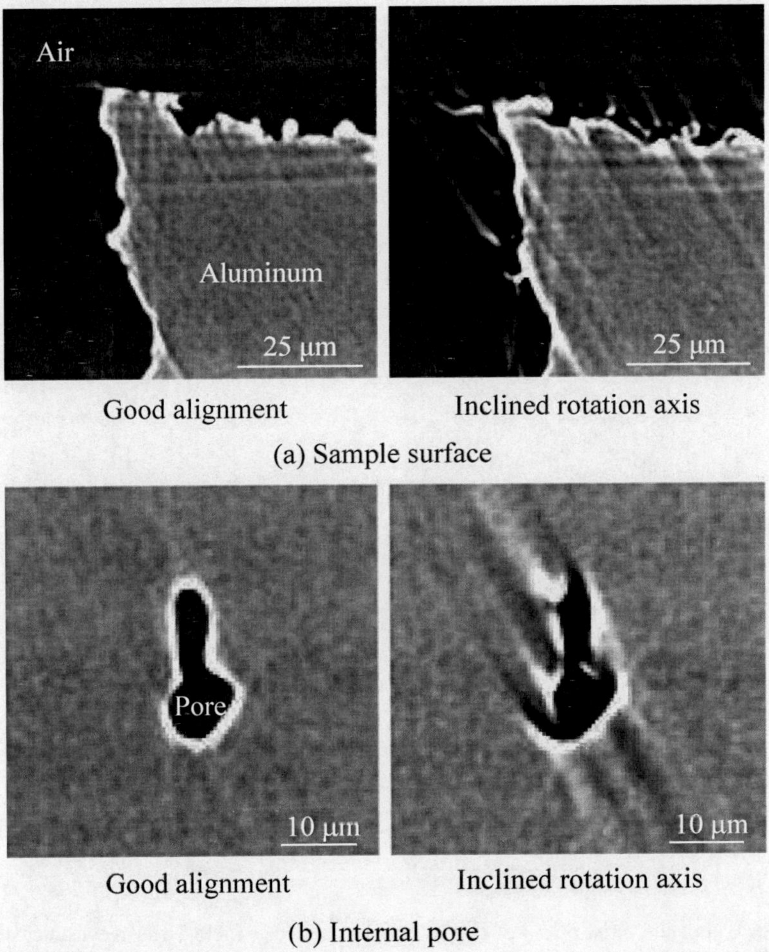

Fig. 7.37 Artifacts when the detector and sample rotation stage axis are misaligned. The detector axis is misaligned by 0.258° (9 pixel) in this case. The sample is an aluminum alloy. **a** sample surface; and **b** an internal pore. Experiments were conducted at BL20XU in SPring-8, the sample-detector distance was 25 mm, and the X-ray energy was 20 keV

being insufficiently fixed. Figure 7.39 shows samples that were intentionally moved vertically or horizontally (i.e., within the detector plane). Examining the directions in which the sample drifted using cross-section images reveals a significant elongation of the pore morphology. Consequently, pores are difficult to confirm on the right-hand side of Fig. 7.39a. Pores can be somewhat identified in cross-section layers that were

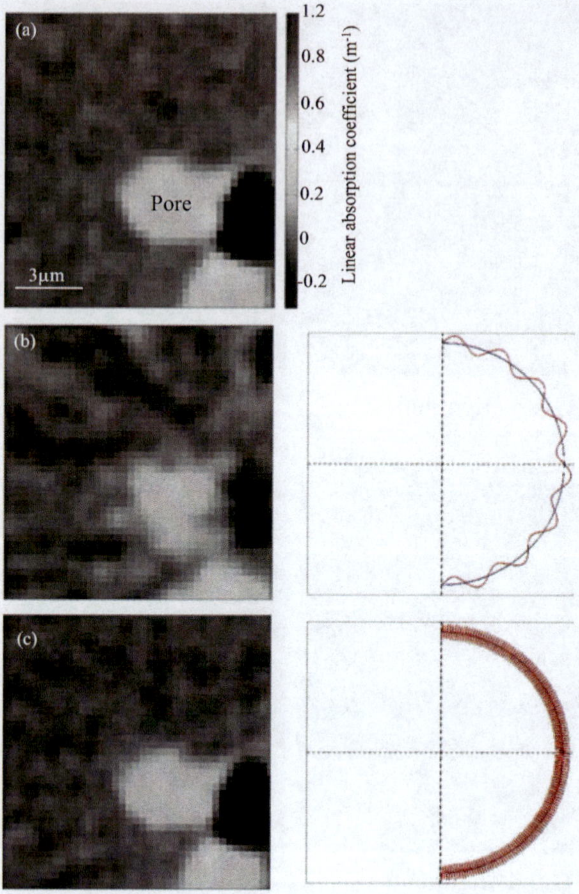

Fig. 7.38 Simulation of the effects of sample rotation stage eccentricity on the reconstructed image [29]. The sample is an aluminum alloy and the figure in **a** shows a magnified image of a particle inside that alloy. Experiments were conducted at BL47XU in SPring-8, the sample-detector distance was 25 mm, and the X-ray energy was 20 keV. The eccentricity patterns are shown in the right-hand sides of **b** and **c**

orthogonal to the sample drift direction like those on the right-hand side of Fig. 7.39b, regardless of whether the sample drift was significant. Comparisons of the 0° and 180° images in the manner described in Sect. 7.2 should be conducted if sample drift is suspected in images. Observations with high spatial resolution require sample fixing methods and sample drift countermeasures that correspond to that spatial resolution. Sufficient consideration must be given to thermal and mechanical stability (creep deformation, stress-relaxation phenomena, etc.) of the sample, particularly with the imaging-type tomography. Furthermore, the reduction of imaging time using fast tomography can be effective in these cases as well.

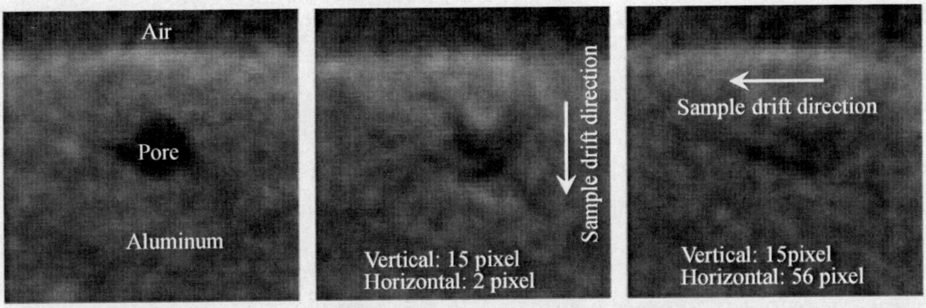

No drift | Sample drift in the vertical direction | Sample drift in the horizontal direction

(a) Comparison in x-y images

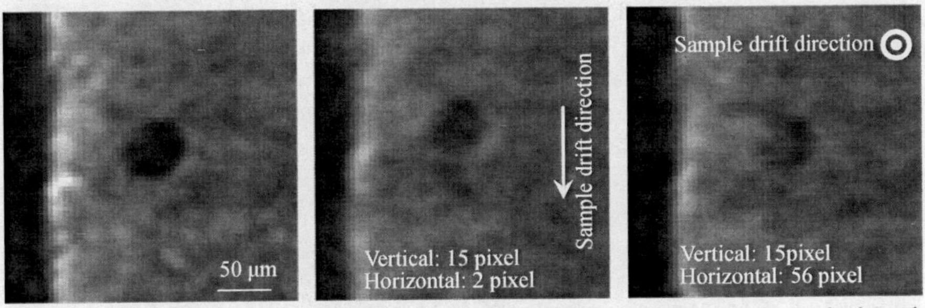

No drift | Sample drift in the vertical direction | Sample drift in the horizontal direction

(b) Comparison in y-z images

Fig. 7.39 Influence of sample movement on image quality. Reconstructed cross-section images of aluminum specimens scanned with a Skyscan 1072. The exposure time was 380 ms and the voxel size was 4.6 µm. **a** comparison in x-y plae; and **b** comparison in y-z plane. The specimen was fixed with scotch tape and left to drift with its own weight in either the vertical (z) or horizontal (x) directions. The movement amounts written in the figures are the total movement when imaged from 0 to 180°

(5) Cone Beam Artifacts

As discussed in Sect. 3.3.4, cone beam-based imaging exhibits shadows for which data are unobtainable with further distance from the origin in the z-axis direction. The image reconstruction accuracy decreases due to this missing data. This influence is more pronounced in industrial X-ray CT scanners as the cone angle increases (i.e. larger magnification) and as the cross-sections are vertically distanced from the center.

7.6.3 *Artifacts Due to Imaging Conditions*

(1) Exposure Time, Projection Number, and Missing Wedge Influences

Similar to Fig. 7.32, Figs. 7.40 and 7.41 show aluminum alloys imaged using industrial X-ray CT scanners, where both the exposure time and projection number were changed to an extreme. The magnification is considerable in these images and the appropriate values for both are an exposure time of 380 ms and a projection number of approximately 1800 projections based on the sampling theorem. Although not in the form of artifacts, large effects are observed in the image quality of both images. Decreasing projection number and exposure time result in large decreases in S/N ratio, spatial resolution, and contrast. The internal structure morphologies in these cases also change when they drop below half of the optimal values. Although not observed in the experiments of Fig. 7.41, artifacts referred to as *aliasing artifacts* occur when the projection number is decreased. All cross-sections of the sample or under-sampled regions show a sequence of streaks that expand outward starting from regions with high absorption coefficients. These can be resolved by simply

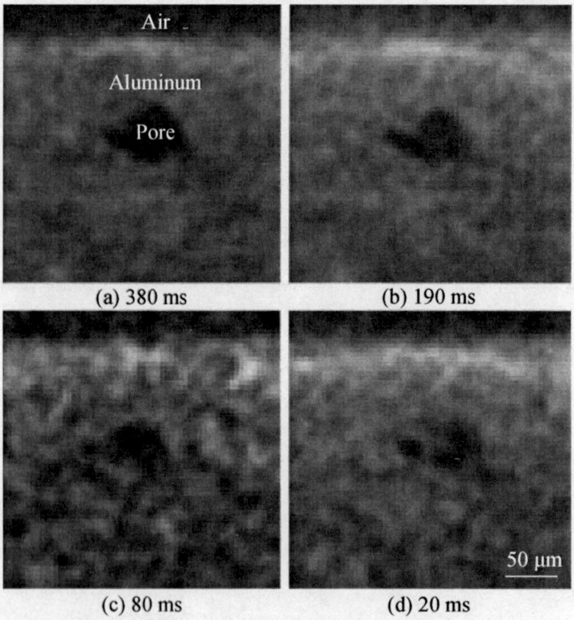

Fig. 7.40 Influence of exposure time on image quality. Reconstructed cross section images where aluminum specimens were scanned with a Skyscan 1072. The number of projections was 1800 images/180° and the voxel size was 4.6 μm. Exposure time is 380, 190, 80 and 20 ms in **a**, **b**, **c** and **d**, respectively

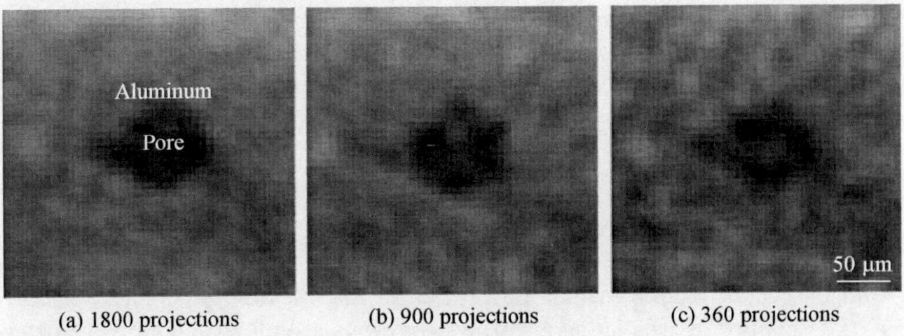

Fig. 7.41 Influence of the number of projections on image quality. Reconstructed cross section images where aluminum specimens were scanned with a Skyscan 1072; The exposure time was 380 ms and voxel size 4.6 μm. The number of projections is 1800, 900 and 360 in **a**, **b** and **c**, respectively

increasing the number of projections to the optimal value; these artifacts are not a major problem with standard X-ray tomography.

Next, Fig. 7.42 shows a reconstructed image where a certain range of projections could not be obtained with a 180° rotation angle (the "missing wedge" defect). This corresponds to when the sample size is extremely large in a given direction or when devices for *in situ* observations create a shadow over a certain portion. The data in

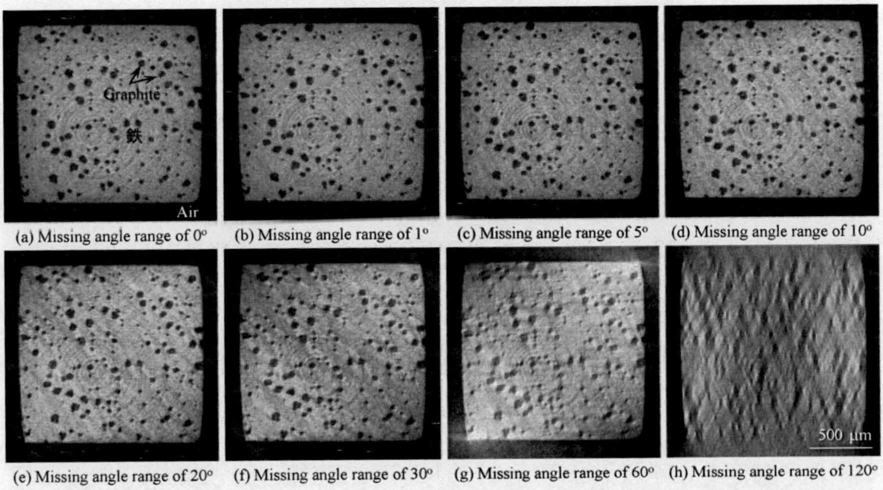

Fig. 7.42 Influence of missing angle on image quality. Sample is a spherical graphite cast iron. Experiments were conducted at BL20XU in SPring-8; the sample-detector distance was 120 mm, X-ray energy 37.7 keV, and voxel size 1.4 μm. Missing angle range is 0, 1, 5, 10, 20, 30, 60 and 120 in **a**, **b**, **c**, **d**, **e**, **f**, **g** and **h**, respectively

Fig. 7.42 are reconstructed by substituting the data from 1 to 120° with I_0 images. Structural observations can be conducted visually when the missing angle is 30° or less. Microstructure shapes like that of graphite do not change considerably if below 10°. However, further inspection shows a decrease in image quality even with slight angle defects. These effects are evident when the microstructure is barely at the spatial resolution or when observing structures with weak contrast.

(2) X-Ray Beam Diameter Influences

The image reconstruction method of a region of interest from incomplete projection data obtained by irradiating a narrow X-ray on a sample with a large diameter was explained in the discussion of specialized image reconstruction methods in Sect. 3.3.5. Artifacts appear when standard reconstruction methods are applied to samples with a large diameter that unintentionally protrude out of the field-of-view. Figure 7.43 is an example of this. A white fringe can be seen on the edges of the field-of-view. If these have large effects on image assessment, measures such as the application of a specialized reconstruction method, expansion of the field-of-view, or reduction in sample size are necessary. However, the locations at which the artifacts appear are limited, hence, frequently; this may not have adverse effects on assessment.

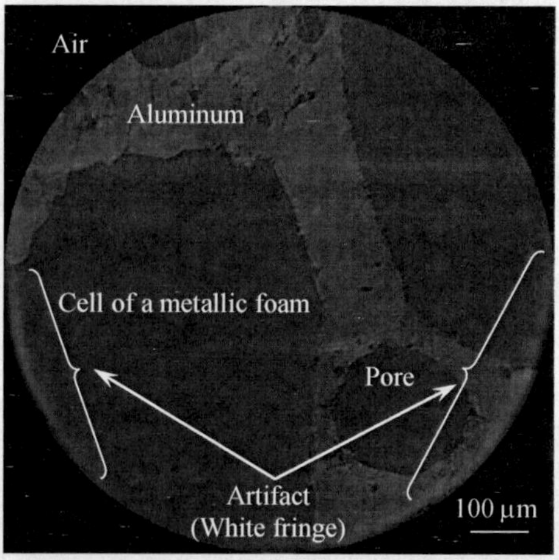

Fig. 7.43 Artifacts appearing in an image reconstructed with standard reconstruction methods when a specimen that is considerably larger than the field of view size is used. The sample was a porous aluminum alloy with a diameter of 7 mm, using a beam width of 1 mm. Experiments were conducted at BL47XU in SPring-8, the sample-detector distance was 55 mm, and the X-ray energy was 20 keV

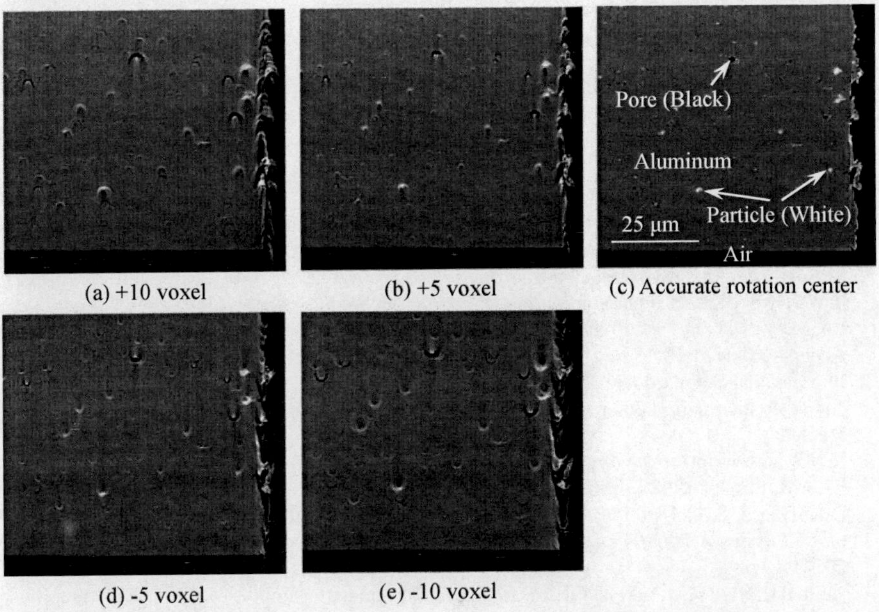

Fig. 7.44 Artifacts due to misalignment of the reconstruction center. Reconstructed cross section image where an aluminum specimen was imaged with a Skyscan 1072. Exposure time was 380 ms and the voxel size was 4.6 μm. The reconstruction center was accurately aligned in **c** and other figures were reconstructed with a misalignment specified in each figure. Misalignment value is + 10, +5, -5 and -10 in **a**, **b**, **d** and **e**, respectively

7.6.4 Artifacts Due to Reconstruction

Figure 7.43 shows artifacts formed when the reconstruction center was misaligned. Focusing on particles, pores, and protrusions on the surface generates artifacts in the shape of a tuning fork when the center of rotation is misaligned. The size of these artifacts increases with larger misalignment. The direction of this tuning fork inverts when misaligned in the reverse direction; therefore, this can be used to determine the direction towards which the center of rotation is misaligned. In either case, specifying the approximate coordinate of the center of rotation and conducting reconstruction, regardless of image quality, is the most fundamental element of X-ray tomography. Cross-sections, where structures such as particles or pores can serve as a reference point, should be sought when conducting an accurate reconstruction (Fig. 7.44).

References

1. Japanese Industrial Standards, Industrial X-ray CT devices—terminology. JIS **B7442** (2013)
2. Japanese Industrial Standards, Medical radiation devices—defined terminology. JIS **Z4005** (2012)
3. Medical Imaging Technology Handbook, Japanese Society of Medical Imaging Technology (2012). ISBN-13:978-4990666705
4. For example, Hamamatsu Photonics High-Performance Image Control System HiPic User Manual. http://www-bl20.spring8.or.jp/detectors/manual/. Accessed June 2018
5. SPring-8 BL20: How to make a transmission image movie (2010.07.06 version, txt format), http://www-bl20.spring8.or.jp/xct/manual/make-movie.txt. Accessed June 2018
6. ImageJ website, https://imagej.nih.gov/ij/docs/guide/146-26.html#toc-Subsection-26.6. Accessed June 2018
7. International Standard IEC, 62220-1, First edition, 2003–10, Medical electrical equipment—characteristics of digital X-ray imaging devices—Part 1: determination of the detective quantum efficiency
8. K.M. Hanson, in *Proceedings of SPIE 3336, Physics of Medical Imaging*, ed. by J.T. Dobbins III, J.M. Boone (1998), pp. 243–250
9. D.R. Dance, S. Christofides, A.D.A. Maidment, I.D. McLean, K.H. Ng, in *Diagnostic Radiology Physics A Handbook for Teachers and Students* (International atomic energy agency, 2014), p. 78
10. I. Elbakri, SU-GG-I-121: JDQE: A user—friendly ImageJ Plugin for DQE Calculation. Med. Phys. **37**, 3129 (2010)
11. CT Image Measurement Program CT measure Basic ver. 0.97b2 2016.8.22, Japanese Society of CT Technology, http://www.jsct-tech.org/index.php?page=member_ctmeasurefree. Accessed June 2018
12. Japanese Industrial Standards, Phantoms for X-Ray CT devices. JIS Z **4923** (2015)
13. K. Ichikawa, S. Muramatsu, Standard X-Ray CT image measurements (Ohmsha, 2018), p. 119
14. H. Toda, M. Kobayashi, T. Kubo, K. Moizumi, D. Sugiyama, Y. Yamamoto, T. Harada, K. Hayashi, Y. Hangai, Y. Murakami, J. Jpn. Instit. Light Metals **63**, 343–349 (2013)
15. F.R. Verdun, D. Racine, J.G. Ott, M.J. Tapiovaara, P. Toroi, F.O. Bochud, W.J.H. Veldkamp, A. Schegerer, R.W. Bouwman, I.H. Giron, N.W. Marshall, S. Edyvean, Physica Med. **31**, 823–843 (2015)
16. 10 spatial resolution in CT. J. ICRU, **12**, 1 (2012), Report 87. https://doi.org/10.1093/jicru/ndt001 (Oxford University Press)
17. Micro-CT Wire Phantom: QRM Gmbh, http://www.qrm.de/content/pdf/QRM-MicroCT-Wire.pdf. Accessed July 2018
18. T.M. Buzug, *Computed Tomography: From Photon Statistics to Modern Cone-Beam CT* (Springer, Berlin, Germany, 2008)
19. X-Ray CT Resolution Specimen Au Absorbent (JIMA RT CT-01) Catalog: Japan Inspection Instruments Manufacturers' Association (JIMA), http://www.jima.jp/content/pdf/catalog_rt_ct01_j.pdf. Accessed July 2018
20. MicroCT Bar Pattern Phantom: QRM Gmbh, http://www.qrm.de/content/pdf/QRM-MicroCT-Barpattern-Phantom.pdf. Accessed July 2018
21. D. Seo, F. Tomizato, H. Toda, K. Uesugi, A. Takeuchi, Y. Suzuki, M. Kobayashi, Appl. Phys. Lett. **101**, 261901 (2012)
22. K. Ichikawa, H. Kunitomo, T. Sakurai, K. Ohashi, M. Sugiyama, T. Miyaji, H. Fujita, Jpn. J. Radiologic. Technol. **58**, 1261–1267 (2002)
23. S.M. Bentzen, Med. Phys. **10**, 579–581 (1983)
24. F.F. Yin, M.L. Giger, K. Doi, Med. Phys. **17**, 960–966 (1990)
25. Photography—electronic still-picture cameras. Resolution measurements (2000(E)). ISO Standard 12233
26. P. Hammersberg, M. Mangard, J. X-ray Sci. Technol. **8**, 75–93 (1998)

27. P.M. Jenneson, R.D. Luggar, E.J. Morton, O. Gundogdu, U. Tuzun, J. Appl. Phys. **96**, 2889–2894 (2004)
28. Image J Plug-in Xlib, https://imagej.net/Xlib. Accessed Sept 2018
29. M. Kobayashi, Y. Sugihara, H. Toda, K. Uesugi, J. Jpn. Instit. Light Metals **63**, 273–278 (2013)

Chapter 8
3D Image Processing and 3D Image Analysis

Let us assume that the latest hardware was suitably selected based on a deep foundational knowledge of X-rays and a 2D transmission image set was obtained as required using the ground-breaking applied imaging methods we have previously examined. Furthermore, assuming a high-quality 3D image, where the optimal spatial resolution, detectability, noise, contrast, and artifacts were obtained, while image reconstruction was conducted with fine attention to detail; if there is an insufficient understanding regarding image processing and the 3D images were haphazardly processed using the software packaged with the industrial X-ray CT scanner or specialized software that is commercially available, all of this accumulated knowledge, preparation, and consideration of the latest software and hardware technology is wasted. For example, if a faintly imaged microstructure is deemed noise and unintentionally filtered or segmented out of the image, then two structures with slight differences in contrast would be processed as identical. Countless pitfalls like this can be encountered. The author has, unfortunately, seen these kinds of examples numerous times both inside and outside of the laboratory.

The various image processing and analysis techniques available are never a universal magic wand and it is extremely rare to use these to stumble upon an entirely novel discovery. Researchers should carefully inspect raw images with their own eyes and determine beforehand what may be shown in the image. Further confirming these visual inspections, quantitatively assessing them, and effectively expressing them with figures are supplemental methods that comprise the bulk of image analysis and image processing techniques. Even at this stage, the researcher should be firmly cognizant of the five factors of spatial resolution, detectability, noise, contrast, and artifacts.

H. Toda, *X-Ray CT*,
https://doi.org/10.1007/978-981-16-0590-1_8

8.1 Filtering

Noise and artifacts are unavoidable components in images obtained by X-ray tomography. Such problems also include realistic issues such as insufficient contrast or spatial resolution. *Removing the noise, etc. from an image whose quality is worse than expected, identifying specific structures and facilitating their qualitative evaluation, enabling segmentation (to be explained later), and prompting qualitative evaluation is the process known as "applying a filter" or "filtering."* Filtering involves 3D images or 2D images as subjects. Furthermore, there are two types of *spatial filtering* and *frequency filtering*. Applying some type of operation on an aggregation of pixels in an image and re-distributing the pixel or voxel values based on the operation results is referred to as spatial filtering. Meanwhile, conducting filtering in the frequency space after applying a Fourier transform to eliminate specified frequency components or noise with a certain periodicity, obtaining the output image after applying an inverse Fourier transform is referred to as frequency filtering.

Representative filter types and overviews are provided in Table 8.1. This section will provide the basic algorithms of representative filters and their effects according to the filtering objective. As filtering a 3D image is fundamentally the same process as that of a 2D image, this section will introduce examples of 2D filtering that are easier to formulate and whose application examples are easier to see.

8.1.1 Averaging Filter

In contrast to the gray-level transformation (detailed in Sect. 7.1) that examines the pixel values of the pixels to be processed only and changes their degree of contrast, the *averaging filter* calculates a new pixel value by including the information surrounding the subject pixel and assigns it to the pixel value of the subject. As shown in Table 8.1, averaging filters include *linear filters* such as a *moving-average filter* and weighted-average filter, as well as nonlinear filters such as a *median filter* or*bilateral filter*.

As shown in the following equation, linear filters apply a weighted coefficient matrix (also referred to as a *kernel*) $h(m, n)$ with a size $(2w + 1) \times (2w + 1)$ on the original image $f(x, y)$ to obtain a filtered image $g(x, y)$:

$$g(x, y) = \sum_{m=-w}^{w} \sum_{n=-w}^{w} f(x, y)h(m, n) \tag{8.1}$$

Mathematically, this corresponds to convoluting a mask function to the original image. The filtering of the entire image is conducted by applying this operation to convert all pixel values through a raster scan. As shown in the following equation, the moving average filter indicates a case among linear filters where the weighting coefficients of the filter are all identical. The brightness after filtering, in this

Table 8.1 Overview of primary spatial/frequency filters and their functions

Type	Designation	Major functions
Averaging filter	Moving average filter	Replacement to an average value in a certain region
	Weighted average filter	Suppression of blurring using a weighted coefficient for gray value in the voxel of interest
	Gaussian filter	Weighting to follow the Gaussian distribution with the voxel of interest as its center
	Median filter	Replacement to a median in a certain region preserving edges to some extent
	Bilateral filter	Averaging filter that places an importance in edge preservation
Edge detection/enhancement filter	Differential filter	First differential affected by edge detection and noise
	Prewitt filter	Combination of the first differential and averaging treatment to extract the edges that align in a specific direction
	Sobel filter	Combination of the first differential and weighted averaging treatment to extract the edges that align in a specific direction
	Laplacian filter	Secondary differential to extract edges independently of direction
	Laplacian of Gaussian filter	Combination of the Gaussian and Laplacian filters
	Sharpening filter	Enhancement of high-frequency components by subtracting averaged data from raw data
Frequency filter	Low-pass filter	High attenuation for high-frequency component allowing low-frequency component to pass through
	High-pass filter	High attenuation for low-frequency component allowing high-frequency component to pass through
	Band-pass filter	Removal of the frequency components above or below a specific range

case, is preserved, so the image can be normalized by making the sum of coefficients equal to 1.

$$h_{m.a.} = \frac{1}{9}\begin{bmatrix} 1 & 1 & 1 \\ 1 & 1 & 1 \\ 1 & 1 & 1 \end{bmatrix} \tag{8.2}$$

This processing is equivalent to taking the average of the local region where the pixel value changes in the overall image become smoother. Furthermore, any noise included in the image is effectively reduced. However, the filtering process also results in the blurring of the image. Furthermore, a larger mask size results in significant blurring of the image. Image deterioration can also occur, such as the elimination of point or line structures or the reduced sharpness of an edge.

As shown in the following equation, the weighted-average filter exhibits a stronger tendency for preserving the original information than the moving-average filter as it increases the coefficients of weighting coefficient matrix multiplied on the central pixel to be processed. This controls the blurring tendency to some extent as well.

$$h_{w.a.} = \frac{1}{16}\begin{bmatrix} 1 & 2 & 1 \\ 2 & 4 & 2 \\ 1 & 2 & 1 \end{bmatrix} \tag{8.3}$$

Furthermore, among the weighted-average filters in Table 8.1, the *Gaussian filter* assigns the coefficients in the weighting coefficient matrix so that their values follow a bell-shaped Gaussian function, with the middle pixel at the center. The filter-coefficient weights follow a Gaussian function with an average of 0 and a variation of σ^2. A 2D Gaussian function is shown in the following equation.

$$h_{Gauss}(x, y) = \frac{1}{2\pi\sigma^2} exp\left(\frac{-x^2 + y^2}{2\sigma^2}\right) \tag{8.4}$$

The shape of the Gaussian function can be changed as shown in Fig. 8.1 by changing the variation σ^2. Furthermore, Eq. (8.3) takes the form of a Gaussian filter. Figure 8.2 shows the reduced-noise image when a Gaussian filter is applied to a high-noise image and its pixel value histogram. Both the salt-and-pepper noise, which takes values of 0 and 255 when pixel values are 8-bit, and noise with intermediate pixel values were reduced, and the peak for air, which was not visible before the filter application, appear alongside the peak for the sample metal. Furthermore, Fig. 8.3 shows changes in the image when Gaussian filters with kernels of various sizes were applied once or twice. Strong filter applications eliminate considerable amounts of noise, but at the same time makes spot-like microstructures and sharp edges difficult to maintain.

Linear filters can significantly eliminate microstructures and can be considered low-pass filters in this sense. The usage of nonlinear filters should be considered

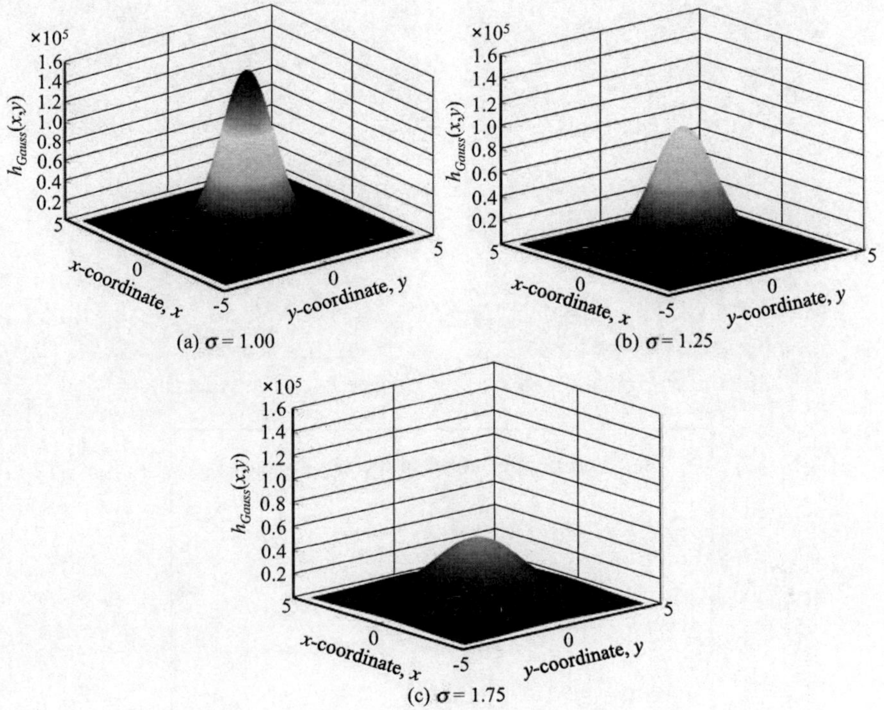

Fig. 8.1 Shape of a 2D Gaussian function. Comparisons of function shapes when three different standard deviation values were used

when the negative effects of linear filters may be problematic. Of these, the median filter outputs the median value of the pixel value distribution in the local region. This can remove spike-like noise with sharp pixel value changes effectively without significantly changing structures such as edges. However, narrow line structures or corners are difficult to preserve.

Meanwhile, bilateral filters can effectively prevent the blurring of edges. This is expressed with the following equation:

$$h_{BL}(x, y) = \frac{\sum_{m=-w}^{w}\sum_{n=-w}^{w} f(x+m, y+n) exp\left(\frac{-m^2+n^2}{2{\sigma_1}^2}\right) exp\left(\frac{-\{f(x,y)-f(x+m,y+n)\}^2}{2{\sigma_2}^2}\right)}{\sum_{m=-w}^{w}\sum_{n=-w}^{w} exp\left(\frac{-m^2+n^2}{2{\sigma_1}^2}\right) exp\left(\frac{-\{f(x,y)-f(x+m,y+n)\}^2}{2{\sigma_2}^2}\right)} \tag{8.5}$$

The term σ_1 is the same as the coefficient of the Gaussian filter and σ_2 is the standard deviation, which expresses the difference from the central pixel value. Both are changed to control the filtering results. Thus, the first exponential function in the numerator of Eq. (8.5) reflects the distance between pixels in the Gaussian function weight. Meanwhile, the second exponential function similarly reflects the difference

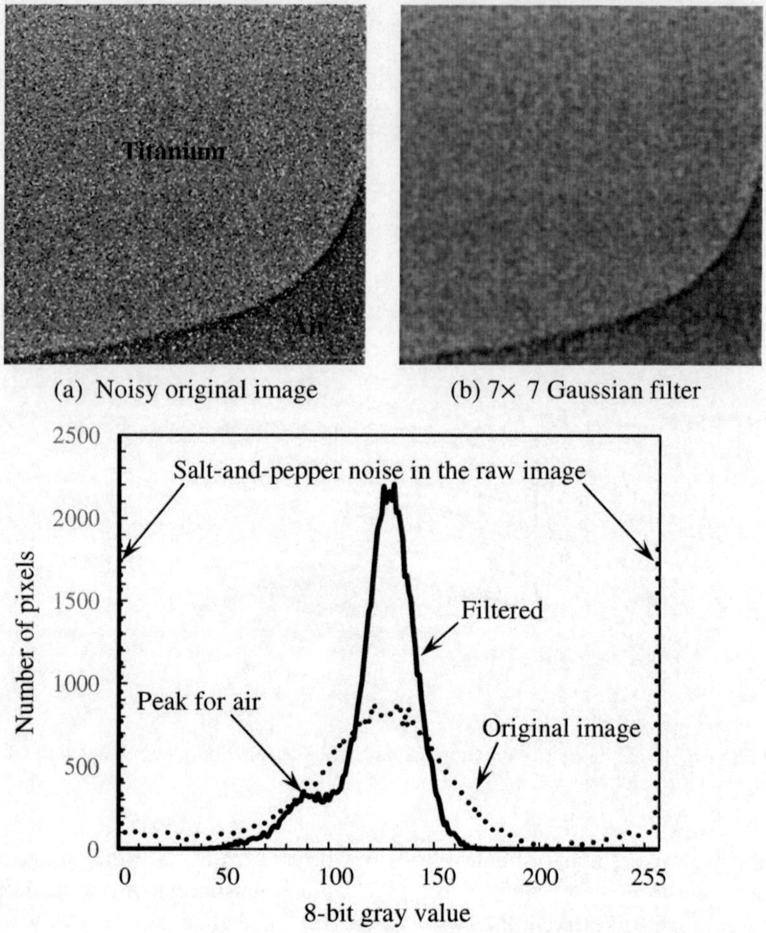

(a) Noisy original image (b) 7× 7 Gaussian filter

(c) Histogram of gray value distribution

Fig. 8.2 Differences in a cross-section image obtained with X-ray tomography when various averaging filters were applied. The air peak appeared with filtering and a general segmentation procedure can be conducted

from the central pixel value to be processed in the Gaussian function. The former approaches 0 as the absolute value of the brightness difference $|f(x, y) - f(x + m, y + n)|$ is larger; the weight conversely becomes larger as the pixel values are similar. In other words, the weight stipulated by the pixel value difference becomes smaller at the edge even if the spatial weight stipulated by the first Gaussian function is large, reducing the likelihood of blurring. Furthermore, the denominator of Eq. (8.5) is included so that the sum of coefficients equals 1. Recently, although the bilateral filter has been widely used, it is problematic in that it requires a lengthy computational time.

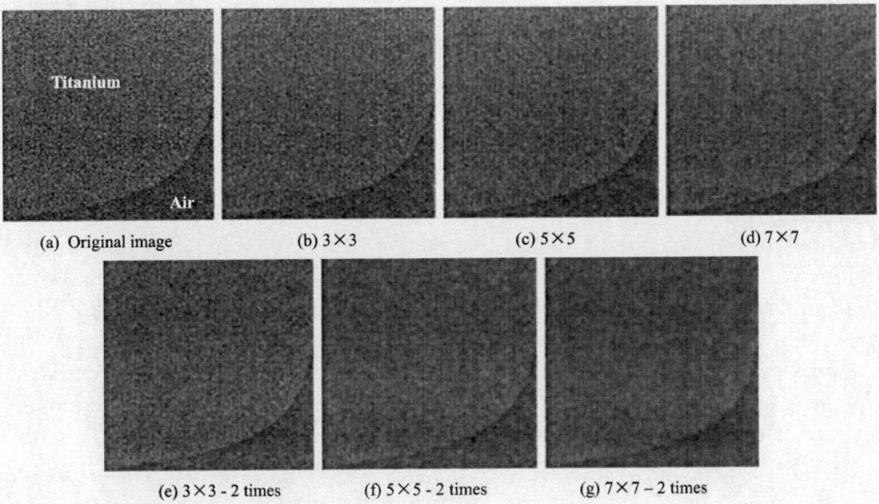

Fig. 8.3 Differences in cross-section images obtained with X-ray tomography when Gaussian filters with kernels of various sizes were applied. The bottom row shows images when each filter was applied twice

Finally, the results of applying various averaging filters introduced in this chapter to actual X-ray tomography images are shown in Fig. 8.4. The noise removal and edge preservation characteristics of each filter can be compared in actual images. Changing the filter parameters results in different effects, therefore, these should be viewed only as a reference. The structure (dual-phase) can be faintly observed in the

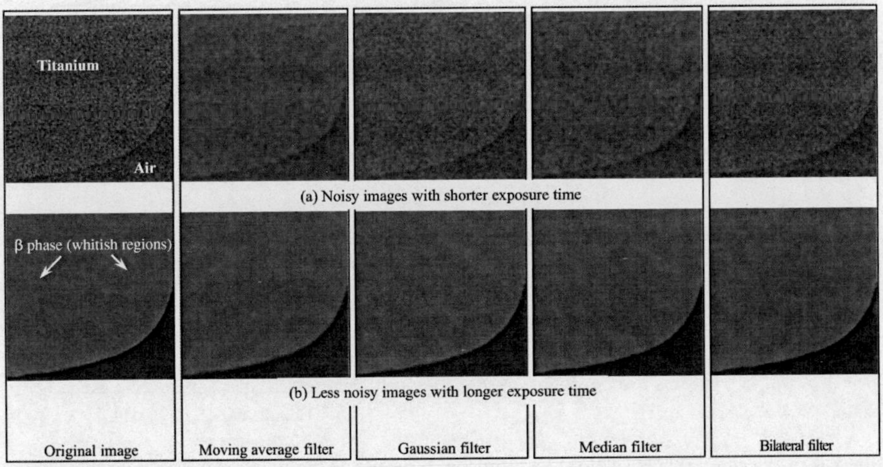

Fig. 8.4 Differences in cross-section images obtained with X-ray tomography when various averaging filters were applied

metal when the exposure time is long. Not only does the blurring of the air–titanium interface change, but the visualization of the dual-phase structure also changes.

8.1.2 *Edge Detection/Enhancing Filter*

As shown in Fig. 8.5, taking the first derivative of a pixel value distribution when regions with different linear absorption coefficients are present in the matrix can result in an image whose boundaries (edges) are enhanced. This is the *differential filter* referred to in Table 8.1. Digital images are composed of discrete data, so this is approximated in practice by taking finite differences. These finite differences are separately taken in the two x- and y- directions, where the edges of these directions are detected. The first-order linear forward difference approximation for a first derivative is expressed as follows:

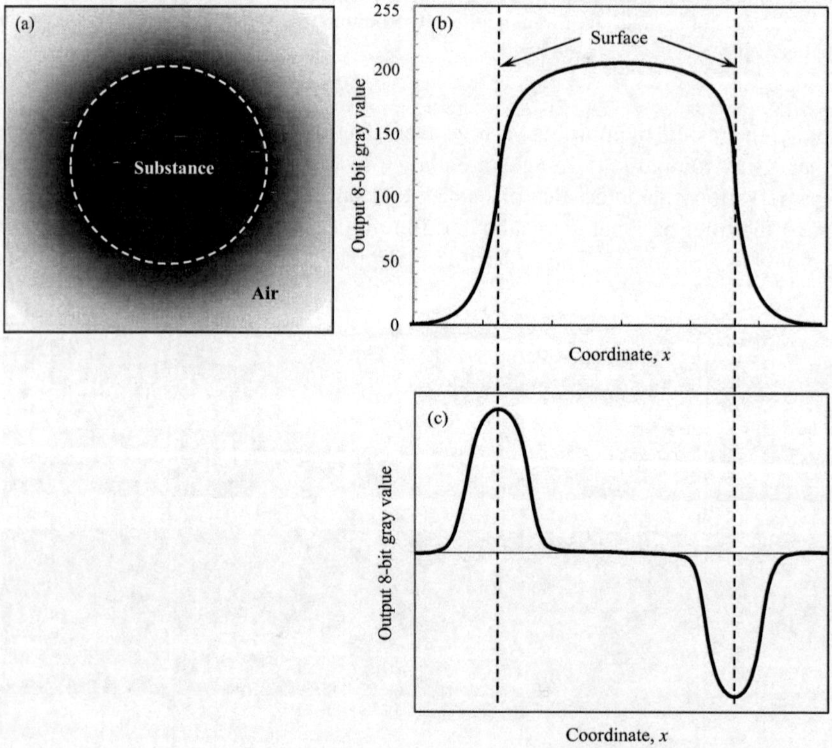

Fig. 8.5 **a** Structure (black section) in an image, **b** profile of the pixel values, and **c** schematic of its first derivative. The edge can be detected by taking the first derivative

$$\frac{\partial f}{\partial x}(x, y) \approx f(x+1, y) - f(x, y) \quad (x\text{-direction}) \tag{8.6}$$

$$\frac{\partial f}{\partial y}(x, y) \approx f(x, y+1) - f(x, y) \quad (y\text{-direction}) \tag{8.7}$$

These can be simply calculated by applying the linear filters $h^x{}_{\text{diff}}$ and $h^y{}_{\text{diff}}$ as follows:

$$h^x_{diff} = \begin{bmatrix} 0 & 0 & 0 \\ 0 & -1 & 1 \\ 0 & 0 & 0 \end{bmatrix} \quad (x\text{-direction}) \tag{8.8}$$

$$h^y_{diff} = \begin{bmatrix} 0 & 1 & 0 \\ 0 & -1 & 0 \\ 0 & 0 & 0 \end{bmatrix} \quad (y\text{-direction}) \tag{8.9}$$

However, it can be easily considered that the first derivatives would enhance noise-filled sections where the pixel values rapidly change simultaneously with the edge sections. The addition of an averaging process to suppress noise effects is referred to as the *Prewitt filter* in Table 8.1. Its linear filters $h^x{}_{\text{Prwt}}$ and $h^y{}_{\text{Prwt}}$ are expressed as follows:

$$h^x_{Prwt} = \begin{bmatrix} -1 & 0 & 1 \\ -1 & 0 & 1 \\ -1 & 0 & 1 \end{bmatrix} \quad (x\text{ - direction}) \tag{8.10}$$

$$h^y_{Prwt} = \begin{bmatrix} -1 & -1 & -1 \\ 0 & 0 & 0 \\ 1 & 1 & 1 \end{bmatrix} \quad (y\text{ - direction}) \tag{8.11}$$

The application of a weighted average to the pixel for this averaging process is referred to as the *Sobel filter* in Table 8.1. The $h^x{}_{\text{Sobel}}$ in the x-direction is as follows:

$$h^x_{Sobel} = \begin{bmatrix} -1 & 0 & 1 \\ -2 & 0 & 2 \\ -1 & 0 & 1 \end{bmatrix} \quad (x\text{ - direction}) \tag{8.12}$$

The edge detection sensitivity of the Sobel filter is higher than that of the Prewitt filter, but the former is more likely to experience noise issues.

Furthermore, the *Laplacian filter* in Table 8.1 uses a second derivative. A second-order forward difference approximation is then used for this.

$$\frac{\partial^2 f}{\partial x^2}(x, y) \approx f(x+1, y) - f(x, y) - \{f(x, y) - f(x-1, y)\} \quad (x \text{ direction}) \tag{8.13}$$

$$\frac{\partial^2 f}{\partial y^2}(x, y) \approx f(x, y+1) - f(x, y) - \{f(x, y) - f(x, y-1)\} \quad (y \text{ direction}) \tag{8.14}$$

Taking the second derivative shows pairings of an upward convex peak and a downward convex peak on either side of the edge, as shown in Fig. 8.6. The edge positions correspond to the points where the values become 0 at the boundaries of both peaks (*zero-crossing points*). The edge can be detected using these even when the pixel transition is gradual. The Laplacian s expressed as follows in two-dimensional orthogonal coordinates:

$$\nabla^2 f(x, y) \approx f(x-1, y) + f(x, y-1) - 4f(x, y) + f(x+1, y) + f(x, y+1) \tag{8.15}$$

The linear filter $h^2{}_{\nabla}{}^2$, in this case, is expressed as follows:

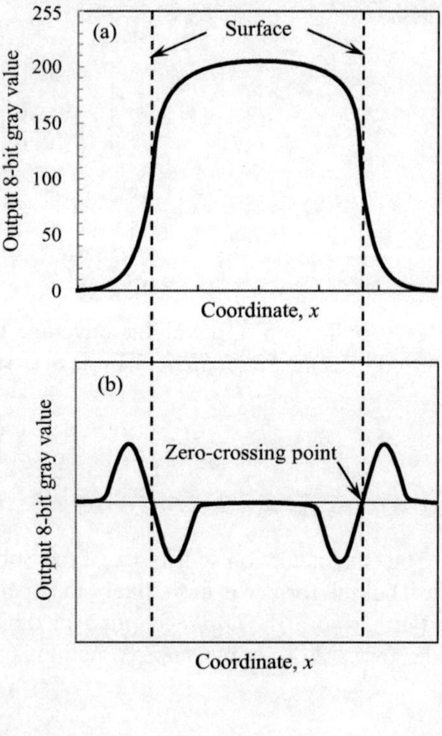

Fig. 8.6 **a** Pixel value profile in the edge section, and **b** schematic of its second derivative. The edge can be detected by taking the second derivative

$$h^2_{\nabla^2} = \begin{bmatrix} 0 & 1 & 0 \\ 1 & -4 & 1 \\ 0 & 1 & 0 \end{bmatrix} \tag{8.16}$$

This is a 4-neighbor filter that calculates the second derivative of the pixel to be processed relative to the four pixels in the vertical and horizontal directions for a 2D image. This equation is as shown below when the four pixels in the diagonal directions are considered (8-neighbor filter):

$$h^4_{\nabla^2} = \begin{bmatrix} 1 & 1 & 1 \\ 1 & -8 & 1 \\ 1 & 1 & 1 \end{bmatrix} \tag{8.17}$$

However, it is frequently the case that the outer contour of noise is recognized as an edge when simply applying a Laplacian filter to an X-ray tomography image, resulting in the insufficient detection of the interface. A *Laplacian of Gaussian (LoG) filter*, which is mentioned in Table 8.1 and applies a Laplacian filter after reducing noise with a Gaussian filter, is used with this potential issue in mind.

An image where the pixel value changes in the edge of the original image are enhanced can be obtained as shown in the following equation by subtracting the second-derivative image, determined by applying the Laplacian filter on the image, from the original image:

$$g(x, y) = f(x, y) - \nabla^2 f(x, y) \tag{8.18}$$

This is the *sharpening filter* in Table 8.1. The 8-neighbor sharpening filter is as follows:

$$h^8_{Sharpng} = \begin{bmatrix} -1 & -1 & -1 \\ -1 & 9 & -1 \\ -1 & -1 & -1 \end{bmatrix} \tag{8.19}$$

The application of the sharpening filter results in the undershoot and overshoot of the pixel values in the interface area occurring in pairs as shown in Fig. 8.7; the inclination angle of the edge also becomes larger. Figure 8.8 shows the application of the 8-neighbor sharpening filter on a cross-section image from Fig. 3.18d, which was considerably blurred due to deliberate reconstruction without any filter correction. Internal structures, which were difficult to confirm in the original image, were thus somewhat confirmed.

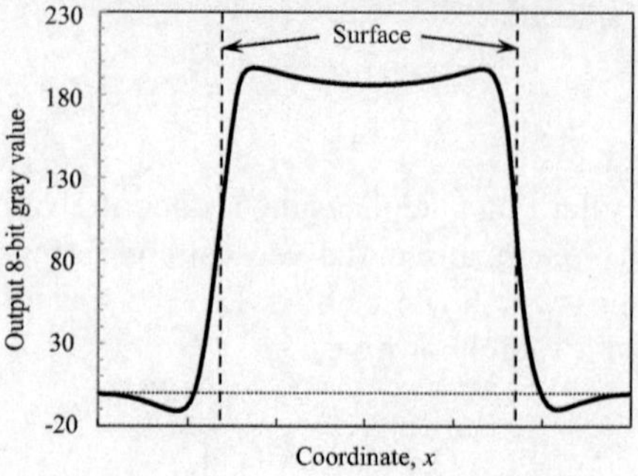

Fig. 8.7 Pixel value profile in an edge-enhanced image obtained by subtracting the second derivative in Fig. 8.6b from the original image in Fig. 8.6a. The grayscale range in the vertical axis has been changed from Figs. 8.5 and 8.6

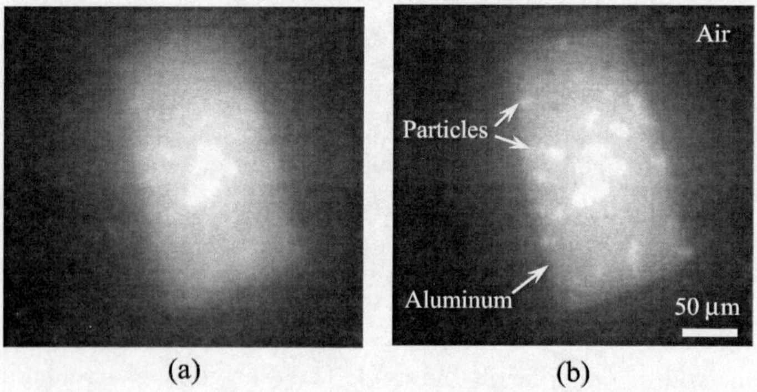

Fig. 8.8 Changes in the cross-section image in Fig. 3.18d after applying a sharpening filter. The sample is aluminum with dispersed particles in it. **a** Blurred original image in Fig.3.18d, **b** After applying the sharpening filter

8.1.3 Frequency Filter

Figure 8.9 shows a *low-pass filter*, where an inverse Fourier transform is applied after shielding all frequencies above a specified level in the frequency space. Filters that shield all frequencies below a specified level are referred to as *high-pass filters*; those that only allow a certain frequency range to transmit are referred to as *band-path filters*. Furthermore, the simple multiplication of $F(u,v)$ after the Fourier transform to the Fourier transform of the filter function $H(u, v)$ enables the creation of various filters, such as high-pass enhancement filters, which emphasize the high-frequency range.

$$G(u, v) = F(u, v)H(u, v) \tag{8.20}$$

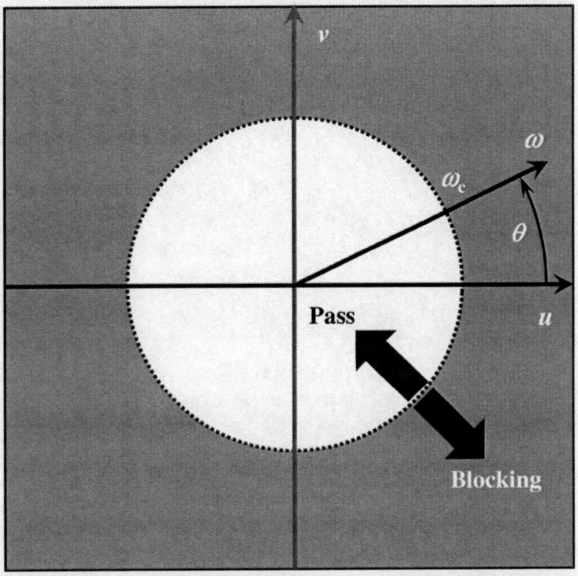

Fig. 8.9 Schematic of a low-pass filter in a frequency space where all the frequencies above a given value are blocked

Figure 8.10 shows the application of a low pass filter in a 3D image of an aluminum material microstructure after salt-and-pepper noise was deliberately added. Only the noise components were effectively removed without largely altering the microstructure. The matrix section was also smoothed out to some extent.

Other filters, based on optimal filtering theory, which seek to restore a deteriorated image, include the *Wiener filter*. Restoring an original image based on blur retrieval is possible by convoluting a so-called reverse filter. However, this tends to simultaneously create wave-like artifacts referred to as ringing and emphasize noise. On a fundamental level, it is paramount to take the various devices that compose the X-ray CT scanner into consideration and gain an understanding of the level of spatial resolution that can reasonably be obtained at the raw-data stage. However, these filters can be used if blur cannot be avoided for a given reason. Please refer to technical books for further investigation [1].

8.2 Segmentation

Segmentation is the first fundamental step in image processing when qualitatively assessing the morphology of the internal structure in 3D images and when quantitatively handling image data through image analysis. This refers to imaging operations that distinguish different structures or tissues into separate regions. The simplest segmentation is assigning an appropriate threshold value and separating multiple regions, structures, tissues, etc. A number of methods can be selected if this does not work. In either case, it is important not to be distracted by just the segmentation

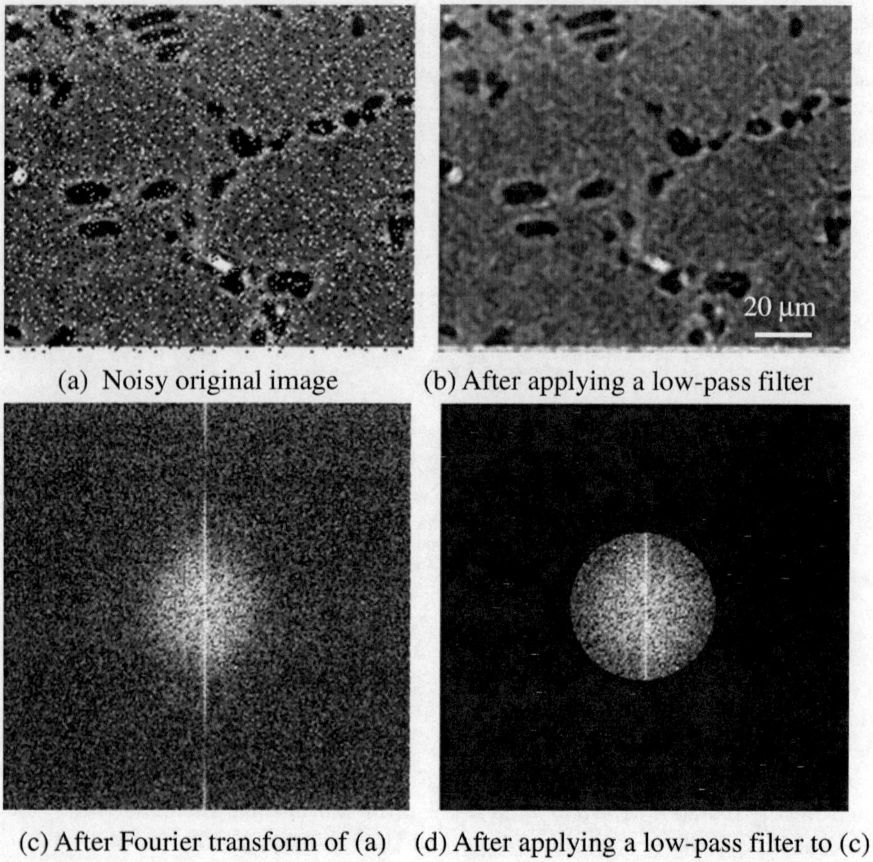

Fig. 8.10 Cross-section image obtained by X-ray tomography (aluminum—silicon alloy, Fig. 2.22) with salt-and-pepper noise deliberately added and where a low-pass filter was applied

results and to be cognizant of the segmentation validity by gaining an understanding of the pixel value distributions in the original image.

Many of the methods shown here can be easily conducted with software such as ImageJ. Users can test them with 2D images to determine which processing method is the most suited for their images. Furthermore, the image processing library "scikit-image" is available for the programming language Python, which can conduct not only the segmentations here but also the filtering, mentioned in the previous section, the coordinate transformation in Chap. 9, and the image analysis detailed in Sect. 8.5 [2].

8.2.1 Simple Threshold-Based Segmentation

The selection of an optimal threshold is key to segmentation using a threshold for distinguishing interfaces in internal structures. This point is discussed below.

Figure 8.11 shows a simple example where a 3D image of a sample with two internal structures was represented using an 8-bit grayscale image. Figure 8.12 shows a schematic of the pixel value distributions and segmentation results in these cases. The pixel value distributions were assumed as normal distributions for each phase and randomly distributed in the image. Figure 8.11a shows that the pixel value distribution curves of the two structures were separate from one another and that segmentation could be easily conducted by applying threshold 1 at the location indicated by the arrow. Figure 8.11b shows some overlap in the curves for the pixel value distributions of the two structures; however, applying threshold 2 corresponding to the valley of the histogram can enable a segmentation with a high degree of accuracy. In this case, only 0.5% of the pixels in structure 2 are included among those below threshold 2. Meanwhile, Fig. 8.11c shows a large overlap between the curves of the pixel value distributions of the two structures. Although the valley of the histogram can be identified, the two structures cannot be effectively distinguished even if threshold 3 was applied at the pixel value corresponding to that valley for segmentation. In this

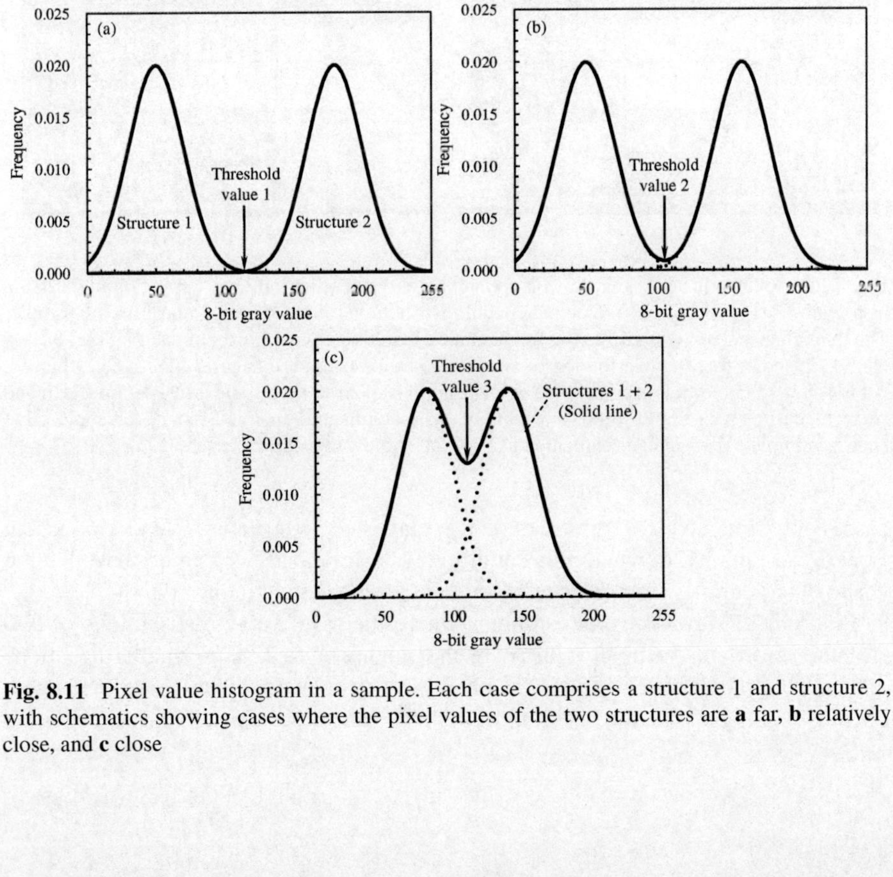

Fig. 8.11 Pixel value histogram in a sample. Each case comprises a structure 1 and structure 2, with schematics showing cases where the pixel values of the two structures are **a** far, **b** relatively close, and **c** close

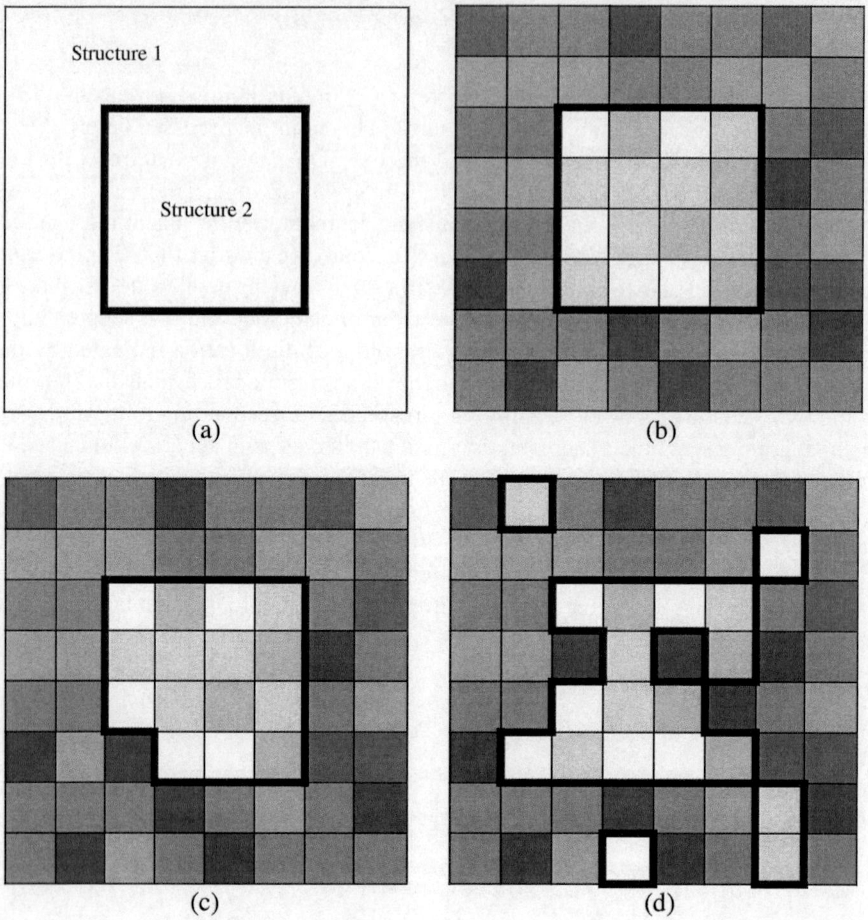

Fig. 8.12 Schematic showing the 3D image pixel value distributions corresponding to the histograms in Fig. 8.11, and the actual spatial distributions of structures 1 and 2 in the sample. The thick lines in (**b**)–(**d**) indicate the segmentation result when the threshold value in Fig. 8.11 is applied. **c** shows that the distribution of structure 2 in the original image is effectively achieved, but (**d**) shows large differences with (**a**). **a** Distributions of structures 1 and 2 in a sample. **b** Image segmented using a threshold value shown in Fig. 8.11a. **c** Image segmented using a threshold value shown in Fig. 8.11b. **d** Image segmented using a threshold value shown in Fig. 8.11c

case, approximately 10% of structure 1 is classified as structure 2 and vice versa. Looking at Fig. 8.12d, where this condition was simulated, we can observe that the shape of structure 1 was considerably altered after segmentation.

Figure 8.13 shows several examples where the pixel value distributions of two structures overlap. The pixel values of each structure were set as normal distributions and only the standard deviation of the pixel value distribution of structure 1 changed. As shown in Fig. 8.13a, the positions of both the valley of the pixel value distribution

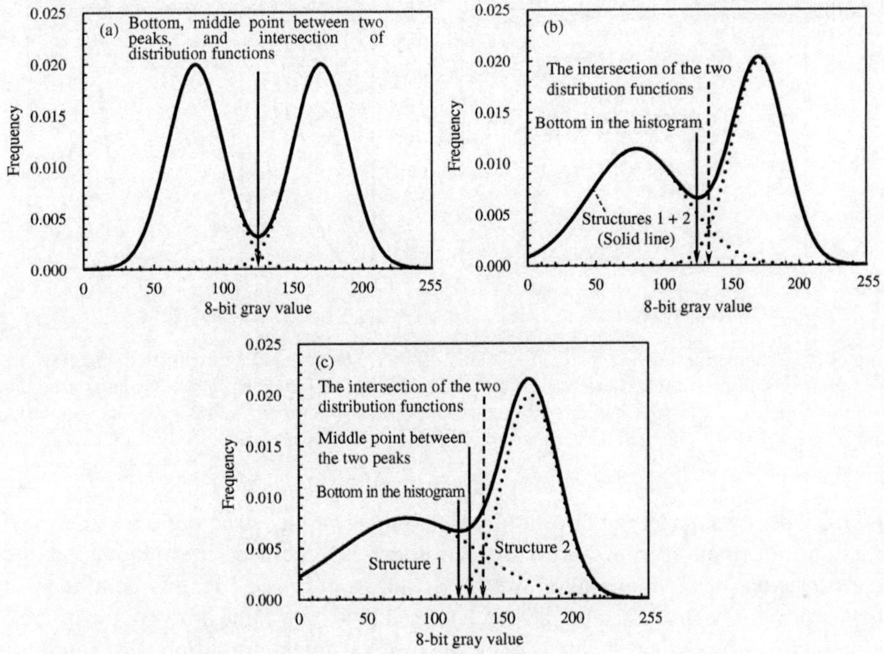

Fig. 8.13 Pixel value histograms in a sample. Each case comprises a structure 1 and structure 2 and the pixel value distributions of each structure are overlapping. Schematics of cases where the pixel value distribution variation of structure 1 is **a** small, **b** somewhat large, and **c** large are shown

and the midpoint between the peaks of structures 1 and 2 are virtually the same when the spread and height of the pixel value distributions are nearly identical. Meanwhile, as the pixel value distribution of structure 1 has a larger spread from Fig. 8.13b through c, a significant portion of structure 1 is classified as structure 2 when the valley position is used for segmentation. Moreover, many of the pixels in structure 1 would not be classified as structure 1. Curve fitting should be conducted to determine the pixel value distributions of the two structures (dotted lines) from the pixel value distribution of the 3D image (solid line). Using that intersection point (dashed line in the vertical direction) would yield a better segmentation. If fitting is difficult, the intersection point between the two peaks should be used instead of the histogram valley, as shown in Fig. 8.13c. However, care must be taken in selecting the optimal threshold depending on the segmentation objective, for example, when it is necessary to classify without excessively changing the fraction of each structure from reality or prioritizing the correct extraction of one structure over another.

The *percentile method* uses information on the fractions of specified phase or structure to estimate a threshold. Structures 1 and 2 in Fig. 8.14a considerably overlap as each has pixel value distributions with wide ranges to some extent. The fraction of structure 1 is measured using either the same sample or a different sample of the same material or component using an optical microscope or scanning electron

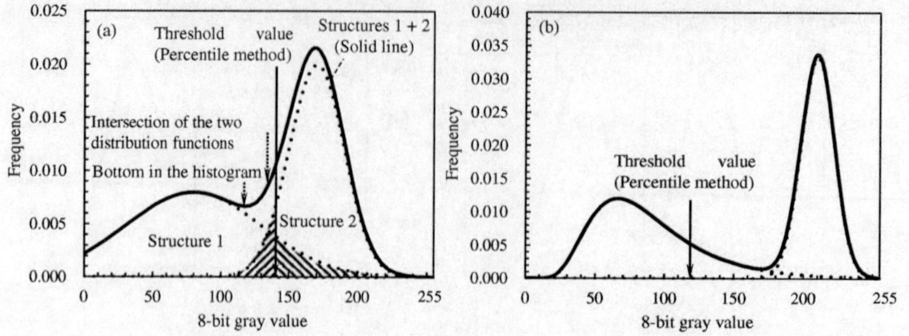

Fig. 8.14 Schematic showing percentile method-based segmentation effects with pixel value histograms; examples where there is a second phase with a wide range in the background (structure 1 in this case), **a** case where the pixel value distributions of structures 1 and 2 are both symmetric, and **b** case where the pixel value distribution of structure 1 is asymmetric

microscope in advance; the threshold is determined so that structure 1 has the same fraction. Information from X-ray diffraction can also be used if structure 1 is the microstructure of the material. However, as shown in Fig. 8.14a, this is influenced by the threshold determined by the percentile method (solid line arrow in the figure), which only ensures that the areas of erroneously segmented portions are equal for each structure and has minimal benefits. The percentile method is most effective when curve fitting the pixel value distributions of the two structures from the pixel value distribution of the 3D image is difficult for any given reason. Meanwhile, an asymmetric pixel value distribution of structure 1 is considered, as shown in Fig. 8.14b. This corresponds to the case where there are noise, heterogeneous element distributions, and microstructure distributions below the spatial resolution (particles, pores, etc.) that predominantly influence only the upper end of the pixel value distribution of structure 1. The percentile method is somewhat effective in removing the factors that make the pixel value distribution of one structure asymmetric. Furthermore, this method could be deemed particularly effective when correcting for the X-ray refraction and divergence (discussed in Sect. 2.2.2 (2)) and magnification projection due to the cone beam, all of which can result in a loss of quantitative dimensions.

The *Otsu method*, which considers the variation of pixel values, is also often used. The threshold in this method is selected so that, for multiple regions separated when using a given threshold, the pixel variation σ_i^2 in region *i* are minimized and the pixel variation σ_{ij}^2 between regions *i* and *j* (variation of the average pixel value determined for each separated region) is maximized [3]. Here, the pixel value variation σ_t^2 of the total image is set as $\sigma_t^2 = \sigma_i^2 + \sigma_{ij}^2$. The *degree of separation S* is defined as follows for this assessment:

$$S = \frac{\sigma_{ij}^2}{\sigma_i^2} = \frac{\sigma_{ij}^2}{\sigma_t^2 - \sigma_{ij}^2} \tag{8.21}$$

Here, $\sigma_t{}^2$ is fixed and the threshold should be determined so that $\sigma_{ij}{}^2$ is maximized; $\sigma_{ij}{}^2$ is expressed using the number of pixels N_i in region i and average pixel value $g^m{}_i$ [3]. This is expressed as follows when considering division into two regions:

$$\sigma_{ij}^2 = \frac{N_1 N_2 \left(g_1^m - g_2^m\right)^2}{(N_1 + N_2)^2} \quad (8.22)$$

Care must be taken so that this method is based on the assumption that the pixel value distributions are normal.

Single-threshold methods do not yield favorable results when there is a gradient in image brightness. Numerous methods, which use variable thresholds depending on the location, have been proposed for these cases. Progress has been made in the field of machine vision—used in applications such as automatic inspection during the industrial product manufacturing process—for this type of processing. Some examples include methods that calculate the average pixel value in a small region several pixels wide and then in turn use that value to calculate the local threshold value [4, 5], or those that make the pixel value in adjacent pixels and the threshold value linearly proportional to one another [6–8]. Among these, Niblack used the average and standard deviation of pixel values in adjacent regions to determine a threshold t as shown below:

$$t = k\sigma_i + g_i^m \quad (8.23)$$

Here, k is a constant, for which Niblack assigns a value of 0.18 [5]. Methods, which use a different threshold value depending on the location, are based on the assumption that the brightness gradient within the small region can be ignored. Numerous references where these have been applied to actual 3D images in X-ray tomography are listed; please refer to them as necessary [9–12].

8.2.2 Edge Detection Filter Usage

Instead of segmenting based on pixel value, it is possible to apply the edge detection filters introduced in Sect. 8.1.2 to conduct segmentations by specifying the interfaces of multiple structures. Segmentation can be conducted by applying hole-filling processing methods (discussed later) if the edge detection is perfect. Meanwhile, watershed processing (also discussed later) can be applied when the edges are discontinuous.

This method is suitable when there is a clear contrast in the internal structure. However, accuracy cannot be assured when there are a large number of edges, such as with microscopic structures.

8.2.3 Region Growing Technique

The *region growing technique* is a processing method that manually or automatically selects a given reference pixel (seed), applies a growth standard (e.g. based on pixel value differences between adjacent pixels and the reference pixel), and extracts an internal structure, which belongs only to the reference pixel, by sequentially growing the region from that reference pixel. The upper/lower limit of the pixel value can be set in these cases, and only one side can be selected as the standard. The latter case is substantially equivalent to the simple threshold-based segmentation from Sect. 8.2.1. However, the extracted regions vary as shown schematically in Fig. 8.15 due to differences in the threshold application method. As shown in Fig. 8.15, the

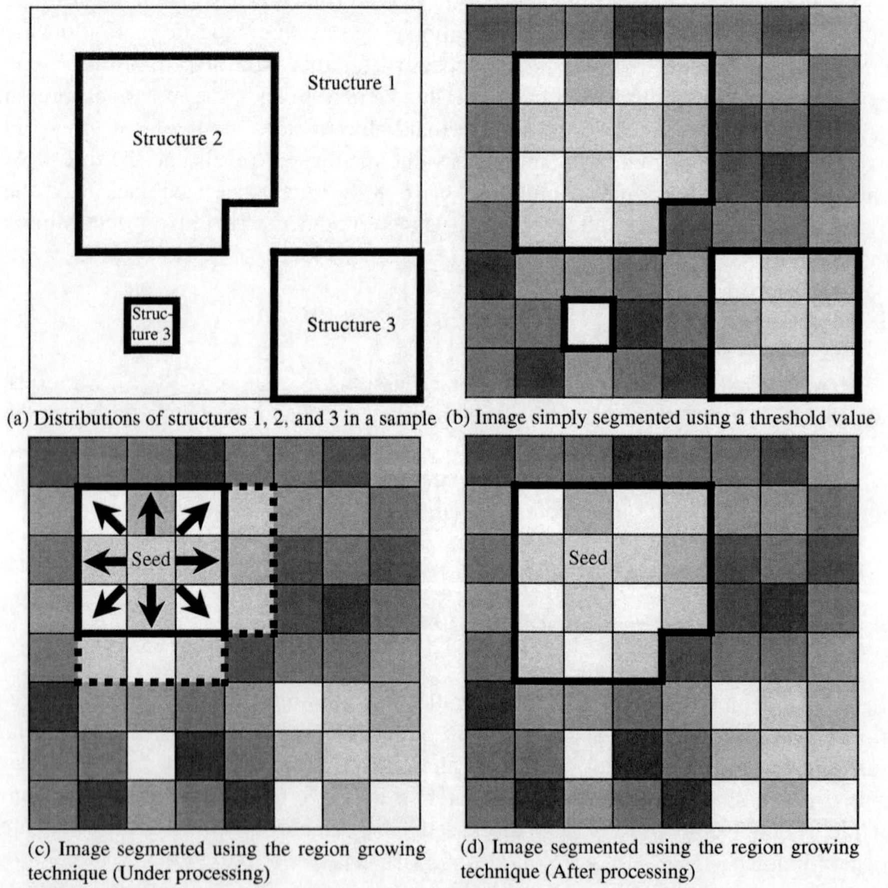

Fig. 8.15 Diagrams of the pixel value distributions in 3D images corresponding to the histogram in Fig. 8.11a and the spatial distribution of the internal structure of the sample. The heavy lines in **b** denote the results of simple segmentation using the thresholds in Fig. 8.11. **c** and **d** show results before and after processing using the region growing technique

simple threshold application results in the extraction of all four internal structures, whereas the region growing technique only results in the extraction of structure 2, which enclosed the specified pixel as a seed. The selection of the pixel to be used as the seed and growth standards are linked and the segmentation result naturally changes when a pixel with a different pixel value is selected. Furthermore, noise located in the extracted region having a pixel value that does not satisfy the threshold criterion will be removed from the extracted region. Hole-filling techniques will need to be used when this is an issue.

8.2.4 Watershed Method

As shown in the 1D example in Fig. 8.16, the *watershed method* views the center of an extracted region as the valley of an undulating landscape and the boundaries between regions as ridges, similar to when precipitation in a mountainous region is divided into two regions with the watershed as the boundary, using these edges to separate regions.

An application of the watershed method is shown in Fig. 8.17. This is the gallium decoration applied on crystallographic grain boundaries of an aluminum alloy using the liquid–metal wetting technique introduced in Sect. 5.5.1, visualized using projection X-ray tomography at SPring-8. The binarized image in Fig. 8.17 shows white points due to noise or dispersion particles in the crystallographic grains and discontinuous crystallographic grain boundaries due to insufficient gallium penetration. Applying the watershed method to these images as-is will not result in the accurate extraction of the crystallographic grains. Appropriate pre-processing and post-processing are, therefore, required. Generally, pre-processing is first conducted. For example, noise and particles were removed, as shown in Fig. 8.17b, by using the dilation/erosion processing discussed in the next section. Various filters are used with the same objective in mind. Afterward, the distance transform shown in Fig. 8.17c is applied. This converts the pixel values to the minimum distance from that pixel to the background pixel. As shown in Fig. 8.17c, there is an excessive amount of gray value valleys forming in the elongated crystallographic grains at this stage such as in crystallographic grain A, as shown with the four arrows for that crystallographic grain. Applying an *H-minima transform* on this, as shown in Fig. 8.17d, can reduce the pixel value valleys, which have a value less than the assigned H-value. This can ultimately allow the polycrystalline structure to be accurately separated by each crystallographic grain, as shown in Fig. 8.17e. Figure 8.17f shows an example where an inappropriate value was deliberately set for the H-minima transform, which resulted in the over-segmentation of crystallographic grain A. Verifications by comparing with the original image should be conducted to check for over-segmentation.

The watershed method is particularly effective when the edge extraction is incomplete as shown in Fig. 8.17 or as shown schematically in Fig. 8.18, where internal structures comprising similar materials are adjacent or nearby each other and where

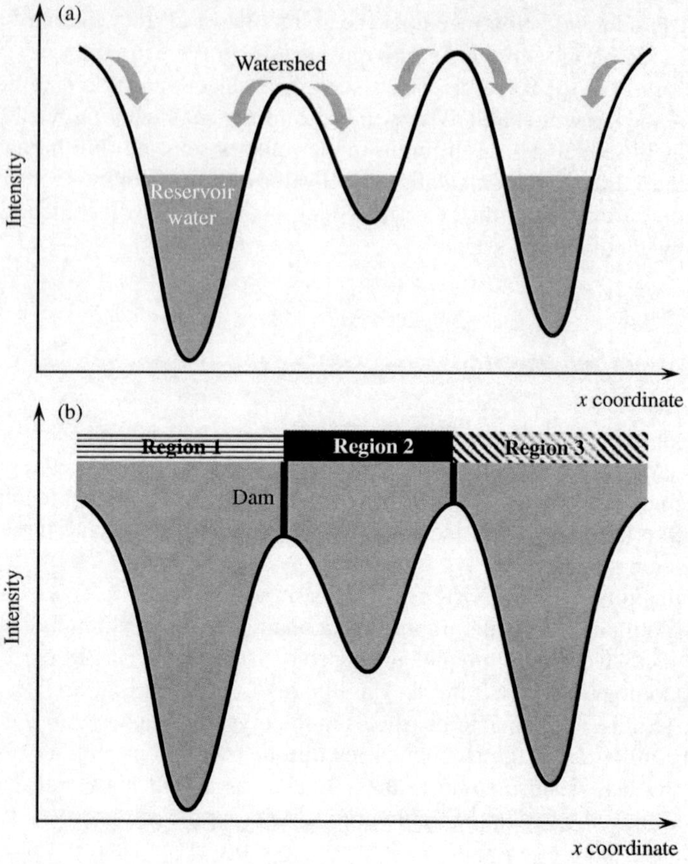

Fig. 8.16 Schematic of the watershed method; **a** During processing with the watershed method and **b** after processing with the watershed method

their interfaces cannot be visualized with X-ray tomography (see Fig. 8.22 in the next section for reference).

Note that the distance transform is an image processing step that is applied to a binarized image. The distance transform is also used for measuring dimensions such as the thickness of a 3D structure, and a pre-processing step for the watershed method mentioned above and skeleton transformation. Figure 8.19 shows the application of the distance transform in a branching structure and Fig. 8.19c shows the skeletonization of this structure; Fig. 8.19c is a thin line with a thickness of a single pixel. This can be combined with the distance values of that structure to retrieve the original structure from inverse distance transforms. The distance transform is resistant to noise and can be effective when its effects need to be avoided.

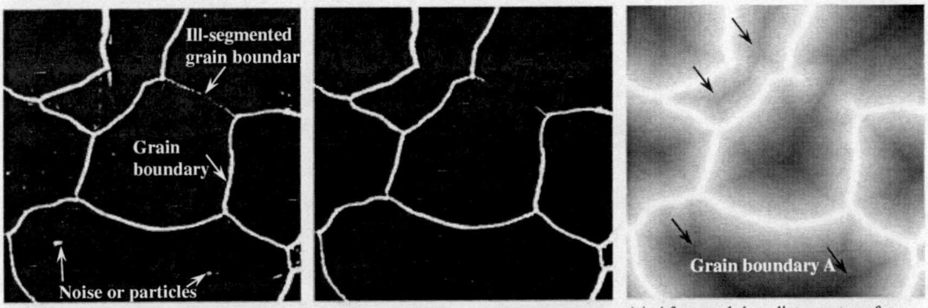

(a) Segmented image

(b) After applying noise removal processing to (a)

(c) After applying distance transform to (b)

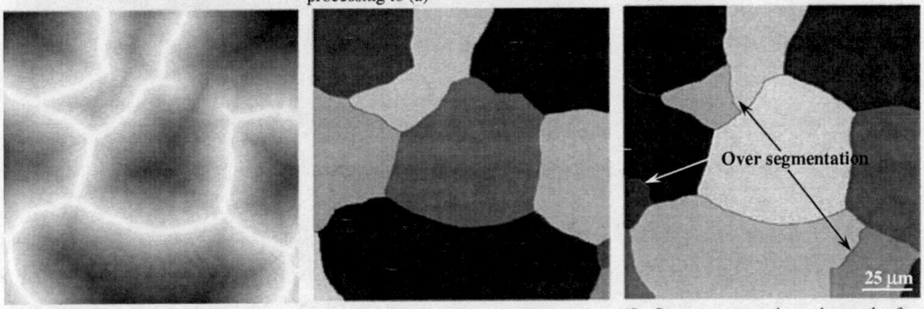

(d) After applying H-minima transform to (c)

(e) After applying the watershed method to (d)

(f) Over segmentation observed after applying the watershed method to (d) with inappropriate conditions

Fig. 8.17 Examples where 3D images of crystallographic grain boundaries in aluminum alloys were obtained and segmentation was conducted according to each crystallographic grain by processing discontinuous grain boundaries and noise, applying the watershed method; **f** shows an example of the over-segmentation of crystallographic grains when the processing conditions were not suitable

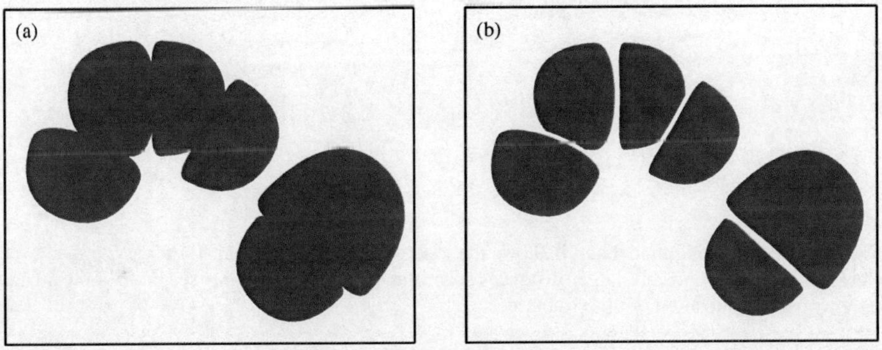

Fig. 8.18 Examples where the application of the watershed method is effective; **a** prior to processing with the watershed method and **b** after processing with the watershed method

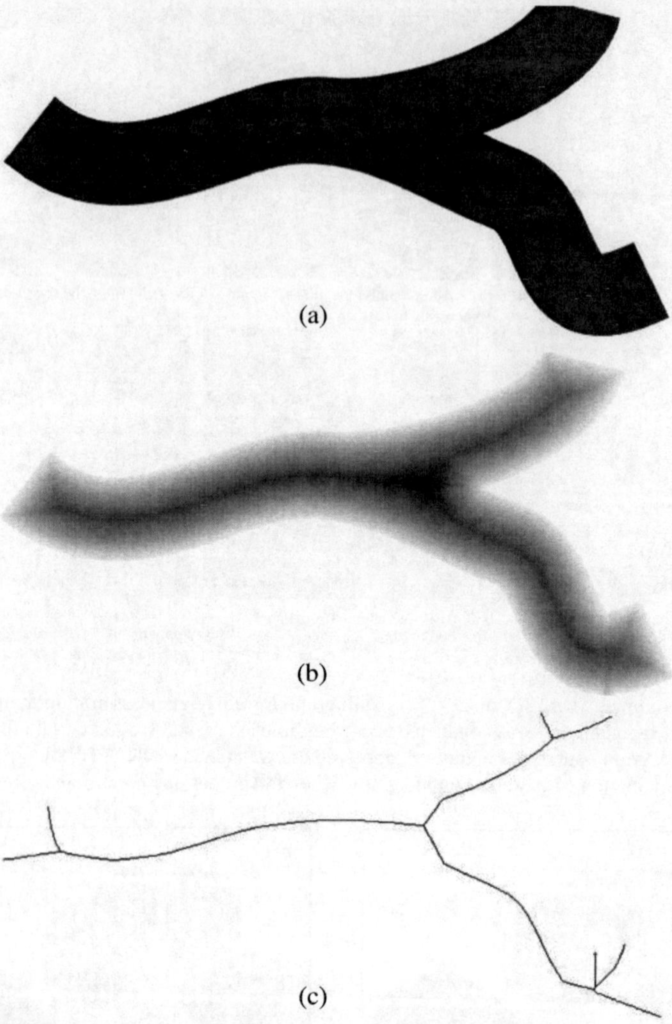

Fig. 8.19 **a** Branching model and **b** shows the case where a distance transform was applied to it. Here, darker pixels indicate larger distances from the closest 255-value pixel in an 8-bit image; **c** shows a skeletonization of this structure

8.2.5 Segmentation Using Machine Learning

Segmentation based on machine learning has been increasingly used in X-ray tomography in recent years as segmentation based on the use of higher-grade information. This is a method that effectively applies the amount of information provided in X-ray tomography. Although details are omitted here, the WEKA segmentation developed by researchers at the University of the Basque Country in 2017 is a well-known ImageJ plug-in capable of analyzing both 2D and 3D data [13]. For example, conducting segmentation in complex images with a large amount of noise and artifacts after learning with data based on scanning electron microscopy, optical microscopy, or EBSD can increase the reliability of that segmentation process.

8.3 Various Types of Image Processing

Representative examples of image processing that can be applied to binarized images are introduced here.

8.3.1 Dilation and Erosion

Dilation and *erosion* processing are mathematically based on *Minkowski sums* and *Minkowski differences*. We consider a set of pixels A, which correspond to an internal structure in an image, and a set of pixels B, which correspond to a structural element with a simple shape (e.g., symmetrical figure). The Minkowski sum and Minkowski difference for the elements *a* and *b* of the former and latter, respectively, are expressed as follows:

$$A \oplus B = \{a + b : a \in A, b \in B\} \tag{8.24}$$

$$A \ominus B = \{a - b : a \in A, b \in B\} \tag{8.25}$$

Here, $\oplus$ and $\ominus$ indicate the Minkowski sum and Minkowski difference, respectively. These correspond to the union and intersection, respectively, of set A translated in all directions by the length of element *b*. Holes and gaps between internal structures that are smaller than the structural element are filled in with the Minkowski sum and elements smaller in diameter or width than the structural element are removed in the Minkowski difference.

Figure 8.20 shows a schematic of the dilation and erosion effects in 2D. In this case, 4-neighbor dilation/erosion was conducted in the vertical and horizontal directions in 2D. However, the 8-neighbor method, which includes diagonal directions, is more generally used. Figure 8.21 shows 4-neighbor and 8-neighbor structural elements. The original image in Fig. 8.20a shows complex particles with holes and noise. First applying a dilation to this results in the internal holes being filled in, as shown in Fig. 8.20b, and the adjacent particles and noise being connected. Applying erosion to this as shown in Fig. 8.20c fills the holes and leaves the associated noise in place. This is expressed as (A $\oplus$ B) $\ominus$ B in operational terms. Figure 8.20d shows that, where erosion was first applied, the closed particles opened up and the noise

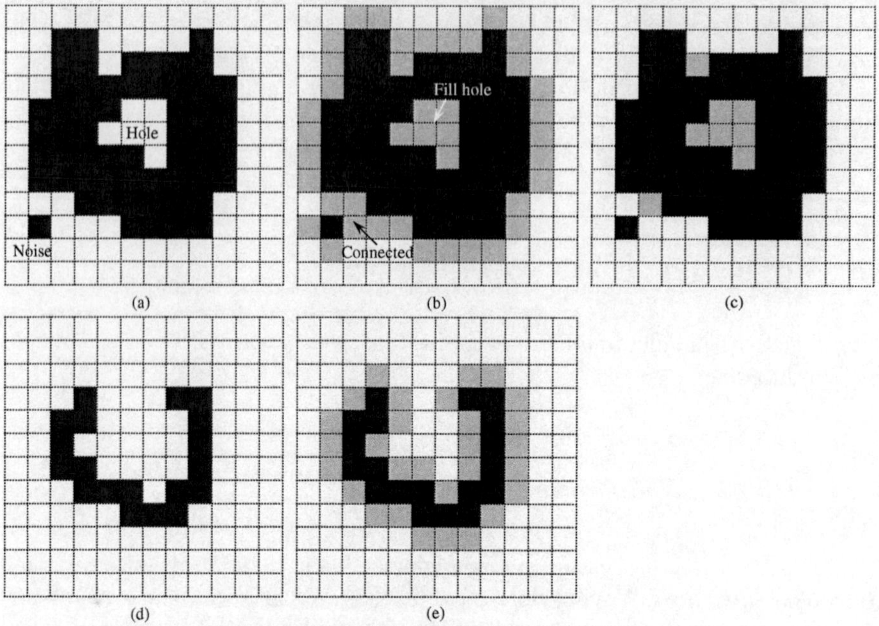

Fig. 8.20 **a** Original image, **b** dilation processing of the original image, **c** erosion processing of the image in **b**, **d** erosion processing of the original image, and **e** dilation processing of the image in **d**; the gray pixels indicate those which newly became a part of the internal structure as a result of dilation processing

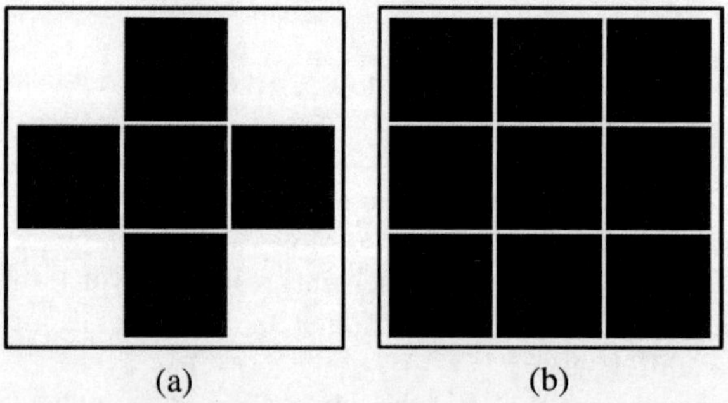

Fig. 8.21 **a** 4-neighbor and **b** 8-neighbor structuring elements used in dilation and erosion processing

was removed. This is expressed by $(A \ominus B) \oplus B$. Next, in Fig. 8.20e, which shows the subsequent application of dilation, the noise and complex structures like protrusions in the original image have been removed. Furthermore, Fig. 8.22 shows the application of the operation $(A \ominus B) \oplus B$ on an image with only slightly connected internal structures. The combination of dilation and erosion resulted in the clean separation

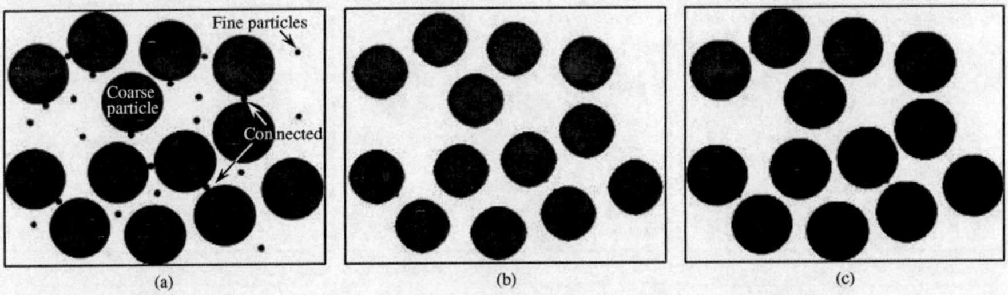

Fig. 8.22 **a** Original image with fine particles and coarse particles mixed; **b** shows the application of erosion to remove the connected sections and fine particles and **c** is the application of dilation on image (**b**)

of the internal structure. The separation of the internal structure enables measurements of the true size and shape of coarse particles that were difficult to measure in Fig. 8.22a.

Noise removal, opening, hole-filling, and internal structure separation can be conducted by applying multiple dilation and erosion steps for the same number of repetitions in this manner. Furthermore, taking the differences between images in Fig. 8.20 allows for the detection of noise or holes. However, the sizes of the examples in Fig. 8.20 vary significantly, with the particle area being 57 pixels in the original image, 65 pixels when dilation followed by erosion was applied, and 47 pixels when erosion followed by dilation was applied. Results can be completely different in this way depending on the order of dilation and erosion, and sufficient care must be taken with regard to the change in morphology or size.

8.3.2 Differences Between Dilation/Erosion-Processed Images

The outer shell of an internal structure can be expressed by taking the pixel difference of an erosion-processed image from a dilation-processed image. Furthermore, taking the pixel difference of an image, which has undergone erosion after dilation ($(A \oplus B)\ominus B$), or an image, which has undergone dilation after erosion ($(A\ominus B) \oplus B$) from the original image, enables the extraction of detailed internal structures such as defects or cracks. The former is referred to as a *bottom-hat transformation* and is expressed as $(A \oplus B)\ominus B - A$. The latter is referred to as a *top-hat transformation* and expressed as $A - (A\ominus B) \oplus B$.

Figure 8.23a shows an image that is primarily composed of spherical particles and includes elongated particles and cracks within the spherical particles. Applying a top-hat transformation allows for the extraction of only the elongated particles. Moreover, applying a bottom-hat transformation extracts only the cracks within the particle interior. Furthermore, the perimeters of the elongated particles, spherical particles,

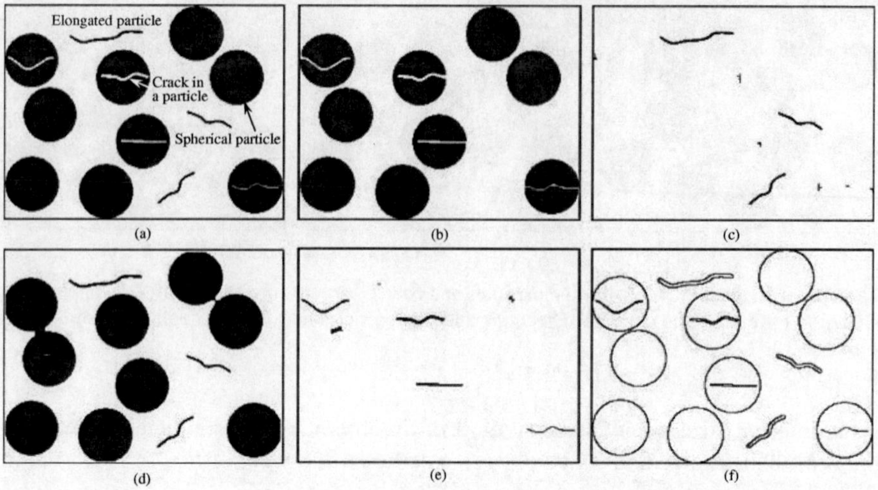

Fig. 8.23 **a** shows the original image with elongated particles and spherical particles mixed; **c** is the result of the application of erosion and then dilation to the original image (**b**), which is then subtracted from the original image in (**a**) (top-hat transformation); **e** is the result of the application of dilation and then erosion to the original image (**d**), which is then subtracted from the original image in (**a**) (bottom-hat transformation); and **f** shows the pixel subtraction of the erosion image from the dilation image

and cracks are all extracted when taking the pixel difference of the erosion-processed image from the dilation-processed image in Fig. 8.23f.

8.3.3 *Thinning Processing*

There are numerous *thinning/skeletonization* algorithms other than distance transforms that have been proposed; *Hilditch thinning* is a representative example [14]. This successively determines whether the pixels at the interfaces between the internal structure and the matrix satisfy a number of conditions and ultimately creates a structure with a line-width of a single pixel by converting the gray values of pixels, which satisfy the conditions from 1 to the background value of 0. Other references are listed here; please refer to them as needed [15, 16].

8.3.4 *Spatial Tessellation*

As shown in Fig. 8.24a, a *Voronoi tessellation* is when isolated characteristic points (e.g. particles) in a given space are used for spatial tessellation. Convex polygons and polyhedrons are generated in 2D and 3D, respectively, without overlap or gaps in a

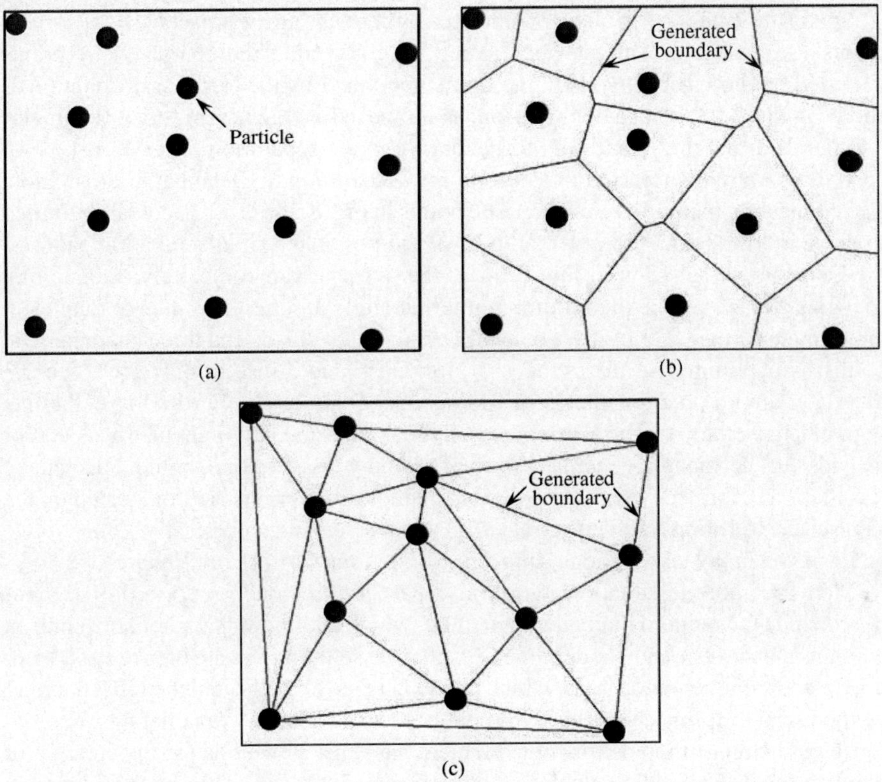

Fig. 8.24 **a** Original image with fine particles dispersed in space, **b** shows the application of a Voronoi tessellation to divide the space, and **c** shows the application of a Delaunay tessellation

given space, as shown in Fig. 8.24b. Convex polygons refer to those where all diagonal lines exist completely in the interior of that polygon. Furthermore, convex polyhedrons refer to those with no self-intersection, where any two face angles across a given section are less than π. The number of generated figures in a Voronoi tessellation is equivalent to the number of characteristic points, and labeling is conducted based on the original characteristic points. Furthermore, each of the separated domains become convex domains. Another particular characteristic is that when three adjacent characteristic points are selected, as shown in Fig. 8.24b, the triple junction point of the polygon that fits those points is equidistant to each characteristic point.

There have been numerous reports on algorithms for Voronoi tessellation, for example, where a line is drawn from each point to every other point and planes are drawn to bisect these lines. A Voronoi tessellation can be made by dividing the spaces so that the smallest figure enclosed by the planes is set as the cell, which corresponds to that characteristic point [17]. There have also been reports on simple algorithms with applied dilation processing [18].

Spatial tessellation can also be conducted by connecting the adjacent characteristic points in Fig. 8.24a. This can be created by connecting the characteristic points enclosed in two adjacent cells in the figure generated by the Voronoi tessellation as shown in Fig. 8.24b with a line segment, as shown in Fig. 8.24c. This creates triangles and tetrahedrons in 2D and 3D, respectively, as the separation units. This type of spatial separation is referred to as *Delaunay tessellation* or Delaunay triangulation. The connection method of characteristic points in Fig. 8.24a is not necessarily limited to those in Fig. 8.24c. However, among all the possible spatial separation patterns, the Delaunay tessellation in Fig. 8.24c is the one that can completely exclude thin triangles and maximize the minimum internal angle among the 3 internal angles in the triangle formed. Here, if we consider a circle that sits on the three characteristic points, which comprise the vertices of a given triangle in Fig. 8.24c, we can see that there are no other characteristic points inside that circle. Furthermore, the line segments generated by the Voronoi tessellation act as the perpendicular bisectors of the sides of the triangle generated by the Delaunay tessellation. A unique Delaunay tessellation is formed when four adjacent characteristic points are not located on the same circle, forming a convex tetrahedron. Conversely, it can be easily seen that two types of Delaunay tessellations can form for the four vertices of a square.

Many Delaunay tessellation algorithms have been previously reported [19]. Examples include the random incremental method, which adds new characteristic points in domains that have already undergone Delaunay tessellation and are then re-separated; divide-and-conquer algorithms, which create 2D convex hulls; and the gift-wrapping method, which forms each figure in sequence [19]. Delaunay tessellations are used for the creation of models for finite element analysis, as well as for the creation of tetrahedrons used for the calculation of strains in 3D strain mapping introduced in the next chapter. As shown in Fig. 8.24, the outer edges of the spatially separated domains in the Voronoi tessellation are equivalent to the outer edges of the sample, whereas the outer edges become the sides or planes of the outermost characteristic points; additionally, there are unseparated domains in the outer edge. This signifies that there are strain values that cannot be obtained in areas up to a given thickness from the surface when conducting 3D strain mapping; care should be taken with this aspect.

8.4 3D Rendering

Readers are likely to use 3D rendering software affiliated with their X-ray CT device or those that are commercially available rather than develop home-made software to do so. Furthermore, the 3D rendering functionalities of numerical analysis software like Mathematica or MATLAB can be used. References are provided for those who would like to learn 3D rendering in earnest [20] but the focus here is on 3D rendering algorithms that are important for X-ray tomography. Maximum-intensity or average-intensity projection methods used for the display of 3D images obtained from medical-use X-ray CT scanners are omitted from this book.

8.4.1 Virtual Cross Section Display

As shown in Fig. 8.25, the more secure way to confirm images from X-ray tomography is to assess a virtual cross-section of an image whose grayscale range has been suitably modified and where the bit count has been reduced. Aside from the *x–y*, *y–z*, and *z–x* planes based on the coordinates of the detector plane, the virtual cross-section can be used for an arbitrary plane and curved surfaces. Images that have undergone various filtering, segmentation, and image processing steps may potentially have lost important structures, particularly internal structures with fine detail

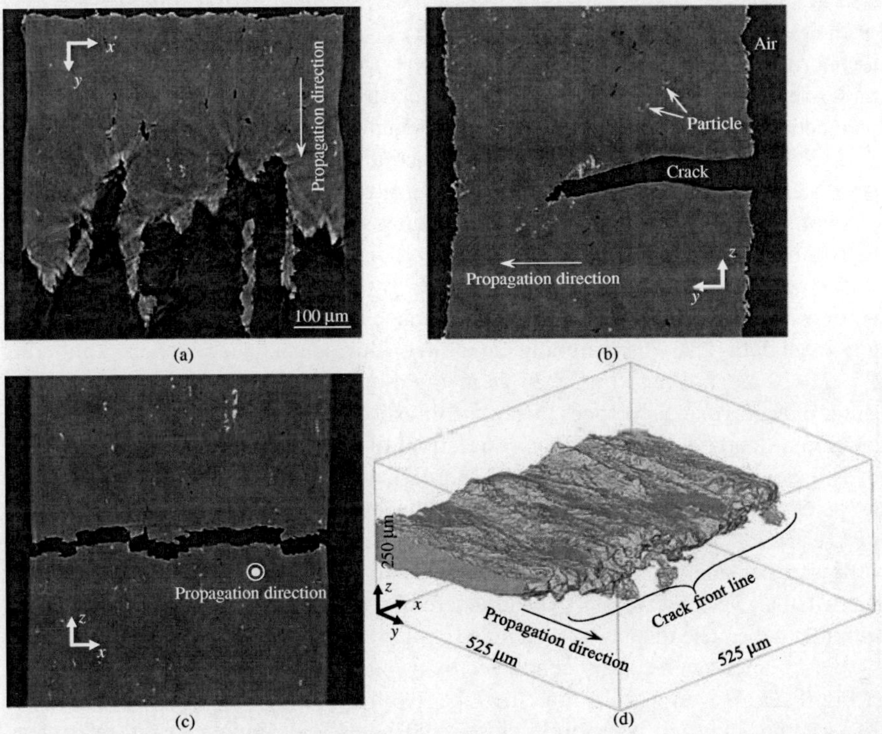

Fig. 8.25 3D image of in situ observations of tensile cracks in A7075 aluminum alloy. Images of the virtual **a** *x–y*, **b** *y–z*, and **c** *z–x* cross-sections, as well as (**d**) the display of the 3D image with 3D rendering software. Other than the cracks, the arrangement of white and dark-gray particles can be seen on the crystallographic grain boundary. The artifacts in the *x–y* cross-section are due to X-ray refraction (courtesy of Kazuyuki Shimizu of Kyushu University)

or low contrast. Instead of merely conducting 3D rendering from the start, a virtual cross-section with a physically meaningful direction should be selected, after which all cross-sections should be observed continuously in that given direction, providing a complete view of the internal structure. Close examinations of images using MPR would enable the rough visualization of the 3D image in the mind of the observer. Particularly for 3D images being viewed for the first time, 3D rendering should be used as a supplement, an aid in representing the object effectively.

8.4.2 Surface Rendering

Surface rendering refers to when the surface of the sample or the interface between internal structures in the sample interior of a 3D image are extracted, polygonized, and rendered in 3D. Surface rendering is effective for 3D images whose surface or interface contours are clear. Surface rendering is also used for many of the 3D images used in applied examples introduced in Chap. 6. Segmentation techniques are used for the extraction of surfaces or interfaces. The surfaces of the extracted sample of the internal structure are converted to polygons. Triangles are primarily used in these cases to express curved surfaces. Surface rendering involves the determination of the intersection points where light from a hypothetical light source first interacts with the surface of a sample or an internal structure and displays the surface of the object by using the morphology of the pixels at the intersection point. Furthermore, the spatial sense of the 3D structure is displayed by adding shadows induced by the light from the specified direction.

The *marching cubes algorithm* published by Lorensen and Cline in 1987 is a well-known algorithm for converting 3D images comprising isotropic voxels into polygonal data [21]. The marching cubes algorithm sets adjacent voxels numbering $2 \times 2 \times 2$ as shown in Fig. 8.26 as a single unit sets a cubic structure with the center of each voxel as vertices. Next, the threshold to be set as the surface/interface extraction standard and magnitude correlations of the pixel values of the 8 voxels are investigated, with a value of 1 (internal) or 0 (external) assigned to the vertex when the pixel value is larger or smaller than the threshold, respectively. It may appear as if there are 256 combinations of 0s and 1s for the eight vertices, but these are reduced to 15 after considering inversion or rotational symmetry, as shown in Fig. 8.27. Type 0 is when all vertex values are 0 and where the surface/interface does not intersect with that cube. The display of a sample or internal structure using the isosurface of the pixel values can be easily conducted by using the library of 15 standard patterns in Fig. 8.27. However, care must be taken with the marching cubes algorithm as the structure may not be entirely closed and holes may remain in the interior, as understood from Fig. 8.27.

The standard marching cubes algorithm is widely used even today but there have been various proposed surface rendering methods that are faster and more accurate. However, computational times rapidly increase as the 3D rendering technique becomes more advanced.

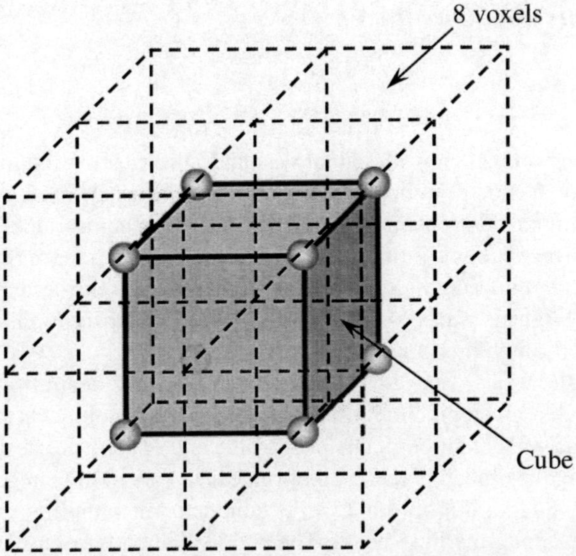

Fig. 8.26 Eight voxels (dotted lines) comprising the cube, which is the basis for surface determination in the marching cubes algorithm (gray). The center of the eight voxels corresponds to the eight vertices of the cube. The basic pattern for determining the polygon is based on the pixel values of the eight voxels

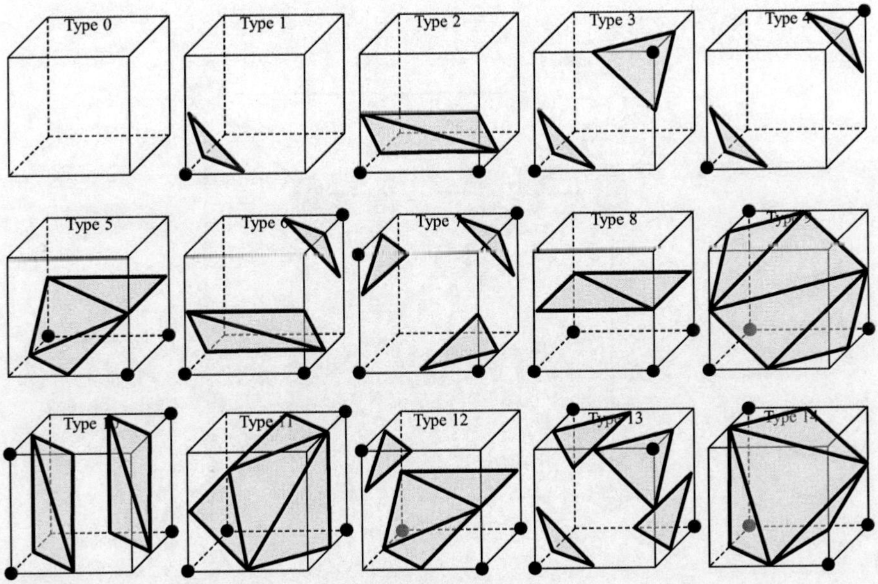

Fig. 8.27 Fifteen basic patterns for determining the surface shape with the marching cubes algorithm. The black circles indicate the interior side and the vertices without circles indicate the external side

8.4.3 *Volume Rendering*

(1) Ray Casting

Volume rendering refers to a method that visualizes the information in the interior of an object by transforming the image data distributed in a 3D space as a perspective image. The positional relationships between an internal structure and its surroundings or the sample surface become intuitively easier to understand if the region of interest is emphasized while rendering its surrounding domains in a semi-transparent manner. For this reason, light is traced while considering X-ray absorption when the light beam transmits through the sample.

Figure 8.28 shows an overview of *ray casting*. The light beam from the observer transmits through a given pixel on a screen, incident to the object at position s_0 and continues by exiting at position s. The pixel value and opacity are designated for the voxel, expressing the object; both the pixel value and *opacity* are integrated along the light beam. The integration, in this case, is conducted at sampling points arranged at fixed intervals along the light beam. The pixel value and opacity at the sampling points can be determined with linear interpolation methods from the voxel data.

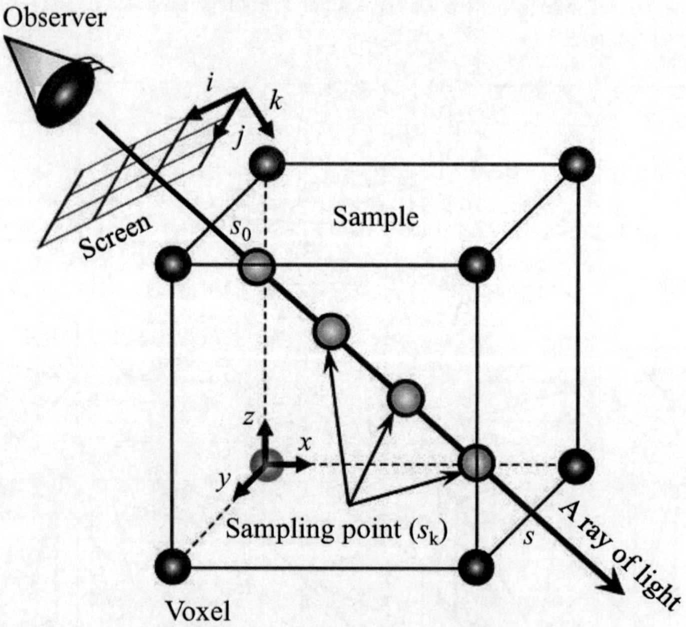

Fig. 8.28 Explanation of ray casting. The light beam is projected from the pixel (i, j) on the screen to the sample stipulated by sample coordinates (x, y, z). The brightness arriving at the viewpoint is determined by sequentially accumulating the brightness at the discretely arranged sampling points

The following differential equation is used to express the *equation of radiative transfer* when an X-ray with intensity I is emitted from a light source with flux q, transmitted through a sample with true absorption κ, and when X-ray scattering is ignored [22].

$$\frac{\partial}{\partial s} I = -\kappa I + q \tag{8.26}$$

This can be analytically solved as follows [22]:

$$I(s) = I(s_0)e^{-\tau(s_0,s)} + \int_{s_0}^{s} q(s')e^{-\tau(s',s)}ds' \tag{8.27}$$

Here, s' is on the light beam within the sample interior and τ is the *optical depth*, which is expressed as follows using positions s_1 and s_2 within the sample [22]:

$$\tau(s_1, s_2) = \int_{s_1}^{s_2} \kappa(s)d \tag{8.28}$$

Discretizing Eq. (8.27) with the sampling points arranged at equal intervals as shown in Fig. 8.28 results in the following equation[22]:

$$I(s_k) = I(s_{k-1})e^{-\tau(s_{k-1},s_k)} + \int_{s_{k-1}}^{s_k} q(s)e^{-\tau(s,s_k)}ds \tag{8.29}$$

Here, $e^{-\tau(s_{k-1},s_k)}$ is the transparency and $1-e^{-\tau(s_{k-1},s_k)}$ is the opacity. Generally, the first term in the right-hand side of the above equation is the transparency multiplied by the light coming from the light source; the second term indicates the X-ray absorption of the sample. This equation serves as the basis of volume rendering.

Bartz et al. reported on direct comparisons between the ray casting method and marching cubes algorithm [23]. This report indicated that the ray casting method effectively brought out a depth feeling and allowed for easier understanding of the shape but the marching cubes algorithm was more capable of retaining microscale morphologies. The microscale morphologies tend to be lost with standard ray casting methods and must be supplemented with oversampling.

(2) Texture-Based Volume Rendering

The recent use of GPUs in high-speed image reconstruction was introduced in Sect. 3.4. The use of GPUs in various image processing/display functionalities for graphics displays is extremely effective in 3D rendering. Texture-based methods are used for GPU-based volume rendering. *Texture-based methods* first convert a 3D

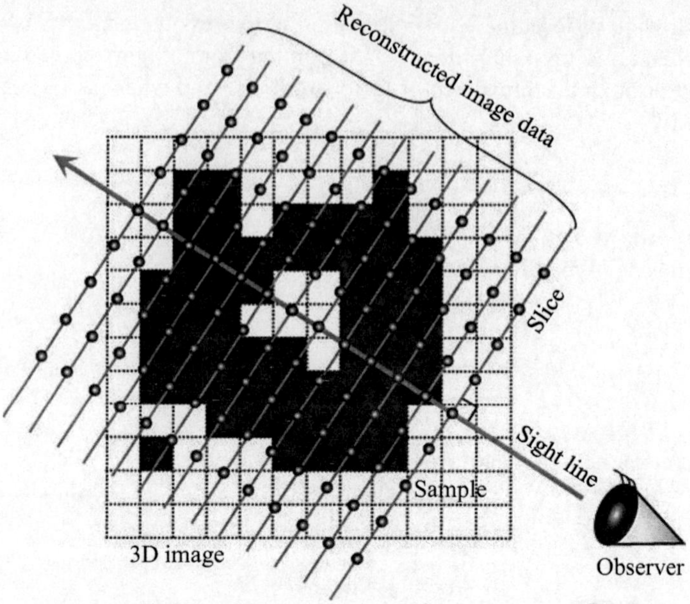

Fig. 8.29 Explanation of 2D texture-based volume rendering. Slices perpendicular to the line of sight are prepared and 3D rendering is conducted by adding the pixel values in order once the 3D image is reconstructed as a stack of slices

image into a texture and read this into GPU video memory. Next, as shown schematically in Fig. 8.29, the texture is mapped onto slices, which are polygons that are perpendicular to the line of sight. The 3D images are newly reconstructed as stacks of slices, after which the 3D rendering is conducted by adding the pixel values in order of the slices furthest from the line of sight. This enables a 3D rendering based on high-speed image processing using a GPU. There are 2D or 3D variations for texture-based volume rendering. Several relevant references are listed here; please refer to them as required [24].

8.5 Quantitative Geometric Analyses

Isaac Newton once said, "*God created everything by number, weight, and measure.*" Being able to conduct image processing on not just 3D images but various image data to numerically express them and assess the results is a fundamental goal of science and technology. As shown in Fig. 8.30a, sampling-based measurement and assessments have been standard academic approaches for surfaces, cross-sections, and thin films. Furthermore, analyzing the internal structure or microstructure by simplifying and abstracting them using the obtained average values has also been standard practice. However, there are many examples of applying conventional 2D-based analyses

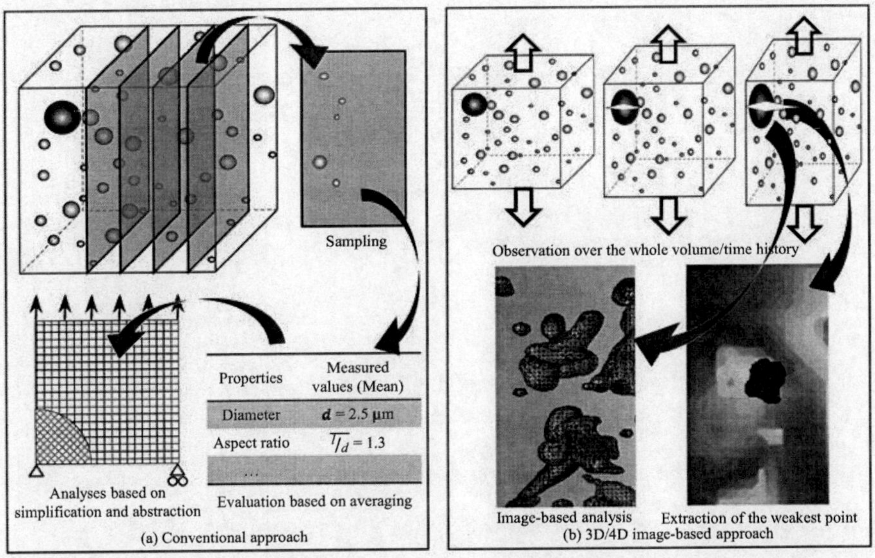

Fig. 8.30 Schematic showing the differences between an **a** assessment analysis based on conventional 2D cross-section/surface observations and **b** academic approach involving 3D images and its continuous scanning (4D observation). The conventional approach cannot assess the relationship between internal tissue/structure and macroscale characteristics particularly in cases where relatively large microstructures (black: particles, manufacturing defects, etc.), seen in the left-hand side of the top figure in **a**, dominate the macroscale characteristics of the material

while still obtaining 3D images from X-ray tomography; this is a waste of valuable information and is highly unfortunate. For example, Fig. 8.31 shows comparisons of 2D and 3D shape/morphology measurements between relatively spherical graphite nodules in a ductile cast iron and extremely elongated silicon particles in a cast aluminum alloy. Measuring the virtual 3D cross-sections of the elongated eutectic silicon particles in the aluminum in 2D results in a dramatically higher number of particles assessed as simple circular morphologies, which do not exist. Furthermore, the graphite size is measured considerably smaller in the spherical graphite nodules. These tendencies become more pronounced as the 3D morphology of the internal structure becomes more complex.

As shown in Fig. 8.30b, the merits of 3D observations are (1) being able to assess complex structures and heterogeneous distributions that exist in reality, (2) determining the total number of internal structures or microstructures, and (3) grasping how changes occur under external disturbances over time. Merit (1) can be thought of as the polar opposite of the "simplification," "abstraction," and "averaging" in Fig. 8.30a, which is based on 2D observations; whereas (2) and (3) are the opposites of "sampling." Sampling-based approaches are not effective when a limited number of internal structures or microstructures affect the overall system or in extreme cases, when a single structure/microstructure stipulates the behavior and characteristics of the entire system. An example of this includes fatigue cracking, which propagates

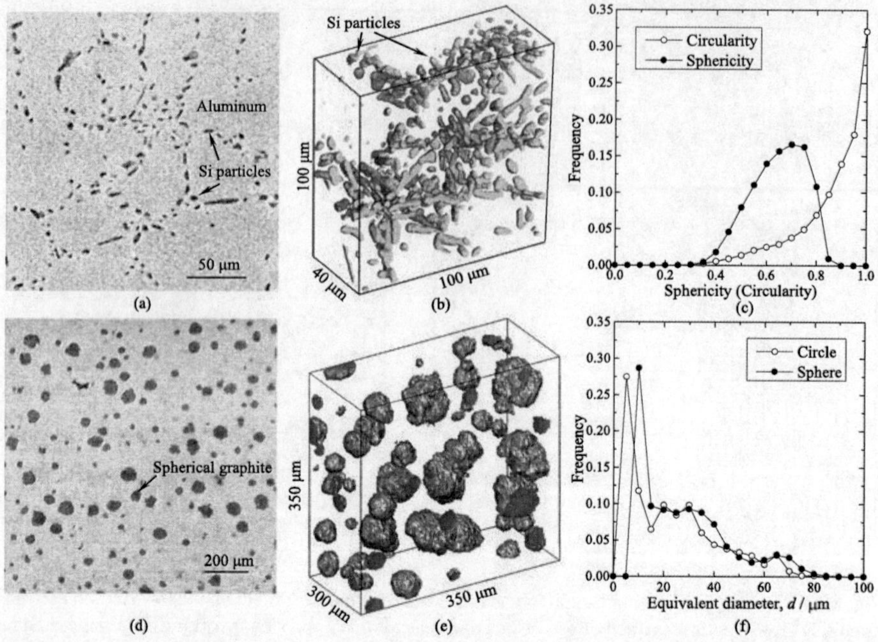

Fig. 8.31 Differences when conducting analysis of a 3D image in 3D and when applying conventional 2D analyses to virtual cross-sections of 3D images; **a** and **b** show the virtual cross-section and 3D image (same sample, same image data), respectively, of the Al-7% Si alloy; **d** and **e** show the virtual cross-section and 3D image (same sample, same image data), respectively, of the spherical graphite cast iron; **c** shows a comparison of the shape (sphericity, circularity) of the former; and **f** shows a comparison of the size (sphere, circle equivalent diameter) of the latter (courtesy of Kazuyuki Shimizu of Kyushu University)

as a result of the cracks that form from a single source (e.g. microscale defect in the surface or interior). There are also no effective approaches other than the visualization/assessment of the total number/volume in cases where the internal defect size has a normal distribution and where products with defects of a given size need to be detected and removed as defective products.

Polygonization techniques such as the marching cubes algorithm introduced in Sect. 8.4.2 serve as the basis of 3D image analysis. Volume, surface area, and length measurements can be accurately conducted once the surface morphology has been determined. Recently, 3D image analysis has been included in 3D rendering software; the Image Processing Toolbox in MATLAB and the various plug-in libraries in ImageJ can be used to conduct them as well.

Table 8.2 lists microstructure parameters that can be analyzed with MATLAB-based home-made software used by the author. These include various parameters relating to the size, shape, and spatial distribution of microstructures dispersed in a material interior. These exclude various parameters (several dozen in total) relating to crystallographic orientation such as the Schmid factor due to their special nature.

Table 8.2 List of categories that can be analyzed with MATLAB-based home-made software used by the author

Category	Microstructural parameter	Remark
Category I size (9 parameters in total)	D; V; S	Equivalent diameter, volume, and surface area of a particle, respectively
	B; L; W; T	Minimum volume bounding box; length of a particle (measured along x axial-direction); width of a particle (measured along y axial-direction); thickness of a particle (measured along z axial-direction)
	$O = \frac{O_1+O_2+O_3}{3}$	Average length of three principal axes
	G	Geodesic distance
Category II shape (24 parameters in total)	$O_4 = \frac{O_3}{O_1}$; $O_5 = \frac{O_2}{O_1}$; $O_6 = \frac{O_3}{O_2}$	Aspect ratio of length of principal axes
	θ	Angle that is according to the first principal axes
	$f_1 = \frac{L}{W}$; $f_2 = \frac{L}{T}$; $f_3 = \frac{W}{T}$	Aspect ratio of Bounding box
	f_4, f_5, f_6	Deviations from spherical shape
	f_7, f_8, f_9	Deviations from cubic shape
	$f_{10} = \frac{\pi G^3}{216V}$	Elongation index
	p_2, p_3, p_4	First moment invariant of a particle. This defines the center of objects
	p_5, p_6, p_7	Second moment invariant of a particle. This represents the deviation from the center of object
	p_8	Third moment invariant of a particle
	p_9	Forth moment invariant of a particle
	C	Mean Curvature of a particle
	E	Euler Number of a particle. It describes the connectivity of a particle

(continued)

Table 8.2 (continued)

Category	Microstructural parameter	Remark
Category III distribution (16 parameters in total)	d_2, d_3, d_5	Average distance within each k-nearest neighborhood (k = 2, 3, 5)
	$De = \frac{N}{V}$	Density
	$K(t) = \frac{\lvert B\rvert}{N^2} \sum_{i=1}^{N} \sum_{i \neq j} w(i, j)^{-1} I\left(x_i - x_j \leq t\right)$	K-function: average number of cells within distance t with the univariate K-function defined in 3D; for example, $w(i, j) = \exp\left(-\lVert x_i - x_j \rVert^2\right)$
	$G(t) = \frac{1}{N} \sum_{i=1}^{N} I(y_i \leq t)$	G-function: cumulative nearest neighbor function, measuring the fraction of nearest-neighboring distances that are less than or equal to t
	$F(t) = \frac{1}{g} \sum_{i=1}^{g} I(y_i \leq t)$	F-function: the distance from each grid point to its nearest neighboring cell is measured. g grid points placed at regular intervals
	$I = \sum_{i=1}^{N} \sum_{j=1}^{N} a_{ij}(x_i - \overline{x})(x_j - \overline{x}) / \left(S^2 \sum_{i=1}^{N} \sum_{j=1}^{N} a_{ij}\right)$	Global spatial auto-correlation: $S^2 = \frac{1}{N} \sum_{i=1}^{N} (x_i - \overline{x})^2$, x_i is the ith pore, $\overline{x} = \frac{1}{N} \sum_{i=1}^{N} x_i$ is the average of the x_i . a_{ij} is the adjacency criterion
	$I = \frac{N}{2\sum_{i=1}^{N} \sum_{j=1}^{N} a_{ij}} \frac{\sum_{i=1}^{N} \sum_{j=1}^{N} a_{ij}(x_i - \overline{x})(x_j - \overline{x})}{\sum_{i=1}^{N} (x_i - \overline{x})^2}$	Moran's I statistics
	$C = \frac{N}{2\sum_{i=1}^{N} \sum_{j=1}^{N} a_{ij}} \frac{\sum_{i=1}^{N} \sum_{j=1}^{N} a_{ij}(x_i - x_j)^2}{\sum_{i=1}^{N} (x_i - \overline{x})^2}$	Geary's C statistics: based on the weighted sum of square difference between observations, which is defined by:
	$G(t) = \sum_{i=1}^{N} \sum_{j=1}^{N} a_{ij}(t) x_i x_j / \sum_{i=1}^{N} \sum_{j=1}^{N} x_i x_j$	Getis's global statistics
	l_2, l_3, l_5	Local volume fraction
	d_{m}	Average distance within each cluster
	l_{m}	Local volume fraction

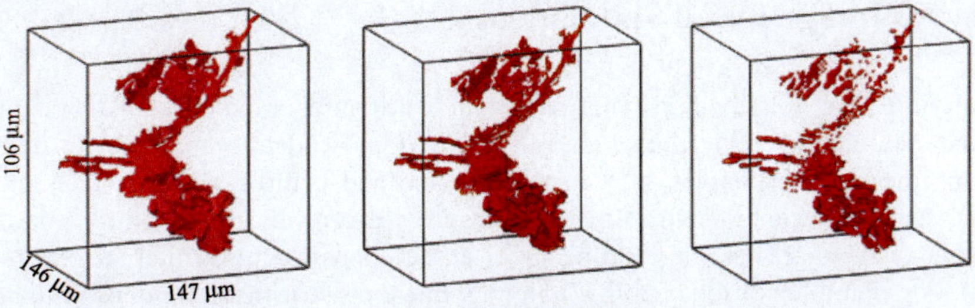

(a)	(b)	(c)
$V_{Voxel} = 18.4 \times 10^3\ \mu m^3$	$V_{Voxel} = 15.8 \times 10^3\ \mu m^3$	$V_{Voxel} = 15.1 \times 10^3\ \mu m^3$
$V_{BB} = 2.26 \times 10^6\ \mu m^3$	$V_{BB} = 2.25 \times 10^6\ \mu m^3$	$V_{BB} = 2.19 \times 10^6\ \mu m^3$
(a) Voxel size: 0.5 µm	(b) Voxel size: 1.0 µm	(c) Voxel size: 2.0 µm

Fig. 8.32 Imaging of particles in an aluminum alloy during synchrotron radiation X-ray tomography experiments. The same particle in the same material was imaged at three different spatial resolutions to simulate the visualization of smaller particles. The two measurement method types of the precision volume measurement (V_{Voxel}) and the substitution with the volume of a rectangular box (bounding box), which exactly encloses the particle (V_{BB}), are compared

Table 8.2 includes multiple parameters used for measuring the size, shape, and spatial distribution. As reiterated throughout this book, this is because X-ray tomography involves the visualization of internal structures that are close to its effective spatial resolution due to limitations from the spatial resolution capacity of the devices used, whether from industrial-use scanners or synchrotron radiation X-ray tomography.

Figure 8.32 shows the quantitative analysis of size when visualizing the same structure in the same material as the spatial resolution of the 3D image is reduced. The size of the microstructure shown in these images in the longitudinal and width directions greatly exceeds the spatial resolution of the 3D image but has a minimal thickness. For this reason, a low spatial resolution relative to the microstructure thickness results in significant decreases in the value calculated by volume and converted to diameter and decreasing measurement accuracy. However, although applying a bounding box that barely encloses the microstructure and setting its long sides as the microstructure size may result in inaccurate measurements at first glance, this also shows that the decrease in measurement accuracy is effectively controlled even when measuring under low spatial resolution. The relationship between internal structure and macroscale characteristics can be clarified in this manner by brainstorming parameters, from which significantly correlated ones are selected statistically.

An overview of the foundational knowledge relating to modern geometry (e.g. classic geometry, differential geometry, and topology) is omitted here due to space limitations. Please refer to specialized sources as required [25–27].

8.6 3D Image-Based Simulations

Observing the relationships between microstructures/internal structures and macroscale characteristics/behavior is an important academic approach in determining the basic principles of that phenomenon and controlling those characteristics. An ideal scenario is to directly measure a given physical quantity relating to macroscale characteristics and behavior at the microstructures/internal structure level. For example, various applied imaging/image processing techniques affiliated with 3D imaging, such as the 3D strain mapping introduced in the next chapter or the 3D mapping of elemental concentrations introduced in Sect. 5.4, correspond to this. However, combining 3D imaging and various numerical simulations to calculated local physical quantities is also extremely effective when this is not possible. This is referred to as a *3D image-based simulation*. Various calculation subjects can be considered in addition to structural analysis of deformation or fracture, including casting/solidification analysis, fluids, heat transfer, electromagnetic fields, and elemental diffusion.

Figure 8.33 shows the process for creating an analysis model where 3D image-based simulations of elastoplastic finite element analysis were conducted to observe the interactions between cracks and silicon particles in 3D images obtained by scanning aluminum–silicon alloys at BL47XU in SPring-8 [28]. There was only a single crack in this case, but there are several tens of thousands of particles for a single 3D image. This is far too many for the element number in the model, so the model was created by restricting this to approximately 100 particles in a given domain to observe the damage at the crack tip. First, the internal structure, which is the subject of numerical analysis, was extracted, after which its surface shape was approximated as an aggregate of triangular meshes. The file format for this is referred to as *standard triangulated language* (STL). STL files comprise the coordinates and normal vectors of all the triangular mesh vertices. There has been a steadily increasing need for image data obtained by X-ray tomography to be converted into STL formats with the recent widespread use of 3D printers. For these reasons, a wide range of commercially sold and free software can be used in addition to various 3D rendering software or MATLAB/ImageJ for converting from 3D images to STL files. Furthermore, various pre-processors for numerical analysis can be used to create a 3D model for numerical analysis from STL files for example. The sample in Fig. 8.33 is separated with 126,000 tetrahedral elements, but numerical analyses have been conducted with approximately 1 million elements.

The size and shape of the surface mesh play a significant role in achieving numerical analysis accuracy. A high-quality and defect-free surface mesh is required to produce a 3D mesh for numerical analysis from a surface mesh [29]. Furthermore, the mesh resolution must be fine enough so that sufficient numerical analysis accuracy can be obtained and coarse enough so that the computational time can be reasonably shortened. The most laborious and time-consuming processes, in particular, involve correcting triangular mesh connectivity [30] or distorted element shapes [31], as

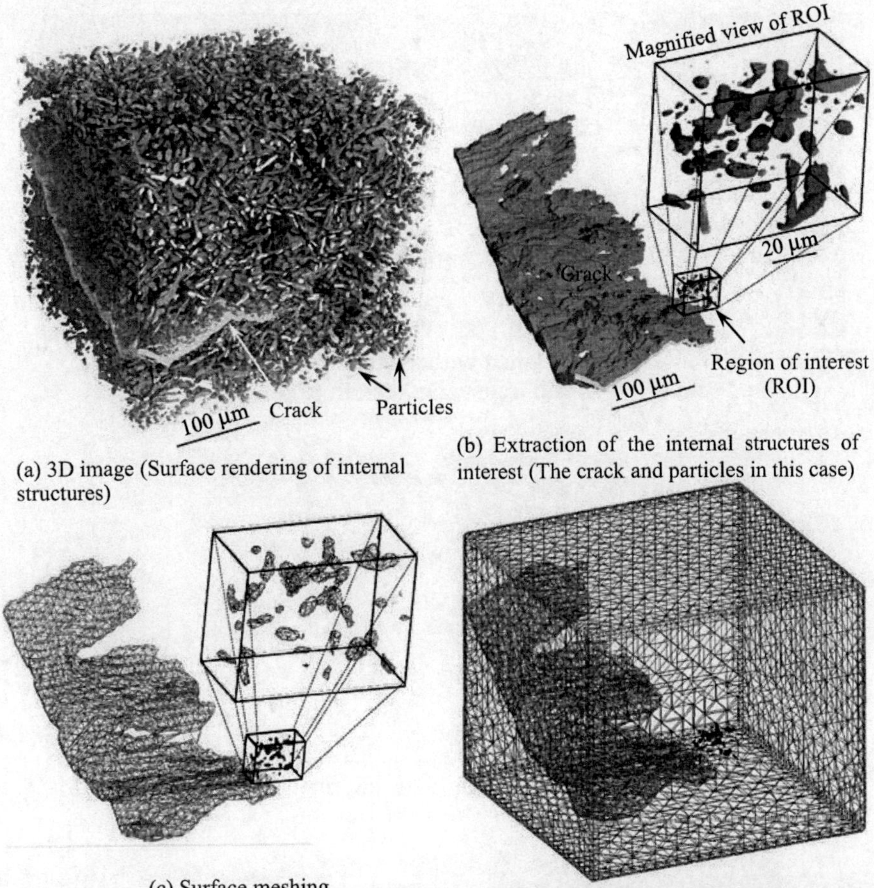

(a) 3D image (Surface rendering of internal structures)

(b) Extraction of the internal structures of interest (The crack and particles in this case)

(c) Surface meshing

(d) 3D model for numerical simulation

Fig. 8.33 **a** Imaging of the particles and cracks in an aluminum alloy during synchrotron radiation X-ray tomography experiments. From this, a process for creating 3D models for numerical analysis is shown, where **b** the internal structures of interest to be used for numerical analysis were extracted, **c** their surface shapes were approximated with triangular meshes, and **d** the internal structure was also separated using 3D elements [28]. The original figure is in color; please refer to the figure in the original publication for details

well as controlling the number of elements. Commercially available software for correcting mesh shapes has recently been developed.

Apart from this, there have been methods developed for directly using voxels as elements for numerical analysis [32]. Though this has the disadvantages of retaining a rough surface shape or lack of ability to control analysis accuracy with mesh resolution such as with the triangular meshes, this does not require mesh correction and its model can be easily created. Figure 8.34 shows an example of research comparing the particle stress and fracture behavior in a material similar to Fig. 8.33

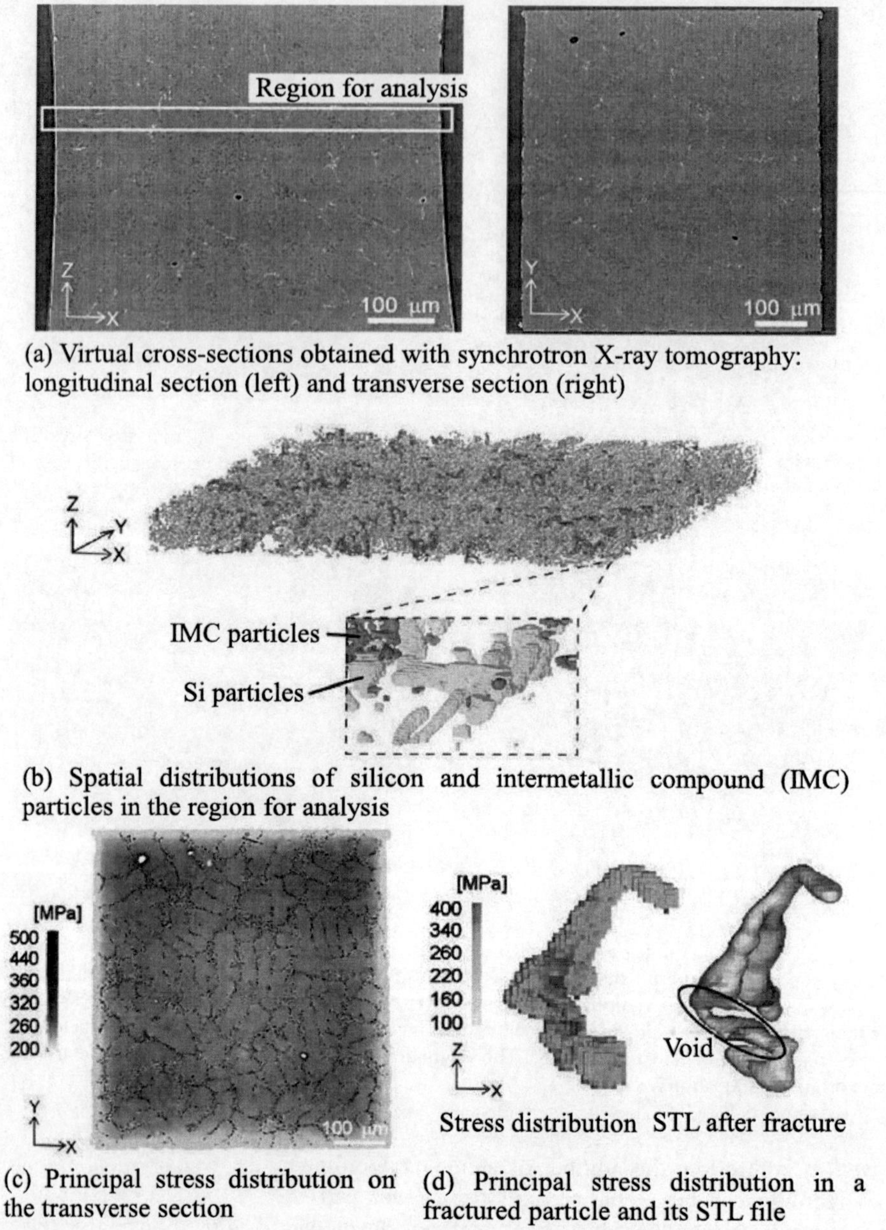

(a) Virtual cross-sections obtained with synchrotron X-ray tomography: longitudinal section (left) and transverse section (right)

(b) Spatial distributions of silicon and intermetallic compound (IMC) particles in the region for analysis

(c) Principal stress distribution on the transverse section

(d) Principal stress distribution in a fractured particle and its STL file

Fig. 8.34 Example of image-based finite element analysis where voxels are directly used as elements for numerical analysis. The relationships between the stress and fracture of silicon and intermetallic compounds in AC_4CH aluminum were analyzed (courtesy of Osamu Kuwazuru and Masaki Teranishi of Fukui University)

[32]. Images in Fig. 8.34b and d show that particles with a size of several to 10 μm are expressed by 0.5-μm voxels. Effective numerical analyses are possible with these types of methods as the voxel size, in this case, is considerably smaller than a particle and the curvature of the particle surface is relatively low.

Finally, several implementation examples of 3D image-based simulations are introduced here. Figure 8.35 shows representative analysis models and the calculation results of various 3D image-based simulations conducted using 3D images from industrial-use X-ray CT scanners and synchrotron radiation X-ray tomography conducted at the author's laboratory [28, 33–37]. These are all structural analyses, with Fig. 8.35b being a viscoplastic analysis and all others being elastoplastic analyses. Figure 8.35a and f shows calculations of local stress/strain distributions, which cannot be directly measured at the microscale (cell walls with porous metal thickness less than 10 μm) and macroscale (external shape of the sample) level, respectively; the former has been useful as analysis data to elucidate complex fracture behavior in porous metal, and the latter for verifying nanovoid formation due to hydrogen. Figure 8.35b shows the use of inverse analysis to determine gas pressures, which can explain the blister (production defect) growth behavior observed with X-ray tomography by varying gas pressure in a blister interior and conducting multiple calculations as a parametric study. These results not only clarified the contribution of hydrogen but also that of pressurized carbon dioxide or nitrogen gas dissolved during the casting process. Figure 8.35c assesses heterogeneous crack tip deformation due to both cracks and the microstructure by simplifying the modeling of the crack shape without considering the microstructures. Figure 8.35d assesses the mixed-mode crack propagation driving force of opening—in-plane shear—out-of-plane shear caused by complex crack shapes. In each case, this has contributed to our current understanding of complex fracture behavior in structural materials. Figure 8.35d separates only the central section (region A in the figure) near the crack tip with fine meshes and sets the rest with gradually coarsening meshes to increase analysis accuracy and strike a balance between analysis accuracy and computational cost. Finally, Fig. 8.35e is slightly different in that local damage indicators, which cannot be measured/assessed with X-ray tomography (e.g. triaxiality of stress–strain or intraparticle elastic stress), are calculated and where original data, which assist the microstructure optimization of structural material, are obtained. This achieves the optimization of microstructure in structural materials due to reverse 4D material engineering introduced in the next chapter.

X-ray tomography-based observations and assessments can be supplemented effectively using 3D image-based simulations, and high-level analyses, which acutely arrive at the fundamental principles of various physical phenomena, suggesting that guiding principles are possible.

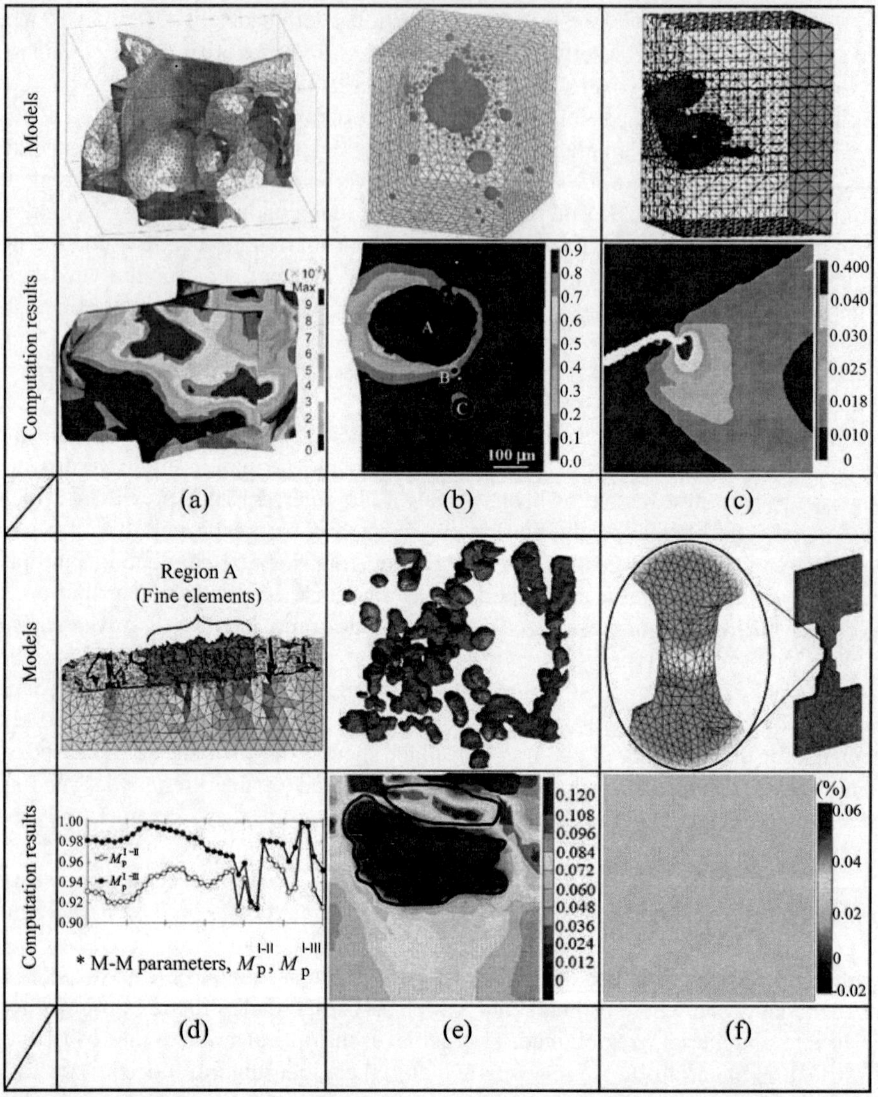

Fig. 8.35 Numerous application examples of 3D image-based simulations and examples of structural analysis, such as material deformation/fracture and manufacturing defect growth [28, 33–37]. The original figure is in color; please refer to the figure in the original publication for details. **a** Deformation behavior of cell walls in an aluminum foam [33]. **b** Inverse analysis of blistering in an aluminum alloy [34]. **c** Plastic zone shape at a crack-tip [28]. **d** Mixed-mode (M-M) deformation behavior of a crack-tip region [35]. **e** Damage behavior of particles and pores in metals [36]. **f** Inhomogeneous strain distribution due to macroscopic shape of a specimen [37]

8.7 3D Expression

Devices like 3D printers or 3D displays are also effective for assessing 3D images obtained with X-ray tomography. As this discipline has significantly developed in recent years, this section, indicating its importance due to concerns provided here, will likely become obsolete.

There are a several 3D models in the author's office (Fig. 8.36), but unlike observing normal displays, the ability to sense microstructures or complex 3D structures in a tactile manner greatly aids in visually grasping the structure. It is important to have a firm grasp of the relationship between the spatial resolution of lamination intervals in a 3D printer and the spatial resolution of the 3D images during 3D printing. There is a wide range of materials that can currently be used in this application as well, including ABS resin, polypropylene, nylon, acrylic, epoxy resin, plaster, and metal. Some of these include transparent materials for easily observing the interior or others that can be dyed and enable a wide range of assessments. Furthermore, complex curved surfaces created by clay are used not only for visual appreciation but are scanned with X-ray tomography as image data from which an STL file is produced and then used to fabricate this shape in a 3D printer. This digital manufacturing method has been widely anticipated as a new manufacturing process for complex shape/small-quantity production. Furthermore, the ongoing relationship with 3D printers and X-ray tomography, for example, using high-resolution X-ray CT scanners to assess the fabrication accuracy of products made with a 3D printer, is expected to become increasingly important in the future.

The use of 3D displays has become widespread in standard TVs as well, by combining active 3D glasses with liquid crystal shutters and passive polarizing 3D glasses. However, this has slowed down with regard to private use. Like with 3D printers, this has large effects in supplementing the visual understanding of complex 3D structures, and its effective use is anticipated. Figure 8.37 shows the observation of high-accuracy 3D images from X-ray tomography using a proper high-resolution 3D display, which connects two LCDs with a half-mirror system.

8.8 Effective Presentation Applications

The importance of 3D displays was briefly discussed in the previous section. The continuous display of 2D cross-section images in order and the use of continuous animations, which rotate, translate, magnify, and shrink 3D images, are effective methods to assess and demonstrate complex internal structures. Notebook computers, which have high-performance graphic boards have recently been commercially available and effective presentations, which use these types of 3D image displays, can be easily conducted. For researchers, as academic journals become increasingly digital, many academic papers include supplemental materials such as animation files in

(a) 3D model of a crack

(c) Fracture surface and internal void distribution after a tensile test in a DP steel

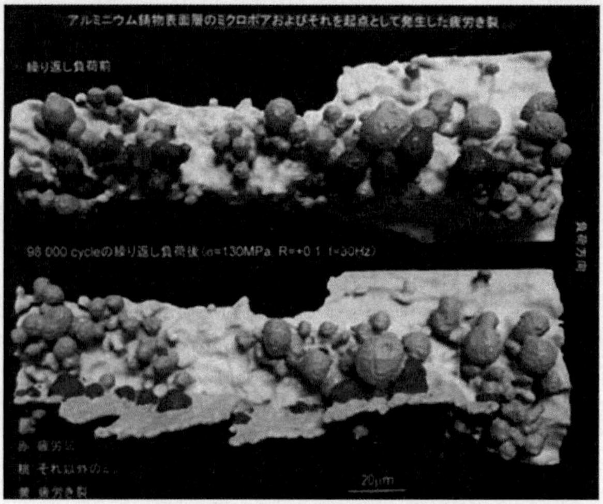

(b) Fatigue crack initiation from surface defects in a cast aluminum

Fig. 8.36 Example of a 3D-printed sample of a 3D image obtained from X-ray tomography; **a** and **c** are models made from plaster, whereas **b** is made from transparent resin

addition to the article figures, which are displayed alongside the article. In addition, researchers who handle 3D images should actively take advantage of these opportunities.

Fig. 8.37 Assessing 3D images obtained from X-ray tomography with high-resolution 3D displays using a halfmirror system, as conducted in the author's laboratory

References

1. K. Nishiyama, *Optimal Filtering (System Control Series), Baifukan* (2001)
2. Scikit-image website. https://scikit-image.org/. Accessed September 2018
3. N. Otsu, IEEE Trans. Syst. Man and Cybern. **9**, 62–66 (1979)
4. C.K. Chow, T. Kaneko, Comput. Biomed. Res. **5**, 388–410 (1972)
5. W. Niblack, *An Introduction to Digital Image Processing* (Prentice-Hall, Englewood Cliffs, NJ, USA, 1986), pp. 115–116
6. W. Oh, W.B. Lindquist, IEEE Trans. Pattern Anal. Mach. Intell. **21**, 590–602 (1999)
7. K.V. Mardia, T.J. Hainsworth, IEEE Trans. Pattern Anal. Mach. Intell. **6**, 919–927 (1988)
8. O. Wirjadi, Berichte des Fraunhofer ITWM, Nr. 123 (2007), p. 2
9. O. Brunke, S. Oldenbach, F. Beckmann, Eur. Phys. J. Appl. Phys. **29**, 73–81 (2005)
10. A.J. Burghardt, G.J. Kazakia, S. Majumdar, Ann. Biomed. Eng. **35**, 1678–1686 (2007)
11. D.S. Feeney, J.W. Crawford, T. Daniell, P.D. Hallett, N. Nunan, N.K. Ritz, M. Rivers, I.M. Young, Microb. Ecol. **52**, 151–158 (2006)
12. S. Ramaswamy, M. Gupta, A. Goel, U. Aaltosalmi, M. Kataja, A. Koponen, B.V. Ramarao, Colloids Surf. A **241**, 323–333 (2006)
13. I. Arganda-Carreras, V. Kaynig, C. Rueden, W. Kevin, J. Eliceiri, A. Schindelin, H. Cardona, S. Sebastian, *Bioinformatics* (Oxford University Press) (2917). https://doi.org/10.1093/bioinformatics/btx180 (on Google Scholar)
14. C.J. Hilditch, *Machine Intelligence*, ed. by B. Meltzer, D. Mitchie, Chap. 22, (Edinburgh University Press, Edinburgh, Scotland, 1969), pp. 403–420
15. M.V. Nagendraprasad, P.S.P. Wang, A. Gupta, Digital Signal Process. **3**, 97–102 (1993)
16. P.S.P. Wang, Y.Y. Zhang, IEEE Trans. Comput. **C-38**, 741–745 (1989)
17. K.J. Kurzydlowski, B. Ralph, *The Quantitative Description of the Microstructure of Materials* (CRC Press, Boca Raton, Florida, 1995), pp. 184–189
18. L. Vincent, In: *Proceedings of the 1991 IEEE Computer Society Conference on Computer Vision and Pattern Recognition, Maui, Hawaii* (1991), pp. 520–525
19. P. Su, R. Drysdale, Comput. Geometry Theor. Appl. **7**, 361–386 (1997)
20. N. Kurachi, CG Magic: Rendering, Ohmsha (2007)
21. W.E. Lorensen, H.E. Cline, ACM SIGGRAPH Comput. Graph. **21**, 163–169 (1987)
22. J. Ohser, K. Schladitz, *Advanced Tomographic Methods in Materials Research and Engineering*, ed. John Banhart (Oxford University Press, 2008), Section 3
23. D. Bartz, M. Meißner, *Proceedings of Volume Graphics '99, Swansea* (1999), pp. 33–48
24. K. Engel, M. Hadwiger, J. Kniss, J.M. Lefohn, C.R. –Salama, D. Weiskopf, Real-time volume graphics. In: *ACM SIGGRAPH Course 28* (2004)
25. G. Lohmann, *Volumetric Image Analysis* (Wiley and Teubner Publishers, New Jersey, 1998)

26. E. Kreyszig, *Differential Geometry (Dover Books on Mathematics)*, 1st edn. (Dover Publications, New York, 1991)
27. W. Thurston, *Three-Dimensional Geometry and Topology* (Princeton University Press, Princeton, NJ, 1997)
28. L. Qian, H. Toda, K. Uesugi, M. Kobayashi, T. Kobayashi, Phys. Rev. Lett. **100**, 115505 (2008)
29. H. Date, J. Jpn Soc. Precis. Eng. **74**, 1264–1268 (2008)
30. C.S. Chong, A.S. Kumar, H.P. Lee, Finite Element Anal. Design **43**, 1109–1119 (2007)
31. E. Bechet, J.C. Cuilliere, F. Trochu, Comput. Aided Des. **34**, 1–17 (2002)
32. M. Teranishi, O. Kuwazuru, M. Kobayashi, H. Toda, Trans. Jpn. Soc. Mechanic. Eng. **84**, 18–00028 (2018)
33. H. Toda, M. Takata, T. Ohgaki, M. Kobayashi, T. Kobayashi, K. Uesugi, K. Makii, Y. Aruga, Adv. Eng. Mater. **8**, 459–467 (2006)
34. H. Toda, P.C. Qu, S. Ito, K. Shimizu, K. Uesugi, A. Takeuchi, Y. Suzuki, M. Kobayashi, Int. J. Cast Met. Res. **27**, 369–377 (2014)
35. H. Toda, E. Maire, S. Yamauchi, H. Tsuruta, T. Hiramatsu, M. Kobayashi, Acta Mater. **59**, 1995–2008 (2011)
36. A. Hosokawa, T. Kurino, H. Toda, K. Uesugi, A. Takeuchi, Y. Suzuki, J. Jpn. Inst. Light Metals **64**, 611–612 (2014)
37. H. Su, H. Toda, R. Masunaga, K. Shimizu, H.Y. Gao, K. Sasaki, M.S. Bhuiyan, K. Uesugi, A. Takeuchi, Y. Watanabe, Acta Materialia, **159** (2018). Accepted

Chapter 9
4D Image Analysis

The 2D visualization devices that we have grown accustomed to using in laboratories, such as scanning electron microscopes and transmission electron microscopes, have constraints of only being able to use thin-films that are considerably less than the deep sub-micron scale or only being able to assess material surfaces. Thin-films and surfaces occasionally have behaved completely differently from that of the bulk material. These differences can be explained mechanically through terms like plane stress conditions or stress singularities, which are at the intersection of different phase interfaces and the surface. Meanwhile, X-ray tomography can use the hardware discussed in Sect. 4.5.1 such as various material testing machines, heating/cooling devices, and imaging techniques including fast tomography, as described in Sect. 5.3, to record how the interior of samples with sufficient size change with external disturbances or over time in the form of multiple sets of 3D images. This is referred to as 4D observation. The *fourth dimension* (*4D*) in this case refers to adding the time axis to 3D. As the term 3D has become increasingly well-known among the general population, so has the term 4D become steadily familiar among gynecology departments in hospitals, as well as a projection method for fanciful films. Industrial X-ray CT scanners are capable of 4D observations, and synchrotron radiation X-ray tomography which uses high-brilliance X-rays even more so. The obtained 4D images can be viewed as often as required when using the 3D rendering methods in Sect. 8.4. Furthermore, applying pixel subtraction to 3D images of the same samples can reveal subtle changes in the internal structures and enable the quantitative assessment of time evolution behavior and various behaviors that evolve with time via 4D image analysis.

The first step for this type of 4D image use is to *align* (*register*) the 3D images of the same samples taken consecutively to enable its assessment and analysis. This chapter will discuss the basics of registration, after which *particle tracking* and 4D techniques for mapping mechanical strain and various fracture mechanics parameters will be discussed. Finally, applications of machine learning techniques referred to as reverse 4D material engineering by the author to 4D images will be discussed. These may be somewhat tedious for researchers and technicians who are in fields

H. Toda, *X-Ray CT*,
https://doi.org/10.1007/978-981-16-0590-1_9

that are different from the material engineering or mechanical engineering disciplines engaged by the author. However, it is the author's hope that these techniques will be learned and serve as a reference for applications to the readers' respective disciplines.

9.1 Registration

The registration described here can be applied to not only the registration of 3D images that have been consecutively captured by X-ray tomography, but also to the registration between different analysis models used for 3D image-based simulations (discussed in Sect. 8.6), between an analysis model and a 3D image, or between discrete 3D images obtained from various applied methods discussed in Chap. 8. Coordinate transforms serve as their basis for this, with linear and nonlinear variations. This section will introduce these methods. The accuracy of this registration is key to the success of 4D image analysis. For these reasons, it should be clear that this must be completed as carefully and optimally as possible, however, it is also important in assessing registration results and improving registration based on these assessments.

9.1.1 Affine Transformation

When registering two images, one of them is set as the fixed image and the other as the moving image; the moving image can be translated, rotated, magnified/minimized, and shear-deformed to ensure an accurate registration with the fixed image. An *affine transformation* refers to a linear transformation where a given point P (x, y) is mapped to a point P' (X, Y) by multiplying it with a matrix A and adding a vector t. The physical object is translated, rotated, magnified/minimized, sheared, and otherwise deformed as a combination of these processes through the affine transformation. A 2D affine transformation can be expressed using the six parameters shown in the equation below:

$$\begin{pmatrix} X \\ Y \end{pmatrix} = \begin{pmatrix} a_{11} & a_{12} \\ a_{21} & a_{22} \end{pmatrix} \begin{pmatrix} x \\ y \end{pmatrix} + \begin{pmatrix} b_x \\ b_y \end{pmatrix} \tag{9.1}$$

Here, the matrix A is $\begin{pmatrix} a_{11} & 0 \\ 0 & a_{22} \end{pmatrix}$ when applying a magnification/minimization. The matrix is $\begin{pmatrix} 1 & a_{12} \\ 0 & 1 \end{pmatrix}$, $\begin{pmatrix} 1 & 0 \\ a_{21} & 1 \end{pmatrix}$, or $\begin{pmatrix} 1 & a_{12} \\ a_{21} & 1 \end{pmatrix}$ when shearing, depending on the direction. Shear deformation refers to deformation when an angular change is induced as shown in Fig. 9.1. Furthermore, the rotation matrix is $\begin{pmatrix} cos\theta & -sin\theta \\ sin\theta & cos\theta \end{pmatrix}$ when the

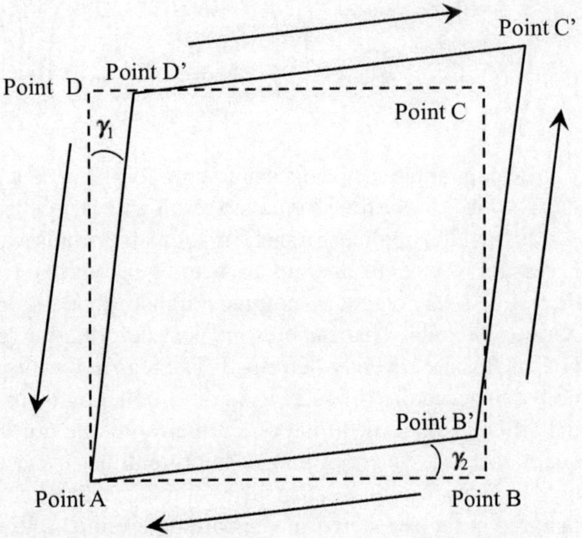

Fig. 9.1 Schematic explaining shear deformation. What was originally a square was assumed to have undergone shear deformation into a rhomboid (AB'C'D'). The distortion of the shape in this case is expressed by angles γ_1 and γ_2 on both sides, and the shear strain is expressed as $\gamma = \gamma_1 + \gamma_2$

rotation angle is θ. Basic deformations should be multiplied to create a composite mapping when combining these deformations.

The point can be expressed as a product of the deformation matrix R as shown in the following equation by introducing the homogeneous coordinates from Eq. (9.1).

$$\begin{pmatrix} X \\ Y \\ 1 \end{pmatrix} = \begin{pmatrix} a_{11} & a_{12} & b_x \\ a_{21} & a_{22} & b_y \\ 0 & 0 & 1 \end{pmatrix} \begin{pmatrix} x \\ y \\ 1 \end{pmatrix} \tag{9.2}$$

Furthermore, expressions based on homogeneous coordinates for 3D cases can be expressed with 12 parameters, as shown in the following equation:

$$\begin{pmatrix} X \\ Y \\ Z \\ 1 \end{pmatrix} = \begin{pmatrix} a_{11} & a_{12} & a_{13} & b_x \\ a_{21} & a_{22} & a_{23} & b_y \\ a_{31} & a_{32} & a_{33} & b_z \\ 0 & 0 & 0 & 1 \end{pmatrix} \begin{pmatrix} x \\ y \\ z \\ 1 \end{pmatrix} \tag{9.3}$$

Registration can be accomplished using rigid-body transformations for both translations in three directions and rotations on three axes when the sample does not deform in the 3D image based on X-ray tomography. The transformation matrix R, in this case, when rotating in the x, y, and z-axes is as follows:

$$\begin{pmatrix} 1 & 0 & 0 & 0 \\ 0 & cos\theta & -sin\theta & 0 \\ 0 & sin\theta & cos\theta & 0 \\ 0 & 0 & 0 & 1 \end{pmatrix}, \begin{pmatrix} cos\theta & 0 & sin\theta & 0 \\ 0 & 1 & 0 & 0 \\ -sin\theta & 0 & cos\theta & 0 \\ 0 & 0 & 0 & 1 \end{pmatrix}, \begin{pmatrix} cos\theta & -sin\theta & 0 & 0 \\ sin\theta & cos\theta & 0 & 0 \\ 0 & 0 & 1 & 0 \\ 0 & 0 & 0 & 1 \end{pmatrix} \tag{9.4}$$

The commutative law does not apply for the products of matrices; therefore, care must be taken as the transformation result will vary depending on the order of rotation.

Meanwhile, nonlinear transformation algorithms such as the B-spline method are necessary when deformations are heterogeneous [1]. However, the author's research discipline does not require nonlinear transformations with consecutive 3D images in X-ray tomography, barring pre- and post-deformation registration between images of objects that are severely deformed. The reason for this is because extreme deformations can be handled by increasing the number of 3D image projections and reducing the difference between images. Furthermore, the objective of registration is not the qualitative assessment of images but oftentimes extracting the identical microstructural features in a consecutive series of images. In other words, once the characteristic features in the registered images are determined, the mechanical analysis of local deformation in a sample would be conducted in the image before the application of transformations such as magnification or minimization. For this reason, dividing the overall 3D images into sub-regions, registering each sub-region with linear transformations, and extracting the characteristic features from them is considered more efficient than applying a nonlinear transformation algorithm to the entire 3D image.

9.1.2 Various Registration Methods

Various registration methods are first shown in Table 9.1. These assume scenarios where the same device is used to image the same sample consecutively. In other words, registration of the absorption contrast and phase-contrast images of the same sample, or those of different samples, needs to be considered separately.

The most basic registration method of 3D images obtained with synchrotron radiation X-ray tomography or industrial X-ray CT scanners is to identify the centroid of regions that are closed within a sample, with clear boundaries (e.g. particles, embedded components, holes like pores), and to register them using them as *landmarks*. Metals, ceramics, and polymers, as well as components, substrates, and industrial products, which combine those materials, often have several to several dozen characteristic features with clear corresponding relationships that can be used as landmarks. The centroid positions of the characteristic features can be easily calculated using the quantitative geometric analysis from Sect. 8.5. As such, registration can be conducted by simply minimizing the misalignment in their positions, so long as the local deformation is not too large. Thirion defined line segments referred to as crest lines that are like mountain ridges in the undulations of the internal structure surface, as well as characteristic points referred to as extreme value points, and used

Table 9.1 Overview of methods for registering 3D images successively scanned with X-ray tomography

Type	Obvious landmarks	Usage of image information	Remarks	Characteristics
Use of landmarks	Used	Characteristic features only	Registration using landmarks	– Accurate and rapid – Easy verification – Careful attention required when features change or deform
	Not used (Extraction of characteristic lines or dots by analyzing internal structures)	Characteristic features only	Registration by defining and extracting landmarks	– Relatively accurate and rapid – Unavailable when shape changes
Disuse of landmarks	Not used	Only the surface of a sample or internal structures	Usage of a sample itself and/or the shape of internal structures	– Relatively rapid – Careful attention required when shape changes
	Not used (*Used)	Entire image	Registration on the basis of the similarity functions of grey value distribution (Mutual information, etc.)	– Unavailable when internal structures drastically change – Occasionally long computation time – Effective for inhomogeneous deformation even when landmarks are available*
Combination of the above two types	Used	It depends	Preliminary registration using landmarks followed by the usage of similarity functions	– Reduction of computation time possible – Unnecessary when landmarks do not exhibit position change nor deformation

these to conduct registration between 3D images [3]. The authors have defined the valleys and peaks of the concave and convex portions, respectively, of surface undulations in the internal structures as changes in the Gaussian curvature and used them for registration and 3D strain mapping [4]. These can also be said to be variations of registration using landmarks. A commonly used method when an appropriate landmark cannot be used is to attach particles on the external surface and use this for registration.

On the other hand, there have been numerous methods proposed for when the boundary between the internal structure and background is unclear and landmark identification cannot be reproduced, as represented by medical images. A superior handbook, which provides a comprehensive discussion of registration methods for images obtained by medical-use CT scanners, is cited here [2]. The surface or internal structure can be segmented when landmarks cannot be used. These can be used for registration after being elastically deformed as necessary. Furthermore, registration can be conducted by determining a principal axis and using that alongside the centroid if the sample or internal structure has clear directionality (e.g. a thin elongated shape).

The mainstream registration method when landmarks cannot be used is to define some type of *similarity function* and use the gray values of all the pixels in the two 3D images to calculate this similarity function while conducting the registration. Examples include calculating the variation of the pixel value ratios in corresponding positions of two 3D images, the entropy of the pixel value histograms, or iteratively minimizing the absolute value or sum of squares of residuals [2]. Registration based on mutual information is what is important for practical use here. As defined by information theory, if the probability of a given event i ($i = 1 - n$) occurring was set as p_i, then the expected value of the amount of information gained from a given information source (average information amount) is referred to as the entropy H, which is defined as follows:

$$H = -\sum_{i=1}^{n} p_i log p_i \tag{9.5}$$

In the case of images, the pixel value of the pixel is the information, the pixel size is the amount of information, and the pixel value histogram p_i is used for the probability of a pixel with a given pixel value. Furthermore, $n = 256$ for an 8-bit image. Here, we define a joint entropy $H(X, Y)$ when two events occur simultaneously. This is defined as follows, where the probability that the pixel value combination in two images (information source X, Y) becomes (x, y) or, in other words, the joint probability $p(x, y)$.

$$H(X, Y) = -\sum_{x,y} p(x, y) log p(x, y) \tag{9.6}$$

Furthermore, defining the *mutual information* $I(X, Y)$ makes this an index for the overlap extent of information source X, Y.

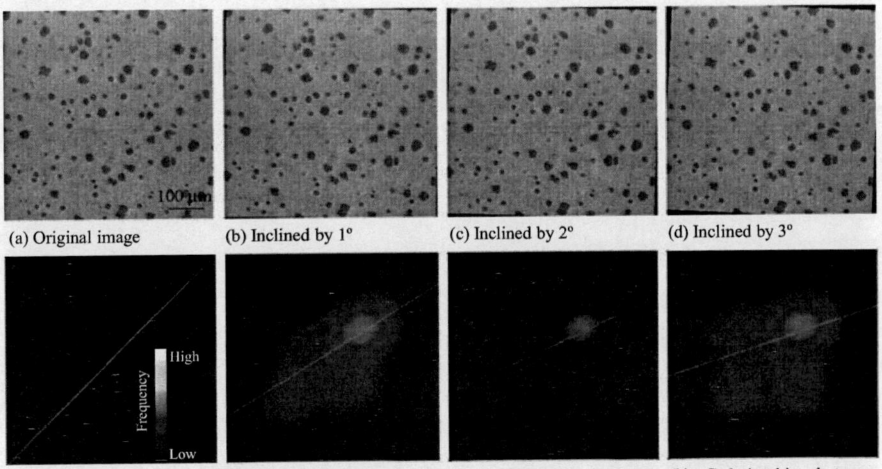

(a) Original image (b) Inclined by 1° (c) Inclined by 2° (d) Inclined by 3°

(e) Relationship between original images (f) Relationship between image (b) and the original image (g) Relationship between image (c) and the original image (h) Relationship between image (d) and the original image

Fig. 9.2 Relationships between the pixel values of the original image and rotated image are assessed to visualize the effects of successful registration by deliberately rotating the 3D image obtained by X-ray tomography in one direction overlapping this directly with the original image. The top row shows the images used and the bottom row shows the 2D histograms

$$I(X, Y) = H(X) + H(Y) - H(X, Y) \tag{9.7}$$

Here, $I\ (X, Y) = 0$ when X and Y are independent. The statistical dependency of two images can be quantitatively expressed in the registration of 3D images by calculating the mutual information.

Figure 9.2b shows a 2D histogram of the relationship between the original image in Fig. 9.2a and an image where the original is combined with a deliberately inclined version of that image. Accurate registrations (i.e., without rotation) result in a histogram alignment on a single line. Meanwhile, slight inclinations in both images result in large changes in the 2D histogram and fine changes in mutual information values. Methods that use mutual information can be used not only for registration but also for assessing the accuracy of other registration methods that use landmarks.

9.2 Particle Tracking

As shown schematically in Fig. 9.3, identifying the same characteristic features in consecutive 3D images and matching them after conducting the quantitative geometric analyses in Sect. 8.5 and the registration in Sect. 9.1 allows for the calculation of a variety of mechanical information. This type of extracting and mapping of the same characteristic feature is referred to as *tracking* and often involves tracking

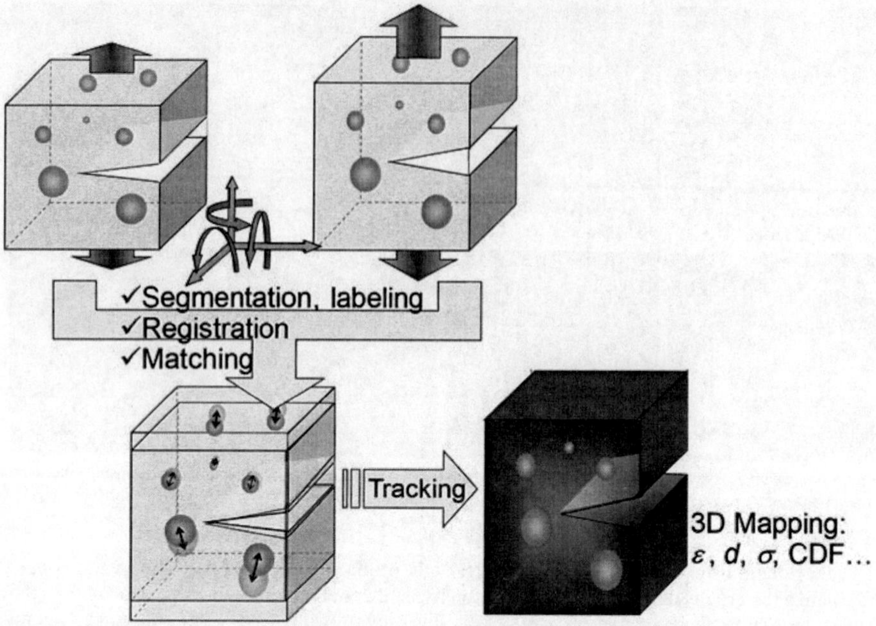

Fig. 9.3 Schematic explaining 4D image analysis conducted by finding an identical characteristic point in continuous 3D images, conducting particle tracking for all particles, and using these results to conduct 3D mapping of various mechanical quantities [7]

independent internal structures that exhibit a certain particle shape, which is referred to as *particle tracking*.

Incidentally, an extremely large number of internal structures can be observed in structural metals with high-spatial-resolution X-ray CT scanners like synchrotron radiation X-ray tomography. A number of microstructures (e.g. particles, pores) with a volume larger than 9 voxels in various aluminum alloys is shown in Table 9.2. Several thousand micropores can be observed in the field of view of a single 3D image even in 99.999% pure aluminum. This increases up to several hundred thousand in high-strength aluminum alloys with high alloying element concentrations. Conducting particle tracking of these types of microstructural features—or, in the case of industrial products, the many small components that comprise them—facilitates the individual identification of changes due to various external disturbances, such as high-temperature exposure, deformation or fracture, and change over time. The fracture of materials and the failure of industrial products are dominated not by the simultaneous fracture or functional cessation of several tens of thousands of internal structures but rather the weakest few, or in some extreme cases, a single internal structure and its effects on the overall characteristics or functionalities. Particle tracking is thus extremely effective in extracting truths from a series of 3D images, which have phenomena dominated by local events like this. Furthermore, analyzing particle tracking data and conducting 3D mapping by calculating

Table. 9.2 Number density of metal microstructures obtained from 3D images when observing with X-ray microtomography at SPring-8. The materials used were aluminum alloys, which have many micropores with precipitated hydrogen and intermetallic compound particles due to their chemical activity and supersaturated hydrogen relative to other metals. A wide range of materials, from high-purity aluminum to various alloys for industrial use, was listed. The right column shows the number density of internal structures that are visible in a standard-sized field of view in synchrotron radiation X-ray tomography

Material	Microstructural feature	Number density per 1 mm^3
Pure aluminum (99.999%)	Pore	3,036
Commercial pure aluminum (A1050)	Pore	10,375
Al–Cu–Mg alloy (A2024)	Particle	150,185
	Pore	63,111
Al–Mg alloy (A5XXX)	Pore	20,160
Al–Mg-Si alloy (A6061)	Pore	21,184
Al–Zn–Mg–Cu alloy (A7075)	Pore	28,852

mechanical quantities would enable the quantitative comprehension of these types of local phenomena. Methods and application examples for this are introduced from this section onward.

The various patterns of change in particles considered necessary for particle tracking are shown in Table 9.3. Depending on conditions, particle tracking can be conducted in chronological order, in reverse chronological order, or using the procedure for the cases when particles cannot be observed.

9.2.1 Particle Tracking Between Two Frames

(1) Basic Methods

First, the microstructures of structural and functional materials (such as those listed in Table 9.2) or the structural components of industrial products are analyzed based on the quantitative geometric analyses from Sect. 8.5, and various parameters based on classical geometry, differential geometry, and modern geometries such as topology are calculated as necessary. Although these can be classified between size, shape, and spatial distribution, the parameters associated with size and shape are used here. Methods by Nielsen et al. use the center of gravity and volume [5]. The author's group uses the center of gravity, volume, and surface area due to the consideration of shape effects [6]. Furthermore, the internal structure is assumed to have either size distributions, shape distributions, both, or neither, due to characteristics of the sample conducting tracking. For these reasons, the authors defined the matching probability parameter mentioned below, determined the optimal values of the interior parameters α, β, γ ($\alpha + \beta + \gamma = 1$) for each sample and image pair, and assured the accuracy of particle tracking [6].

Table. 9.3 Summary of growth and annihilation patterns, applications, and actual examples of internal structures (e.g., particles, pores, or regions when particle tracking is needed). The black arrows in the rightward direction indicate particle tracking that should be conducted in the forward direction of time and the black arrows in the leftward direction indicate particle tracking that should be conducted back in time. The methods in Sect. 9.2.1 (1) are used for both cases. The large gray arrows refer to steps where particle tracking methods for "non-visible particles" (Sect. 9.2.1 (5)) need to be conducted

Pattern	Imaging steps for obtaining 3D images (Time series) Step *n* - 2 | Step *n* - 1 | Step *n* | Step *n* + 1 | Step *n* + 2	Examples
A: Growth or expansion	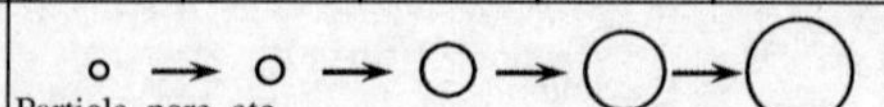	- Growth/expansion of particles - Migration of particles
B: Nucleation and growth	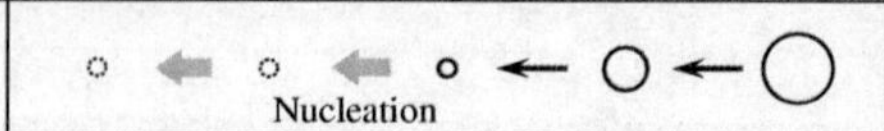	- Precipitation of particles - Initiation of damage or cracking - Assembly of micro parts
C: Annihilation	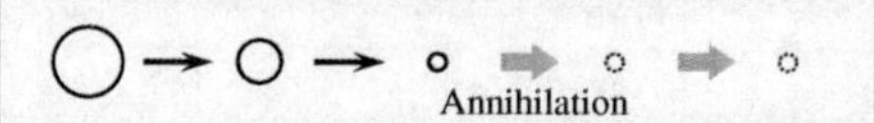	- Pore collapse or particle dissolution - Crack closure - Falling-off of a part
D: Growth and coalescence	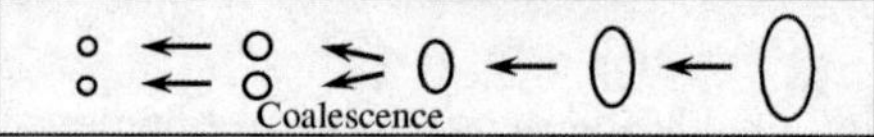	- Coalescence of particles due to growth - Particle coalescence due to migration - Imaging of neighboring parts
E: Division		- Fracture of particles or parts
F: Renucleation after annihilation	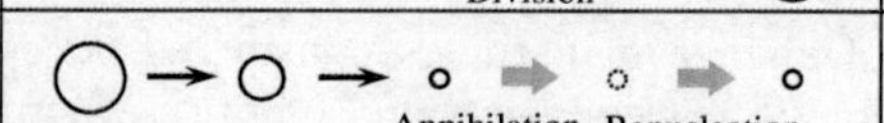	- Precipitation after particle dissolution - Collapse of pores and subsequent renucleation
G: Observation before nucleation or invisible		- Prediction of fracture path before fracture using subsequently formed fracture surface - Invisible nucleation process

$$M_P^{i,j} = \alpha M_L^{i,j} + \beta M_S^{i,j} + \gamma M_V^{i,j} \tag{9.8}$$

Here, $M_P{}^{i,j}$ is set as the matching probability parameter between the particle labeled i in the 3D image at the mth stage and the particle labeled j in the 3D image at the (m + 1)th stage. Furthermore, the parameters relating to the center of gravity, surface area, and volume are set as $M_L{}^{i,j}$, $M_S{}^{i,j}$, and $M_V{}^{i,j}$, respectively. $M_S{}^{i,j}$ and $M_V{}^{i,j}$ are normalized by taking the surface area and volume ratios, respectively, of particles i and j. $M_L{}^{i,j}$ is defined as follows [6]:

$$M_L^{i,j} = \begin{cases} (r_s - d_{i,j})/r_s (d_{i,j} \le r_s) \\ 0(d_{i,j} > r_s) \end{cases} \tag{9.9}$$

Here, $d_{i,j}$ is the distance between particles i and j after registration, r_s is the search range for particle tracking, using optimal values determined in advance based on the given sample [6].

(2) Cluster Matching

Without exceptions, particle tracking accuracy must be as close to 100% as possible for the 3D mapping of mechanical quantities. The reason for this is clear with 3D strain mapping as an example; here scenarios of two different particles in two consecutive images erroneously identified as the same would result in calculations of massive strain at these locations, which would then be assessed as the weakest spots. Particle tracking must be conducted without error even in areas where particles are densely packed together to avoid this issue. The authors use *cluster-matching* to achieve these objectives [6].

The *modified spring model*, which is a type of cluster matching, uses a total of N_{sp} hypothetical springs that connect a large cluster of particles, as shown in Fig. 9.4. The sum of elastic energy E_{sp} of all the hypothetical springs is calculated from the relative vectors determined from the positions of particles in the anterior and posterior 3D images (b_k and $b_k{}'$, respectively).

$$E_{sp} = \frac{1}{N_{sp}} \sum\nolimits_{k=1}^{N_{sp}} \frac{|b'_k - b_k|}{|b_k|} \tag{9.10}$$

E_{sp} is then calculated by sequentially changing the correspondence relation of particles between 3D images before and after deformation with data suspected to be matched in error due to relatively close values of $M_{\mathrm{P}}{}^{i,j}$. The particle mapping is ultimately determined so that E_{sp} is minimized. This enables the mapping of the same particle between two consecutive 3D images even if local heterogeneous deformation occurred in locations where the particles are clustered.

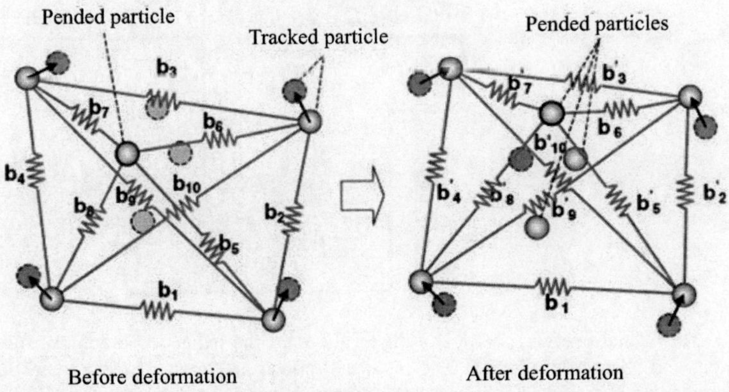

Fig. 9.4 Schematic of a modified spring model that is locally applied to regions where particles are clustered. Adjacent particles are connected with a hypothetical spring and are mapped to minimize the total elastic energy before and after the deformation of the sample in this region [6]

(3) Particle Tracking Prediction

Particle tracking can be difficult in the presence of large and heterogeneous sample deformation represented by necking deformation of plastically deforming metals. In these cases, conducting particle tracking prediction, in addition to the methods mentioned in the previous section, is effective [7]. The simplest tracking prediction is to apply the deformation experienced at the macroscale by the sample (e.g. tension, shearing, or torsion) in the moving image. Other reports have indicated methods that predict particle deformation using the elastic and plastic stress field solutions of the crack tip [7], as well as particle tracking methods not constrained by micro-/macroscale sample deformation systems, individually formulating the trajectories of each particle in a nonlinear manner [7].

(4) Application Examples

A schematic of an experiment where in situ observations of expanded graphite were conducted after compression is shown in Fig. 9.5 [8]. Expanded graphite derives its name from how sulfuric acid penetrates between the layers of graphite, causing the interlamellar spaces to expand in the c-axis direction to form worm-like shapes.

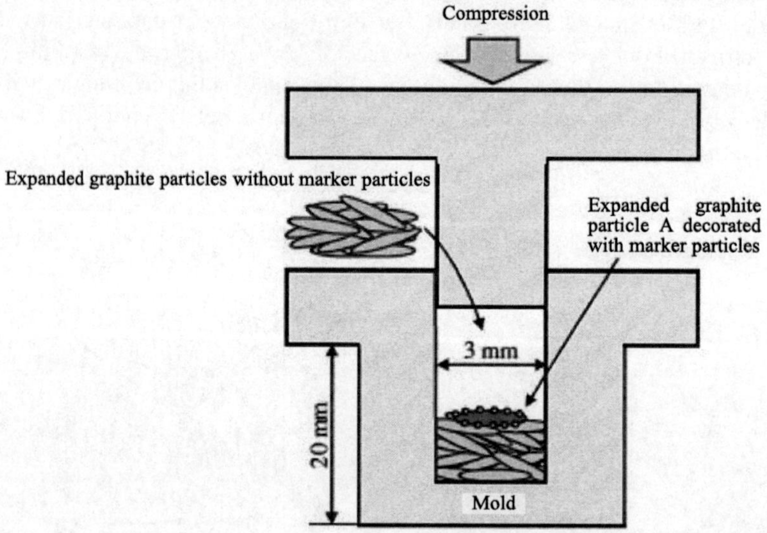

Fig. 9.5 Experimental overview of in-situ observations of the expanded graphite solidification, compression, and restoration process. The mold and punch are made of a polymer. Successive in-situ observations of the process were taken with projection-type X-ray tomography at SPring-8. There are no structures/microstructures that can be identified in the expanded graphite interior or surface with X-ray tomography, so the deformation behavior was quantitatively determined by attaching marker particles to the expanded graphite surface (A in the top figure) [8]

The elastic repulsion when this material is compressed is widely used as a sealant in various industrial fields. However, its shape repulsion mechanism has not clarified until now, as the internal structure of the material is complex and in situ observations of its deformation processes could not be conducted. With this in mind, the authors created a mold and punch with polymers that have high X-ray transmissivity. In situ observations of a series of processes comprising solidification, compression, and recovery of the expanded graphite packed inside were conducted with projection-type X-ray tomography at SPring-8 [8]. This corresponds to Pattern A in Table 9.3. As there are no structures/microstructures that can be identified with X-ray tomography on the interior or surface of the expanded graphite, not only can the deformation be quantified but the shape of the expanded graphite particles cannot be visualized, particularly after solidification. As shown in Fig. 9.6a, b, marker particles were artificially affixed beforehand on the expanded graphite surface, its deformation behavior visualized as shown in Fig. 9.6c, d, and quantitatively determined as shown in Fig. 9.6e [8].

As shown in Fig. 9.6a, the shapes of expanded graphite particles after solidification do not have a worm-like shape as reported up until now but rather in the shape of finely divided sheets. The expanded graphite sheets curved significantly during the solidification process due to adjacent worms, after which it deformed to recover the curved section during compression. Simple compression and recovery in the c-axis direction are conventionally considered to be the expanded graphite particle recovery mechanism; however, strains formed in a complex manner within the expanded graphite particles due to 3D deformation and strain distributions, showing that interactions such as the expansion, curving deformation, and friction acting between the expanded graphite particles combined to bring about its recovery properties [8].

Next, let us consider the analysis of material fracture. There are indications present before the formation of macroscale fractures in the material, as well as slight signs before this. The signs during the initial stage are no more than slight changes generally difficult to detect or assess. However, it should be possible to qualitatively determine the signs in the initial stage in a set of 3D images, which accurately record this sign by repeatedly reviewing these images. Taking this a step further, tracking microscale cracks and damage back in time to the stage immediately before those fractures were clearly observed allows for the specification of cracking and damage initiation points and the correct assessment of coalescence. This corresponds to Patterns B and D in Table 9.3.

Figure 9.7 shows the damage visible when Al–Mg alloys were tensioned at relatively high speeds and temperatures, as well as their development [9]. An extremely large number of defects can be observed immediately before fracture but the coarse defects, which dominated sample fracture, occurred in the initial stage. Tracking damage back in time enables the accurate specification of determining when all defects occurred. Figure 9.8 shows a summary of these types of analysis results. In this case, the liquid–metal wetting technique introduced in Sect. 5.5.1 was applied immediately before fracture and a 3D image of a crystallographic grain boundary,

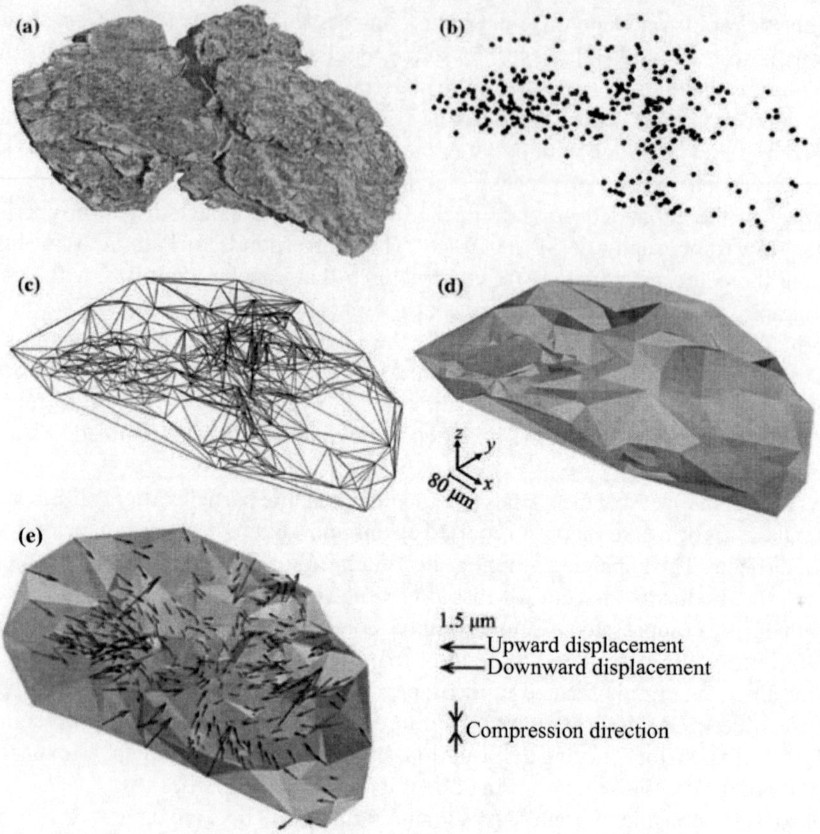

Fig. 9.6 **a** A single expanded graphite particle and the **b** marker particles affixed on its surface (center of gravity locations indicated with black circles) with the expanded graphite particle shape reconstructed by enclosing these dots with surfaces, shown in **c** and **d**; **e** shows the vectors indicating the displacement of each marker point following compaction, compression, and recovery. The original figure is in color, so please refer to the figure of the original publication for details. Bending deformation exceeded the recovery (expansion) in the plane-normal direction and the deformation during the recovery of the expanded graphite particles is shown to be highly complex [8]

as shown in Fig. 9.7b, was obtained. This is used to investigate all voids, specifically to check whether the damage initiation point was at the crystallographic grain boundary or within the grains, and whether they were caused by dispersed particles. The behavior of actual materials is extremely complex, observing a 3D image. It is often the case that the predominant factor cannot be determined. However, these analyses showed that voids initiated due to the presence of dispersed particles on the crystallographic grain boundaries expanded significantly and were directly linked to the ultimate fracture of the material. The validity of these types of results can be easily verified by reviewing 3D images as shown in Fig. 9.7b. This can be considered the main strength of image-based analysis.

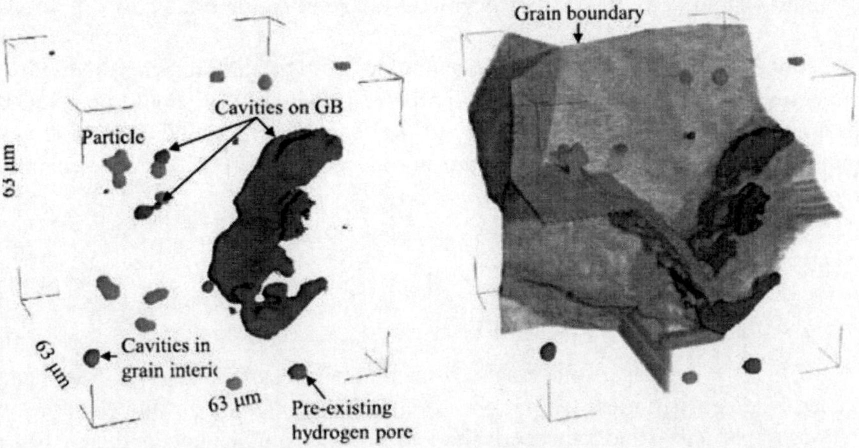

(a) Voids and particles observed during deformation (b) Grain boundary image superposed on (a)

Fig. 9.7 Observation results of damage nucleation and growth behavior during Al–Mg alloy tension tests at temperatures of 500 °C and a strain rate of 10^{-2} s^{-1} [9]; **a** shows the voids and particles at a given observed region during deformation. The right-hand figure showed the superposition of a 3D image of crystallographic grain boundaries obtained from the liquid–metal wetting technique. The original figure is in color; please refer to the figure in the original publication for details

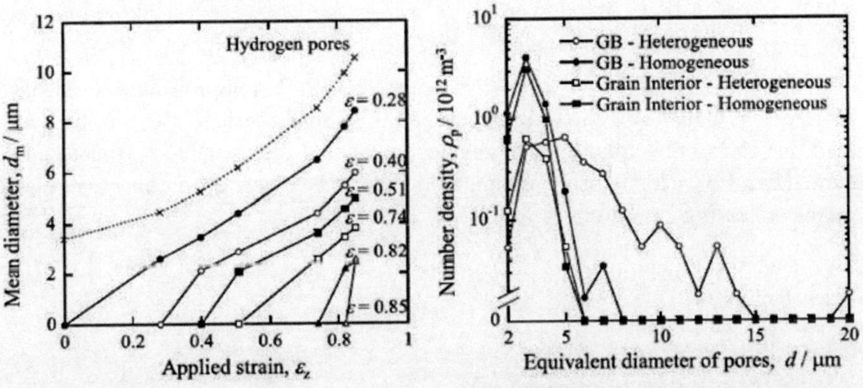

(a) Growth behavior of voids nucleated at each loading step

(b) Histograms of void diameter at an applied strain of 13%, which have been classified according to its nucleation sites (grain boundary or grain interior) and modes (heterogeneous nucleation (i.e. nucleation on particles) or homogeneous nucleation independent of particles).

Fig. 9.8 Analysis results of the 3D image sets in Fig. 9.7 [9]. In **a**, all observed voids were tracked back through time, the initiation stage was specified and the growth behavior through each initiation stage was summarized; **b** shows whether the initiation points were due to crystallographic grain boundaries or the particles, in the form of histograms. Voids on crystallographic grain boundaries, which were generated due to particles, have coarse growth

(5) Non-visible Particle Tracking Methods and Application Examples

Detailed methods among those corresponding to "(3) Particle Tracking Prediction" can continuously track the position of particles to where they should be, even if they are no longer visible. This corresponds to Patterns B, C, F, and G in Table 9.3. Examples of these types of methods are introduced here [7]. These are generally referred to as *3D scattered data interpolation*.

$$u_i = \sum_{j=1}^{N} c^j g(\|p_i - p_j\|) \tag{9.11}$$

Here, u_i is the particle displacement; N the number of particles; $\|\cdot\|$ the Euclidean distance between particles; g $(\|\cdot\|)$ the *radial basis function*; p_i and p_j are the positions of particles i and j, respectively; and c^j is the complement coefficient to be determined. The radial basis function has a form similar to that in Fig. 9.9, is a function of distance r from the center, and is dependent on the coefficient k. The displacement field is generally expressed for all particles using the radial basis function, where each particle is at the center.

$$\begin{pmatrix} u_x^1 \cdots u_x^N \\ u_y^1 \cdots u_y^N \\ u_z^1 \cdots u_z^N \end{pmatrix} = \begin{pmatrix} c_x^1 \cdots c_x^N \\ c_y^1 \cdots c_y^N \\ c_z^1 \cdots c_z^N \end{pmatrix} \begin{pmatrix} g(\|p_1 - p_1\|) & \cdots & g(\|p_1 - p_N\|) \\ \vdots & \ddots & \vdots \\ g(\|p_N - p_1\|) & \cdots & g(\|p_N - p_N\|) \end{pmatrix} \tag{9.12}$$

Equation (9.12) is always a positive constant using an appropriate radial basis function (e.g. Gaussian function) and the complement coefficient c^j can be determined by solving the equation. The displacement u of the invisible particles can be assumed to follow a local displacement field if c^j can be determined; this is calculated with the following equation:

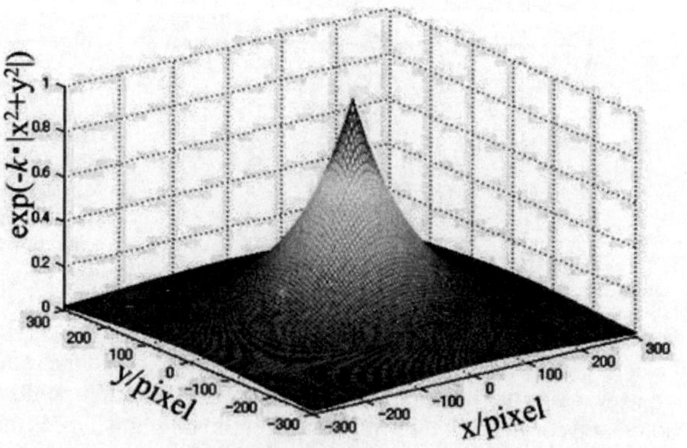

Fig. 9.9 Radial basis function shape. This shape varies widely depending on the coefficient k

$$u = \sum_{i=1}^{N} c^i g(\|P - p_i\|) \tag{9.13}$$

Here, P is the initial position of the non-visible particles. This method has been used for assessments on whether hydrogen micropores temporarily eliminated from shot peening [10] or hot rolling [11] re-precipitate as hydrogen gas, as well as in research specifying the fracture initiation points in 3D images before loading using fracture surface trajectory predictions back in time [12]. Figure 9.10 shows pores that were temporarily eliminated due to hot rolling and regenerated as a result of hydrogen gas precipitation in the same location with re-heating (pore C in the figure) [11]. Equation (9.13) can be used to identify whether the pores prior to annihilation and after regeneration are identical.

Figure 9.11 is a research example specifying microstructures that became fracture initiation points in a high-strength aluminum alloy [12]. The fracture surface first appears only after fracture and does not exist in the material prior to this event. Weak microstructures steadily receive damage during plastic deformation of the material

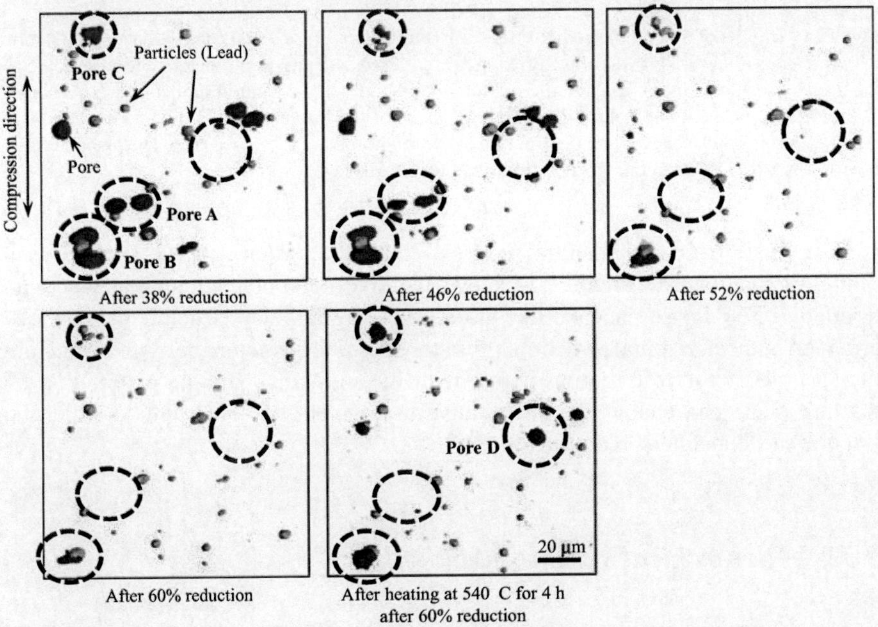

Fig. 9.10 Annihilation, remaining, and regeneration behavior of internal pores when an Al-4 Mg-0.05Pb alloy was rolled up to a maximum reduction rate of 60% at room temperature and then subsequently held at high temperature [11]. Pore A was eliminated through rolling and did not regenerate even with heat treatment. Pore B was not annihilated even with rolling and remained throughout the experiment. Pore C was temporarily annihilated through rolling and later reinitiated in the same position with heat treatment. Pore D was newly formed after heat treatment. The original reference has the figure in color; please refer to this for details

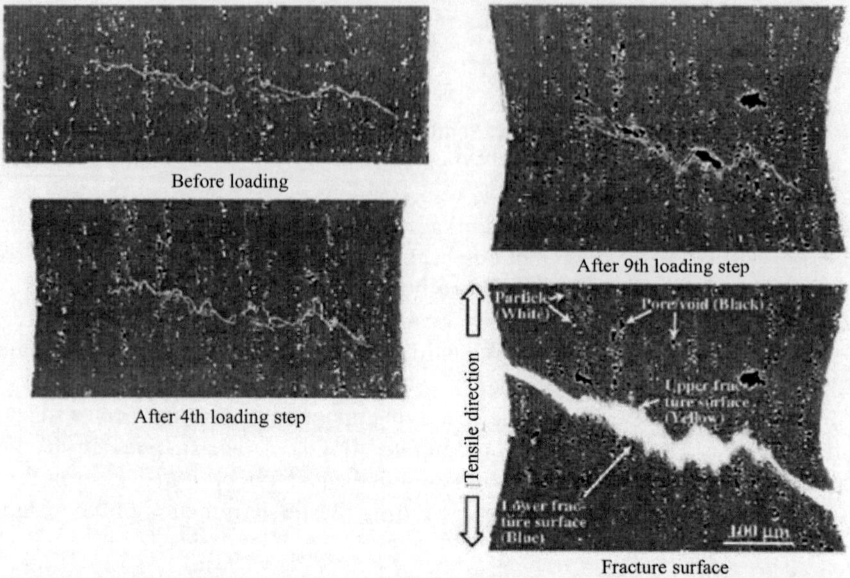

Fig. 9.11 Results of trajectory prediction by following the fracture surface coordinates back in time to the pre-load images. This was done by using 3D scattered data interpolation after the coordinates of the fracture surface (crack) from ductile fracture in the high-strength aluminum alloy were recorded as high-density coordinate points. The microstructure of the fracture plane could be matched with that of the pre-loading image and the fracture initiation point could be specified. The original figure is in color; please refer to the original reference for details [12]

and eventually result in fracture. It is extremely important to specify these weak initial microstructures. Figure 9.11 shows that hydrogen pores in regions where the terminals of elongated pancake-like grains in the initial microstructure happen to be arranged in the horizontal direction ultimately dominate fracture behavior. Matching the specified fracture initiation points with the individual dimple patterns on the fracture plane can enable the quantitative assessment of the extent to which the fracture mechanism is predominant.

9.2.2 Hierarchical Tracking Methods

Aoki and Nakazawa have used approximately 1% of all particles as landmarks to conduct matching between 3D images using relaxation matching [7, 13]. Furthermore, they referred to the remaining particles as non-landmarks, the landmark matching results are used for matching the remaining particles from the estimation of mobile vectors using radial basis functions. This enables 3D imaging and particle tracking to be conducted simultaneously [7, 13].

The landmark particles in the image before and after the present time are set as $A = \{a_1, a_2, \ldots, a_N\}$ and $B = \{b_1, b_2, \ldots, b_M\}$, respectively. The probability vector of the particles a_i in A and B is expressed as $p_i = \left[p_{i(1)}, p_{i(2)}, \ldots, p_{i(M)}\right]^T \hat{\text{I}}\text{R}^M$. Here, $\sum_{k'} p_{i(k')} = 1$. The probability vector is initialized as follows when the volume of particle a_i and particle b_k (V_i and V_k, respectively) are under the allowable value RC [7]:

$$p_{i(k)}^{0'} = \begin{cases} 1 \text{ if } \left|\frac{V_k - V_i}{V_i}\right| \leq \text{RC} \\ 0 \text{ otherwise} \end{cases} \qquad p_{i(k)}^{0} = \frac{p_{i(k)}^{0'}}{\sum p_{i(k')}^{0'}} \tag{9.14}$$

Next, the probability matrix at time t is updated with the updating coefficient $q_i(k)$ [7]:

$$p_{i(k)}^{t+1} = \frac{q_{i(k)} p_{i(k)}^{t}}{\sum q_{i(k')} p_{i(k')}^{t}}, \quad q_{i(k)} = \sum_{j \in \zeta_i} \max\left(\gamma_{\text{ij}}(k, l) p_{j(l)}^{t}\right) \tag{9.15}$$

Here, $\gamma_{ij}(k, l)$ is the assessment function that determines matching success; $e_{ij}(k, l)$ is expressed as follows by setting the distance between two landmark particles as d [7]:

$$\gamma_{\text{ij}}(\text{k}, \text{l}) = \max\left(1 - \tfrac{1}{W} e_{\text{ij}}(\text{k}, \text{l}), 0\right) \quad e_{\text{ij}}(k, l) = \frac{\|d_{\text{kl}} - d_{\text{ij}}\|}{\|d_{\text{ij}}\|} \tag{9.16}$$

As shown in Eq. (9.15), the most certain match between particles A and B is the combination where the probability vector at the current time multiplied by $\gamma_{ij}(k, l)$ is maximized; $\gamma_{ij}(k, l) = 1$ if $e_{ij}(k, l) = 0$, and $\gamma_{ij}(k, l) = 0$ if $e_{ij}(k, l)$ is above its critical value W.

The non-landmark matching method conducted afterwards is the same as that in Sect. 9.2.1 (5) [7].

9.2.3 3D Strain Mapping

(1) Method

Several dozen to a hundred thousand particles can typically be tracked in an aluminum alloy for a single field of view in a single 3D image using methods from Sects. 9.2.1 and 9.2.2. Next, the sample interior can be divided into multiple tetrahedrons with the tracked particles as vertices by using the Delaunay tessellation introduced in Sect. 8.3.4. Assuming the deformation is small, the vertical strain (ε_x, ε_y, ε_z) and shear strain (γ_{xy}, γ_{yz}, γ_{zx}) in a single tetrahedron can be calculated from the particle deformation (u_i, v_i, w_i), (u_j, v_j, w_j), (u_k, v_k, w_k), and (u_l, v_l, w_l) at four vertices *i, j, k,* and *l* [6, 7].

$$\{\varepsilon_x \ \varepsilon_y \ \varepsilon_z \ \gamma_{xy} \ \gamma_{yz} \ \gamma_{zx}\}^t = [\mathrm{B}_i \ \mathrm{B}_j \ \mathrm{B}_k \ \mathrm{B}_l]\{u_i \ v_i \ w_i \ u_j \ v_j \ w_j \ u_k \ v_k \ w_k \ u_l \ v_l \ w_l\}^t \tag{9.17}$$

Here, $[\mathbf{B}_i]$ from Eq. (9.17) is defined as follows:

$$[B_i] = \frac{1}{6V}\begin{bmatrix} o_i & 0 & 0 \\ 0 & p_i & 0 \\ 0 & 0 & q_i \\ p_i & o_i & 0 \\ 0 & q_i & p_i \\ q_i & 0 & o_i \end{bmatrix} \tag{9.18}$$

Here, V is the volume of the tetrahedron and o_i, p_i, and q_i are as follows:

$$o_i = y_k z_j + y_l z_k + y_j z_l - y_k z_l - y_j z_k - y_l z_j \tag{9.19}$$

$$p_i = x_l z_j + x_j z_k + x_k z_l - x_j z_l - x_l z_k - x_k z_j \tag{9.20}$$

$$q_i = x_l y_k + x_j y_l + x_k y_j - x_j y_k - x_k y_j - x_l y_k \tag{9.21}$$

Here, $x_j, y_j, z_j, x_k, y_k, z_k, x_l, y_l$, and z_l are the coordinates of the tetrahedron vertices j, k, and l. $[\mathbf{B}_j]$, $[\mathbf{B}_k]$, and $[\mathbf{B}_l]$ can be similarly determined by substituting $i \to j$, $j \to k$, $k \to l$, and $l \to i$, respectively.

Furthermore, *digital volume correlation* can also be used for 3D strain mapping. Commercially available software is also sold for this. Details can be found in the referenced article [7].

(2) Application Examples

There are virtually no methods that visualize plastic strains in 3D, therefore, it is often the case that applications of 3D strain mapping enable the observation of specific behaviors in the material interior that have not been identified until now. Figure 9.12 shows ex-situ *observations* of the aluminum alloy hot-rolling process [11]. Compared to adding external disturbances on a sample rotation stage during the in-situ observations introduced in Sect. 4.5.1, ex-situ observations involve the application of external disturbances after the sample is removed from the sample rotation stage and then returning it to that stage, after which observations are repeated. Absorption contrast tomography is used in this case, but there are not many particles with which particle tracking can be conducted in these materials. For these reasons, fine and soft lead particles, which do not influence the compression deformation that occurs during rolling in the material preparation process, are artificially added and dispersed at a microscale and high-density level. This enables 3D strain mapping. The aluminum alloy is not homogeneously compressed during rolling and instead has the shear band run centrally through the alloy, which induces local tension; a representative

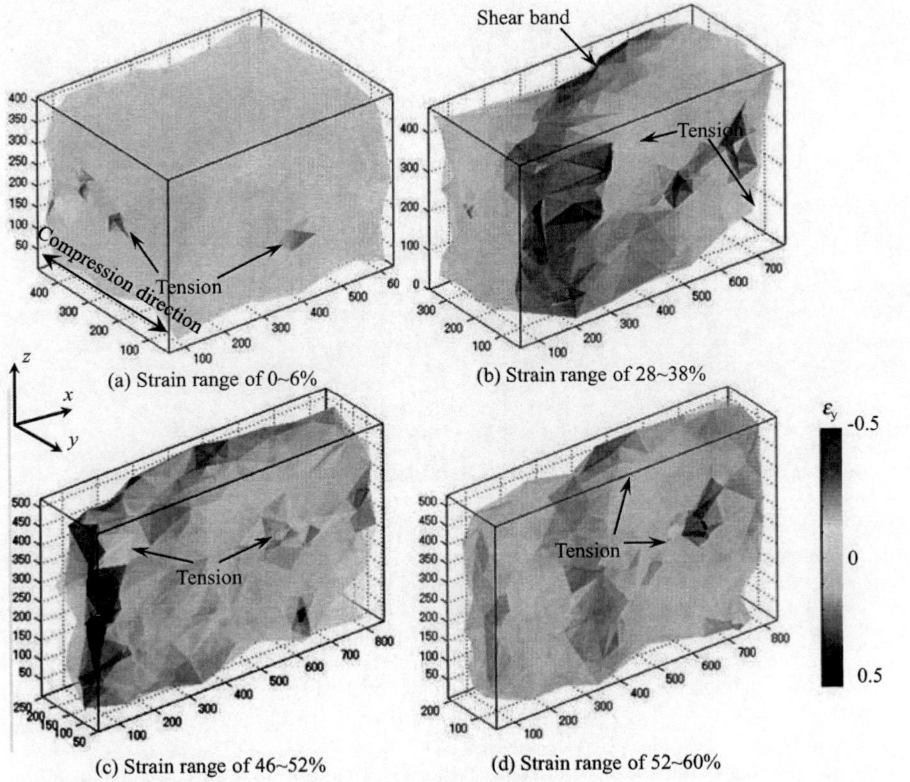

Fig. 9.12 Application of 3D strain mapping to assess the annihilation, remaining, and recovery behavior of internal pores when an Al-4 Mg-0.05 Pb alloy was rolled to a maximum reduction ratio of 60% at room temperature; it was then held at high temperatures. Normal strains in the compression direction due to rolling are shown. The reference has a color figure for the same observation; please refer to this for details [11]

example is shown in Fig. 9.12b [11]. This causes manufacturing defects to remain in the material without collapsing, regardless of whether the material is sufficiently rolled.

Strain mapping has a variety of uses that are not merely limited to the visualization of plastic deformation. Figure 9.13 is an example of 3D strain mapping during tensile experiments of Al–Zn–Mg alloys with a high Zn content (approximately 10%), which are likely to exhibit hydrogen embrittlement [14]. The localization of plastic strains in a gauge section can be clearly observed from the early stages of tension through this 3D mapping. Furthermore, high-density nanovoids with diameters of several nm to approximately 10 μm are present in the material. This is clarified with multi-scale observations that combined the three techniques of transmission electron microscopy observations, imaging X-ray nanotomography using the Fresnel zone plate introduced in Sect. 5.1.1 and standard X-ray microtomography [14]. The material should undergo slight expansion with the formation of nanovoids.

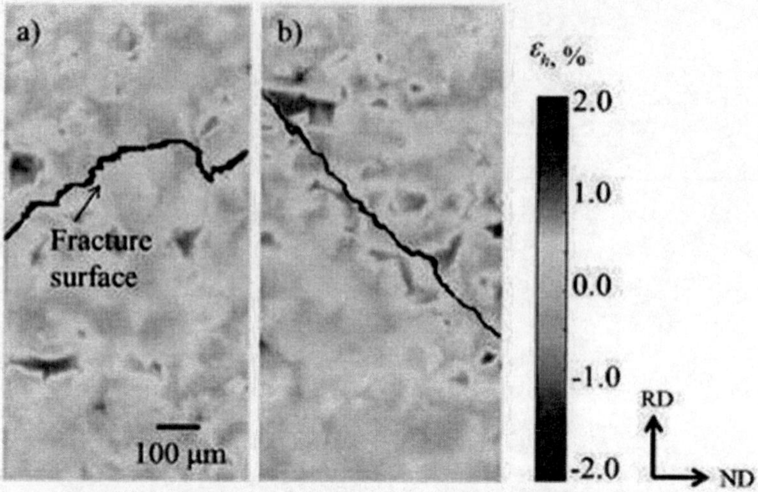

Fig. 9.13 Hydrostatic strain distribution when Al–Zn–Mg alloys with high Zn content are hydrogen-charged with electric discharge machining during tensile tests; **a** shows when the load strains are between 2.1–6.8% (holding time of 0.53 ks) and **b** shows when the load strains are between 1.1–5.5% (holding time of 3.34 ks). Both show hydrostatic tensile strain with a maximum of over 1% and an average of around 0.2%. These are due to nanovoids with a diameter of several nm to μm that are generated at a high density in the material interior. The original figure is in color; please refer to the figure in the original publication for details [14]

Figure 9.13 shows the spatial distribution of this expansion in the form of hydrostatic strains using 3D strain mapping.

Finally, dislocation density measurements using 3D strain mapping are introduced. Plastic strains occur due to dislocation activity and are distinct only in their nano- or micro-scale perspectives. Figure 9.14 shows the calculations of *statistically stored dislocation (SSD)* and *geometrically necessary dislocation (GND)* using 3D strain mapping [15]. For example, the SSD density ρ_{SSD} is calculated as follows from the equivalent plastic strain ε_{p} [16]:

$$\rho_{SSD} = \frac{\sqrt{3}\varepsilon_p}{bl} \tag{9.22}$$

Here, l is the mean free path of the dislocation and b is the Burgers vector. The GND density ρ_{GND} is as follows [17]:

$$\rho_{GND} = \bar{r}\frac{\eta_p}{b} \tag{9.23}$$

Here, $\bar{r}$ is the Nye coefficient (1.9) and η_{p} is the plastic strain gradient. Figure 9.14 shows the increases in atomic vacancy concentrations due to the motion of screw dislocation, including jogs. As dislocation density in the material can be calculated in the form of 3D maps, assessments of how the hydrogen is being trapped in hydrogen

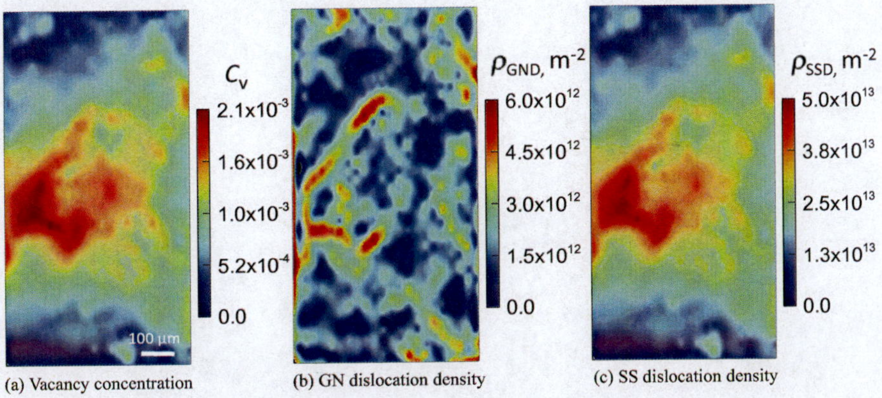

Fig. 9.14 Calculations of dislocation density and atomic vacancy concentration distribution using the data for 3D strain mapping when the Al–Zn–Mg alloys with high Zn content were hydrogen-charged with electric discharge machining during tensile experiments. The load strain is 4.4%. The densities of screw dislocation and edge dislocations in both SS and GN dislocations were assumed to be equivalent for calculations. The original figure is in color; please see the original reference for details [15]

trap sites (e.g. dislocations, crystallographic grain boundaries, precipitates, atomic vacancies, and pores) can be conducted at the local region level, such as at the crack tip [15].

9.2.3.1 Local Fracture Resistance Mapping

(1) Particle Tracking-Based Methods

If the particle tracking in Sect. 9.2 is achieved, then this can be applied to other methods besides 3D strain mapping. One example of this is the direct measurement of fracture mechanics parameters such as the stress intensity factor and the J-integral. To the best of the authors' knowledge, no methods exist that directly measure local fracture resistance along the crack front or driving forces of crack propagation, other than those that combine X-ray tomography and particle tracking. Figure 9.15 shows the distribution of fatigue cracks and micropores near the crack front [18]. The authors sought to conduct high-density mapping of the stress intensity factor by tracking these pores during early synchrotron radiation X-ray tomography experiments [18]. Part of these results is shown in Fig. 9.16 [18]. This is the mapping of the driving force of mode I (opening) crack propagation. Elastic fracture mechanics generally relates modes' I, II, and III stress intensity factors k_{I}, k_{II}, and k_{III} with the displacements u_x, u_y, and u_z in the x, y, and z directions, respectively, as follows:

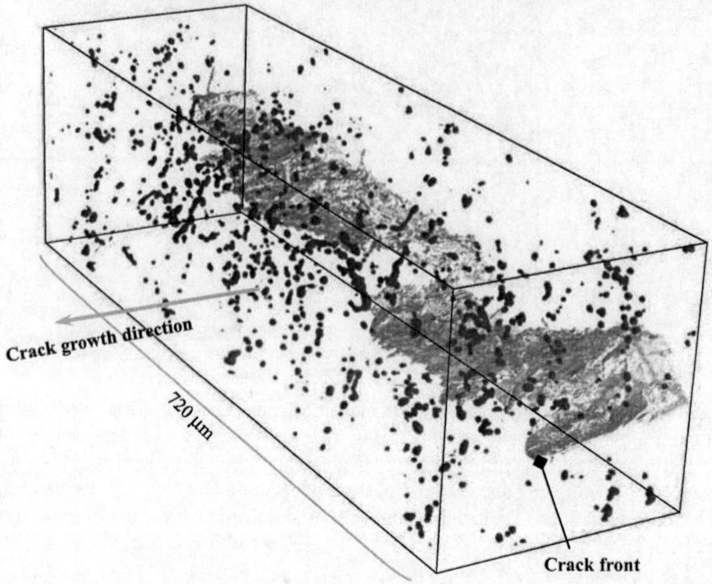

Fig. 9.15 Magnification of the crack tip section when a fatigue crack was induced in an A2024 aluminum alloy [18]. The black dots are pores, the average diameter 3.6 µm, and there are 805 pores total in the above enclosure. These are the earliest experiments where the fracture mechanics parameters were determined with particle tracking

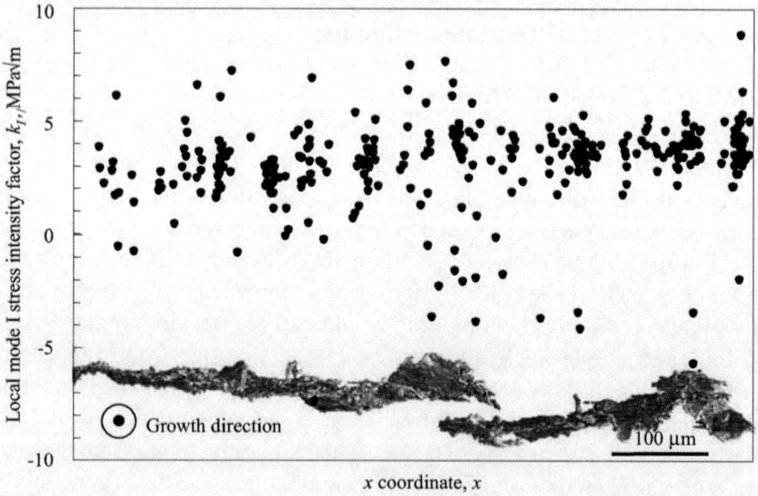

Fig. 9.16 Determination of the fracture mechanics parameters through inverse analysis, using the physical displacements obtained by tracking the pores near the crack tip seen in Fig. 9.15 [18]. This is the first research of its kind conducting high-density 3D mapping of fracture mechanics parameters like the stress intensity factor along the crack front

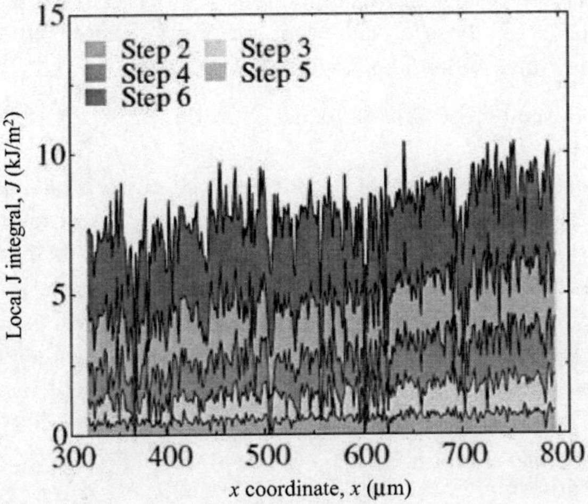

Fig. 9.17 Results of preparing fatigue pre-cracked A2024 aluminum alloy materials, where the J integral distribution along the crack front was mapped in 3D using in-situ observations during fracture toughness experiments [19]. Changes in the distribution of driving force for crack propagation can be seen as the crack propagated

$$u_x = \frac{1+\nu}{E}\sqrt{\frac{2r}{\pi}}\left\{k_I\cos\frac{\theta}{2}\left[1-2\nu+\sin^2\left(\frac{\theta}{2}\right)\right]+k_{II}\sin\frac{\theta}{2}\left[2(1-\nu)+\cos^2\left(\frac{\theta}{2}\right)\right]\right\}+\left(1-\nu^2\right)\frac{\mathrm{Trcos}\theta}{E} \tag{9.24}$$

$$u_y = \frac{1+\nu}{E}\sqrt{\frac{2r}{\pi}}\left\{k_I\sin\frac{\theta}{2}\left[2(1-\nu)-\cos^2\left(\frac{\theta}{2}\right)\right]+k_{II}\cos\frac{\theta}{2}\left[1-2\nu-\sin^2\left(\frac{\theta}{2}\right)\right]\right\}-\nu(1+\nu)\frac{\mathrm{Trcos}\theta}{E} \tag{9.25}$$

$$u_z = 2\frac{1+\nu}{E}\sqrt{\frac{2r}{\pi}}k_{III}\sin\frac{\theta}{2} \tag{9.26}$$

Here, r and θ are the distance and angle from the crack tip to the particle, E is Young's modulus, ν is Poisson's ratio, and T is the T stress. A single pair of local stress intensity factors can be calculated from the displacement of a single particle. The driving force for crack propagation along the crack front is extremely heterogeneous; this results in areas where crack propagation is greatly advanced or has outpaced other areas, becoming more intricate and yielding complex crack front morphologies. Furthermore, mapping of the driving force of crack propagation was conducted for mode II (in-plane shear) and mode III (antiplane shear) in reference [18]. This research visualized the local behavior of the crack closure processes, which are extremely important to the understanding of fatigue cracks. This research also directly linked the generation of the roughness-induced crack closure with the source of this phenomenon, i.e., the changes in the local driving force of crack propagation in both modes II and III [18].

Figure 9.17 shows these analyses in the range of elastic fracture mechanics [19]. The equations used for these calculations are omitted here, but the principles of derivation for the J-integral are the same as in Fig. 9.16.

(2) Analyses Based on the Crack Opening Profile

The crack-tip opening displacement (CTOD) is a parameter that assesses the extent of plastic deformation by using the opening amount of the crack tip, whose crack opening occurs due to plastic deformation near the tip. As is the case with the stress intensity factor and J-integral, the CTOD is used as a fracture criterion by determining its critical value. However, there are no experimental measurement techniques for directly measuring the CTOD aside from the specimen surface, therefore, this must be estimated from the opening amount of the crack mouth, which can be measured. However, CTOD distributions near the crack front can be directly and accurately measured if X-ray microtomography is used and the spatial resolution of the 3D image is sufficiently smaller than the opening amount.

Figure 9.18 shows examples of CTOD measurements in fracture toughness experiments of aluminum alloys [20]. Considering crack deflection, a linear segment close to the crack tip is detected and an algorithm that measures CTOD is prepared by specifying the crack tip based on this detection. Figure 9.19 shows the driving force for crack propagation along the crack front determined in this manner. This material comprises a dual-phase microstructure, where the eutectic structure in the right half of the figure has a low fracture resistance that allows for cracks to progress ahead of others, while the α - aluminum phase in the center has an extremely high fracture resistance that arrests cracks. CTOD measurements showed that the fracture resistance in the two phases varied by a factor of 10 in practice.

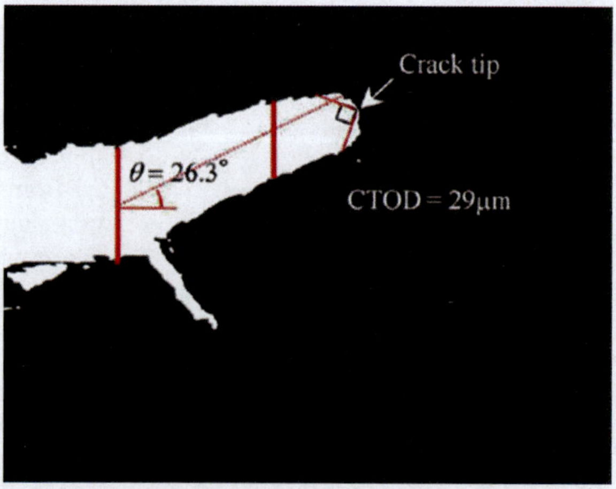

Fig. 9.18 Example of measuring the crack-tip opening displacement (CTOD) in a cross-section when conducting in-situ observations of the fatigue pre-cracked Al-6.7% Si-0.31% Mg aluminum alloy during fracture toughness experiments [20]

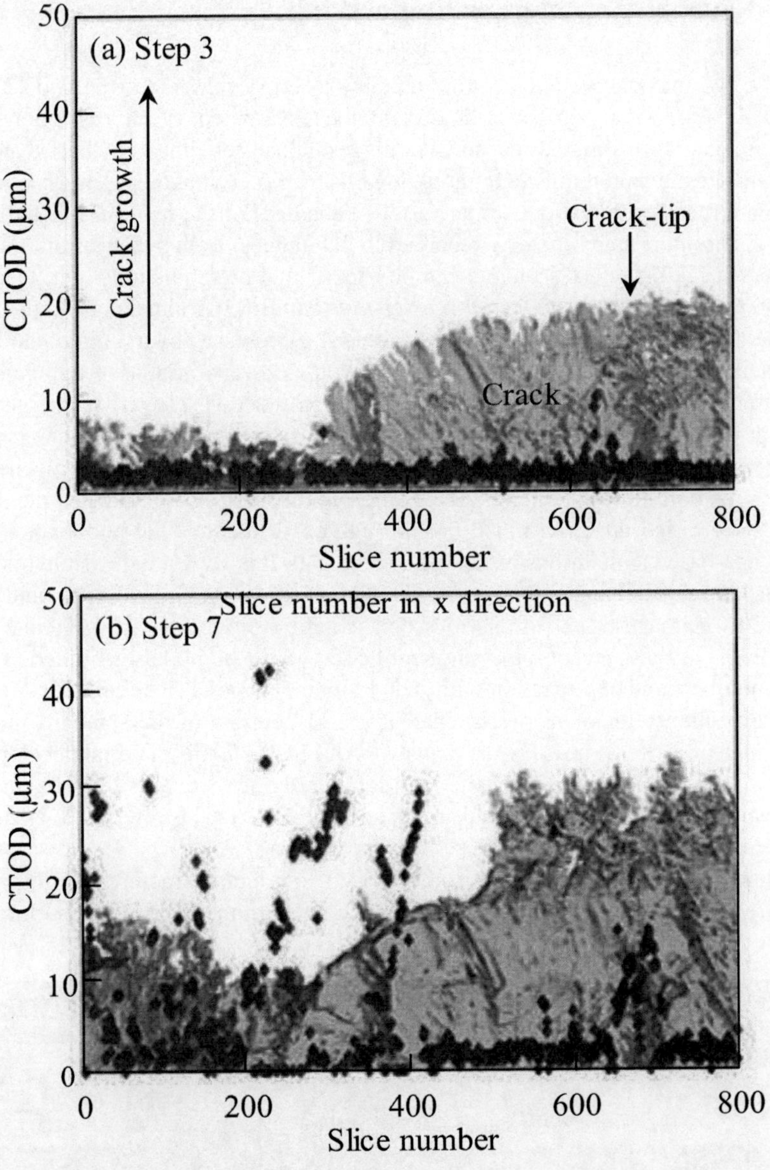

Fig. 9.19 Results of 3D mapping the distribution of CTOD along the crack front when conducting in-situ observations of fatigue pre-cracked Al-6.7% Si-0.31% Mg aluminum alloy during fracture toughness experiments [20]. Changes in the distribution of the driving force for crack propagation can be seen as the crack propagates. The original figure is in color; please refer to the reference publication for details

9.3 Reverse 4D Materials Engineering

Finally, we introduce an X-ray tomography-based development method [21] for materials design that combines the quantitative assessment of internal microstructures, image-based simulations, and 4D image analyses obtained in Chap. 8 and this chapter. The main prerequisite for applying this method is that the microstructures or internal structures that stipulate the various behaviors of the materials or industrial products of interest are properly recorded in 3D images. Both synchrotron radiation and industrial X-ray CT scanners can be used if this prerequisite is met. Once the complex 3D microstructure was expressed with multiple parameters from the quantitative geometric analysis in Sect. 8.5, this can be expressed by a limited number of parameters through coarsening. This becomes the design variable of the material's design. Next, the material characteristic to be optimized is selected as a target variable, after which predictions of the optimal value, worst value, and characteristic value for a given microstructure morphology are made. This is shown schematically in the bottom row of Fig. 9.20. Changes in the target variable when the design variable is varied are determined not through experimental trial and error but the 3D image-based simulations discussed in Sect. 8.6. It is currently possible to obtain high-accuracy 3D images, however, there are high computational costs and work required for image-based simulations that use these images as is, particularly for modeling. For these reasons, the number of computational models is limited to some small numbers and optimization with such limited data points is required. Currently, these aspects are major technical challenges. In contrast to the standard material development process shown in the upper row of Fig. 9.20, this material development process takes a path that is opposite to the standard one that starts from the internal microstructure of existing materials to material design. This is referred to as *reverse 4D materials engineering* (*R4ME*).

Figure 9.21 shows the application of R4ME on ductile fracture due to particle fracture and hydrogen micropore growth in A2024 aluminum alloys. After ultimately

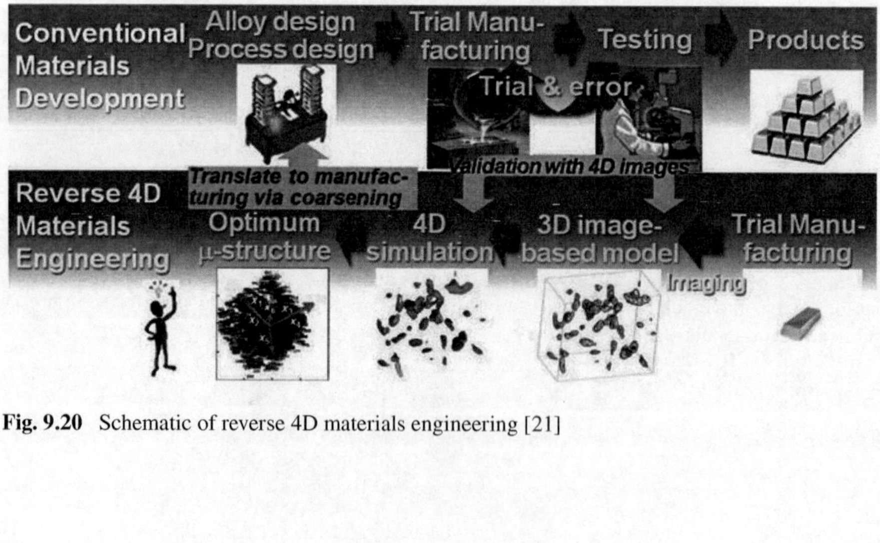

Fig. 9.20 Schematic of reverse 4D materials engineering [21]

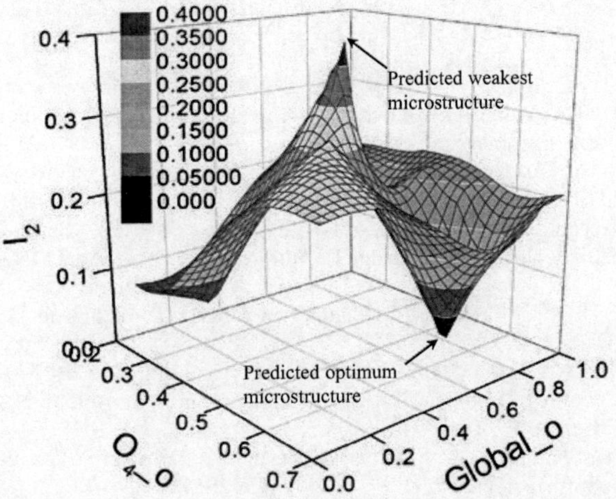

Fig. 9.21 Application of reverse 4D materials engineering with regards to ductile fracture due to the particle fracture and growth of the internal hydrogen micropores in an A2024 aluminum alloy. The volume ratio (I_2) of areas in the aluminum section, which exceed a total equivalent plastic strain value of 0.05, was set as an indicator for how easily the material would fracture (objective function). A total of 45 geometric parameters expressing the particle and pore size, shape, and spatial distribution were used; ultimately five parameters were selected through statistical analysis

coarsening into a combination of particles and pores comprising five design variables, the relationship between these variables and target variables, which express the ease of fracture, are shown with a metamodel. Figure 9.21 shows the selection of two parameters among the five design variables whose correlations with the target variable are strongest and where the response surface was drawn as a 3D graph. The fracture resistance changes significantly with changes in the microstructure; both the optimal and weakest microstructures can be clearly observed. Utilizing X-ray tomography during material development to obtain the optimal structures makes it possible to obtain materials and industrial products with superior performance without relying on specialized equipment or raw materials.

The 3D/4D images obtained from X-ray tomography can solidly reflect the microstructures and internal structures of material and industrial products; phenomena can also be accurately understood through 4D observations and its various accompanying analyses. However, materials, industrial products, and physical phenomena are in reality highly complex, and their true natures tend to be difficult to understand by using the 3D/4D observations. As shown in Fig. 9.21, the relationships between various phenomena and materials or industrial products can be accurately assessed and more easily understood using methods such as R4ME. Furthermore, material design guidelines with higher accuracy can be achieved and their performance increased in this manner.

References

1. B. Zitová, J. Flusser, Image Vis. Comput. **21**, 977–1000 (2003)
2. J.B.A. Maintz, M.A. Viergever, An overview of medical image registration methods, Utrecht University. Inform. Comput. Sci (1998)
3. J.-P. Thirion, Int. J. Comput. Vision **18**, 121–137 (1996)
4. H. Toda, A. Takijiri, M. Azuma, S. Yabu, K. Hayashi, D. Seo, M. Kobayashi, K. Hirayama, A. Takeuchi, K. Uesugi, Acta Mater. **126**, 401–412 (2017)
5. S.F. Nielsen, H.F. Poulsen, F. Beckmann, C. Thorning, J.A. Wert, Acta Mater. **51**, 2407–2415 (2003)
6. M. Kobayashi, H. Toda, Y. Kawai, T. Ohgaki, K. Uesugi, D.S. Wilkinson, T. Kobayashi, Acta Mater. **56**, 2167–2181 (2008)
7. H. Toda, E. Maire, Y. Aoki, M. Kobayashi, J. Strain Anal. Eng. Des. **46**, 549–561 (2011)
8. H. Toda, K. Tsubone, K. Shimizu, K. Uesugi, A. Takeuchi, Y. Suzuki, M. Nakazawa, Y. Aoki, M. Kobayashi, Carbon **59**, 184–191 (2013)
9. H. Toda, Z.A.B. Shamsudin, K. Shimizu, K. Uesugi, A. Takeuchi, Y. Suzuki, M. Nakazawa, Y. Aoki, M. Kobayashi, Acta Mater. **61**, 2403–2413 (2013)
10. H. Toda, T. Yamaguchi, M. Nakawaza, Y. Aoki, K. Uesugi, Y. Suzuki, M. Kobayashi, Mater. Trans. **51**, 1288–1295 (2010)
11. H. Toda, K. Minami, K. Koyama, K. Ichitani, M. Kobayashi, K. Uesugi, Y. Suzuki, Acta Mater. **57**, 4391–4403 (2009)
12. H. Toda, H. Oogo, K. Horikawa, K. Uesugi, A. Takeuchi, Y. Suzuki, M. Nakawawa, Y. Aoki, M. Kobayashi, Metall. Mater. Trans. A **45A**, 765–776 (2014)
13. M. Nakazawa, M. Kobayashi, H. Toda, Y. Aoki, IEEE J. Trans. Fund. Mater. **131**, 1–9 (2011)
14. H. Su, H. Toda, R. Masunaga, K. Shimizu, H. Gao, K. Sasaki, Md.S. Bhuiyan, K. Uesugi, A. Takeuchi, Y. Watanabe, Acta Mater. **159**, 332–343 (2018)
15. H. Su, H. Toda, K. Shimizu, K. Uesugi, A. Takeuchi, Acta Mater (2018) (under review)
16. S. Brinckmann, T. Siegmund, Y. Huang, Int. J. Plast **22**, 1784–1797 (2006)
17. E. Martínez-Pañeda, C.F. Niordsonam, R.P. Gangloff, Acta Mater. **117**, 321–332 (2016)
18. H. Toda, I. Sinclair, J.-Y. Buffiere, E. Maire, K.H. Khor, P. Gregson, T. Kobayashi, Acta Mater. **52**, 1305–1317 (2004)
19. H. Toda, E. Maire, S. Yamauchi, H. Tsuruta, T. Hiramatsu, M. Kobayashi, Acta Mater. **59**, 1995–2008 (2011)
20. H. Toda, S. Yamamoto, M. Kobayashi, K. Uesugi, H. Zhang, Acta Mater. **56**, 6027–6039 (2008)
21. H. Toda, R. Batres, O. Kuwazuru, M. Kobayashi, A. Kosokawa, J. Japan Inst. Light Metals **64**, 518–524 (2014)

Chapter 10
Dimensional and Morphological Measurements

Industrial X-ray CT scanners have been increasingly used to measure the internal and external dimensions and shapes of various industrial products and components, being used as *coordinate measuring machines* (*CMM*). There are currently products referred to as dimensional X-ray CT scanners, which are commercially available. Moreover, accuracy assessment methods have also been standardized.

Incidentally, modern industrial products have external shapes whose complexity cannot be fully expressed with a drawing alone and have fine internal structures, which can be controlled to a high degree of precision. As such, there is an increasing requirement for measuring and managing the dimensions and shape of end products, assuring product quality by determining its internal defects. Being able to conduct dimension and shape measurements of products in 3D enables a clear assessment of whether a product was created as designed and whether it conforms to standards. The internal structure of products and components, which otherwise have no means of measurement, are primarily assessed through cutting; however, a proper assessment is difficult to conduct for its required cost and effort. The role of X-ray tomography in this context is extremely significant. X-ray tomography also has the benefit of being able to assess external shapes accurately without being affected by various factors, which would otherwise be affected by conventional contact-based or optical-based dimensional/morphological measurement methods (e.g. light reflection from glossy surfaces, overhanging structures, surface roughness). Meanwhile, one cannot ignore the presence of various issues unique to X-ray tomography, such as the trade-off between sample size and spatial resolution, which was not an issue with conventional measurement methods (Sect. 7.5.1), as well as various noise unique to X-rays and artifacts. The various issues presented in this book should be duly considered while conducting 3D imaging to prevent them from becoming actual weaknesses of the X-ray tomography—a lesson that holds true for dimensional and morphological measurements as well.

H. Toda, *X-Ray CT*,
https://doi.org/10.1007/978-981-16-0590-1_10

(a) Shape designed by means of the 3D CAD

(b) Polygon data of a casting obtained from the X-ray tomography

(c) Contour map representing shape error between the designed shape and a real product

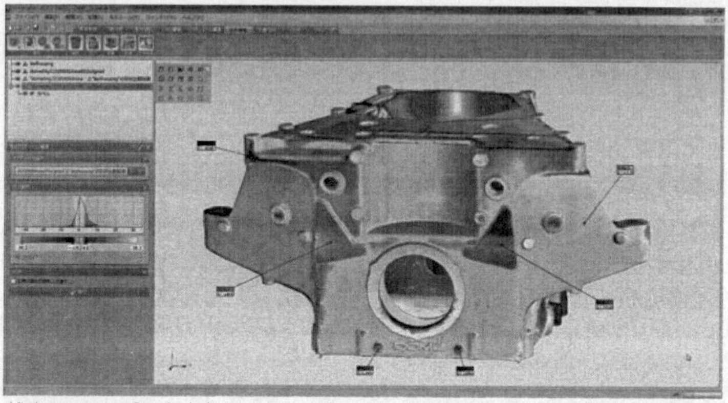

(d) Appearance of analysis in (c) using a software for the reverse engineering (Computer screen)

Fig. 10.1 Example of reverse engineering using an X-ray CT scanner; analysis example using a Nihon Visual Science PointMaster V5.5 (courtesy of Katsuhiko Taki of Nihon Visual Science)

The representative application example of dimensional and morphological measurements is probably the measurement of the wall thickness and bore diameter of hollow components. The registration of 3D images of the obtained components and products and 3D CAD models using the methods discussed in Sect. 9.1 allows for the quantitative assessment of how much both components vary and in which locations. Figure 10.1 shows an implementation example of this. The product is a bell housing casting in an automobile. This type of assessment is particularly important during the trial manufacturing stage of the component or product, also when transitioning from trial production to mass production. Furthermore, the types of assessments mentioned above can be conducted when several components are assembled, as shown in Fig. 6.13, using X-ray tomography. Furthermore, the relationship between the performance and actual shape/production method of the product can be clarified by conducting various simulations using the measured 3D component morphology, which can, in turn, be used to optimize the morphology and control performance variability. In this manner, the 3D data provided by X-ray tomography matches extremely well with the design, analysis, and manufacturing process of industrial products based on current computer-aided design (CAD), computer-aided manufacturing (CAM), and computer-aided engineering (CAE).

A representative example, which requires internal defect measurements, is the detection of packing deficiencies, shrinkage cavities, porosity, and cracks in cast materials such as aluminum and injection-molding plastic materials. This enables efficient determinations, which not only eject components with the above-mentioned properties from the production line but also leave products with harmless defects in the production line. X-ray tomography-based dimensional/morphological measurements are also important in the configuration of new manufacturing processes, such as the production of components from 3D printers.

Industrial products with high accuracy and reliability have been pursued in the various fields of manufacturing industry. The active and efficient use of X-ray tomography should be promoted to ensure that these properties are maintained and developed in the future.

10.1 Device Technology

Figure 10.2 shows the characteristics of dimensional X-ray CT scanner device technology. Temperature control in the device (20 ± 0.5 °C [1]), vibration control, and positional stability due to a high-stiffness/low-thermal expansion stage [1] can be thought of as considerations unique to dimensional X-ray CT scanners. Temperature control is operated continuously over 24 h to regulate sample drift during startup. Furthermore, vibration control is included not only as control of the housing interior and exterior but at times in the assessment of the installation environment and its control. Next, the use of high-spatial-resolution detectors, microfocus radiation sources, and high-accuracy positioning stages is standard practice in obtaining high spatial resolution. As detailed in Sect. 7.5μ, the maximum spatial resolution of an X-ray CT scanner is constrained by the lowest accuracy among the various factors in X-ray tomography including the X-ray source, sample rotation stage, X-ray focusing elements in the case of an imaging optical system, and detectors. However, *accuracy* and *traceability* are more important than spatial resolution for the X-ray CT scanners as coordinate-measuring machines. In these cases, there are two factors associated with measurement accuracy: the deviation from the true value and the extent of data variation for each measurement. The positional stability of the focal spot and attachment of 2D measurement devices for correction and confirmation (e.g. laser interferometer) are important to ensure traceability. The voxel size must be accurately corrected for these cases. There are commercially available devices that automatically correct values and reflect this in the measured parameters. Some devices control the focal spot position in real-time so that it has no variation. High-energy X-ray sources and large-scale, high-accuracy, high-load-capacity positioning stages must be used particularly when large products or components are to be measured.

Two situations must be considered with regard to dimensional/morphological measurement: when relatively high-energy X-ray sources are used to measure large-scale components and products and when small-scale components and products must be measured with microtomography. As shown in Fig. 10.2, the device technology, in either case, is based on that described in Chap. 4, with no special cases in the basic

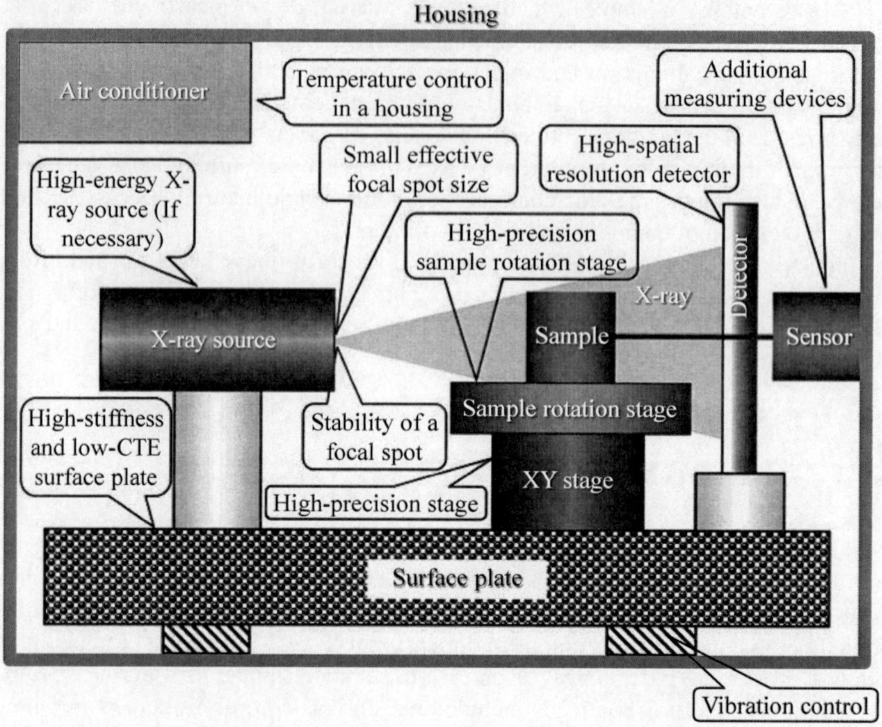

Fig. 10.2 Summary of the important elements regarding the device technologies in a dimensional X-ray CT scanner according to component machine (ten categories in the balloons)

principles and rules of X-ray tomography or its various component devices. In other words, dimensional and morphological measurements according to the performance of each X-ray CT scanner can be conducted without necessarily using a machine referred to as a dimensional X-ray CT scanner if the relationships between the methods (i.e., various constitutive devices, reconstructions, various image processing techniques) and the image quality of 3D images (i.e., spatial resolution, noise, and contrast) are sufficiently understood through this book and if some additional considerations are made with regard to the heat drift and vibrations of the sample. In addition to the aforementioned consideration of the hardware, a major point of difference between dimensional and imaging X-ray CT scanners is perhaps the support from various user-friendly software and the assurance of accuracy. The former refers to support when simultaneously and rapidly assessing large numbers of measurement points, or when creating assessment reports. Furthermore, fast imaging and imaging analysis are necessary to support high throughput and the handling of samples by industrial robots when setting up production lines for in-line inspection to conduct a total inspection of products and components. The latter is discussed in the next section.

10.2 Measurement Accuracy

10.2.1 Standardization

International testing standards are present when using X-ray CT scanners for measurements. The reliability of an X-ray CT scanner as a coordinate measuring machine can be increased with the listing of the characteristics of the scanner as per international testing standards or with the regular maintenance and correction of the X-ray CT scanner. At present, we are still in a transitional stage concerning these types of viewpoints.

First, with regard to the testing standards of X-ray tomography itself, ISO 15,708–1:2017 provides a definition of the terminology used [2]; ISO 15,708–2:2017 provides general principles, devices, and samples [3]; ISO 15,708–3:2017 provides the operation of industrial X-ray CT scanners and the interpretation of the obtained images [4]; and ISO 15,708–4:2017 provides inspections of X-ray tomography including dimensional/morphological measurements and assessments of the device [5]. ISO 15,708–4:2017, in particular, also has descriptions relating to the accuracy of dimensional/morphological measurements [5].

Meanwhile, VDI/VDE 2617–13 [6] and VDI/VDE 2630–1.3 [7], which are the German test standards relating to dimensional/morphological measurements as of 2018, have also been applied to X-ray CT scanners outside of Germany. These are used as guidelines for applying ISO10360, which are the ISO standard for accuracy assessments of coordinate measuring machines, to dimensional X-ray CT scanners. Furthermore, VDI/VDE 2630–1.1:2016–05 [8] includes entries on basic items and definitions, and VDI/VDE 2630–1.2:2016–07 [9] includes entries on influential quantities relating to the morphological measurements conducted with X-ray CT scanners. Using these standards enables the comparison of different X-ray CT scanners based on the definitions in performance assessments of dimensional X-ray CT scanners and standard devices. According to Matsuzaki, the issue of whether there are any problems with the application of these standards has been a subject of international debate [1].

Length measurement errors and probing errors representing local measurement errors are the subjects of the ISO (international standardization) relating to performance assessments of dimensional X-ray CT scanners [10]. The latest standardization trends should be studied and used.

10.2.2 Uncertainty in Measurement Accuracy

The assessment of *uncertainty in measurement accuracy* relating to the various above-mentioned component devices, post-measurement reconstruction, and 3D image handling is discussed here. In addition, standard devices for the assessment of dimensional and shape measurement characteristics in X-ray CT scanners are briefly discussed. Various factors relating to imaging hardware, measurement environment, sample, measurement conditions, reconstruction, and various image processing that may influence measurement accuracy are comprehensively listed in Table 10.1. It is important to consider in advance, what the major influences are and what can be ignored in measurements conducted by the researcher.

Traceability can be guaranteed if there is a standard device whose shape and size are known in advance and its measurements periodically conducted and reflected in the dimensional/morphological measurements. This refers to whether the measurement results can be verified to be the same each time and whether the X-ray CT scanner accurately revises based on measurement results or corrects the measurement data. This requires that the shape and size of the standard device are measured and confirmed using another coordinate measuring machine or another X-ray CT scanner with an effective spatial resolution and accuracy that is considerably higher than the X-ray CT scanner being used for dimensional/morphological measurements. Furthermore, measurements must be repeatedly conducted and the variation in results must be statistically assessed to assure reproducibility.

Standard devices of various forms are used. Representative examples, shown in Fig. 10.3, include the forest gage, where multiple spheres are supported by stays and standing close together, and step cylinder, which is a hollow cylinder with multiple steps. The spheres can generally be measured to assess probing error. The dimensional error can be assessed by measuring between the two points whose distances are known in advance. The forest gage is designed so that multiple data can be simultaneously obtained. Furthermore, the step cylinder considers how changes in material thickness influence measurements, as well as the verification of internal structure assessment by measuring the opened holes in the interior.

The difference between the maximum radius $R_{\max}$ and minimum radius $R_{\min}$ when the sphere diameters are measured are assessed as a *form probing error* in VDI/VDE 2630–1.3. Furthermore, the deviation from the true diameter value determined in advance is assessed as the *size probing error*. Sphere size and measurement methods, as well as error calculation specifications, are stipulated here. A standard device in which four spheres are arranged and attached to a cylinder is used to measure the distance between spheres with regard to distance error.

VDI/VDE 2630–2.1 describes a method for assessing error U during measurement using the corrected standard device. Here, U is expressed as follows, based on various causes of error [11]:

$$U = k\sqrt{u_{cal}^2 + u_p^2 + u_w^2 + u_b^2} \tag{10.1}$$

Here, u_{cal} is the uncertainty when calibrating the standard device using the 3D measurement method, u_{p} the uncertainty related to the reproducibility of the measurement method, u_{w} the uncertainty caused by the sample and the variation in its production process, and u_{b} the uncertainty due to the measurement method procedure. Furthermore, k is a coefficient that varies with the confidence interval, with $k = 2$ at a 95.45% confidence interval.

Table 10.1 Various factors which can influence the measurement accuracy of dimensional X-ray CT scanners and their classifications

Category	Sub-category	Cause
Apparatus	X-ray source	Drift
		Ageing
		Poor fixation
		Beam hardening
		Tube voltage
	Positioning stage	Alignment of a rotation axis
		Shortage of load bearing capacity
		Eccentricity
		Surface runout
		Cyclic positioning accuracy
	Detector	Alignment in beam direction
		Angular alignment
		Planarity
		Sensitivity/Dynamic range
		Spatial resolution
		Noise characteristic
		Pixel size (Poor calibration)
Environment	Housing/surface plate	Temperature
		Stiffness
	Installation environment	Vibration
Observation object	Sample	Chemical composition
		Size
		Deformation
		Drift
		Surface roughness
		Poor fixing
		Beam hardening
Condition	Measuring condition	Number of projections/exposure time
		Magnification and its calibration
		Selection of a filter
Post processing	Reconstruction	Cone beam artifact
		Reconstruction filter
	Image processing	Filtering
		Various image processing

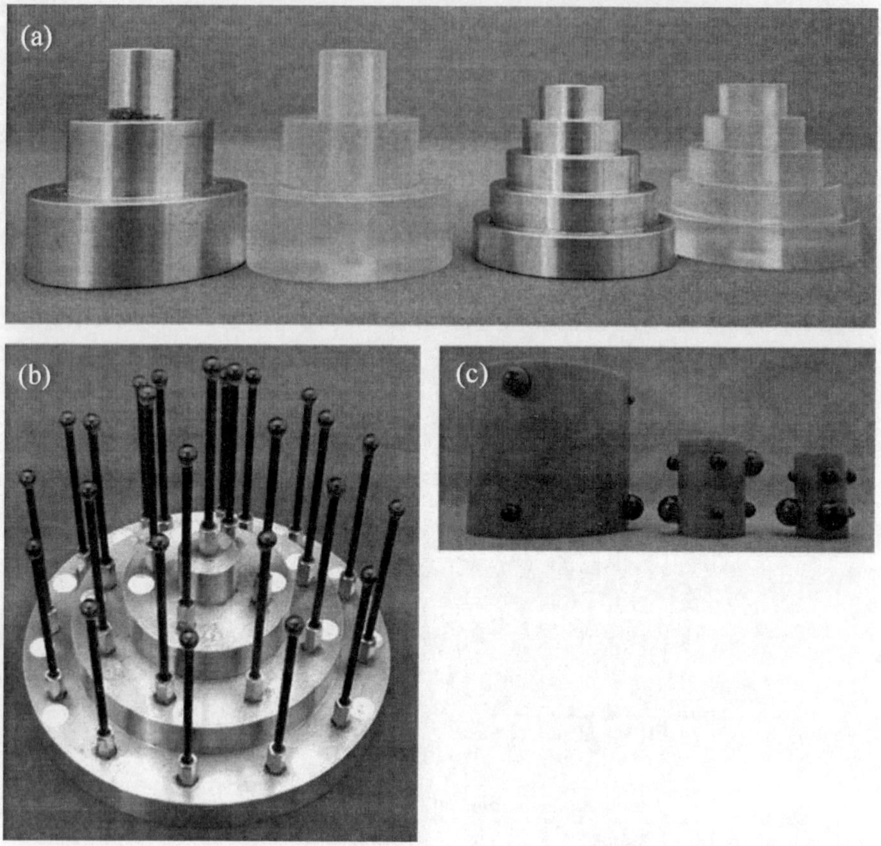

Fig. 10.3 Standard devices used for measurement accuracy assessments of the dimensional X-ray CT scanner: **a** step gage, **b** forest gage, and **c** is a device made independently by Mr. Takahashi (courtesy of Yuichi Takahashi of the Gunma Industrial Technology Center)

Table 10.1 does not include u_{cal}. Meanwhile, other factors cover a lot of ground in actual practice, as shown in Table 10.1, and the substantial meaning of Eq. (10.1) must be sufficiently understood by carefully examining the X-ray CT scanner, measurement conditions, sample, and data processing methods. For example, Jiménez et al. applied the method in Eq. (10.1) on microtomography devices [12]. They expressed u_{w} and u_{b} as follows [12]:

$$u_w = \sqrt{{u_{w_1}}^2 + {u_{w_2}}^2} \tag{10.2}$$

$$u_b = \sqrt{{u_{b_1}}^2 + {u_{b_2}}^2} \tag{10.3}$$

Here, u_{w_1} is the uncertainty related to the variation in the mechanical properties of the sample and u_{w_2} is the uncertainty related to the variation in the thermal expansion coefficients of the sample [12]. Furthermore, u_{b_1} is the uncertainty related to temperature fluctuations during measurement and u_{b_2} is the uncertainty related to surface position specification [12]. According to them, u_{cal} is equal to approximately 2.5–2.7 μm when an optical coordinate measuring machine is used, accounting for the largest share of U. This is followed by u_{p}, which is measured 10 times and is approximately 0.5–1.8 μm. Meanwhile, Kraemer et al. similarly investigated measurement uncertainty using a microtomography device with a focal spot size of 8 μm [13]. These results showed that a result similar to Jiménez was obtained when distances between spheres with a diameter of 2 mm were measured; however, u_{b} was considerably larger at 8.2–13.2 μm when the sphere diameters were measured.

Finally, we summarized this from the viewpoint of the relationship between uncertainty and spatial resolution of the 3D image. The effective pixel size of the 3D image in the case of Jiménez was 8 μm and the spatial resolution of the 3D image was estimated to be at most 16 μm based on the sampling theorem. In contrast, the uncertainty was 5.6–7.0 μm, which is less than half of the spatial resolution. The effective pixel size in the case of Kraemer et al. was 13 μm and the maximum spatial resolution value was 26 μm. The uncertainty at this point was 2–5 μm for the distance between the spheres, which is 10–20% of the spatial resolution but 16–26 μm for the sphere diameter, which is at a level similar to the spatial resolution. In this manner, it should be noted that the uncertainty could vary by a factor of 10 even with the identical sample by changing the measurement location.

These studies show that the standard device must first be calibrated to a high level of precision. Next, it must be understood that the factors, which dominate uncertainty vary for each case, and the physical elements, which bring about these factors, should be understood at the level shown in Table 10.1. Finally, the uncertainty level brought about by these factors should be understood if possible. Care must also be taken as this uncertainty can vary considerably from the effective spatial resolution of the X-ray CT scanner or 3D image, which is relatively easy to determine.

10.3 Reverse Engineering

Reverse engineering is a production method that involves decomposing and analyzing industrial products including software; understanding the material and thermomechanical heat treatment of that product, the shape and functionality of constituent components, and the materials specifications; and actively applying these aspects for production. For example, a business can purchase a product from another company, decompose it and thoroughly analyze it to create a similar product. Alternatively, it can use the product's technology to design or develop an original product. However, this chapter focuses on reverse engineering in the narrow sense of conducting the 3D imaging of industrial products with X-ray tomography, applying the obtained digital data to production. For example, the outer shells of automobiles are created with multiple complex curved surfaces, unlike those from several decades ago. Even if these are rendered with CAD, humans cannot completely capture this shape in a monitor. With this in mind, the external appearance of the automobile is created using a clay model and once the designer corrects this with human perception, this is measured to create 3D data, from which a metal mold is created. Production is enabled if this metal mold is used, and CAE-based analysis is enabled if 3D data are used.

Figure 10.4 shows a flowchart of reverse engineering and the necessary elements in X-ray tomography for conducting reverse engineering; Fig. 10.5 shows application examples. Among these, CAE can conduct a variety of measurements, including structural analysis, solidification analysis, fluids, heat transfer, electromagnetic fields, and diffusion of elements. Explanations relating to STL files and structural analyses were provided in Sect. 8.6; however, further discussions on CAE or CAM are beyond the scope of this book and are therefore omitted.

The same consideration afforded to the thought processes and points of caution regarding 3D imaging, which have been discussed up until now mainly in Sect. 6.7, is necessary for 3D imaging. In other words, the size of the industrial products and components to be examined, and their average atomic numbers (taking voids into account) must be considered when selecting the necessary X-ray CT scanner and X-ray energy to be used during imaging, taking field of view and transmissivity into account. X-ray CT scanners equipped with the compact electron accelerators

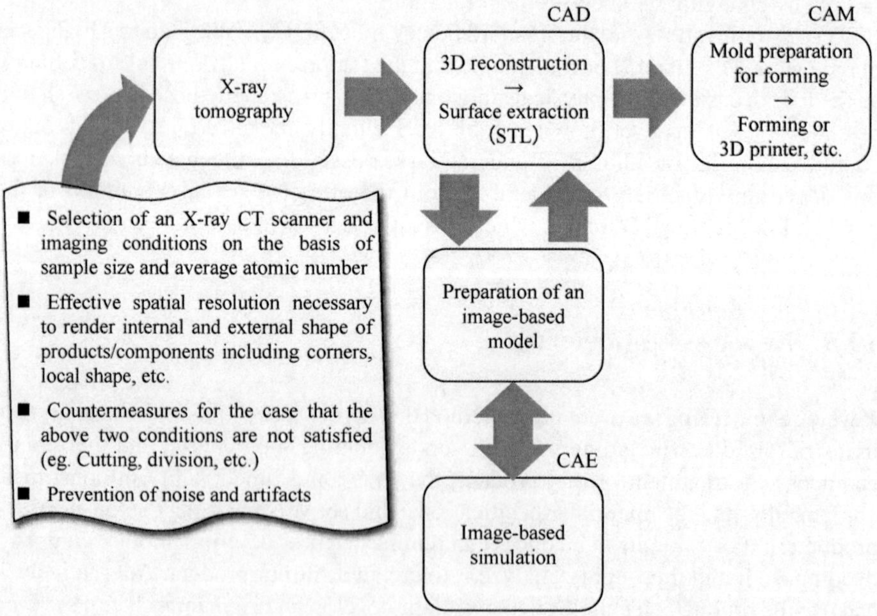

Fig. 10.4 Flowchart of reverse engineering using an X-ray CT scanner and a schematic summarizing the major points of 3D imaging

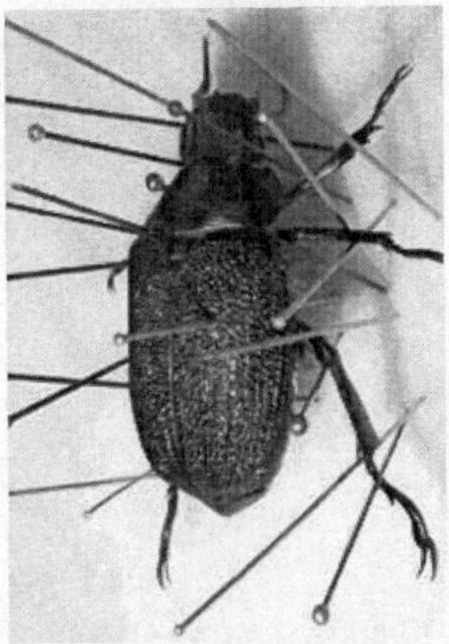

(a) Real object such as industrial products (In this case an insect specimen)

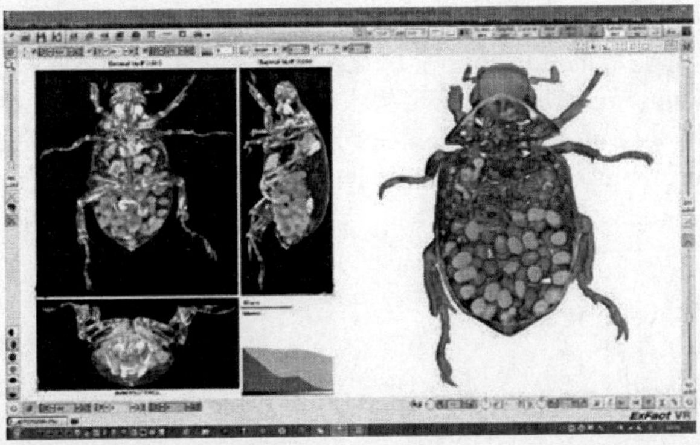

(b) Conversion into image data using the X-ray tomography. MPR (Left) and a 3D image (Right) using a 3D image rendering software

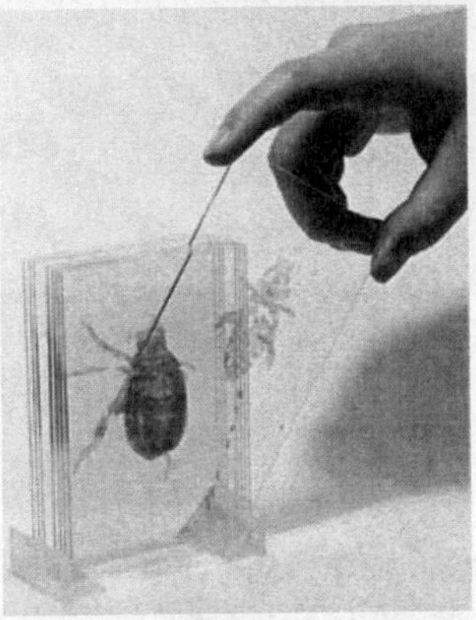

(c) A solid model based on the 3D image data that consists of stacked transparent thin plates

Fig. 10.5 Example of reverse engineering. Here, 3D imaging of an insect specimen instead of an industrial product or component is conducted and its surface is extracted, after which a 3D model was created using the 3D image data. Various production processes can be conducted, including metal forging and casting, plastic injection, and 3D printing (courtesy of Katsuhiko Taki of Nihon Visual Science)

discussed in Sect. 4.1.3 become necessary when the product or component size is large or when the material has a large average atomic number (e.g., iron). Next, the effective spatial resolution necessary for a 3D image is determined by considering the internal and external shape of the product/component, reproducibility of the pin-angle section, size and shape of the micro-section with the narrowest width, and the separability between the micro-sections. When transmissivity and spatial resolution are incompatible, some type of plan is required prior to 3D imaging such as cutting or assembly separation, as discussed in Sect. 6.7. Furthermore, as briefly discussed in

Sect. 8.6, the connectivity of the triangular mesh, the correction of distorted element shapes, and the control of the number of elements require immense amounts of time and effort. Other than this, the noise and artifacts that always accompany 3D imaging should be sufficiently determined in advance and should be eliminated as much as possible during the 3D image processing or segmentation step.

References

1. K. Matsuzaki, AIST Bull. Metrol. **9**, 311–321 (2016)
2. ISO 15708–1:2017: Non-destructive testing—Radiation methods for computed tomography—Part 1: Terminology (2017)
3. ISO 15708–2:2017: Non-destructive testing—Radiation methods for computed tomography—Part 2: Principles, equipment and samples (2017)
4. ISO 15708–3:2017: Non-destructive testing—Radiation methods for computed tomography—Part 3: Operation and interpretation (2017)
5. ISO 15708–4:2017: Non-destructive testing—Radiation methods for computed tomography—Part 4: Qualification (2017)
6. VDI/VDE 2617–13: Accuracy of coordinate measuring machines characteristics and their testing—Guideline for the application of DIN EN ISO 10360 for coordinate measuring machines with CT-sensors (2011)
7. VDI/VDE 2630–1.3: Computed tomography on dimensional measurement—Guideline for the application of DIN EN ISO 10360 for coordinate measuring machines with CT sensors (2011)
8. VDI/VDE 2630–1.1: Computed tomography in dimensional measurement—Fundamentals and definitions (2016)
9. VDI/VDE 2630–1.2: Computed tomography in dimensional measurement Influencing variables on measurement results and recommendations for computed tomography dimensional measurements (2018)
10. METI industrial science and technology policy and environment Bureau—Metrology policy office, AIST, Hitachi, Ltd.: Foundational assessment technology development project assessment document for next-generation 3D internal and external measurement (2016)
11. VDI/VDE 2630–2.1 Computed tomography in dimensional measurement: Determination of the uncertainty of measurement and the test process suitability of coordinate measurement systems with CT sensors (2015)
12. R. Jiménez, M. Torralba, J.A. Yagüe-Fabra, S. Ontiveros, G. Tosello, Sensors **17**, 1–21 (2017)
13. A. Kraemer, G. Lanza, Procedia CIRP, 14th CIRP Conference on Computer Aided Tolerancing (CAT), vol. 43 (2016), pp. 362–367

Correction to: X-Ray CT

Correction to:
H. Toda, *X-Ray CT*,
https://doi.org/10.1007/978-981-16-0590-1

The original version of the book was inadvertently published with the wrong copyright holder text in the book. This has now been updated and approved by the author.

The updated version of the book can be found at
https://doi.org/10.1007/978-981-16-0590-1

H. Toda, *X-Ray CT*,
https://doi.org/10.1007/978-981-16-0590-1_11

Index

H. Toda, *X-Ray CT*,
https://doi.org/10.1007/978-981-16-0590-1